Electrical Technology

Other works by the same author

Fundamentals of Electrical Engineering

with Arthur Morley
Basic Engineering Science
Elementary Engineering Science
Mechanical Engineering Science
Principles of Electricity
Electrical Engineering Science

with Christopher Hughes
Engineering Science in SI Units

Electrical Technology

Edward Hughes
D.Sc.(Eng.), Ph.D., C.Eng., F.I.E.E.
Fellow of the Heriot-Watt College, Edinburgh.
Formerly Vice-Principal and Head of the Engineering Department,
Brighton College of Technology

FOURTH EDITION IN SI UNITS

Longman

LONGMAN GROUP LIMITED
London

Associated companies, branches and representatives
throughout the world

© Edward Hughes 1960
New Editions © Edward Hughes 1963, 1967, 1969

First published 1960
Second Edition 1963
Third Edition 1967
Fourth Edition 1969
Third impression (*SI Units*) 1972
Fourth impression 1973

ISBN 582 44449 7 cased   0 582 44453 5 paper

Made and printed in Great Britain by
Richard Clay (*The Chaucer Press*), *Ltd, Bungay, Suffolk*

Preface to First Edition

This volume covers the electrical engineering syllabuses of the Second and Third Year Courses for the Ordinary National Certificate in Electrical Engineering and of the First Year Course leading to a Degree in Engineering.

The rationalized M.K.S. system of units has been used throughout this book, but for the convenience of those who may have to refer to data expressed in the C.G.S. electromagnetic or electrostatic system, a conversion table is given on page 687. The symbols, abbreviations and nomenclature are in accordance with the recommendations of the British Standards Institution and the Institution of Electrical Engineers; and, for the convenience of students, the symbols and abbreviations used in this book have been tabulated on pages xiii–xvii. The 'per-unit' notation has been used wherever possible so that the student may have ample opportunity of becoming familiar with this term.

It is impossible to acquire a thorough understanding of electrical principles without working out a large number of numerical problems; and while doing this, students should make a habit of writing the solutions in an orderly manner, attaching the name of the unit wherever possible. When students tackle problems in examinations or in industry, it is important that they express their solutions in a way that is readily intelligible to others, and this facility can only be acquired by experience. Guidance in this respect is given by the 106 worked examples in the text, and the 670 problems afford ample opportunity for practice.

Most of the questions have been taken from examination papers; and for permission to reproduce these questions I am indebted to the University of London, the East Midland Educational Union, the Northern Counties Technical Examination Council, the Union of Educational Institutions and the Union of Lancashire and Cheshire Institutes. The answers given to the problems are of slide-rule accuracy—in general, they are given to three significant figures. The greatest care has been taken to eliminate errors in both the text and the answers; but if any mistakes are found, I shall be very grateful to have them brought to my attention.

I wish to express my thanks to Dr. F. T. Chapman, C.B.E., M.I.E.E., and Mr. E. F. Piper, A.M.I.E.E., for reading the manuscript and making valuable suggestions.

Hove,   EDWARD HUGHES
April 1959

Preface to Third Edition

Recent changes in Electrical Engineering syllabuses for the first year of degree courses and for the Ordinary National Certificate in Engineering have made it necessary to revise the contents of this volume. An appreciable portion of the original text dealing with d.c. machines has been deleted and chapters involving Electronics have been expanded to cover practically all aspects included in the O.N.C. syllabuses of the various examining institutions. In the chapter on Semiconductor Devices, the thyristor is included in view of its rapidly growing importance.

A departure from previous practice is the use of the term *phasor* instead of *vector* in connection with a.c. theory. This change is in line with continental and American practice.

Most of the questions from S2 and S3 papers of the examining institutions mentioned in the Preface to the first edition have been replaced by questions from recent O.1 and O.2 papers of these institutions. Questions have also been taken from O.1 and O.2 papers of the Scottish Association for National Certificates and Diplomas and from those of the Welsh Joint Education Committee. To all these institutions I express my thanks for permission to reproduce their questions.

I am grateful to Mr Norman L. White, B.Sc.(Eng.), for reading the revised manuscript for chapters 19 and 20 and for his helpful suggestions.

January 1966   E. H.

Preface to Fourth Edition

The principal changes in this edition are concerned with nomenclature in order that the terms may be in agreement with the International System of Units (SI) adopted by the International Organization for Standardization and by the International Electrotechnical Commission. These changes have involved the replacement of *mho* by *siemens*, of *weber per square metre* by *tesla*, of *cycle per second* by *hertz* and of *lumen per square foot* by the metric unit *lux*.

I wish to express my thanks to the various examining institutions for their permission to modify questions reproduced from their examination papers in accordance with the changes listed above.

September 1968   E. H.

In the 1972 impression, references to British units have been deleted except for a conversion table on page 695. The terms *magnetizing force* and *electric force* have been replaced by *magnetic field strength* and *electric field strength* respectively; and an appendix on *The generalized theory of electrical machines* has been substituted for the trigonometrical and log tables. Also, for ease of reference, definitions of SI units are listed on page xviii.

October, 1971   E. H.

In the 1973 impression, new matter has been included dealing with speed control of d.c. motors by thyristors, imperfect capacitor and loss angle, waveform of transformer magnetizing current, tungsten-halogen and high-pressure mercury and sodium lamps, phototransistor and rectifier voltmeter. The terms *illumination* and *luminous efficiency* have been replaced by *illuminance* and *luminous efficacy* respectively. An extension of the Appendix has replaced the conversion table on page 695.

January, 1973   E. H.

Contents

*

currents—displacement current in a dielectric—energy in charged capacitor—attraction between charged plates—dielectric strength.

bias—power amplifier—Miller effect—tetrode and pentode
—thyratron—cathode-ray oscilloscope—electrostatic and
magnetic deflections on screen of cathode-ray tube—photoelectric cells.

SYMBOLS, ABBREVIATIONS AND DEFINITIONS

Based upon PD 5686: 1972 (BSI) and *The International System of Units* (1790), prepared jointly by the National Physical Laboratory, UK, and the National Bureau of Standards, USA, and approved by the International Bureau of Weights and Measures.

Notes on the use of symbols and abbreviations

1. A unit symbol is the same for the singular and the plural: for example, 10 kg, 5 V; and should be used only after a numerical value: for example, m kilograms, 5 kg.

2. Full point should be omitted after a unit symbol: for example, 5 mA, 10 μF, etc.

3. Full point should be used in a multi-word abbreviation: for example, e.m.f., p.d.

4. In a compound unit-symbol, the product of two units is preferably indicated by a dot, especially in manuscript. The dot may be dispensed with when there is no risk of confusion with another unit symbol: for example, N·m or N m, but not mN. A solidus (/) denotes division: for example, J/kg, rev/min.

5. Only one multiplying prefix should be applied to a given unit: for example, 5 megagrams, not 5 kilokilograms.

6. A prefix should be applied to the numerator rather than the denominator: for example, MN/m^2 rather than N/mm^2.

7. The abbreviated forms a.c. and d.c. should be used only as adjectives: for example, d.c. motor, a.c. circuit.

8. A hyphen is inserted between the numerical value and the unit when the combination is used adjectivally: for example, a 240-volt (or 240-V) motor, a 2-ohm (or 2-Ω) resistor.

Units of length, volume, mass and time

Quantity	Unit	Symbol	Quantity	Unit	Symbol
Length	metre	m	Volume	cubic metre	m^3
	kilometre	km		litre (0·001m³)	l
Mass	kilogram	kg	Time	second	s
	megagram	Mg		minute	min
	or tonne (1 Mg)	t		hour	h

Electricity and Magnetism

Quantity	Quantity symbol	Unit	Unit symbol
Admittance	Y	siemens	S
Angular velocity	ω	radian per second	rad/s
Capacitance	C	farad	F
		microfarad	μF
		picofarad	pF
Charge or Quantity of electricity	Q	coulomb	C
Conductance	G	siemens	S
Conductivity	σ	siemens per metre	S/m
Current			
Steady or r.m.s. value	I	ampere	A
		milliampere	mA
		microampere	μA
Instantaneous value	i		
Maximum value	I_m		
Current density	J	ampere per metre²	A/m^2
Difference of potential			
Steady or r.m.s. value	V	volt	V
		millivolt	mV
		kilovolt	kV
Instantaneous value	v		
Maximum value	V_m		
Electric field strength	E	volt per metre	V/m
Electric flux	Ψ	coulomb	C
Electric flux density	D	coulomb per square metre	C/m^2
Electromotive force			
Steady or r.m.s. value	E	volt	V
Instantaneous value	e		
Maximum value	E_m		
Energy	W	joule	J
		kilojoule	kJ
		megajoule	MJ
		watt hour	W·h
		kilowatt hour	kW·h
		electronvolt	eV
Force	F	newton	N

Quantity	Quantity symbol	Unit	Unit symbol
Frequency	f	hertz	Hz
		kilohertz	kHz
		megahertz	MHz
Impedance	Z	ohm	Ω
Inductance, self	L	henry (plural, henrys)	H
Inductance, mutual	M	henry (plural, henrys)	H
Magnetic field strength	H	ampere per metre	A/m
Magnetic flux	Φ	weber	Wb
Magnetic flux density	B	tesla	T
Magnetomotive force	F	ampere	A
Permeability of free space or Magnetic constant	μ_0	henry per metre	H/m
Permeability, relative	μ_r		
Permeability, absolute	μ	henry per metre	H/m
Permittivity of free space or Electric constant	ϵ_0	farad per metre	F/m
Permittivity, relative	ϵ_r		
Permittivity, absolute	ϵ	farad per metre	F/m
Power	P	watt	W
		kilowatt	kW
		megawatt	MW
Reactance	X	ohm	Ω
Reactive voltampere	—	var	VAr
Reluctance	S	ampere per weber	A/Wb
Resistance	R	ohm	Ω
		microhm	$\mu\Omega$
		megohm	$M\Omega$
Resistivity	ρ	ohm metre	$\Omega\cdot m$
		microhm metre	$\mu\Omega\cdot m$
Susceptance	B	siemens	S
Torque	T	newton metre	N·m
Voltampere	—	voltampere	VA
		kilovoltampere	kVA
Wavelength	λ	metre	m
		micrometre	μm

Light

Quantity	Quantity symbol	Unit	Unit symbol
Illuminance	E	lux	lx
Luminance (objective brightness)	L	candela per square metre	cd/m²
Luminous flux	Φ	lumen	lm
Luminous intensity	I	candela	cd
Luminous efficacy	—	lumen per watt	lm/W

ELECTRICAL MACHINES

Term	Symbol
Number of armature conductors	Z
„ commutator segments or bars	C
„ pairs of poles	p
„ parallel circuits	c
„ phases	m
„ turns	N

ABBREVIATIONS FOR MULTIPLES AND SUB-MULTIPLES

T	tera	10^{12}
G	giga	10^{9}
M	mega or meg	10^{6}
k	kilo	10^{3}
d	deci	10^{-1}
c	centi	10^{-2}
m	milli	10^{-3}
μ	micro	10^{-6}
n	nano	10^{-9}
p	pico	10^{-12}

GREEK LETTERS USED AS SYMBOLS

Letter	Capital	Small
Alpha	—	α (angle, temperature coefficient of resistance, current amplification factor for common-base transistor)
Beta	—	β (current amplification factor for common-emitter transistor)
Delta	Δ (increment, mesh connection)	δ (small increment)
Epsilon	—	ϵ (permittivity)
Eta	—	η (efficiency)
Theta	—	θ (angle, temperature)
Lambda	—	λ (wavelength)
Mu	—	μ (micro, permeability, amplification factor)
Pi	—	π (circumference/diameter)
Rho	—	ρ (resistivity)
Sigma	Σ (sum of)	σ (conductivity)
Phi	Φ (magnetic flux)	ϕ (angle, phase difference)
Psi	Ψ (electric flux)	—
Omega	Ω (ohm)	ω (solid angle, angular velocity, angular frequency)

MISCELLANEOUS

Term	Symbol	Term	Symbol
Approximately equal to	$\simeq$	Much less than	$\ll$
Proportional to	$\propto$	Base of natural logarithms	e
Infinity	∞	Common logarithm of x	$\log x$
Sum of	Σ	Natural logarithm of x	$\ln x$
Increment or finite differ-		Complex operator $\sqrt{-1}$	j
ence operator	Δ, δ	Temperature	θ
Greater than	$>$	Time constant	T
Less than	$<$	Efficiency	η
Much greater than	$\gg$	Per unit	p.u.

ELECTRONIC VALVES AND SEMICONDUCTOR DEVICES
Based on B.S. 3363:1961

Amplification factor	μ
Mutual conductance	g_m
Anode a.c. (or slope) resistance	r_a
Voltage amplification	A
Current amplification factor for common-base transistor circuit	α
Current amplification factor for common-emitter transistor circuit	β

Subscripts for quantity symbols:

Anode	A, a
Cathode	K, k
Control terminal	G, g
Emitter	E, e
Base	B, b
Collector	C, c

D.C. and average values are indicated by upper-case subscripts, e.g.

I_A = direct anode current (no signal),
V_G = direct grid-cathode p.d.,
I_C = direct collector current (no signal),
V_{CB} = direct collector-base p.d. in a transistor circuit (no signal).

Values of varying components are indicated by lower-case subscripts, e.g.

i_a and I_a = instantaneous and r.m.s. values respectively of the varying component of anode current,

v_{cb} and V_{cb} = instantaneous and r.m.s. values respectively of the varying component of the p.d. between collector and base in a transistor circuit.

Sequence of subscripts: the first subscript denotes the terminal at which the current is measured, or where the terminal potential is measured with respect to the reference terminal; e.g. V_{CB} represents the direct potential of the collector with reference to the base in a transistor circuit.

Where the use of the reference terminal is not required to preserve the meaning of the symbol, the second subscript may be omitted; e.g. V_A represents the direct potential of the anode with reference to the cathode of a thermionic valve.

Definitions of electric and magnetic SI units

The *ampere* (A) is that constant *current* which, if maintained in two straight parallel conductors of infinite length, of negligible circular cross-section, and placed 1 m apart in vacuum, would produce between these conductors a force equal to 2×10^{-7} N per metre of length.

The *coulomb* (C) is the *quantity of electricity* transported in 1 s by 1 A.

The *volt* (V) is the *difference of electrical potential* between two points of a conductor carrying a constant current of 1 A, when the power dissipated between these points is equal to 1 W.

The *ohm* (Ω) is the *resistance* between two points of a conductor when a constant difference of potential of 1 V, applied between these points, produces in this conductor a current of 1 A, the conductor not being a source of any electromotive force.

The *henry* (H) is the *inductance* of a closed circuit in which an e.m.f. of 1 V is produced when the electric current in the circuit varies uniformly at the rate of 1 A/s.

(*Note:* this also applies to the e.m.f. in one circuit produced by a varying current in a second circuit, i.e. mutual inductance.)

The *farad* (F) is the *capacitance* of a capacitor between the plates of which there appears a difference of potential of 1 V when it is charged by 1 C of electricity.

The *weber* (Wb) is the *magnetic flux* which, linking a circuit of one turn, produces in it an e.m.f. of 1 V when it is reduced to zero at a uniform rate in 1 s.

The *tesla* (T) is the *magnetic flux density* equal to 1 Wb/m².

Definitions of other derived SI units

The *newton* (N) is the *force* which, when applied to a mass of 1 kg, gives it an acceleration of 1 m/s².

The *pascal* (Pa) is the *stress* or *pressure* equal to 1 N/m².

The *joule* (J) is the *work done* when a force of 1 N is exerted through a distance of 1 m in the direction of the force.

The *watt* (W) is the *power* equal to 1 J/s.

The *hertz* (Hz) is the unit of *frequency*, namely the number of cycles per second.

The *lumen* (lm) is the *luminous flux* emitted within unit solid angle by a point source having a uniform intensity of 1 candela.

The *lux* (lx) is an *illumination* of 1 lm/m².

CHAPTER 1

The Electric Circuit

1.1 The International System of units

The International System of Units, known as SI in every language, derives all the units in the various technologies from the following *seven* base units:

Quantity	Unit	Symbol
length	metre	m
mass	kilogram	kg
time	second	s
electric current	ampere	A
temperature	kelvin	K
luminous intensity	candela	cd
amount of substance	mole	mol

The *metre* is the length equal to 1 650 763·73 wavelengths of the orange line in the spectrum of an internationally-specified krypton discharge lamp.

The *kilogram* is the mass of a platinum-iridium cylinder preserved at the International Bureau of Weights and Measures at Sèvres, near Paris.

The *second* is the interval occupied by 9 192 631 770 cycles of the radiation corresponding to the transition of the caesium-133 atom.

The *ampere* is defined on page 6.

The *kelvin* is 1/273·16 of the thermodynamic temperature of the triple point of water. On the Celsius scale, the temperature of the triple point of water is 0·01°C,

hence, $0°C = 273·15$ K.

A temperature interval of 1°C = a temperature interval of 1 K.
The *candela* is defined on page 606.
The *mole* does not affect the units used in this book.

1

1.2 Unit of force

The SI unit of force is the *newton* (N), namely the *force which, when applied to a mass of 1 kg, gives it an acceleration of 1 m/s². Hence the force F required to give a mass m an acceleration a is:

$$F \text{ [newtons]} = m \text{ [kilograms]} \times a \text{ [metres/second}^2\text{]} \qquad (1.1)$$

Weight. The weight of a body is the gravitational force exerted by the earth on that body. Owing to the variation in the radius of the earth, the gravitational force on a given mass, at sea level, is different at different latitudes, as shown in fig. 1.1. It will be seen that the weight of a 1-kg mass at sea level in the London area is practically 9·81 N. For most purposes we can assume:

$$\text{the weight of a body} \simeq 9{\cdot}81 \ m \text{ newtons} \qquad (1.2)$$

where m is the mass of the body in kilograms.

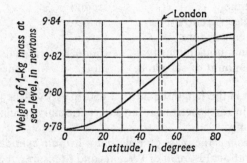

Fig. 1.1. Variation of weight with latitude.

Example 1.1 *A force of 50 N is applied to a mass of 200 kg. Calculate the acceleration.*

Substituting in expression (1.1), we have:

$$50 \text{ [N]} = 200 \text{ [kg]} \times a$$
$$\therefore \qquad a = 0{\cdot}25 \text{ m/s}^2.$$

Example 1.2 *A steel block has a mass of 80 kg. Calculate the weight of the block at sea level in the vicinity of London.*

Since the weight of a 1-kg mass $\simeq$ 9·81 N,

∴   weight of the steel block = 80 [kg] × 9·81 [N/kg]*

$$= 784·8 \text{ N.}$$

1.3 Unit of turning moment or torque

If a force F, in newtons, is acting at right angles to a radius r, in metres, from a point,

turning moment or torque about that point
$$= Fr \text{ newton metres} \quad (1.3)$$

1.4 Unit of work or energy

The SI unit of energy is the *joule*† (after the English physicist, James P. Joule, 1818–89). *The joule is the work done when a force of* 1 *N acts through a distance of* 1 *m in the direction of the force.* Hence, if a force F acts through distance d in its own direction,

$$\text{work done} = F \text{ [newtons]} \times d \text{ [metres]}$$

$$= Fd \text{ joules} \quad (1.4)$$

If a body having mass m, in kilograms, is moving with velocity v, in metres/second,

$$\text{kinetic energy} = \tfrac{1}{2}mv^2 \text{ joules} \quad (1.5)$$

If a body having mass m, in kilograms, is lifted vertically through height h, in metres, and if g is the gravitational acceleration, in metres/second², in that region,

potential energy acquired by body = work done in lifting the body

$$= mgh \text{ joules} \quad (1.5A)$$
$$\simeq 9·81 \ mh.$$

---

* This expression can alternatively be stated thus:

weight of block $\simeq$ 80 [kg] × 9·81 [m/s²] = 784·8 N.

When the body under consideration is *stationary*, [N/kg] is better since it emphasizes the relationship between mass and weight, namely that a mass of 1 kg has a weight of approximately 9·81 N.

† *Joule* is pronounced 'jool', rhyming with 'tool'.

Example 1.3 *A body having a mass of 30 kg is supported 50 m above the earth's surface. What is its potential energy relative to the ground?*

If the body is allowed to fall freely, calculate its kinetic energy just before it touches the ground. Assume gravitational acceleration to be 9·81 m/s².

Weight of body = 30 [kg] × 9·81 [N/kg] = 294·3 N,

∴   potential energy = 294·3 [N] × 50 [m] = 14 715 J.

If v be the velocity of the body after it has fallen a distance s with an acceleration g,

$$v = \sqrt{(2\,gs)} = \sqrt{(2 \times 9\cdot81 \times 50)} = 31\cdot32 \text{ m/s,}$$

and   kinetic energy = $\frac{1}{2} \times$ 30 [kg] × (31·32)² [m/s]² = 14 715 J.

Hence the whole of the initial potential energy has been converted into kinetic energy. When the body is finally brought to rest by impact with the ground, practically the whole of this kinetic energy is converted into heat.

Specific heat capacity. Different substances absorb different amounts of heat to raise the temperature of a given mass of the substance by one degree. The heat required to raise the temperature of 1 kg of a substance by 1 degree is termed the *specific heat capacity* of that substance. Hence, if c represents the specific heat capacity of a substance in joules per kilogram kelvin, the heat required to raise the temperature of m kilograms of the substance by t degrees

$$= mct \text{ joules} \tag{1.6}$$

The following table gives the approximate values of the specific heat capacity of some well-known substances for a temperature range between 0°C and 100°C:

Substance	Specific heat capacity
Water	4190 J/kg·K
Copper	390 ,,
Iron	500 ,,
Aluminium	950 ,,

Example 1.4 *Calculate the mass of copper that can be heated from 10°C to 80°C by the absorption of 100 kJ of thermal energy.*

Increase of temperature = 80 − 10 = 70°C = 70K.

Quantity of heat = 100 kJ = 100 000 J.

Assuming the specific heat capacity of copper to be 390 J/kg·K,

$$100\ 000\ [\text{J}] = m \times 390\ [\text{J/kg·K}] \times 70\ [\text{K}]$$

∴ $m = 3 \cdot 66$ kg.

1.5 Unit of power

Since power is the rate of doing work, it follows that the SI unit of power is the *joule/second* or *watt* (after the Scottish engineer, James Watt, 1736–1819). In practice, the watt is often found to be inconveniently small and so the *kilowatt* is frequently used. Similarly, when we are dealing with a large amount of energy, it is often convenient to express the latter in *kilowatt hours* rather than in joules.

$$\begin{aligned} 1\ \text{kW·h} &= 1000\ \text{watt hours} \\ &= 1000 \times 3600\ \text{watt seconds or joules} \\ &= 3\ 600\ 000\ \text{J} = 3 \cdot 6\ \text{MJ}. \end{aligned}$$

If T is the torque, in newton metres, due to a force acting about an axis of rotation, and if n is the speed in revolutions/second,

$$\begin{aligned} \text{power} &= \text{torque in newton metres} \times \text{speed in radians/second} \\ &= T \times 2\pi n \ \text{joules/second or watts} \\ &= \omega T \ \text{watts} \end{aligned} \tag{1.7}$$

where ω = angular velocity in radians/second.

Example 1.5 *A stone block, having a mass of 120 kg, is hauled 100 m in 2 min along a horizontal floor. The coefficient of friction is 0·3. Calculate* (a) *the horizontal force required,* (b) *the work done and* (c) *the power.*

(a)   Weight of stone $\simeq 120\ [\text{kg}] \times 9 \cdot 81\ [\text{N/kg}] = 1177 \cdot 2$ N,

∴   force required $= 0 \cdot 3 \times 1177 \cdot 2\ [\text{N}] = 353 \cdot 16$ N.

(b)   Work done $= 353 \cdot 16\ [\text{N}] \times 100\ [\text{m}] = 35\ 316$ J

                     $= 35 \cdot 3$ kJ.

(c) $\text{Power} = \dfrac{35\ 316\ [\text{J}]}{(2 \times 60)\ [\text{s}]} = 294 \cdot 3$ W.

Example 1.6 *An electric motor is developing 10 kW at a speed of 900 rev/min. Calculate the torque available at the shaft.*

$$\text{Speed} = \frac{900\ [\text{rev/min}]}{60\ [\text{s/min}]} = 15\ \text{rev/s}.$$

Substituting in expression (1.7), we have:

$$10\ 000\ [W] = T \times 2\pi \times 15\ [rev/s]$$
$$\therefore \qquad\qquad T = 106\ \text{N·m}.$$

Example 1.7 *An electric heater is required to heat 15 litres of water from 12°C to the boiling point (100°C). Calculate* (a) *the electrical energy consumed* (i) *in megajoules,* (ii) *in kilowatt hours, and* (b) *the cost of the energy consumed if the charge is* 2 *p/kW·h. Assume the specific heat capacity of water to be* 4190 *J/kg·K,* 1 *litre of water to have a mass of* 1 *kg and the efficiency of the heater to be* 85 *per cent.*

(a)   Mass of water = 15 kg.

         Rise of temperature = 100 − 12 = 88°C = 88 K.

From expression (1.6),

useful heat = 15 [kg] × 4190 [J/kg·K] × 88 [K]
$$= 5\ 530\ 000\ J = 5{\cdot}53\ \text{MJ}.$$

Since   efficiency $= \dfrac{\text{useful energy}}{\text{energy absorbed}}$

i.e. $0{\cdot}85 = \dfrac{5{\cdot}53\ [\text{MJ}]}{\text{energy absorbed}}$

$\therefore$ energy absorbed = 6·51 MJ.

Since   1 kW·h = 3·6 MJ

$\therefore$ energy absorbed $= \dfrac{6{\cdot}51\ [\text{MJ}]}{3{\cdot}6\ [\text{MJ/kW·h}]} = 1{\cdot}8\ \text{kW·h}.$

(b)   Cost of energy = 1·8 [kW·h] × 2 [p/kW·h] = 3·6 p.

1.6 Electrical units

(a) *Current.* The unit of current is the *ampere* and is one of the SI base units mentioned in section 1.1.

The *ampere* is defined as that *current which, if maintained in two straight parallel conductors of infinite length, of negligible circular cross-section, and placed* 1 *metre apart in a vacuum, would produce between these conductors a force of* 2×10^{-7} *newton per metre of length*. The conductors are attracted towards each other if the currents are in the same direction, whereas they repel each other if the currents are in opposite directions.

The value of the current in terms of this definition can be determined by means of a very elaborately constructed balance in which the force between fixed and moving coils carrying the current is balanced by the force of gravity acting on a known mass.

(b) *Quantity of electricity.* The unit of electrical quantity is the *coulomb*, namely the *quantity of electricity passing a given point in a circuit when a current of* 1 *ampere is maintained for* 1 *second*. Hence,

$$Q \text{ [coulombs]} = I \text{ [amperes]} \times t \text{ [seconds]}.$$
1 ampere hour = 3600 coulombs.

When a current is passed through an electrolyte, chemical decomposition takes place, and the mass of an element liberated by 1 coulomb is termed the *electrochemical equivalent* of that element.

If z = electrochemical equivalent of an element in milligrams per coulomb,
and I = current, in amperes, for time t, in seconds,
  mass of element liberated = zIt milligrams.

The electrochemical equivalents of copper and silver are 0·3294 mg/C and 1·1182 mg/C respectively.

The value of a direct current can be determined with considerable accuracy by passing the current through either a copper or a silver voltameter (i.e. two copper plates in a copper sulphate solution or two silver plates in a silver nitrate solution) for a given time and noting the increase in the mass of the negative plate (or cathode).

(c) *Potential difference.** The unit of potential difference is the *volt*, namely *the difference of potential between two points of a conducting wire carrying a current of* 1 *ampere, when the power dissipated between these points is equal to* 1 *watt*.

(d) *Resistance.* The unit of electric resistance is the *ohm*, namely *the resistance between two points of a conductor when a potential difference of* 1 *volt, applied between these points, produces in this conductor a current of* 1 *ampere, the conductor not being a source of any electromotive force*.

---

* The term *voltage* originally meant a difference of potential expressed in volts; but it is now used as a synonym for potential difference irrespective of the unit in which it is expressed. For instance, the voltage between the lines of a transmission system may be 400 kV, while in communication and electronic circuits, the voltage between two points may be 5μV.

Alternatively, the *ohm* can be defined as *the resistance of a circuit in which a current of* 1 *ampere generates heat at the rate of* 1 *watt*.

If V represents the p.d., in volts, across a circuit having resistance R, in ohms, carrying a current I, in amperes, for time t, in seconds,

$$V = IR \text{ or } I = V/R \text{ [by Ohm's Law]}$$
$$\text{power} = IV = I^2R = V^2/R \text{ watts}$$
and $$\text{heat generated} = I^2Rt = IVt \text{ joules.}$$

(*e*) *Electromotive force.* An electromotive force is that which tends to produce an electric current in a circuit, and the *unit of e.m.f.* is the *volt*. The principal sources of e.m.f. are:

(i) the electrodes of dissimilar materials immersed in an electrolyte, as in primary and secondary cells (section 23.1);

(ii) the relative movement of a conductor and a magnetic flux, as in electric generators and transformers; this source can, alternatively, be expressed as the variation of magnetic flux linked with a coil (sections 2.11 and 2.12);

(iii) the difference of temperature between junctions of dissimilar metals, as in thermo-junctions (section 22.7).

Consider a circuit of resistance R ohms to be connected across a cell having an internal resistance r ohms, as in fig. 1.2, and suppose I amperes to be the current, then:

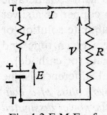

$$\text{p.d. across } R = IR \text{ volts}$$
and $$\text{p.d. across } r = Ir \text{ volts.}$$
$$\left.\begin{array}{l}\text{Total power dis-}\\\text{sipated in } R \text{ and } r\end{array}\right\} = I^2(R + r) \text{ watts.}$$

Fig. 1.2 E.M.F. of a cell.

The e.m.f. of the cell is responsible for maintaining the electric current in the circuit; and the value of the e.m.f., E volts, must be such that the electrical power generated by chemical action in the cell is equal to that dissipated as heat in the resistance of the circuit,

i.e. $$IE = I^2(R + r)$$
∴ $$E = I(R + r)$$

It follows that the difference of potential, V volts, across terminals TT (fig. 1.2) is given by:

$$V = IR = E - Ir$$

1.7 Electrical reference standards

It was mentioned in section 1.6 (*a*) that the absolute value of the ampere is determined by means of a current balance. The fixed and moving coils are made to known dimensions and from these dimensions and the relative position of the coils, the value of the current to exert a certain force can be calculated. Also, the absolute value of the ohm can be determined by means of a copper disc rotated in a uniform magnetic field. Both of these methods require such elaborate and expensive equipment that they are only carried out at infrequent intervals in national laboratories. During intervals between the

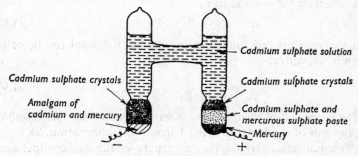

Fig. 1.3 Weston cadmium standard cell.

absolute determinations, use is made of standards of resistance and of e.m.f. in the form of manganin*-wire resistors and Weston cells respectively.

The Weston cadmium standard cell is shown in section in fig. 1.3. The positive electrode is mercury and the negative electrode is an amalgam of mercury and cadmium. The electrolyte is a saturated solution of cadmium sulphate in water slightly acidulated with sulphuric acid. The mercurous sulphate acts as a depolarizer. When this cell is manufactured to a specification prescribed by the International Electrotechnical Commission, its e.m.f. is exactly 1·018 59 volts at 20°C. Consequently the volt can be taken as 1/1·018 59 of the e.m.f. of a cadmium cell at 20°C. The e.m.f. of the cell falls by about 40 microvolts per degree rise of temperature. Since the internal resistance is about 1000 Ω, this cell is intended only as a standard of e.m.f. and not as a source of electrical energy.

* Manganin is an alloy of copper, manganese and nickel, with sometimes a trace of iron; its temperature coefficient of resistance may be as low as 0·000 003 per degree. (See *Dictionary of Applied Physics*, Vol. II, p. 710.)

1.8 Resistance

A resistor (i.e. a wire or other form of material used simply because of its resistance) is said to be *linear* if the current through the resistor is proportional to the p.d. across its terminals. If the resistance varies with the magnitude of the current or voltage, the resistor is said to be *non-linear*. A semiconductor junction rectifier (section 20.7) is an example of a non-linear resistor. In this book, resistors will normally be assumed to be linear, i.e. their resistance will be assumed to remain constant when the temperature is maintained constant.

If resistors of resistance R_1, R_2, R_3, etc., are connected in series, the total resistance R is given by:

$$R = R_1 + R_2 + R_3 + \ldots \tag{1.8}$$

If the resistors are connected in parallel, the total resistance is given by R, where:

$$\frac{1}{R} = \frac{1}{R_1} + \frac{1}{R_2} + \frac{1}{R_3} + \ldots \tag{1.9}$$

The reciprocal of the resistance is termed the *conductance* (symbol G), the unit of conductance being 1 *siemens** (abbreviation, S).

For conductances G_1, G_2, G_3, etc., in parallel, the total conductance G is given by:

$$G = G_1 + G_2 + G_3 + \ldots \tag{1.10}$$

If l = length of an electrical circuit, in metres
and a = cross-sectional area of the circuit, in square metres

$$\text{resistance of circuit} = R = \rho l/a \text{ ohms} \tag{1.11}$$

where ρ = resistivity of the material.

$$= \frac{R \text{ [ohms]} \times a \text{ [metres}^2]}{l \text{ [metres]}}$$

$$= Ra/l \text{ ohm metres,}$$

e.g. the resistivity of annealed copper at 20°C is $1 \cdot 725 \times 10^{-8} \, \Omega$ m.

1.9 Temperature coefficient of resistance

The resistance of all pure metals increases with increase of temperature, whereas the resistance of carbon, electrolytes and insulating

---

* The brothers Werner von Siemens (1816–92) and Sir William Siemens, F.R.S. (1823–83) were pioneers of electrical engineering, scientifically and industrially. William Siemens was, in 1872, the first President of the I.E.E.

materials decreases with increase of temperature. Certain alloys, such as manganin (section 1.7) show practically no change of resistance for a considerable variation of temperature. For a moderate range of temperature, such as 100°C, the change of resistance is usually proportional to the change of temperature; and the ratio of the change of resistance per degree change of temperature to the resistance at some definite temperature, adopted as standard, is termed *the temperature coefficient of resistance* and is represented by the Greek letter α.

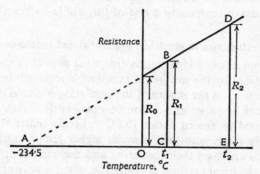

Fig. 1.4 Variation of resistance of copper with temperature.

The variation of resistance of copper for a range over which copper conductors are usually operated is represented by the graph in fig. 1.4. If this graph is extended backwards, the point of intersection with the horizontal axis is found to be −234·5°C. Hence, for a copper conductor having a resistance of 1 Ω at 0°C, the change of resistance for 1°C change of temperature is $(1/234·5)$ Ω, namely 0·004 264 Ω,

$$\therefore \quad a_0 = \frac{0·004\ 264\ [\Omega/°C]}{1\ [\Omega]} = 0·004\ 264/°C.$$

In general, if a material having a resistance R_0 at 0°C has a resistance R_1 at t_1 and R_2 at t_2, and if a_0 is the temperature coefficient of resistance at 0°C,

$$R_1 = R_0\left(1 + a_0 t_1\right) \text{ and } R_2 = R_0\left(1 + a_0 t_2\right)$$

$$\therefore \qquad \frac{R_1}{R_2} = \frac{1 + a_0 t_1}{1 + a_0 t_2} \qquad\qquad (1.12)$$

It is often the practice to assume the standard temperature to be

20°C, which is roughly the average atmospheric temperature. This involves using a different value for the temperature coefficient of resistance, e.g. the temperature coefficient of resistance of copper at 20°C is 0·003 92/°C. Hence, for a material having a resistance R_{20} at 20°C and temperature coefficient of resistance a_{20} at 20°C, the resistance R_t at temperature t is given by:

$$R_t = R_{20}\{1 + a_{20}(t - 20)\} \qquad (1.12A)$$

Hence if the resistance of a coil, such as a field winding of an electrical machine, is measured at the beginning and at the end of a test, the mean temperature rise of the whole coil can be calculated.

1.10 Temperature rise in electrical apparatus and machines

The maximum power which can be dissipated as heat in an electrical circuit is limited by the maximum permissible temperature, and the latter depends upon the nature of the insulating material employed. Materials such as paper and cotton become brittle if their temperature is allowed to exceed about 100°C; whereas materials such as mica and glass can withstand a much higher temperature without any injurious effect on their insulating and mechanical properties.

When an electrical machine is *loaded* (i.e. when supplying electrical power in the case of a generator or mechanical power in the case of a motor), the temperature rise of the machine is largely due to the I^2R loss in the windings; and the greater the load, the greater is the loss and therefore the higher the temperature rise. The *full load* or *rated output* of the machine is the maximum output obtainable under specified conditions, e.g. for a specified temperature rise after the machine has supplied that load continuously for several hours.

The temperature of a coil may be measured by (i) a thermometer, (ii) the increase of resistance of the coil and (iii) thermo-junctions embedded in the coil. The third method enables the distribution of temperature throughout the coil to be determined but is only possible if the thermo-junctions are inserted in the coil when the latter is being wound. Since the heat generated at the centre of the coil has to flow outwards, the temperature at the centre may be considerably higher than that at the surface.

1.11 Maximum power transfer

Let us consider a source, such as a battery or a d.c. generator, having an e.m.f. E and an internal resistance r, as shown enclosed by the dotted rectangle in fig. 1.5. A variable resistor is connected across terminals A and B of the source. If the value of the load resistance is R, then

$$I = E/(r + R)$$

and power transferred to load $= I^2R$

$$= \frac{E^2R}{(r + R)^2} = \frac{E^2R}{r^2 + 2rR + R^2}$$

$$= \frac{E^2}{(r^2/R) + 2r + R} \tag{1.13}$$

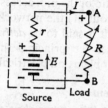

Fig. 1.5 Resistance matching.

This power is a maximum when the denominator of (1.13) is a minimum,

i.e. when

$$\frac{d}{dR}\{(r^2/R) + 2r + R\} = 0$$

$$\therefore \qquad - (r^2/R^2) + 1 = 0$$

or $\qquad R = r \tag{1.14}$

To check that this condition gives the minimum and not the maximum value of the denominator in expression (1.13), expression $\{-(r^2/R^2) + 1\}$ should be differentiated with respect to R, thus:

$$\frac{d}{dR}\{1 - (r^2/R^2)\} = 2r^2/R^3.$$

Since this quantity is positive, expression (1.14) is the condition for the denominator of (1.13) to be a minimum and therefore the output power to be a maximum. Hence the power transferred from the source to the load is a maximum when the resistance of the load is equal to the internal resistance of the source. This condition is referred to as *resistance matching*.

Resistance matching is of importance in communication and

electronic circuits where the source usually has a relatively high resistance and where it is desired to transfer the largest possible amount of power from the source to the load. In the case of d.c. generators and secondary batteries, the internal resistance is so low that it is impossible to satisfy the above condition without overloading the source.

1.12 Network theorems

A network consists of a number of branches or circuit elements, considered as a unit, and is said to be *passive* if it contains no source of e.m.f. The equivalent resistance between any two terminals of a passive network is the ratio of the p.d. across the two terminals to the current flowing into (or out of) the network. When a network contains a source of e.m.f., it is said to be *active*.

We shall now consider the principal theorems that have been developed for solving problems on electrical networks.

1.13 Superposition theorem

In a linear network containing more than one source of e.m.f., the resultant current in any branch is the algebraic sum of the currents that would be produced by each e.m.f., acting alone, all the other sources of e.m.f. being replaced meanwhile by their respective internal resistances.

Let us consider the application of this theorem to the solution of the following problem.

Example 1.8 *Battery A in fig. 1.6 has an e.m.f. of 6 V and an internal resistance of 2 Ω. The corresponding values for battery B are 4 V and 3 Ω respectively. The two batteries are connected in parallel across a*

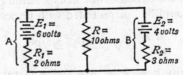

Fig. 1.6 Circuit diagram for example 1.8.

10-Ω resistor R. Calculate the current in each branch of the network.

Fig. 1.7 (*a*) represents the network with battery A only.

$$\left.\begin{array}{l}\text{Equivalent resistance of } R \text{ and } R_2 \\ \text{in parallel}\end{array}\right\} = \frac{10 \times 3}{10 + 3} = 2 \cdot 31 \ \Omega.$$

$$\therefore \qquad I_1 = \frac{6}{2 + 2\cdot31} = 1\cdot392 \text{ A.}$$

Hence,
$$I_2 = 1\cdot392 \times \frac{3}{10 + 3} = 0\cdot321 \text{ A}$$

and
$$I_3 = 1\cdot392 - 0\cdot321 = 1\cdot071 \text{ A.}$$

Fig. 1.7 (*b*) represents the network with battery B only.

Equivalent resistance of R and R_1 in parallel $\left.\right\} = \dfrac{2 \times 10}{2 + 10} = 1\cdot667 \ \Omega.$

$$\therefore \qquad I_4 = \frac{4}{3 + 1\cdot667} = 0\cdot856 \text{ A.}$$

Hence,
$$I_5 = 0\cdot856 \times \frac{2}{2 + 10} = 0\cdot143 \text{ A}$$

and
$$I_6 = 0\cdot856 - 0\cdot143 = 0\cdot713 \text{ A.}$$

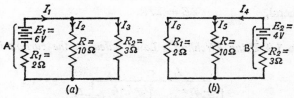

Fig. 1.7 Diagrams for solution of example 1.8.

Superimposing the results for fig. 1.7 (*b*) on those for fig. 1.7 (*a*), we have:

$$\text{resultant current through A} = I_1 - I_6$$
$$= 1\cdot392 - 0\cdot713 = 0\cdot679 \text{ A}$$
$$\text{and resultant current through B} = I_4 - I_3$$
$$= 0\cdot856 - 1\cdot071$$
$$= -0\cdot215 \text{ A,}$$

i.e. battery B is being charged at $0\cdot215$ A.

$$\text{Resultant current through } R = I_2 + I_5$$
$$= 0\cdot321 + 0\cdot143 = 0\cdot464 \text{ A.}$$

1.14 Kirchhoff's Laws

First Law *The total current flowing towards a node* is equal to the total current flowing away from that node,* i.e. *the algebraic sum*

* A *node* of a network is defined as the junction point of two or more branches of that network; e.g. C in fig. 1.8 is a node of three branches.

of the currents flowing towards a node is zero. Thus at node C in fig. 1.8,

$$I_1 + I_2 = I_3 \quad \text{or} \quad I_1 + I_2 - I_3 = 0$$

In general,
$$\Sigma I = 0 \tag{1.15}$$

where Σ represents the algebraic sum.

Second Law* *In a closed circuit, the algebraic sum of the products of the current and the resistance of each part of the circuit is equal to the resultant e.m.f. in the circuit.* Thus for the closed circuit involving E_1, E_2, R_1 and R_2 in fig. 1.8,

$$E_1 - E_2 = I_1 R_1 - I_2 R_2$$

and for the mesh involving E_2, R_2 and R,

$$E_2 = I_2 R_2 + I_3 R$$

In general,
$$\Sigma E = \Sigma IR. \tag{1.16}$$

Example 1.9 *Using Kirchhoff's Laws, calculate the current in each*

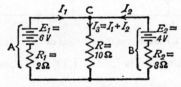

Fig. 1.8 Circuit to illustrate Kirchhoff's Laws.

branch of the circuit shown in fig. 1.8, *the e.m.f.s and resistances of which are the same as those of fig.* 1.6.

Applying Kirchhoff's Laws to the circuit formed by A and R, we have:

$$6 = 2I_1 + 10(I_1 + I_2) = 12I_1 + 10I_2 \tag{1.17}$$

Similarly, for the closed circuit formed by A and B,

$$6 - 4 = 2 = 2I_1 - 3I_2 \tag{1.18}$$

Multiplying (1.18) by 6 and subtracting from (1.17), we have:

$$- 6 = 28I_2$$
$$\therefore \qquad I_2 = - 0 \cdot 2143 \text{ A.}$$

* See Note on p. 38 concerning Kirchhoff's Second Law.

Substituting for I_2 in (1.18), we have:

$$I_1 = 1 - 1.5 \times 0.2143 = 0.6786 \text{ A}$$

and $\qquad I_3 = 0.6786 - 0.2143 = 0.4643 \text{ A}.$

Example 1.10 *Three similar primary cells are connected in series to form a closed circuit as shown in fig. 1.9. Each cell has an e.m.f. of 1·5 V and an internal resistance of 30 Ω. Calculate the current and show that points* A, B *and* C *are at the same potential.**

In fig. 1.9, E and R represent the e.m.f. and internal resistance respectively of each cell.

$$\text{Total e.m.f.} = 1.5 \times 3 = 4.5 \text{ V},$$
$$\text{total resistance} = 30 \times 3 = 90 \text{ Ω}$$
$$\therefore \qquad \text{current} = 4.5/90 = 0.05 \text{ A}.$$

The voltage drop due to the internal resistance of each cell is 0.05×30, namely 1·5 V. Hence the e.m.f. of each cell is absorbed in

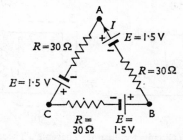

Fig. 1.9 Circuit diagram for example 1.10.

sending the current through the internal resistance of that cell, so that there is no difference of potential between the two terminals of the cell. Consequently the three junctions A, B and C are at the same potential.†

* It can be interesting to draw the circuit of fig. 1.9 on the blackboard and ask students which of points A and B is at the higher potential!

† This result has important practical applications, e.g. in connection with the non-existence of a third harmonic in the terminal voltage of a delta-connected 3-phase machine or transformer. The same principle can be applied to explain why there is no magnetic leakage in a toroid uniformly wound with a magnetizing winding—the m.m.f. per unit length is absorbed in sending the magnetic flux through the reluctance of that length, irrespective of how small that length may be. Hence, all points of the toroid are at the same magnetic potential.

1.15 Thévenin's Theorem

The current through a resistor R *connected across any two points* A *and* B *of an active network* (i.e. a network containing one or more sources of e.m.f.) *is obtained by dividing the p.d. between* A *and* B, *with* R *disconnected, by* $(R + r)$, *where* r *is the resistance of the network measured between points* A *and* B *with* R *disconnected and the sources of e.m.f. replaced by their internal resistances.*

An alternative way of stating Thévenin's Theorem is as follows: *An active network having two terminals* A *and* B *can be replaced by a constant-voltage source having an e.m.f.* E *and an internal resistance* r. *The value of* E *is equal to the open-circuit p.d. between* A *and* B,

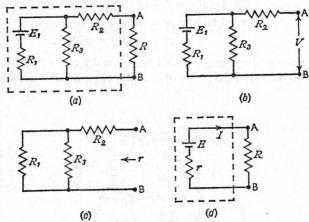

Fig. 1.10 Circuits to illustrate Thévenin's Theorem.

and r *is the resistance of the network measured between* A *and* B *with the load disconnected and the sources of e.m.f. replaced by their internal resistances.*

Suppose A and B in fig. 1.10 (*a*) to be the two terminals of a network consisting of resistors having resistances R_2 and R_3 and a battery having an e.m.f. E_1 and an internal resistance R_1. It is required to determine the current through a load of resistance R connected across AB. With the load disconnected as in fig. 1.10 (*b*),

$$\text{current through } R_3 = \frac{E_1}{R_1 + R_3}$$

and

$$\text{p.d. across } R_3 = \frac{E_1 R_3}{R_1 + R_3}.$$

Since there is no current through R_2,

$$\text{p.d. across AB} = V = \frac{E_1 R_3}{R_1 + R_3}.$$

Fig. 1.10 (c) shows the network with the load disconnected and the battery replaced by its internal resistance R_1.

$$\text{Resistance of network between A and B} = r = R_2 + \frac{R_1 R_3}{R_1 + R_3}.$$

Thévenin's Theorem merely states that the active network enclosed by the dotted line in fig. 1.10 (a) can be replaced by the very simple circuit enclosed by the dotted line in fig. 1.10 (d) and consisting of a source having an e.m.f. E equal to the open-circuit potential difference V between A and B, and an internal resistance r, where V and r have the values determined above. Hence,

$$\text{current through } R = I = \frac{E}{r + R}.$$

Thévenin's Theorem—sometimes referred to as Helmholtz's Theorem—is an application of the Superposition Theorem. Thus, if a source having an e.m.f. E equal to the open-circuit p.d. between A and B in fig. 1.10 (b) were inserted in the circuit between R and terminal A in fig. 1.10 (a), the positive terminal of the source being connected to A, no current would flow through R. Hence, this source could be regarded as circulating through R a current superimposed upon but opposite in direction to the current through R due to E_1 *alone*. Since the resultant current is zero, it follows that a source of e.m.f. E connected in series with R and the equivalent resistance r of the network, as in fig. 1.10 (d), would circulate a current I having the same value as that through R in fig. 1.10 (a); but in order that the direction of the current through R may be from A towards B, the polarity of the source must be as shown in fig. 1.10 (d).

Example 1.11 C *and* D *in fig.* 1.11 (a) (*which is similar to fig.* 1.7) *represent the two terminals of an active network. Calculate the current through R.*

With R disconnected as in fig. 1.11 (b),

$$I_1 = \frac{6 - 4}{2 + 3} = 0 \cdot 4 \text{ A}$$

and $$\text{p.d. across CD} = E_1 - I_1 R_1$$

i.e. $$V = 6 - (0 \cdot 4 \times 2) = 5 \cdot 2 \text{ V}.$$

When the e.m.f.s are removed as in fig. 1.11 (c),

$$\text{total resistance between C and D} = \frac{2 \times 3}{2 + 3}$$

i.e. $r = 1 \cdot 2 \ \Omega.$

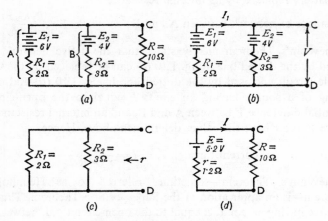

Fig. 1.11 Circuit diagrams for example 1.11.

Hence the network AB in fig. 1.11 (a) can be replaced by a single source having an e.m.f. of 5·2 V and an internal resistance of 1·2 Ω, as in fig. 1.11 (d); consequently,

$$I = \frac{5 \cdot 2}{1 \cdot 2 + 10} = 0 \cdot 4643 \ \text{A},$$

namely the value obtained in examples 1.8 and 1.9.

Example 1.12 *The resistances of the various arms of an unbalanced Wheatstone bridge are given in fig.* 1.12. *The battery has an e.m.f. of 2 V and a negligible internal resistance. Determine the value and direction of the current in the galvanometer circuit* BD, *using* (a) *Kirchhoff's Laws and* (b) *Thévenin's Theorem.*

(a) *By Kirchhoff's Laws.* Let I_1, I_2 and I_3 be the currents in arms AB, AD and BD respectively, as shown in fig. 1.12. Then by Kirchhoff's First Law,

current in BC = $I_1 - I_3$

and   current in DC = $I_2 + I_3$.

Applying Kirchhoff's Second Law to the mesh formed by ABC and the battery, we have:

$$2 = 10I_1 + 30(I_1 - I_3)$$
$$= 40I_1 - 30I_3 \tag{1.19}$$

Similarly for mesh ABDA,

$$0 = 10I_1 + 40I_3 - 20I_2 \tag{1.20}$$

and for mesh BDCB,

$$0 = 40I_3 + 15(I_2 + I_3) - 30(I_1 - I_3)$$
$$= -30I_1 + 15I_2 + 85I_3 \tag{1.21}$$

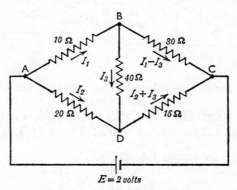

Fig. 1.12 Circuit for example 1.12.

Multiplying (1.20) by 3 and (1.21) by 4, and adding the two expressions thus obtained, we have:

$$0 = -90I_1 + 460I_3$$
$$\therefore \qquad I_1 = 5 \cdot 111 I_3.$$

Substituting for I_1 in (1.19), we have:

$$I_3 = 0 \cdot 0115 \text{ A} = 11 \cdot 5 \text{ mA}.$$

Since the value of I_3 is positive, the direction of I_3 is that assumed in fig. 1.12, namely from B to D.

(*b*) *By Thévenin's Theorem.* Since we require to find the current in the 40-Ω resistor between B and D, the first step is to remove this resistor, as in fig. 1.13 (*a*). Then:

$$\text{p.d. between A and B} = 2 \times \frac{10}{10 + 30} = 0.5 \text{ V}$$

and $\qquad$ p.d. between A and D $= 2 \times \dfrac{20}{20 + 15} = 1.143$ V

$\therefore \qquad$ p.d. between B and D $= 0.643$ V,

B being positive relative to D. Consequently, current in the 40-Ω resistor, when connected between B and D, will flow from B to D.

The next step is to replace the battery by a resistance equal to its

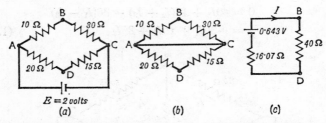

$(a) \qquad\qquad\qquad (b) \qquad\qquad\qquad (c)$

Fig. 1.13 Diagrams for solution of example 1.12 by Thévenin's Theorem.

internal resistance. Since the latter is negligible in this problem, junctions A and C can be short-circuited as in fig. 1.13 (*b*).

Equivalent resistance of BA and BC

$$= \frac{10 \times 30}{10 + 30} = 7.5 \ \Omega$$

and equivalent resistance of AD and CD

$$= \frac{20 \times 15}{20 + 15} = 8.57 \ \Omega,$$

$\therefore \quad$ total resistance of network between B and D

$$= 16.07 \ \Omega.$$

Hence the network of fig. 1.13 (*a*) is equivalent to a source having an e.m.f. of 0·643 V and an internal resistance of 16·07 Ω as in fig. 1.13 (*c*).

$\therefore \qquad$ current through BD $= \dfrac{0.643}{16.07 + 40} = 0.0115$ A

$$= 11.5 \text{ mA from B to D.}$$

1.16 Delta–star transformation

Fig. 1.14 (*a*) shows three resistors R_1, R_2 and R_3 connected in a closed mesh or *delta* to three terminals A, B and C, *their numerical subscripts* 1, 2 *and* 3, *being opposite to the terminals* A, B *and* C *respectively*. It is possible to replace these delta-connected resistors by three resistors R_a, R_b and R_c connected respectively between the same terminals A, B and C and a common point S, as in fig. 1.14 (*b*).

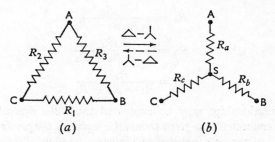

(*a*) (*b*)

Fig. 1.14 Delta–star transformation.

Such an arrangement is said to be *star-connected*. It will be noted that the letter subscripts are now those of the terminals to which the respective resistors are connected. If the star-connected network is to be equivalent to the delta-connected network, the resistance between any two terminals in fig. 1.14 (*b*) must be the same as that between the same two terminals in fig. 1.14 (*a*). Thus, if we consider terminals A and B in fig. 1.14 (*a*), we have a circuit having a resistance R_3 in parallel with a circuit having resistances R_1 and R_2 in series; hence

$$R_{AB} = \frac{R_3(R_1 + R_2)}{R_1 + R_2 + R_3} \qquad (1.22)$$

For fig. 1.14 (*b*), we have:

$$R_{AB} = R_a + R_b \qquad (1.23)$$

In order that the networks of fig. 1.14 (*a*) and (*b*) may be equivalent to each other, the values of R_{AB} represented by expressions (1.22) and (1.23) must be equal,

$$\therefore \qquad R_a + R_b = \frac{R_1 R_3 + R_2 R_3}{R_1 + R_2 + R_3} \qquad (1.24)$$

Similarly, $$R_b + R_c = \frac{R_1 R_2 + R_1 R_3}{R_1 + R_2 + R_3} \qquad (1.25)$$

and
$$R_a + R_c = \frac{R_1 R_2 + R_2 R_3}{R_1 + R_2 + R_3} \qquad (1.26)$$

Subtracting (1.25) from (1.24), we have:

$$R_a - R_c = \frac{R_2 R_3 - R_1 R_2}{R_1 + R_2 + R_3} \qquad (1.27)$$

Adding (1.26) and (1.27) and dividing by 2, we have:

$$R_a = \frac{R_2 R_3}{R_1 + R_2 + R_3} \qquad (1.28)$$

Similarly,
$$R_b = \frac{R_3 R_1}{R_1 + R_2 + R_3} \qquad (1.29)$$

and
$$R_c = \frac{R_1 R_2}{R_1 + R_2 + R_3} \qquad (1.30)$$

These relationships may be expressed thus: *the equivalent star resistance connected to a given terminal is equal to the product of the two delta resistances connected to the same terminal divided by the sum of the delta resistances.*

1.17 Star–delta transformation

Let us next consider how to replace the star-connected network of fig. 1.14 (b) by the equivalent delta-connected network of fig. 1.14 (a). Dividing equation (1.28) by equation (1.29), we have:

$$R_a/R_b = R_2/R_1$$
$$\therefore \qquad R_2 = R_1 R_a/R_b$$

Similarly, dividing (1.28) by (1.30), we have:

$$R_a/R_c = R_3/R_1$$
$$\therefore \qquad R_3 = R_1 R_a/R_c$$

Substituting for R_2 and R_3 in (1.28), we have:

$$R_1 = R_b + R_c + R_b R_c/R_a$$
Similarly,
$$R_2 = R_c + R_a + R_c R_a/R_b$$
and
$$R_3 = R_a + R_b + R_a R_b/R_c$$

These relationships may be expressed thus: *the equivalent delta resistance between two terminals is the sum of the two star resistances connected to those terminals plus the product of the same two star resistances divided by the third star resistance.*

1.18 Two-wire d.c. system of distribution

A d.c. system is usually supplied either from d.c. generators or from rectifiers at a voltage that is maintained approximately constant. In fact, one of the regulations governing the distribution of electrical energy stipulates that the voltage at a consumer's premises must not vary by more than ± 6 per cent; for instance, if the consumer is supplied at a nominal voltage of 240 V, the actual voltage should not exceed 254 V or fall below 226 V.

Fig. 1.15 gives the general arrangement of a distribution system. Two d.c. generators DD are shown connected in parallel to bus-bars BB. The bus-bars are two copper bars that extend the whole length

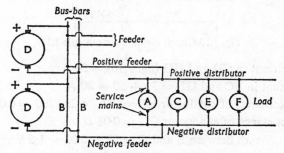

Fig. 1.15 A distribution system.

of the switchboard, 'bus' being an abbreviation of 'omnibus', a Latin word meaning 'for all'. Thus, all the generators at the generating station and all the cables connecting the station to various 'feeding' points are connected to the bus-bars. The cables radiating from the station are called *feeders*, whereas *distributors* are cables to which the *service mains* supplying the individual consumers are connected. A distributor is connected at one or more points to feeders, but no service mains are connected to the latter.

In practice, the distributors are interconnected to form a network that is almost like that of a spider's web. This network is connected to feeders at suitable points. Such an arrangement has the advantage that if for some reason a feeder has to be disconnected, the current to the section normally supplied by that feeder can still be supplied through other feeders and distributors.

Example 1.13 *Two loads, A and B (fig. 1.16), taking 50 A and 30 A respectively, are connected to a two-wire distributor at distances of*

200 *m and* 300 *m respectively from the feeding point, the p.d. at which is* 120·*V. The resistance of the distributor is* 0·01 Ω *per* 100 *m of single conductor. Find:* (a) *the p.d. across each load,* (b) *the cost of the energy wasted in the distributor if the above loads are maintained constant for* 10 *hours. Assume the cost of energy to be* 1·2 *p/kW h.*

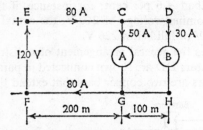

Fig. 1.16 Circuit diagram for example 1.13.

(*a*) Current in CD and HG (fig. 1.16) = 30 A

and current in EC and GF = 30 + 50 = 80 A

Resistance of conductor EC = 0·01 × 200/100 = 0·02 Ω

and resistance of conductor CD = 0·01 Ω

Hence,   p.d. between E and C = 80 × 0·02 = 1·6 V

Similarly,   p.d. between G and F = 1·6 V

But   p.d. between E and F = sum of the p.d.s in circuit ECGF

i.e.   120 = 1·6 + p.d. between C and G + 1·6.

If the voltage drop in the service main be neglected,
     p.d. across load A = 120 − 3·2 = 116·8 V.

Also

     p.d. between C and D = 30 × 0·01 = 0·3 V

and p.d. between H and G = 0·3 V

But p.d. between C and G = sum of p.d.s in circuit CDHG

i.e.   116·8 = 0·3 + p.d. between D and H + 0·3

so that p.d. across load B = 116·8 − 0·6 = 116·2 V.

(*b*) Power wasted in conductors EC and FG

= current × voltage drops in EC and FG

= 80 × 3·2 = 256 W

Power wasted in conductors CD and HG

$$= 30 \times 0.6 = 18 \text{ W}$$

∴   total power wasted in distributor

$$= 256 + 18 = 274 \text{ W}$$

$$= 0.274 \text{ kW}$$

and energy wasted in 10 hours $= 0.274 \times 10 = 2.74$ kW h

∴   cost of this energy $= 2.74 \times 1.2 = 3.29$ p.

Example 1.14 *A two-wire ring distributor (i.e. a distributor in which each conductor forms a complete circuit or loop, as in fig 1.17) is 300 m long and is fed at 240 V at A. At a point B, 150 m from A, there is*

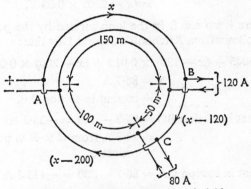

Fig. 1.17 Circuit diagram for example 1.14.

a load of 120 A and at C, 100 m in the opposite direction, there is a load of 80 A. The resistance per 100 m of single conductor is 0.03 Ω. Find: (a) *the current in each section,* (b) *the p.d.s at B and C.*

(a) Let x be the current, in amperes, from A to B in the positive conductor.

From Kirchhoff's First Law, it follows that the current from B to C in positive conductor

$$= x - 120 \text{ A}$$

and current from C to A in positive conductor

$$= x - 120 - 80 = x - 200 \text{ A}.$$

Resistance of positive conductor between A and B

$$= 0.03 \times 150/100 = 0.045 \text{ Ω},$$

resistance of positive conductor between B and C

$$= 0.03 \times 50/100 = 0.015 \ \Omega$$

and resistance of positive conductor between C and A

$$= 0.03 \ \Omega.$$

Hence, voltage drop in positive conductor between A and B

$$= x \times 0.045 \text{ V},$$

voltage drop in positive conductor between B and C

$$= (x - 120) \times 0.015 \text{ V}$$

and voltage drop in positive conductor between C and A

$$= (x - 200) \times 0.03 \text{ V}.$$

Since there is no e.m.f. in the loop formed by the positive conductor, it follows from Kirchhoff's Second Law that:

$$x \times 0.045 + (x - 120) \times 0.015 + (x - 200) \times 0.03 = 0$$

$$\therefore \qquad x = 86.7 \text{ A}$$

$$= \text{current in section AB.}$$

$$\text{Current in section BC} = 86.7 - 120 = -33.3 \text{ A}$$

$$= 33.3 \text{ A from C to B in positive conductor}$$

and current in section CA $= 86.7 - 200 = -113.3$ A

$$= 113.3 \text{ A from A to C in positive conductor.}$$

(*b*) Voltage drop in positive and negative conductors between A and B

$$= 86.7 \times 0.045 \times 2 = 7.8 \text{ V}$$

and voltage drop in positive and negative conductors between A and C

$$= 113.3 \times 0.03 \times 2 = 6.8 \text{ V}$$

$$\therefore \qquad \text{p.d. across load at B} = 240 - 7.8 = 232.2 \text{ V}$$

and „ „ „ C $= 240 - 6.8 = 233.2$ V.

The difference of 1 V between the p.d.s across the loads at B and C should agree with the voltage drop calculated from the current in section BC and the resistance of that section, namely 33·3 ×

$0.015 \times 2 = 0.999$ V. The very slight discrepancy between the two values is due to the fact that the value of x was limited to three significant figures. This degree of accuracy is sufficient for most practical purposes.

Example 1.15. *A distributor, 450 m long, is loaded as shown in fig. 1.18. The p.d. at* AB *is 250 V and that at* CD *is 240 V. The resistance of the distributor is 0.02 Ω per 100 m of single conductor. Calculate the p.d. at each load point.*

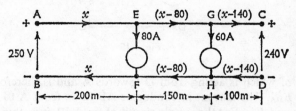

Fig. 1.18 Circuit diagram for example 1.15.

Let x be the current, in amperes, in AE and FB.
From Kirchhoff's First Law,

$$\text{current in EG and HF} = x - 80 \text{ A}$$

and
$$\text{current in GC and DH} = x - 140 \text{ A.}$$

$$\text{Resistance of AE and FB} = 0.02 \times 4 = 0.08 \ \Omega,$$
$$\text{resistance of EG and HF} = 0.02 \times 3 = 0.06 \ \Omega$$

and
$$\text{resistance of GC and DH} = 0.02 \times 2 = 0.04 \ \Omega.$$

∴
$$\text{voltage drop in AE and FB} = 0.08x \text{ V,}$$
$$\text{voltage drop in EG and HF} = 0.06(x - 80) \text{ V}$$

and
$$\text{voltage drop in GC and DH} = 0.04(x - 140) \text{ V.}$$

But voltage drop in AC and DB $= 250 - 240 = 10$ V,
∴ $0.08x + 0.06(x - 80) + 0.04(x - 140) = 10$

so that
$$x = 113.3 \text{ A.}$$

Hence,
$$\text{p.d. across load EF} = 250 - (113.3 \times 0.08)$$
$$= 240.93 \text{ V}$$

and
$$\text{p.d. across load GH} = 240.93 - (33.3 \times 0.06)$$
$$= 238.93 \text{ V.}$$

Alternatively, p.d. across GH $= 240 - (26.7 \times 0.04)$
$$= 238.93 \text{ V.}$$

1.19 Use of double subscripts

Double-subscript notation is used to avoid ambiguity in the direction of current, e.m.f. or p.d. In fig. 1.19, S represents a d.c. source, the

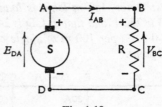

Fig. 1.19

e.m.f. of which is acting from D towards A and is therefore designated E_{DA}. The current in conductor AB flows from A to B and is designated I_{AB}. For the simple circuit of fig. 1.19, it is obvious that:

$$I_{AB} = I_{BC} = I_{CD} = I_{DA}.$$

The p.d. across the load is designated V_{BC} to symbolize that the potential of B is positive with respect to that of C; hence,

$$\text{voltage drop across the load} = V_{BC} = R \times I_{BC}.$$

If an arrow representing the direction of this voltage drop is drawn alongside the load, its head should point towards the end which is at the higher potential, i.e. towards B in fig. 1.19.

If the internal resistance of the source and the resistance of conductors AB and CD are negligible, then:

$$E_{DA} = V_{BC} = R \times I_{BC}$$

i.e. the voltage *rise* E_{DA} in the source is equal to voltage *drop* V_{BC} in the load.

When the double-subscript notation is applied to an a.c. circuit the sequence of the subscripts indicates the direction in which the current, the e.m.f. or the p.d. is assumed to be positive, e.g. in fig. 13.5, E_{CA} represents an alternating e.m.f. having its positive direction from C towards A in fig. 13.4; whereas in fig 13.25, V_{RY} represents the voltage drop between lines R and Y, its instantaneous value being positive when R is positive with respect to Y.

Summary of important formulae and relationships

$$t \ [^{\circ}C] = T \ [K] - 273.15$$
$$F \ [\text{newtons}] = m \ [\text{kg}] \times a \ [\text{m/s}^2] \qquad (1.1)$$
$$\text{Work done [joules]} = F \ [\text{newtons}] \times d \ [\text{metres}] \qquad (1.4)$$
$$\text{Kinetic energy [joules]} = \tfrac{1}{2}m \ [\text{kg}] \times v^2 \ [\text{m/s}]^2 \qquad (1.5)$$
$$\text{Potential energy [joules]} = m \ [\text{kg}] \times g \ [\text{m/s}^2] \times h \ [\text{m}] \qquad (1.5A)$$
$$\simeq 9.81 \ mh$$
$$\text{Power [watts]} = 2\pi T \ [\text{N·m}] \times n \ [\text{rev/s}] \qquad (1.7)$$
$$1 \ \text{kW·h} = 3.6 \ \text{MJ}$$

Heat required to raise the temperature of m kilograms of a substance having specific heat capacity c joules per kilogram kelvin through t degrees $= mct$ joules $\qquad (1.6)$

$$Q \ [\text{coulombs}] = I \ [\text{amperes}] \times t \ [\text{seconds}]$$

Mass of element liberated from electrolyte $\Big\} = zIt = zQ$

Ohm's law: $\qquad I = V/R, \ V = IR \text{ or } R = V/I$
Electrical power $= I^2R = IV = V^2/R$ watts
Electrical energy $= IVt$ joules

For resistors in series:
$$R = R_1 + R_2 + \ldots \qquad (1.8)$$

For resistors in parallel:
$$1/R = 1/R_1 + 1/R_2 + \ldots \qquad (1.9)$$
or $\qquad G = G_1 + G_2 + \ldots \qquad (1.10)$
$$R = \rho l/a \qquad (1.11)$$
$$R_t = R_0 \, (1 + \alpha_0 t) \qquad (1.12)$$
$$= R_0 \, (1 + 0.004\,26t) \text{ for annealed copper}$$
or $\qquad R_t = R_{20} \, \{1 + \alpha_{20} \, (t - 20)\} \qquad (1.12A)$
$$= R_{20} \, \{1 + 0.003\,92 \, (t - 20)\} \text{ for annealed copper}$$

For maximum power transfer,
$$R = r \qquad (1.14)$$

EXAMPLES 1

1. A force of 80 N is applied to a mass of 200 kg. Calculate the acceleration in metres per second squared.
2. Calculate the force, in kilonewtons, required to give a mass of 500 kg an acceleration of 4 m/s^2.

3. A train having a mass of 300 Mg is hauled at a constant speed of 90 km/h along a straight horizontal track. The track resistance is 5 mN per newton of train weight. Calculate (*a*) the tractive effort in kilonewtons, (*b*) the energy in megajoules and in kilowatt hours expended in 10 minutes, (*c*) the power in kilowatts and (*d*) the kinetic energy of the train in kilowatt hours (neglecting rotational inertia).

4. The power required to drive a certain machine at 350 rev/min is 600 kW. Calculate the driving torque in newton metres.

5. A d.c. motor connected to a 240-V supply is developing 20 kW at a speed of 900 rev/min. Calculate the useful torque.

6. If the motor referred to in Q.5 has an efficiency of 88 per cent, calculate (*a*) the current and (*b*) the cost of the energy absorbed if the load is maintained constant for 6 h. Assume the cost of electrical energy to be 1·2 p/kW·h.

7. (*a*) An electric motor runs at 600 rev/min when driving a load requiring a torque of 200 N·m. If the motor input is 15 kW, calculate the efficiency of the motor and the heat lost by the motor per minute, assuming its temperature to remain constant.

   (*b*) An electric kettle is required to heat 0·5 kg of water from 10°C to boiling point in 5 minutes, the supply voltage being 230 V. If the efficiency of the kettle is 80 per cent, calculate the resistance of the heating element. Assume the specific heat capacity of water to be 4·2 kJ/kg·K.   (S.A.N.C., O.1)

8. A pump driven by an electric motor lifts 1·5 m³ of water per minute to a height of 40 m. The pump has an efficiency of 90 per cent and the motor an efficiency of 85 per cent. Determine: (*a*) the power input to the motor; (*b*) the current taken from a 480-V supply; (*c*) the electrical energy consumed when the motor runs at this load for 8 h. Assume the mass of 1 m³ of water to be 1000 kg.

9. An electric kettle is required to heat 0·6 litre of water from 10°C to the boiling point in 5 min, the supply voltage being 240 V. The efficiency of the kettle is 78 per cent. Calculate: (*a*) the resistance of the heating element and (*b*) the cost of the energy consumed at 1·3 p/kW·h. Assume the specific heat capacity of water to be 4190 J/kg·K and 1 litre of water to have a mass of 1 kg.

10. An electric furnace is to melt 40 kg of aluminium per hour, the initial temperature of the aluminium being 12°C. Calculate: (*a*) the power required and (*b*) the cost of operating the furnace for 20 h, given that aluminium has the following thermal properties: specific heat capacity, 950 J/kg·K; melting point, 660°C; specific latent heat of fusion, 450 kJ/kg. Assume the efficiency of the furnace to be 85 per cent and the cost of electrical energy to be 0·8 p/kW·h.

11. A steady current of 5 A is passed through a copper coulometer for 20 min. Calculate the mass of copper deposited on the cathode, assuming the electrochemical equivalent of copper to be 0·33 mg/C.

12. A plate having a total surface area of 20 000 mm² is to be nickel-plated to a thickness of 0·15 mm. If the current available is 6 A, calculate the

time required. Assume the electrochemical equivalent of nickel to be 0·304 mg/C and the density to be 8800 kg/m³.

13. Two coils having resistances of 5 and 8 Ω respectively are connected across a battery having an e.m.f. of 6 V and an internal resistance of 1·5 Ω. Calculate: (a) the terminal voltage and (b) the energy in joules dissipated in the 5-Ω coil if the current remains constant for 4 minutes.

14. A coil of 12-Ω resistance is in parallel with a coil of 20-Ω resistance. This combination is connected in series with a third coil of 8-Ω resistance. If the whole circuit is connected across a battery having an e.m.f. of 30 V and an internal resistance of 2 Ω, calculate (a) the terminal voltage of the battery and (b) the power in the 12-Ω coil.

15. A coil of 20-Ω resistance is joined in parallel with a coil of x-Ω resistance. This combination is then joined in series with a piece of apparatus A, and the whole circuit connected to 100-V mains. What must be the value of x so that A shall dissipate 600 W with 10 A passing through it?   (U.L.C.I., O.1)

16. Two circuits, A and B, are connected in parallel to a 25-V battery, which has an internal resistance of 0·25 Ω. Circuit A consists of two resistors, 6 Ω and 4 Ω, connected in series. Circuit B consists of two resistors, 10 Ω and 5 Ω, connected in series. Determine the current flowing in and the potential difference across each of the four resistors. Also, find the power expended in the external circuit.
(N.C.T.E.C., O.1)

17. A load taking 200 A is supplied by copper and aluminium cables connected in parallel. The total length of conductor in each cable is 200 m, and each conductor has a cross-sectional area of 40 mm². Calculate: (i) the voltage drop in the combined cables; (ii) the current carried by each cable; (iii) the power wasted in each cable. Take the resistivity of copper and aluminium as 0·018 μΩ·m and 0·028 μΩ·m respectively.

18. A circuit, consisting of three resistances 12 Ω ,18 Ω and 36 Ω respectively joined in parallel, is connected in series with a fourth resistance. The whole is supplied at 60 V and it is found that the power dissipated in the 12-Ω resistance is 36 W. Determine the value of the fourth resistance and the total power dissipated in the group.   (U.E.I., O.1)

19. A coil consists of 2000 turns of copper wire having a cross-sectional area of 0·8 mm². The mean length per turn is 80 cm and the resistivity of copper is 0·02 μΩ·m at normal working temperature. Calculate the resistance of the coil and the power dissipated when the coil is connected across a 110-V d.c. supply.

20. An aluminium wire 7·5 m long is connected in parallel with a copper wire 6 m long. When a current of 5 A is passed through the combination, it is found that the current in the aluminium wire is 3 A. The diameter of the aluminium wire is 1 mm. Determine the diameter of the copper wire. Resistivity of copper is 0·017 μΩ·m; that of aluminium is 0·028 μΩ·m.

21. The field winding of a d.c. motor is connected directly across a 440-V supply. When the winding is at the room temperature of 17°C, the current is 2·3 A. After the machine has been running for some hours, the current has fallen to 1·9 A, the voltage remaining unaltered. Calculate the average temperature throughout the winding, assuming the temperature coefficient of resistance of copper to be 0·004 26/°C at 0°C.

22. Define the term *resistance–temperature coefficient*.

    A conductor has a resistance of R_1 ohms at θ_1°C, and consists of copper with a resistance–temperature coefficient a referred to 0°C. Find an expression for the resistance R_2 of the conductor at temperature θ_2°C.

    The field coil of a motor has a resistance of 250 Ω at 15°C. By how much will the resistance increase if the motor attains an average temperature of 45°C when running? Take $a = 0·004\ 28$/°C referred to 0°C. (S.A.N.C., O.2)

23. Explain what is meant by the *temperature coefficient of resistance* of a material.

    A copper rod, 0·4 m long and 4 mm in diameter, has a resistance of 550 μΩ at 20°C. Calculate the resistivity of copper at that temperature.

    If the rod is drawn out into a wire having a uniform diameter of 0·8 mm, calculate the resistance of the wire when its temperature is 60°C. Assume the resistivity to be unchanged and the temperature coefficient of resistance of copper to be 0·004 26/°C at 0°C. (App.El., L.U.)

24. A coil of insulated copper wire has a resistance of 150 Ω at 20°C. When the coil is connected across a 240-V supply, the current after several hours is 1·25 A. Calculate the average temperature throughout the coil, assuming the temperature coefficient of resistance of copper at 20°C to be 0·0039/°C.

25. An aluminium conductor has a resistance of 3·6 Ω at 20°C. What is its resistance at 50°C if the temperature coefficient of resistance of aluminium is 0·004 03/°C at 20°C?

26. Owing to a short circuit, a copper conductor having a cross-sectional area of 25 mm² carries a current of 20 000 A for 30 milliseconds. Neglecting heat loss, calculate the temperature rise of the conductor. Assume the specific heat capacity of copper to be 390 J/kg·K, density to be 8900 kg/m³ and resistivity to be 0·018 μΩ·m.

27. A certain generator has an open-circuit voltage of 12 V and an internal resistance of 40 Ω. Calculate: (*a*) the load resistance for maximum power transfer; (*b*) the corresponding values of the terminal voltage and of the power supplied to the load.

    If the load resistance were increased to twice the value for maximum power transfer, what would be the power absorbed by the load?

28. A battery having an e.m.f. of 105 V and an internal resistance of 1 Ω is connected in parallel with a d.c. generator of e.m.f. 110 V and internal resistance of 0·5 Ω to supply a load having a resistance of

8 Ω. Calculate: (i) the currents in the battery, the generator and the load; (ii) the potential difference across the load.   (U.E.I., O.1)

29. State Kirchhoff's Laws and apply them to the solution of the following problem.

Two batteries, A and B, are connected in parallel, and an 80-Ω resistor is connected across the battery terminals. The e.m.f. and the internal resistance of battery A are 100 V and 5 Ω respectively, and the corresponding values for battery B are 95 V and 3 Ω respectively. Find (a) the value and direction of the current in each battery and (b) the terminal voltage.   (U.E.I., O.1)

30. State Kirchhoff's Laws for an electric circuit, giving an algebraic expression for each law.

A network of resistors has a pair of input terminals AB connected to a d.c. supply and a pair of output terminals CD connected to a load resistor of 120 Ω. The resistances of the network are AC = BD = 180 Ω, and AD = BC = 80 Ω. Find the ratio of the current in the load resistor to that taken from the supply.   (S.A.N.C., O.1)

31. State Kirchhoff's Laws as applied to an electrical circuit.

A secondary cell having an e.m.f. of 2 V and an internal resistance of 1 Ω is connected in series with a primary cell having an e.m.f. of 1·5 V and an internal resistance of 100 Ω, the negative terminal of each cell being connected to the positive terminal of the other cell. A voltmeter having a resistance of 50 Ω is connected to measure the terminal voltage of the cells. Calculate the voltmeter reading and the current in each cell.   (App.El., L.U.)

32. State and explain Kirchhoff's Laws relating to electric circuits. Two storage batteries, A and B, are connected in parallel for charging from a d.c. source having an open-circuit voltage of 14 V and an internal resistance of 0·15 Ω. The open-circuit voltage of A is 11 V and that of B is 11·5 V; the internal resistances are 0·06 Ω and 0·05 Ω respectively. Calculate the initial charging currents.

What precautions are necessary when charging batteries in parallel?   (App.El., L.U.)

33. State Kirchhoff's Laws as applied to an electrical circuit.

Two batteries A and B are joined in parallel. Connected across the battery terminals is a circuit consisting of a battery C in series with a 25-Ω resistor, the negative terminal of C being connected to the positive terminals of A and B. Battery A has an e.m.f. of 108 V and an internal resistance of 3 Ω, and the corresponding values for battery B are 120 V and 2 Ω. Battery C has an e.m.f. of 30 V and a negligible internal resistance. Determine (a) the value and direction of the current in each battery and (b) the terminal voltage of battery A.   (App. El., L.U.)

34. A network is arranged as shown in fig. A. Calculate the value of the current in the 8-Ω resistor by (a) the Superposition Theorem, (b) Kirchhoff's Laws and (c) Thévenin's Theorem.

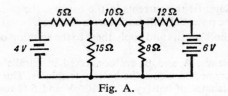

Fig. A.

35. Calculate the voltage across AB in the network shown in fig. B and indicate the polarity of the voltage, using (a) Kirchhoff's Laws and (b) delta-star transformation.

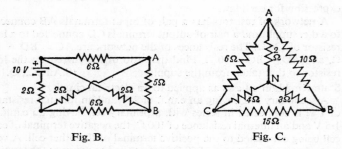

Fig. B.   Fig. C.

36. A network is arranged as in fig. C. Calculate the equivalent resistance between (a) A and B, and (b) A and N.

37. A network is arranged as in fig. D, and a battery having an e.m.f. of 2 V and negligible internal resistance is connected across AC. Determine the value and direction of the current in branch BE.

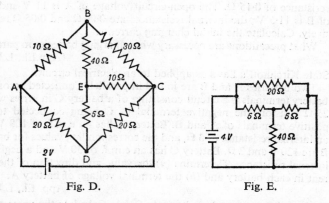

Fig. D.   Fig. E.

38. Calculate the value of the current through the 40-Ω resistor in fig. E.

39. Using Thévenin's Theorem, calculate the current through the 10-Ω resistor in fig. F.

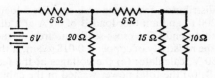

Fig. F.

40. State your own interpretation of Thévenin's Theorem and use it to solve the following problem.

Two batteries are connected in parallel. The e.m.f. and internal resistance of one battery are 120 V and 10 Ω respectively and the corresponding values for the other are 150 V and 20 Ω. A resistor of 50 Ω is connected across the battery terminals. Calculate (*a*) the current through the 50-Ω resistor, and (*b*) the value and direction of the current through each battery.

If the 50-Ω resistor were reduced to 20-Ω resistance, find the new current through it. (E.M.E.U., O.2)

41. Three resistors having resistances 50 Ω, 100 Ω and 150 Ω are star-connected to terminals A, B and C respectively. Calculate the resistances of equivalent delta-connected resistors.

42. Three resistors having resistances 20 Ω, 80 Ω and 30 Ω are delta-connected between terminals AB, BC and CA respectively. Calculate the resistances of equivalent star-connected resistors.

43. With the aid of delta and star connection diagrams, state the basic equations from which the delta-star and star-delta conversion equations can be derived.

A star network, in which N is the star point, is made up as follows: A-N = 70 Ω, B-N = 100 Ω and C-N = 90 Ω. Find the equivalent delta network. If the above star and delta networks were superimposed, what would be the measured resistance between terminals A and C? (E.M.E.U., O.2)

44. A two-wire distributor is fed at 250 V and loads of 60 A, 100 A and 120 A are taken at distances of 100, 200 and 250 metres respectively from the feeding end. If the resistance of the distributor is 0·1 Ω per 1000 metres of single conductor, calculate: (*a*) the current in each section of the distributor; (*b*) the potential difference across each load point; (*c*) the power dissipated in the distributor; (*d*) the energy lost in the distributor in 8 hours. (U.E.I., O.1)

45. A two-wire distributor, 400 m long, is fed at one end at 240 V. At points 250 m and 400 m from the feeding end there are loads of 200 A and 160 A respectively. Calculate the cross-sectional area of each core in order that the voltage at the 160-A load may be 96 per cent of that at the feeding point. Also, determine the cost of the energy loss in the distributor over a period of 6 hours if the above load were maintained constant during that time. Assume the resistivity of the conductor at working temperature to be 0·02 μΩ·m, and the cost of electrical energy to be 0·5 p per kW·h. (App. EL., L.U.)

46. A two-wire distributor, fed at one end, supplies a load A of 70 A at a distance of 200 m and another load B of 50 A at a distance of 300 m from the feeding end. The cross-sectional area of each conductor is 75 mm² and the resistivity of copper is 0·018 μΩ·m. If the voltage across load A is 250 V, calculate: (*a*) the voltages at the feeding point and across load B; (*b*) the total power wasted in the cable.

47. A two-wire ring main, 4 km in length, is fed at 250 V at point A. Respective loads of 80 A and 100 A are applied at point B, 1·5 km from point A, and point C, 2 km in the opposite direction. If the resistance per 100 metres of single conductor is 0·003 Ω, calculate: (*a*) the current in each section; (*b*) the voltages at points B and C.

48. A two-core cable is 400 m long and its resistance is 0·04 Ω per 100 m of single conductor. There are loads of 60 A, 80 A and 30 A at distances 100 m, 150 m and 250 m respectively from one end. If the cable is fed at 240 V at each end, calculate (*a*) the current in each section and (*b*) the voltage across each load. (App. El., L.U.)

49. A two-wire distributor AB, 1200 m long, is fed at end A at 246 V and at end B at 242 V. There are concentrated loads of 70 A and 110 A at points 200 m and 800 m respectively from end A. The resistance of the distributor (go and return) is 0·1 Ω/1000 m. Calculate the p.d. across each load.

Note on Kirchhoff's Second Law (p. 16). In textbooks, this law is expressed in various ways of which the following is typical: *In a closed circuit, the algebraic sum of the e.m.f.'s is equal to the algebraic sum of the voltage drops*. This suggests that there must be differences of potential between various points of the circuit. Actually, there may be *no* p.d. between any two points of the circuit. For instance, if a cylindrical magnet is moved along the axis of a homogeneous metal ring of uniform cross-sectional area, as in fig. 2.18, the e.m.f. induced in the ring circulates a current.

If e and i be the instantaneous values of the e.m.f. and current respectively, and if R and l be the resistance and mean periphery of the ring respectively,

$$e = iR, \quad \text{so that } e/l = i \times R/l.$$

This means that the e.m.f. induced in a length of, say, 1 mm of the ring is absorbed in sending current through the resistance of that millimetre. Consequently, there is no difference of potential across that millimetre length of the ring—in fact, *all points of the ring are at the same potential*.

Similarly, in Ex. 1.10, p. 17, it is shown that when three exactly similar cells are connected in series, there is no p.d. between any two available points of the circuit.

It follows that the definition of Kirchhoff's Second Law given above is applicable *only when the sources of e.m.f. have no internal resistance*. The definition given on p. 16 is applicable to *all* closed circuits.

CHAPTER 2

Electromagnetism

2.1 Magnetic field

If a permanent magnet is suspended so that it is free to swing in a
horizontal plane, as in fig. 2.1, it takes up a position such that a
particular end points towards the earth's North Pole. That end is
therefore said to be the *north-seeking* end of the magnet. Similarly,
the other end is the *south-seeking* end. For short, these are referred

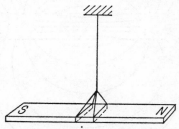

Fig. 2.1 A suspended permanent magnet.

to as the *north* (or N) and *south* (or S) poles respectively of the
magnet.

Let us place a permanent magnet on a table, cover it over with a
sheet of smooth cardboard and sprinkle iron filings uniformly over
the sheet. Slight tapping of the latter causes the filings to set themselves in curved chains between the poles, as shown in fig. 2.2. The
shape and density of these chains enable one to form a mental
picture of the magnetic condition of the space or 'field' around a bar
magnet and lead to the idea of *lines of magnetic flux*. It should be
noted, however, that these lines of magnetic flux have no physical existence; they are purely imaginary and were introduced by
Michael Faraday as a means of visualizing the distribution and
density of a magnetic field. It is important to realize that the
magnetic flux permeates the *whole* of the space occupied by that
flux.

39

2.2 Direction of magnetic field

The direction of a magnetic field is taken as that in which the north-seeking pole of a magnet points when the latter is suspended in the field. Thus, if a bar magnet rests on a table and four compass needles are placed in positions indicated in fig. 2.3, it is found that the needles take up positions such that their axes coincide with the corresponding chain of filings (fig. 2.2) and their N poles are all pointing

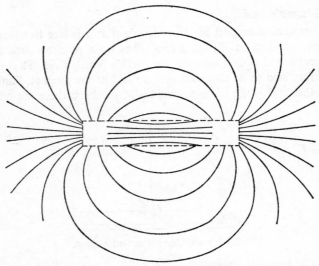

Fig. 2.2 Use of iron filings for determining distribution of magnetic field.

along the dotted line from the N pole of the magnet to its S pole. The lines of magnetic flux are assumed to pass through the magnet, emerge from the N pole and return to the S pole.

2.3 Characteristics of lines of magnetic flux

In spite of the fact that lines of magnetic flux have no physical existence, they do form a very convenient and useful basis for explaining various magnetic effects and for calculating their magnitudes. For this purpose, lines of magnetic flux are assumed to have the following properties:

1. *The direction of a line of magnetic flux at any point in a non-magnetic medium, such as air, is that of the north-seeking pole of a compass-needle placed at that point.*

2. *Each line of magnetic flux forms a closed loop,* as shown by the dotted lines in figs. 2.4 and 2.5. This means that a line of flux emerging from any point at the N-pole end of a magnet passes through the surrounding space back to the S-pole end and is then assumed to con-

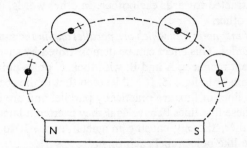

Fig. 2.3 Use of compass needles for determining direction of magnetic field.

tinue through the magnet to the point at which it emerged at the N-pole end.

3. *Lines of magnetic flux never intersect.* This follows from the fact that if a compass needle is placed in a magnetic field, its north-

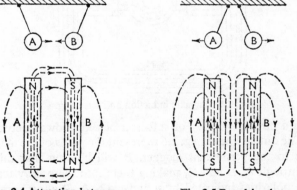

Fig. 2.4 Attraction between magnets.

Fig. 2.5 Repulsion between magnets.

seeking pole will point in one direction only, namely in the direction of the magnetic flux at that point.

4. *Lines of magnetic flux are like stretched elastic cords, always trying to shorten themselves.* This effect can be demonstrated by suspending two permanent magnets, A and B, parallel to each other,

with their poles arranged as in fig. 2.4. The distribution of the resultant magnetic field is indicated by the dotted lines. The lines of magnetic flux passing between A and B behave as if they were in tension, trying to shorten themselves and thereby causing the magnets to be attracted towards each other. In other words, unlike poles attract each other.

5. *Lines of magnetic flux which are parallel and in the same direction repel one another.* This effect can be demonstrated by suspending the two permanent magnets, A and B, with their N poles pointing in the same direction, as in fig. 2.5. It will be seen that in the space between A and B the lines of flux are practically parallel and are in the same direction. These flux lines behave as if they exerted a lateral pressure on one another, thereby causing magnets A and B to repel each other. Hence like poles repel each other.

2.4 Magnetic induction and magnetic screening

In fig. 2.6, N and S are the poles of a U-shaped permanent magnet M, A and B are soft-iron rectangular blocks attached to the magnet

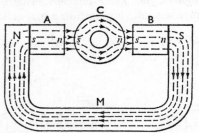

Fig. 2.6 Magnetic induction and screening.

and C is a hollow cylinder of soft-iron placed midway between A and B. The dotted lines in fig. 2.6 represent the paths of the magnetic flux due to the permanent magnet. It will be seen that this flux passes through A, B and C, making them into temporary magnets with the polarities indicated by *n* and *s*, i.e. A, B and C are magnetized by *magnetic induction*. Being of soft-iron, A, B and C lose almost the whole of their magnetism when they are removed from the influence of the permanent magnet M.

Fig. 2.6 also shows that no* flux passes through the air space

* Actually there must be some magnetic flux across the space inside the soft-iron cylinder C, but the density of this flux is so low that, for most purposes, it can be assumed to be zero.

inside cylinder C. Consequently, a body placed in this space would be found to be screened from the magnetic field around it. Magnetic screens are used to protect cathode-ray tubes (section 19.25) and instruments such as moving-iron ammeters and voltmeters (section 22.6) from external magnetic fields.

2.5 Magnetic field due to an electric current

When a conductor carries an electric current, a magnetic field is produced around that conductor—a phenomenon discovered by Oersted at Copenhagen in 1820. He found that when a wire carrying

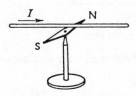

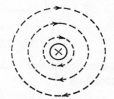

Fig. 2.7 Oersted's experiment.

Fig. 2.8 Magnetic flux due to current in a straight conductor.

an electric current was placed above a magnetic needle (fig. 2.7) and in line with the normal direction of the latter, the needle was deflected clockwise or anticlockwise, depending upon the direction of the current. Thus it is found that if we look along the conductor and if the current is flowing away from us, as shown by the cross* inside the conductor in fig. 2.8, the magnetic field has a clockwise direction and the lines of magnetic flux can be represented by concentric circles around the wire.

A convenient method of representing the relationship between the direction of a current and that of its magnetic field is to place a corkscrew or a woodscrew (fig. 2.9) alongside the conductor carrying the current. In order that the screw may travel in the same direction as the current, namely towards the right in fig. 2.9, it has to be turned clockwise when viewed from the lefthand side. Similarly, the direction of the magnetic field, viewed from the same side, is clockwise around the conductor, as indicated by the curved arrow F.

* It is usual to represent a current receding from the reader by a cross, as in fig. 2.8, and an approaching current by a dot, as on the lefthand conductor in fig. 2.12.

An alternative method of deriving this relationship is to grip the conductor with the *right* hand, with the thumb outstretched parallel to the conductor and pointing in the direction of the current; the

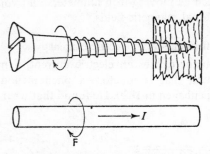

Fig. 2.9 Righthand screw rule.

fingers then point in the direction of the magnetic flux around the conductor.

2.6 Magnetic field of a solenoid

If a coil is wound on an iron rod, as in fig. 2.10, and connected to an accumulator, the iron becomes magnetized and behaves like a per-

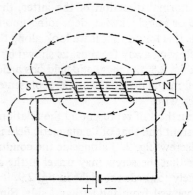

Fig. 2.10 Solenoid with an iron core.

manent magnet. The magnetic field of the electromagnet is represented by the dotted lines and its direction by the arrowheads.

The direction of the magnetic field produced by a current in a solenoid may be deduced by applying either the screw or the grip rule. Thus, if the axis of the screw is placed along that of the solenoid and if the screw is turned in the direction of the current, it travels in the direction of the magnetic field *inside* the solenoid, namely towards the right in fig. 2.10.

The grip rule can be expressed thus: if the solenoid is gripped with the *right* hand, with the fingers pointing in the direction of the current, then the thumb outstretched parallel to the axis of the solenoid points in the direction of the magnetic field *inside* the solenoid.

2.7 Force on a conductor carrying current across a magnetic field

In section 2.5 it was shown that a conductor carrying a current can produce a force on a magnet situated in the vicinity of the conductor. By Newton's Third Law of Motion, namely that to every force there

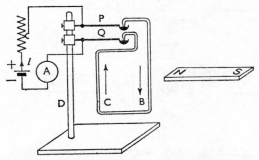

Fig. 2.11 Force on conductor carrying current across a magnetic field.

must be an equal and opposite force, it follows that the magnet must exert an equal force on the conductor. One of the simplest methods of demonstrating this effect is to take a copper wire, about 2 mm in diameter, and bend it into a rectangular loop as repre-

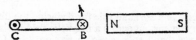

Fig. 2.12 Direction of force on conductor in fig. 2.11.

sented by BC in fig. 2.11. The two tapered ends of the loop dip into mercury contained in cups, one directly above the other, the cups being attached to metal rods P and Q carried by a wooden upright rod D. A current of about 5 A is passed through the loop and the N pole of a permanent magnet NS is moved towards B. If the current in this wire is flowing downwards, as indicated by the arrow in fig. 2.11, it is found that the loop, when viewed from above, turns counterclockwise, as shown in plan in fig. 2.12. If the magnet is reversed and again brought up to B, the loop turns clockwise.

If the magnet is placed on the other side of the loop, the latter

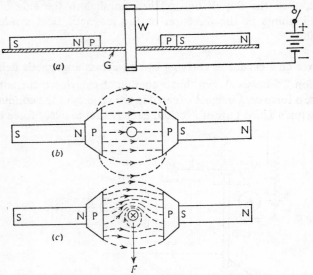

Fig. 2.13 Flux distribution with and without current.

turns clockwise when the N pole of the magnet is moved near to C, and counterclockwise when the magnet is reversed.

These effects can be explained* by the simple apparatus shown in elevation and plan in fig. 2.13. Two permanent magnets NS rest on a sheet of paper or glass G, and soft-iron pole-pieces P are added to increase the area of the magnetic field in the gap between them.

---

* Many textbooks give a *left*hand rule for deducing the direction of the force on a conductor carrying current across a magnetic field. This rule is liable to be confused with the *right*hand rule, given in section 2.10, for determining the direction of a generated e.m.f. The latter rule is extremely useful and should be memorized. Few students can memorize both rules correctly, and it is suggested that the lefthand rule should be forgotten and that the direction of the force on a current-carrying conductor in a magnetic field should be deduced from first principles by drawing separately the magnetic field due to the current in the conductor and that due to the permanent magnet or electromagnet and thus derive the distribution of the resultant magnetic field as in fig. 2.13.

Midway between the pole-pieces is a wire W passing vertically downwards through G and connected through a switch to a 6-V battery capable of giving a very large current for a short time.

With the switch open, iron filings are sprinkled over G and the latter is gently tapped. The filings in the space between PP take up the distribution shown in fig. 2.13 (*b*). If the switch is closed momentarily, the filings rearrange themselves as in fig. 2.13 (*c*). It will be seen that the lines of magnetic flux have been so distorted that they partially surround the wire. This distorted flux acts like stretched elastic cords bent out of the straight; the lines of flux try to return to the shortest paths between PP, thereby exerting a force *F* urging the conductor out of the way.

It has already been shown in section 2.5 that a wire W carrying a current downwards in fig. 2.13 (*a*) produces a magnetic field as shown in fig. 2.8. If this field is compared with that of fig. 2.13 (*b*), it is seen that on the upper side the two fields are in the same direction, whereas on the lower side they are in opposition. Hence, the combined effect is to strengthen the magnetic field on the upper side and weaken it on the lower side, thus giving the distribution shown in fig. 2.13 (*c*).

By combining diagrams similar to figs. 2.13 (*b*) and 2.8, it is easy to understand that if either the current in W or the polarity of magnets NS is reversed, the field is strengthened on the lower side and weakened on the upper side of diagrams corresponding to fig. 2.13 (*b*), so that the direction of the force acting on W is the reverse of that shown in fig. 2.13 (*c*).

On the other hand, if both the current through W and the polarity of the magnets are reversed, the *distribution* of the resultant magnetic field and therefore the direction of the force on W remain unaltered.

2.8 Magnitude of the force on a conductor carrying current across a magnetic field

With the apparatus of fig. 2.11, it can be shown qualitatively that the force on a conductor carrying a current at right-angles to a magnetic field is increased (*a*) when the current in the conductor is increased and (*b*) when the magnetic field is made stronger by bringing the magnet nearer to the conductor. With the aid of more elaborate apparatus, the force on the conductor can be measured for various currents and various densities of the magnetic field, and it is found that:

force on conductor $\propto$ current $\times$ (flux density) $\times$ (length of conductor)

If $F =$ force on conductor in newtons,

    $I =$ current through conductor in amperes

and $l =$ length, in metres, of conductor at right-angles to the magnetic field,

F [newtons] $\propto$ flux density $\times$ l [metres] $\times$ I [amperes].

The *unit of flux density* is taken as *the density of a magnetic field such that a conductor carrying* 1 *ampere at right-angles to that field has a force of* 1 *newton per metre acting upon it.* This unit is termed a *tesla** (T). Hence, for a flux density of B teslas,

$$\text{force on conductor} = BIl \text{ newtons} \tag{2.1}$$

For a magnetic field having a cross-sectional area of a metres^2 and a uniform flux density of B teslas, the *total flux* in *webers†* (Wb) is represented by the Greek capital letter Φ (phi), where

$$\Phi \text{ [webers]} = B \text{ [teslas]} \times a \text{ [metres}^2\text{]}$$

or B [teslas] $= \Phi$ [webers]$/a$ [metres^2]   (2.2)

i.e.   1 tesla = 1 weber/metre^2.

The *weber* may be defined either (i) as that magnetic flux which, when cut at a uniform rate by a conductor in 1 second, generates an e.m.f. of 1 volt (section 2.11), or (ii) as that magnetic flux which, linking a circuit of one turn, induces in it an e.m.f. of 1 volt when the flux is reduced to zero at a uniform rate in 1 second (section 2.12).

Example 2.1 *A conductor carries a current of* 800 *A at right-angles to a magnetic field having a density of* 0·5 *T. Calculate the force on the conductor in newtons per metre length.*

---

* Nikola Tesla (1857–1943), a Yugoslav who emigrated to U.S.A. in 1884, was a very famous electrical inventor. In 1888 he patented 2-phase and 3-phase alternators and motors.

† Wilhelm Eduard Weber (1804–91), a German physicist, was the first to develop a system of absolute electrical and magnetic units.

From expression (2.1),

$$\text{force per metre length} = 0.5 \ [\text{T}] \times 1 \ [\text{m}] \times 800 \ [\text{A}]$$
$$= 400 \ \text{N}.$$

2.9 Electromagnetic induction

On 29 August 1831, Michael Faraday (1791–1867) made the great discovery of *electromagnetic induction*, namely a method of obtaining

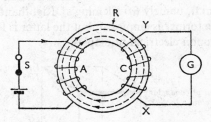

Fig. 2.14 Electromagnetic induction.

an electric current with the aid of magnetic flux. He wound two coils, A and C, on an iron ring R, as in fig. 2.14, and found that when switch S was closed, a deflection was obtained on galvanometer G; and that when S was opened, G was deflected in the reverse direction. A few weeks later he found that when a permanent magnet NS was

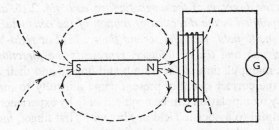

Fig. 2.15 Electromagnetic induction.

moved relative to a coil C (fig. 2.15), galvanometer G was deflected in one direction when the magnet was moved towards the coil and in the reverse direction when the magnet was withdrawn; and it was this experiment that finally convinced Faraday that an electric current could be produced by the movement of magnetic flux relative to a coil. Faraday also showed that the magnitude of the induced e.m.f. is proportional to the rate at which the magnetic flux passing through the coil is varied. Alternatively, we can say that when a conductor cuts or is cut by magnetic flux, an e.m.f. is generated in the conductor and the magnitude of the generated e.m.f. is proportional to the rate at which the conductor cuts or is cut by the magnetic flux.

2.10 Direction of induced e.m.f.

Two methods are available for deducing the direction of the induced or generated e.m.f., namely (*a*) Fleming's* Righthand Rule and (*b*) Lenz's Law. The former is empirical, but the latter is fundamental in that it is based upon electrical principles.

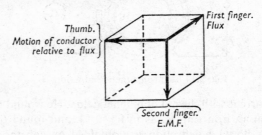

Fig. 2.16 Fleming's Righthand Rule.

(a) **Fleming's Righthand Rule** *If the first finger of the right hand be pointed in the direction of the magnetic flux, as in fig.* 2.16, *and if the thumb be pointed in the direction of motion of the conductor* **relative** *to the magnetic field, then the second finger, held at right-angles to both the thumb and the first finger, represents the direction of the e.m.f.* The manipulation of the thumb and fingers and their association with the correct quantity present some difficulty to many students. Easy manipulation can be acquired only by experience; and it may be helpful to associate *f*ield or *f*lux with *f*irst finger, *m*otion of the conductor relative to the field with the *m* in thu*m*b and e.m.f. with the *e* in second finger. If any two of these are correctly applied, the third is correct automatically.

(b) **Lenz's Law** In 1834 Heinrich Lenz, a German physicist (1804–65), enunciated a simple rule, now known as Lenz's Law, which can be expressed thus: *The direction of an induced e.m.f. is always such*

* John Ambrose Fleming (1849–1945) was Professor of Electrical Engineering at University College, London.

that it tends to set up a current opposing the motion or the change of flux responsible for inducing that e.m.f.

Let us consider the application of Lenz's Law to the ring shown in fig. 2.14. By applying either the screw or the grip rule given in section 2.6, we find that when S is closed and the battery has the polarity shown, the direction of the magnetic flux in the ring is clockwise. Consequently, the current in C must be such as to try to produce a flux in a counterclockwise direction, tending to oppose the growth of the flux due to A, namely the flux which is responsible for the e.m.f. induced in C. But a counterclockwise flux in the ring would require the current in C to be passing through the coil from X to Y (fig. 2.14). Hence, this must also be the direction of the e.m.f. induced in C.

2.11 Magnitude of the generated or induced e.m.f.

Fig. 2.17 represents the elevation and plan of a conductor AA situated in an airgap between poles NS. Suppose AA to be carrying a current, I amperes, in the direction shown. By applying either the screw or the grip rule of section 2.5, it is found that the effect of this current is to strengthen the field on the right and weaken that on the left of A, so that there is a force of BlI newtons (section 2.8) urging the conductor towards the left, where B is the flux density in teslas (or webers per square metre) and l is the length in metres of conductor in the magnetic field. Hence, a force of this magnitude has to be applied in the opposite direction to move A towards the right.

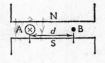

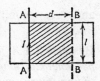

Fig. 2.17 Conductor moved across magnetic field.

The work done in moving conductor AA through a distance d metres to position BB in fig. 2.17 is $(BlI \times d)$ joules. If this movement of AA takes place at a uniform velocity in t seconds, the e.m.f. induced in the conductor is constant at, say, E volts. Hence the electrical power generated in AA is IE watts and the electrical energy is IEt watt-seconds or joules. Since the mechanical energy expended in moving the conductor horizontally across the gap is all converted into electrical energy, then

$$IEt = Blld$$

$$\therefore \qquad E = Bld/t = Blv \text{ volts,}$$

where v is the velocity in metres/second. But Bld = the total flux, Φ webers, in the area shown shaded in fig. 2.17. This flux is cut by the conductor when the latter is moved from AA to BB. Hence

$$E \text{ [volts]} = \frac{\Phi \text{ [webers]}}{t \text{ [seconds]}} \qquad (2.3)$$

i.e. the e.m.f., in volts, generated in a conductor is equal to the rate (in webers/second) at which the magnetic flux is cutting or being cut by the conductor; and the *weber* may therefore be defined as *that magnetic flux which, when cut at a uniform rate by a conductor in* 1 *second, generates an e.m.f. of* 1 *volt.*

In general, if a conductor cuts or is cut by a flux of dΦ weber in dt second,

$$\text{e.m.f. generated in conductor} = \mathrm{d}\Phi/\mathrm{d}t \text{ volts.} \qquad (2.4)$$

Example 2.2 *Calculate the e.m.f. generated in the axle of a car travelling at* 80 *km/h, assuming the length of the axle to be* 2 *m and the vertical component of the earth's magnetic field to be* 40 *microteslas.*

$$80 \text{ km/h} = \frac{(80 \times 1000) \text{ [m]}}{3600 \text{ [s]}}$$

$$= 22 \cdot 2 \text{ m/s}$$

Vertical component of earth's field = 40×10^{-6} T,

$$\therefore \qquad \text{flux cut by axle} = 40 \times 10^{-6} \text{ [T]} \times 2 \text{ [m]} \times$$
$$22 \cdot 2 \text{ [m/s]}$$

$$= 1776 \times 10^{-6} \text{ Wb/s}$$

and   e.m.f. generated in axle = 1776×10^{-6} V

$$= 1776 \text{ } \mu\text{V.}$$

Example 2.3 *A four-pole generator has a magnetic flux of* 12 *mWb per pole. Calculate the average value of the e.m.f. generated in one of the armature conductors while it is moving through the magnetic flux of one pole, if the armature is driven at* 900 *rev/min.*

When a conductor moves through the magnetic field of one pole, it cuts a magnetic flux of 12 mWb.

Time taken for a conductor to move through one revolution $= \frac{60}{900} = \frac{1}{15}$ s.

Since the machine has 4 poles, time taken for a conductor to move through the magnetic field of one pole $= \frac{1}{4} \times \frac{1}{15} = \frac{1}{60}$ s,

∴   average e.m.f. generated in one conductor

$$= (12 \times 10^{-3}) \div (\tfrac{1}{60}) = 0.72 \text{ V}.$$

2.12 Magnitude of e.m.f. induced in a coil

Suppose the magnetic flux through a coil of N turns to be increased by Φ webers in t seconds due to, say, the relative movement of the coil and a magnet (fig. 2.15). Since each of the lines of magnetic flux cuts* each turn, one turn can be regarded as a conductor cut by Φ webers in t seconds; hence, from expression (2.3), the average e.m.f. induced in each turn is Φ/t volts.

The current due to this e.m.f., by Lenz's Law, tries to prevent the increase of flux, i.e. tends to set up an opposing flux. Thus, if the magnet NS in fig. 2.15 is moved towards coil C, the flux passing from left to right through the latter is increased. The e.m.f. induced in the coil circulates a current in the direction represented by the dot and cross in fig. 2.18, where—for simplicity—coil C is represented as one turn. The effect of this

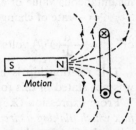

Fig. 2.18 Distortion of magnetic field by induced current.

current is to distort the magnetic field as shown by the dotted lines, thereby tending to push the coil away from the magnet. By Newton's Third Law of Motion, there must be an equal and opposite force tending to oppose the movement of the magnet.

Owing to the fact that the induced e.m.f. circulates a current tending to oppose the increase of flux through the coil, its direction is regarded as negative; hence

average e.m.f. induced in 1 turn $= -\Phi/t$ volts

$$= -\left\{ \begin{array}{l} \text{average rate of } \textit{change} \text{ of} \\ \text{flux in webers per second} \end{array} \right.$$

---

* It is immaterial whether we consider the e.m.f. as being due to change of flux linked with a coil or due to the coil cutting or being cut by lines of flux; the result is exactly the same. The fact of the matter is that we do not know what is really happening; but we can calculate the effect by imagining the magnetic field in the form of lines of flux, some of which expand from nothing when the field is increased or collapse to nothing when the field is reduced. In so doing they may be regarded as cutting the turns of the coil, or alternatively, the effect may be regarded as being due merely to a change in the number of flux-linkages.

and average e.m.f. induced in coil $= -N\Phi/t$ volts $\qquad$ (2.5)

$$= -\left\{ \begin{array}{l} \text{average rate of } \textit{change} \text{ of} \\ \text{flux-linkages per second.} \end{array} \right.$$

The term 'flux-linkages' merely means the product of the flux in webers and the number of turns with which the flux is linked. Thus if a coil of 20 turns has a flux of 0·1 weber through it, the flux-linkages $= 0\cdot1 \times 20 = 2$ weber-turns.

From expression (2.5) it follows that:

instantaneous value of e.m.f., in volts, induced in a coil
$\qquad = -$rate of change of flux-linkages, in weber-turns per second

or $\quad e = -\dfrac{\mathrm{d}}{\mathrm{d}t}(N\Phi)$ volts $\qquad$ (2.6)

This relationship is usually known as *Faraday's Law*, though it was not stated in this form by Faraday.

From expression (2.5) we can define* the *weber as that magnetic flux which, linking a circuit of one turn, induces in it an e.m.f. of 1 volt when the flux is reduced to zero at a uniform rate in 1 second.*

Next, let us consider the case of the two coils, A and C, shown in fig. 2.14. Suppose that when switch S is closed, the flux in the ring increases by Φ webers in t seconds. Then if coil A has N_1 turns,

$$\text{average e.m.f. induced in A} = -N_1\Phi/t \text{ volts.}$$

The minus sign signifies that this e.m.f., in accordance with Lenz's Law, is acting in opposition to the e.m.f. of the battery, thereby trying to prevent the growth of the current.

If coil C is wound with N_2 turns, and if all the flux produced by coil A passes through C,

$$\text{average e.m.f. induced in C} = -N_2\Phi/t \text{ volts.}$$

In this case the minus sign signifies that the e.m.f. circulates a current in such a direction as to tend to set up a flux in opposition to that produced by the current in coil A, thereby delaying the growth of flux in the ring.

In general, if the magnetic flux through a coil increases by $\mathrm{d}\Phi$ weber in $\mathrm{d}t$ second,

$$\text{e.m.f. induced in coil} = -N \cdot \mathrm{d}\Phi/\mathrm{d}t \text{ volts} \qquad (2.7)$$

* This is an alternative to the definition already given on p. 52.

Example 2.4 *A magnetic flux of* 400 μ*Wb passing through a coil of* 1200 *turns is reversed in* 0·1 *s. Calculate the average value of the e.m.f. induced in the coil.*

The magnetic flux has to decrease from 400 μWb to zero and then increase to 400 μWb in the reverse direction; hence the *increase* of flux in the original direction is −800 μWb.

Substituting in expression (2.5), we have:

$$\text{average e.m.f. induced in coil} = -\frac{1200 \times (-800 \times 10^{-6})}{0\cdot 1}$$

$$= 9\cdot 6 \text{ V.}$$

This e.m.f. is positive because its direction is the same as the original direction of the current, at first tending to prevent the current decreasing and then tending to prevent it increasing in the reverse direction.

Summary of Important Formulae

$$\text{Force on conductor} = BlI \text{ newtons} \qquad (2.1)$$

$$\text{E.m.f. generated in conductor} = Blv \text{ volts}$$

$$= d\Phi/dt \text{ volts} \qquad (2.4)$$

$$\text{E.m.f. induced in coil} = -\frac{d}{dt}(N\Phi) \text{ volts} \qquad (2.6)$$

EXAMPLES 2

1. A current-carrying conductor is situated at right-angles to a uniform magnetic field having a density of 0·3 tesla. Calculate the force in newtons per metre length of the conductor when the current is 200 A.
2. Calculate the current in the conductor referred to in Q. 1 when the force per metre length of the conductor is 15 N.
3. A conductor, 150 mm long, is carrying a current of 60 A at right-angles to a magnetic field. The force on the conductor is 3 N. Calculate the density of the field.
4. The coil of a moving-coil loudspeaker has a mean diameter of 30 mm and is wound with 800 turns. It is situated in a radial magnetic field of 0·5 T. Calculate the force on the coil, in newtons, when the current is 12 mA.
5. The armature of a certain motor has 900 conductors and the current per conductor is 24 A. The flux density in the airgap under the poles is 0·6 T. The armature core is 160 mm long and has a diameter of 250 mm. Assume that the core is smooth (i.e. there are no slots and the

*

winding is on the cylindrical surface of the core) and also assume that only two-thirds of the conductors are simultaneously in the magnetic field. Calculate (*a*) the torque in newton metres and (*b*) the mechanical power developed, in kilowatts, if the speed is 700 rev/min.

(*Note.* In the case of slotted cores, the flux density in the slots is very low, so that there is very little torque on the conductors; nearly all the torque is exerted on the teeth.)

6. Explain what happens when a long straight conductor is moved through a uniform magnetic field at constant velocity. Assume that the conductor moves perpendicularly to the field.

   If the ends of the conductor are connected together through an ammeter, what will happen?

   A conductor, 0·6 m long, is carrying a current of 75 A and is placed at right-angles to a magnetic field of uniform flux density. Calculate the value of the flux density if the mechanical force on the conductor is 30 N. (U.L.C.I., O.1)

7. State Lenz's Law.

   A conductor, 500 mm long, is moved at a uniform speed at right-angles to its length and to a uniform magnetic field having a density of 0·4 T. If the e.m.f. generated in the conductor is 2 V and the conductor forms part of a closed circuit having a resistance of 0·5 Ω, calculate: (i) the velocity of the conductor in metres/second; (ii) the force acting on the conductor in newtons; (iii) the work done in joules when the conductor has moved 600 mm. (U.E.I., O.1)

8. A wire, 100 mm long, is moved at a uniform speed of 4 m/s at right-angles to its length and to a uniform magnetic field. Calculate the density of the field if the e.m.f. generated in the wire is 0·15 V.

   If the wire forms part of a closed circuit having a total resistance of 0·04 Ω, calculate the force on the wire in newtons.

9. Give three practical applications of the mechanical force exerted on a current-carrying conductor in a magnetic field.

   A conductor of active length 30 cm carries a current of 100 A and lies at right-angles to a magnetic field of density 0·4 T. Calculate the force in newtons exerted on it. If the force causes the conductor to move at a velocity of 10 m/s, calculate (*a*) the e.m.f. induced in it and (*b*) the power in watts developed by it. (E.M.E.U., O.1)

10. The axle of a certain motor car is 1·5 m long. Calculate the e.m.f. generated in it when the car is travelling at 140 km/h. Assume the vertical component of the earth's magnetic field to be 40 μT.

11. An aeroplane having a wing span of 50 m is flying horizontally at a speed of 600 km/h. Calculate the e.m.f. generated between the wing tips, assuming the vertical component of the earth's magnetic field to be 40 μT. Is it possible to measure this e.m.f.?

12. A copper disc, 250 mm in diameter, is rotated at 300 rev/min about a horizontal axis through its centre and perpendicular to its plane. If the axis points magnetic north and south, calculate the e.m.f. between the

circumference of the disc and the axis. Assume the horizontal component of the earth's field to be 18 μT.

13. A coil of 1500 turns gives rise to a magnetic flux of 2·5 mWb when carrying a certain current. If this current is reversed in 0·2 s, what is the average value of the e.m.f. induced in the coil?

14. A short coil of 200 turns surrounds the middle of a bar magnet. If the magnet sets up a flux of 80 μWb, calculate the average value of the e.m.f. induced in the coil when the latter is removed completely from the influence of the magnet in 0·05 s.

15. The flux through a 500-turn coil increases uniformly from zero to 200 μWb in 3 milliseconds. It remains constant for the fourth millisecond and then decreases uniformly to zero during the fifth millisecond. Draw to scale a graph representing the variation of the e.m.f. induced in the coil.

16. State Lenz's Law. The field coils of a 6-pole d.c. generator, each having 500 turns, are connected in series. When the field is excited, there is a magnetic flux of 0·02 Wb/pole. If the field circuit is opened in 0·02 s and the residual magnetism is 0·002 Wb/pole, calculate the average e.m.f. induced across the field terminals. In which direction is this e.m.f. directed relative to the direction of the field current?

(U.L.C.I., O.1)

17. Two coils, A and B, are wound on the same iron core. There are 300 turns on A and 2800 turns on B. A current of 4 A through coil A produces a flux of 800 μWb in the core. If this current is reversed in 20 ms, calculate the average e.m.f.s induced in coils A and B.

18. A six-pole motor has a magnetic flux of 0·08 Wb per pole and the armature is rotating at 700 rev/min. Calculate the average e.m.f. generated per conductor.

19. A four-pole armature is to generate an average e.m.f. of 1·4 V per conductor, the flux per pole being 15 mWb. Calculate the speed at which the armature must rotate.

CHAPTER 3

Magnetic Circuit

3.1 Introductory

One of the characteristics of lines of magnetic flux is that each line is a closed loop (section 2.3); for instance, in fig. 2.10, the dotted lines represent the flux passing through the iron core and returning through the surrounding air space. The complete closed path followed by any group of lines of magnetic flux is referred to as a *magnetic circuit*. One of the simplest forms of magnetic circuit is shown in fig. 2.14, where the iron ring R provides the path for the magnetic flux.

3.2 Magnetomotive force; magnetic field strength

In an electric circuit, the current is due to the existence of an electromotive force. By analogy, we may say that in a magnetic circuit the

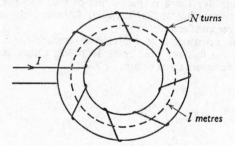

Fig. 3.1 A toroid.

magnetic flux is due to the existence of a *magnetomotive force* (m.m.f.) caused by a current flowing through one or more turns. The value of the m.m.f. is proportional to the current and to the number of turns, and is descriptively expressed in *ampere turns*; but for the purpose of dimensional analysis, it is expressed in *amperes*, since the number of turns is dimensionless. Hence the unit of magnetomotive force is the *ampere*.

If a current of I amperes flows through a coil of N turns, as shown in fig. 3.1, the magnetomotive force is the *total* current linked with the magnetic circuit, namely IN amperes. If the magnetic circuit is homogeneous and of uniform cross-sectional area, the magneto-motive force per metre length of the magnetic circuit is termed the *magnetic field strength* (the term *magnetizing force* is now obsolete), and is represented by the symbol H. Thus, if the mean length of the magnetic circuit of fig. 3.1 is l metres,

$$H = IN/l \text{ amperes/metre} \qquad (3.1)$$

3.3 Permeability of free space or magnetic constant

Suppose A in fig. 3.2 to represent the cross-section of a long straight conductor, situated in a vacuum and carrying a current of one

ampere towards the paper; and suppose the return path of this current to be some considerable distance away from A so that the effect of the return current on the magnetic field in the vicinity of A may be neglected. The lines of magnetic flux surrounding A will, by symmetry, be in the form of concentric circles, as already described in section 2.5, and the dotted circle D in fig. 3.2 represents the path of one of these lines of flux at a radius of 1 metre. Since conductor A and its return

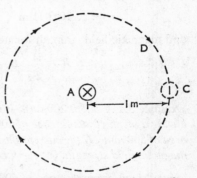

Fig. 3.2 Magnetic field at 1-metre radius due to current in a long straight conductor.

conductor form one turn, the magnetomotive force acting on path D is 1 ampere; and since the length of this line of flux is 2π metres, the magnetic field strength, H, at a radius of 1 m is $1/(2\pi)$ A/m.

If the flux density in the region of line D is B teslas, it follows from expression (2.1) that the force per metre length on a conductor C (parallel to A) carrying 1 ampere at right-angles to this flux is given by:

Force per metre length

$$= B \text{ [T]} \times 1 \text{ [m]} \times 1 \text{ [A]} = B \text{ newtons.}$$

But from the definition of the ampere given in section 1.6 (*a*), this force is 2×10^{-7} newton,

$\therefore$ flux density at 1-m radius from conductor carrying 1 A
$$= B = 2 \times 10^{-7} \text{ tesla.}$$

Hence,

$$\frac{\text{flux density at C}}{\text{magnetic field strength at C}} = \frac{B}{H} = \frac{2 \times 10^{-7} \text{ [T]}}{1/2\pi \text{ [A/m]}}$$
$$= 4\pi \times 10^{-7} \text{ H/m.}^*$$

The ratio B/H for the above condition is termed the *permeability of free space* or *magnetic constant* and is represented by the symbol μ_0. The value of this ratio is almost exactly the same whether the conductor A of fig. 3.2 is assumed to be situated in a vacuum (or free space) or in air or in any other non-magnetic material. Hence,

$$\mu_0 = \frac{B}{H} \text{ for a vacuum and non-magnetic materials}$$
$$= 4\pi \times 10^{-7} \text{ H/m} \tag{3.2}$$

and magnetic field strength for non-magnetic materials

$$= H = \frac{B}{\mu_0} = \frac{B}{4\pi \times 10^{-7}} \text{ amperes/metre} \tag{3.3}$$

Example 3.1 *A coil of* 200 *turns is wound uniformly over a wooden ring having a mean circumference of* 600 *mm and a uniform cross-sectional area of* 500 *mm². If the current through the coil is* 4 *A, calculate* (a) *the magnetic field strength,* (b) *the flux density and* (c) *the total flux.*

(a) Mean circumference $= 600$ mm $= 0.6$ m,
$\therefore$ $H = 4 \times 200/0.6 = 1333$ A/m.

(b) From expression (3.2):
flux density $= \mu_0 H = 4\pi \times 10^{-7} \times 1333$
$= 0.001\ 675 \text{ T} = 1675\ \mu\text{T}.$

(c) Cross-sectional area $= 500$ mm² $= 500 \times 10^{-6}$ m²
$\therefore$ total flux $= 1675\ [\mu\text{T}] \times (500 \times 10^{-6})\ [\text{m}^2]$
$= 0.8375\ \mu\text{Wb}.$

Example 3.2 *Calculate the magnetomotive force required to produce a flux of* 0.015 *Wb across an airgap* 2.5 *mm long, having an effective area of* 200 *cm².*

---

* It is shown in the footnote on p. 112 that the units of absolute permeability are *henrys/metre*; e.g., $\mu_0 = 4\pi \times 10^{-7}$ H/m.

$$\text{Area of airgap} = 200 \times 10^{-4} = 0 \cdot 02 \text{ m}^2$$

$$\therefore \quad \text{flux density} = \frac{0 \cdot 015 \text{ [Wb]}}{0 \cdot 02 \text{ [m}^2]} = 0 \cdot 75 \text{ T.}$$

From expression (3.3),

$$\text{magnetic field strength for gap} = \frac{0 \cdot 75}{4\pi \times 10^{-7}} = 597\ 000 \text{ A/m.}$$

$$\text{Length of gap} = 2 \cdot 5 \text{ mm} = 0 \cdot 0025 \text{ m,}$$

$\therefore$ m.m.f. required to send flux across gap
$$= 597\ 000 \text{ [A/m]} \times 0 \cdot 0025 \text{ [m]} = 1492 \text{ A.}$$

3.4 Relative permeability

In section 2.6 it was shown that the magnetic flux inside a coil is intensified when an iron core is inserted. It follows that if the non-

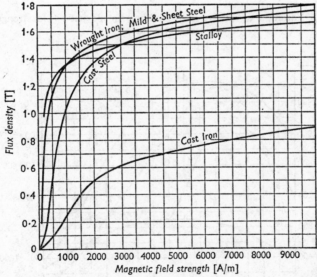

Fig. 3.3 Variation of flux density with magnetic field strength for cast steel, etc.

magnetic core of a toroid, such as that shown in fig. 3.1, is replaced by an iron core, the flux produced by a given m.m.f. is greatly increased; and *the ratio of the flux density produced in a material to the flux density produced in a vacuum* (or in a non-magnetic core)

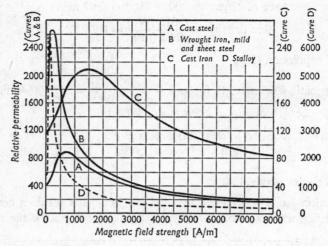

Fig. 3.4 Variation of relative permeability with magnetic field strength for cast steel, etc.

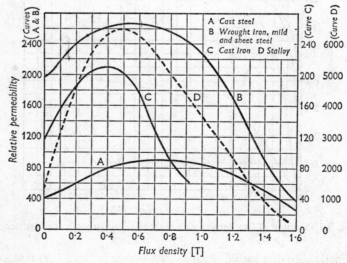

Fig. 3.5 Variation of relative permeability with flux density for cast steel, etc.

by the same magnetic field strength is termed the *relative permeability* and is denoted by the symbol μ_r. For air, $\mu_r = 1$; but for certain nickel-iron alloys, it may be as high as 100 000.

The value of the relative permeability of a ferromagnetic material varies considerably for different values of the magnetic field strength, and it is usually convenient to represent the relationship between the flux density and the magnetic field strength graphically as in fig. 3.3; and the curves in figs. 3.4 and 3.5 represent the corresponding values of the relative permeability plotted against the magnetic field strength and the flux density respectively.

From expression (3.2), $B = \mu_0 H$ for a non-magnetic material; hence, for a material having a relative permeability μ_r,

$$B = \mu_r \mu_0 H$$

$\therefore$ *absolute permeability* $= B/H = \mu_r \mu_0$ (3.4)

3.5 Reluctance

Let us consider an iron ring having a cross-sectional area of a square metres and a mean circumference of l metres (fig. 3.1), wound with N turns carrying a current I amperes; then

total flux $= \Phi =$ flux density $\times$ area $= Ba$ (3.5)

and   m.m.f. $=$ magnetic field strength $\times$ length $= Hl$ (3.6)

Dividing (3.5) by (3.6), we have:

$$\frac{\Phi}{\text{m.m.f.}} = \frac{Ba}{Hl} = \mu_r \mu_0 \times \frac{a}{l}$$

$\therefore$ $\Phi = \dfrac{\text{m.m.f.}}{l/\mu_r \mu_0 a}$ (3.7)

so that $\dfrac{\text{m.m.f.}}{\Phi} = l/\mu_r \mu_0 a$

$= $ *reluctance* of magnetic circuit.

This expression is similar in form to:

$$\frac{\text{e.m.f.}}{I} = \rho l/a$$

for the electric circuit. The denominator, $l/\mu_r \mu_0 a$, in expression (3.7) is similar in form to $\rho l/a$ for the resistance of a conductor except that the absolute permeability, $\mu_r \mu_0$, for the magnetic material corresponds to the reciprocal of the resistivity, namely the conductivity of the electrical material.

Since the m.m.f. is equal to the total number of amperes ($= IN$) acting on the magnetic circuit,

$$\therefore \quad \text{magnetic flux} = \frac{IN}{\text{reluctance}} \tag{3.8}$$

$$\text{where reluctance} = l/\mu_r\mu_0 a \tag{3.9}$$

$$= l/\mu_0 a \text{ for non-magnetic materials.}$$

The symbol for reluctance is S and the quantity is expressed in amperes/weber.

3.6 Comparison of the electric and magnetic circuits

It is helpful to tabulate side by side the various electric and magnetic quantities and their relationships, thus:

Electric circuit		Magnetic circuit	
Quantity	Unit	Quantity	Unit
E.m.f.	volt	M.m.f.	ampere
—	—	Magnetic field strength	ampere/metre
Current	ampere	Magnetic flux	weber
Current density	ampere/m²	Magnetic flux density	tesla
Resistance	ohm	Reluctance	ampere/weber
$\left(= \rho \cdot \dfrac{l}{a}\right)$		$\left(= \dfrac{1}{\mu_r\mu_0} \cdot \dfrac{l}{a}\right)$	
Current = e.m.f./resistance		Flux = m.m.f./reluctance	

One important difference between the electric and magnetic circuits is the fact that energy must be supplied to *maintain* the flow of electricity in a circuit, whereas the magnetic flux, once it is set up, does not require any further supply of energy. For instance, once the flux produced by a current in a solenoid has attained its maximum value, the energy subsequently absorbed by that solenoid is all dissipated as heat due to the resistance of the winding.

Example 3.3 *A mild-steel ring having a cross-sectional area of* 500 *mm² and a mean circumference of* 400 *mm has a coil of* 200 *turns wound uniformly around it. Calculate* (a) *the reluctance of the ring and* (b) *the current required to produce a flux of* 800 *μWb in the ring.*

$$(a) \quad \text{Flux density in ring} = \frac{800 \times 10^{-6} \text{ [Wb]}}{500 \times 10^{-6} \text{ [m}^2\text{]}} = 1 \cdot 6 \text{ T.}$$

From fig. 3.5, the relative permeability of mild steel for a flux density of 1·6 T is about 380.

$$\therefore \quad \text{reluctance of ring} = \frac{0{\cdot}4}{380 \times 4\pi \times 10^{-7} \times 5 \times 10^{-4}}$$

$$= 1{\cdot}677 \times 10^{6} \text{ A/Wb.}$$

(*b*) From expression (3.7),

$$800 \times 10^{-6} = \frac{\text{m.m.f.}}{1{\cdot}677 \times 10^{6}}$$

$$\therefore \quad \text{m.m.f.} = 1342 \text{ A}$$

and   magnetizing current = 1342/200 = 6·7 A.

Alternatively, from expression (3.4),

$$H = \frac{B}{\mu_r \mu_0} = \frac{1{\cdot}6}{380 \times 4\pi \times 10^{-7}}$$

$$= 3350 \text{ A/m,}$$

$$\therefore \quad \text{m.m.f.} = 3350 \times 0{\cdot}4 = 1340 \text{ A}$$

and   magnetizing current = 1340/200 = 6·7 A.

3.7 Composite magnetic circuit

Suppose a magnetic circuit to consist of two specimens of iron, A and B, arranged as in fig. 3.6. If l_1 and l_2 be the mean lengths in metres of the magnetic circuits of A and B respectively, a_1 and a_2 their cross-sectional areas in square metres, and μ_1 and μ_2 their *absolute* permeabilities,

then $$\text{reluctance of A} = \frac{l_1}{\mu_1 a_1}$$

and $$\text{reluctance of B} = \frac{l_2}{\mu_2 a_2}.$$

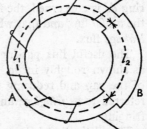

Fig. 3.6 Composite magnetic circuit.

If a coil is wound on core A as in fig. 3.6 and if the magnetic flux is assumed to be confined to the iron core, then

$$\text{total reluctance of magnetic circuit} = \frac{l_1}{\mu_1 a_1} + \frac{l_2}{\mu_2 a_2}$$

and $$\text{total flux} = \Phi = \frac{\text{m.m.f. of coil}}{\text{total reluctance}} = \frac{NI}{\dfrac{l_1}{\mu_1 a_1} + \dfrac{l_2}{\mu_2 a_2}} \qquad (3.10)$$

Fig. 3.7 shows an iron ring with an airgap. If the gap is very short, practically all the flux goes straight across from the one face to the other, so that the area of the gap may be assumed to be the same as that of the iron. Hence, if a is the cross-sectional area and l_1 and l_2 are the lengths of the iron core and gap respectively,

$$\text{total reluctance} = \frac{l_1}{\mu_1 a} + \frac{l_2}{\mu_0 a}$$

3.8 Magnetic leakage and fringing

Suppose dd in fig. 3.8 to represent a metal ring symmetrically situated relative to the airgap in the iron ring, and suppose the magnetizing winding to be concentrated over a short length of the core. As far as

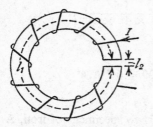

Fig. 3.7 Iron ring with an airgap.

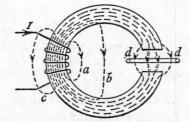

Fig. 3.8 Magnetic leakage and fringing.

ring dd is concerned, the flux passing through it can be regarded as the *useful* flux and that which returns by such paths as a, b and c is *leakage* flux.

The useful flux passing across the gap tends to bulge outwards as shown roughly in fig. 3.8, thereby increasing the effective area of the gap and reducing the flux density in the gap. This effect is referred to as *fringing*; and the longer the airgap, the greater is the fringing.

The distinction between useful and leakage fluxes may be more obvious if we consider an electrical machine. For instance, fig. 3.9 shows two poles of a six-pole machine. The armature slots have, for simplicity, been omitted. Some of the dotted lines do not enter the armature core and thus do not assist in generating an e.m.f. in the armature winding; consequently, they represent leakage flux. On the other hand, some of the flux passes between the pole tips and the armature core, as shown in fig. 3.9, and is referred to as fringing flux. Since this fringing flux is cut by the armature conductors, it forms part of the useful flux.

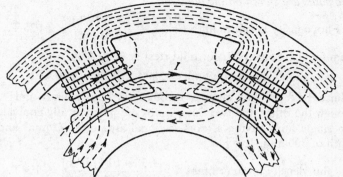

Fig. 3.9 Magnetic leakage and fringing in a machine.

From figs. 3.8 and 3.9 it is seen that the effect of leakage flux is to increase the total flux through the exciting winding,

and leakage factor

$$= \frac{\text{total flux through exciting winding}}{\text{useful flux}} \qquad (3.11)$$

The value of the leakage factor for electrical machines is usually about 1·15–1·25.

Example 3.4 *A magnetic circuit is made of mild steel arranged as in fig. 3.10. The centre limb is wound with 500 turns and has a cross-sectional area of 800 mm². Each of the outer limbs has a cross-sectional area of 500 mm². The airgap has a length of 1 mm. Calculate the current*

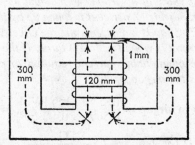

Fig. 3.10 Magnetic circuit for example 3.4.

required to set up a flux of 1·3 *mWb in the centre limb, assuming no magnetic leakage and fringing. The mean lengths of the various magnetic paths are shown on the diagram.*

$$\text{Flux density in centre limb} = \frac{(1\cdot3 \times 10^{-3})\ [\text{Wb}]}{(800 \times 10^{-6})\ [\text{m}^2]} = 1\cdot625\ \text{T}.$$

From fig. 3.3, value of H for mild steel $\simeq$ 3800 A/m,

∴   m.m.f. for centre limb = $3800 \times 0\cdot12 = 456$ A.

Since half the flux returns through one outer limb and half through the other, the two outer limbs are magnetically equivalent to a single limb having a cross-sectional area of 1000 mm² and a length of 300 mm,

∴   flux density in outer limbs $= \dfrac{(1\cdot3 \times 10^{-3})\ [\text{Wb}]}{(1000 \times 10^{-6})\ [\text{m}^2]} = 1\cdot3$ T.

From fig. 3.3, value of H for mild steel $\simeq$ 850 A/m

m.m.f. for outer limbs = $850 \times 0\cdot3 = 255$ A.
Flux density in airgap = 1·625 T

∴   value of H for gap $= \dfrac{1\cdot625}{4\pi \times 10^{-7}}$

$\qquad\qquad\qquad\qquad\quad = 1\cdot292 \times 10^6$ A/m

and   m.m.f. for gap = $1\cdot292 \times 10^6 \times 0\cdot001$
$\qquad\qquad\qquad\qquad\quad = 1292$ A.

Hence,   total m.m.f. = $456 + 255 + 1292$
$\qquad\qquad\qquad\qquad\quad = 2003$ A

and   magnetizing current $= 2003/500 = 4$ A.

Example 3.5 *A magnetic circuit is made up of steel laminations shaped as in fig. 3.11. The width of the iron is 40 mm and the core is built up to a depth of 50 mm, of which 8 per cent is taken up by insulation between the laminations. The gap is 2 mm long and the effective area of the gap is 2500 mm². The coil is wound with 800 turns. If the leakage factor is 1·2, calculate the magnetizing current required to produce a flux of 0·0025 Wb across the airgap.*

$$\text{Flux density in airgap} = \frac{(2\cdot5 \times 10^{-3})\ [\text{Wb}]}{(2500 \times 10^{-6})\ [\text{m}^2]} = 1\ \text{T},$$

∴   value of H for gap $= \dfrac{1\ [\text{T}]}{4\pi \times 10^{-7}\ [\text{H/m}]} = 796\,000$ A/m

and   m.m.f. for gap = 796 000 × 0·002
                              = 1592 A.

Total flux through coil = flux in gap × leakage factor
                       = 0·0025 × 1·2 = 0·003 Wb.

Since only 92 per cent of the cross-section of the core consists of iron,

∴ area of iron in core = 40 × 50 × 0·92 = 1840 mm²
                         = 0·001 84 m²
and flux density in core = 0·003 [Wb]/0·001 84 [m²] = 1·63 T.

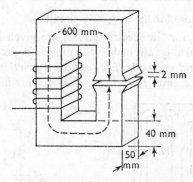

Fig. 3.11 Magnetic circuit for example 3.5.

From fig. 3.3, value of H for laminations ≃ 4000 A/m.

It will be evident from fig. 3.8 that when there is magnetic leakage, the flux density is not uniform over the whole length of the iron core. It is impossible, however, to allow for the variation of magnetic field strength due to this variation of flux density; and the usual practice is to assume that the magnetic field strength estimated for the region of maximum density applies to the whole of the iron core, thereby erring on the safe side.

Hence, m.m.f. for iron core = 4000 × 0·6 = 2400 A
and   total m.m.f. = 1592 + 2400 = 3992 A

∴   magnetizing current = 3992/800 ≃ 5 A.

It should be appreciated that owing to the difficulty of calculating the exact flux densities at various points of the magnetic circuit and to the uncertainty regarding the exact magnetic property of the iron, the results obtained in magnetic calculations are only approximately correct.

When we are dealing with composite magnetic circuits, it is usually helpful to tabulate the results thus:

Part	Area (m²)	Length (m)	Flux (Wb)	Flux density (T)	Amperes/ metre (A/m)	M.M.F. (A)
Iron	0·001 84	0·6	0·003	1·63	4 000	2400
Airgap	0·0025	0·002	0·0025	1·0	796 000	1592

Total m.m.f. = 3992

3.9 Kirchhoff's Laws for the magnetic circuit

Kirchhoff's Laws for the electric circuit are given in section 1.14. These Laws can also be applied to the magnetic circuit thus:

First Law *The total magnetic flux towards a junction is equal to the total magnetic flux away from that junction.* This law follows from

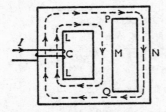

Fig. 3.12 Magnetic circuit to illustrate Kirchhoff's Laws

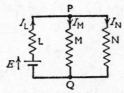

Fig. 3.13 Electric circuit equivalent to fig. 3.12.

the fact that each line of flux forms a closed path; for instance, if an iron core is arranged as shown in fig. 3.12 and if a coil C wound on limb L carries a current, the magnetic flux through C divides at P, some flux passing along limb M and the remainder along limb N, to join again at Q. There is no break or discontinuity in any of the lines of flux at P and Q; consequently the total flux from L towards P is exactly the same as the sum of the fluxes from P towards M and N, i.e.

$$\Phi_L = \Phi_M + \Phi_N$$

or

$$\Phi_L - \Phi_M - \Phi_N = 0.$$

In general,

$$\Sigma\Phi = 0.$$

where Σ represents the algebraic sum.

Second Law *In any closed magnetic circuit, the algebraic sum of the product of the magnetic field strength and the length of each part of the circuit is equal to the resultant magnetomotive force.* For instance, if H_L is the magnetic field strength required for limb L and l_L is the length of the circuit from Q via L to P, and if H_M and l_M are the corresponding values for limb M and H_N and l_N are those for the limb extending from P via N to Q, then:

$$\text{total m.m.f. of coil C} = H_L l_L + H_M l_M$$
$$= H_L l_L + H_N l_N$$
$$\text{and} \qquad 0 = H_M l_M - H_N l_N.$$
$$\text{In general,} \qquad \Sigma \text{ m.m.f.} = \Sigma Hl.$$

It may be helpful to compare the magnetic circuit of fig. 3.12 with the corresponding electric circuit of fig. 3.13.

3.10 Determination of the magnetization curve for an iron ring

(a) *By means of a fluxmeter*

Fig. 3.14 shows an iron ring of uniform cross-section, uniformly wound with a coil P, thereby eliminating magnetic leakage.* Coil P

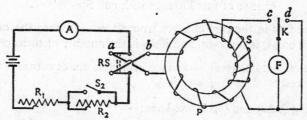

Fig. 3.14 Determination of the magnetization curve for an iron ring.

is connected to a battery through a reversing switch RS, an ammeter A and a variable resistor R_1. Another coil S, which need not be distributed around the ring, is connected through a two-way switch K to fluxmeter F which is a special type of permanent-magnet moving-coil instrument. Current is led into and out of the moving coil of F by fine wires or ligaments so arranged as to exert negligible control over the position of the moving coil. When the flux in the ring is varied, the e.m.f. induced in S sends a current through the fluxmeter and produces a deflection that is proportional to the change

* See footnote on p. 17.

of flux-linkages in coil S. Switch S_2 and resistor R_2 are not required for this test, but will be used when this circuit is employed for determining the hysteresis loop (section 3.12).

The current through coil P is adjusted to a desired value by means of R_1 and switch RS is then reversed several times to bring the iron into a 'cyclic' condition, i.e. into a condition such that the flux in the ring reverses from a certain value in one direction to the same value in the reverse direction. During this operation, switch K should be on *d*, thereby short-circuiting the fluxmeter. With switch RS on, say, *a*, switch K is moved over to *c*, the current through P is reversed by moving RS quickly over to *b* and the fluxmeter deflection is noted.

If N_P = number of turns on coil P,
l = mean circumference of the ring, in metres
and I = current through P, in amperes,
magnetic field strength = $H = IN_P/l$ amperes/metre.

If θ = fluxmeter deflection when current through P is reversed
and c = fluxmeter constant
= no. of weber-turns per unit of scale deflection,
change of flux linkages with coil S = $c\theta$ (3.12)

If the flux in the ring changes from Φ to $-\Phi$ when the current through coil P is reversed, and if N_S is the number of turns on S,

change of flux-linkages ⎫ = change of flux × no. of turns on S
with coil S   ⎬ = $2\Phi N_S$ (3.13)

Equating (3.12) and (3.13), we have:

$$2\Phi N_S = c\theta$$

so that
$$\Phi = \frac{c\theta}{2N_S} \text{ webers.}$$

If a = cross-sectional area of ring in square metres.
flux density in ring = $B = \Phi/a$

$$= \frac{c\theta}{2aN_S} \text{ teslas.}   (3.13A)$$

The test is performed with different values of the current; and from the data, a graph representing the variation of flux density with magnetic field strength can be plotted, as in fig. 3.3.

(b) *By means of a ballistic galvanometer*

A ballistic galvanometer has a moving coil suspended between the poles of a permanent magnet, but the coil is wound on a *non-metallic* former, so that there is very little damping when the coil has a resistor having a high resistance in series. The first deflection or 'throw' is proportional to the number of coulombs discharged through the galvanometer if the duration of the discharge is short compared with the time of one oscillation.

If θ = first deflection or 'throw' of the ballistic galvanometer when the current through coil P is reversed

and k = ballistic constant of the galvanometer

= quantity of electricity in coulombs per unit deflection,

quantity of electricity through galvanometer = $k\theta$ coulombs (3.14)

If Φ = flux produced in ring by I amperes through P

and t = time, in seconds, of reversal of flux,

average e.m.f. induced in S = $N_S \times 2\Phi/t$ volts.

If R = total resistance of the secondary circuit, quantity of electricity through ballistic galvanometer

$$= \text{average current} \times \text{time}$$
$$= \frac{2\Phi N_S}{tR} \times t = 2\Phi N_S/R \text{ coulombs} \quad (3.15)$$

Equating (3.14) and (3.15), we have:

$$k\theta = 2\Phi N_S/R$$

$$\therefore \quad \Phi = \frac{k\theta R}{2N_S} \text{ webers.}$$

The values of the flux density, etc., can then be calculated as already described for method (*a*).

3.11 Hysteresis

If we take a closed iron ring which has been completely demagnetized* and measure the flux density with increasing values of the magnetic field strength, the relationship between the two quantities is

* The simplest method of demagnetizing is to reverse the magnetizing current a large number of times while, at the same time, gradually reducing the current to zero.

represented by curve OAC in fig. 3.15. If the value of H is then reduced, it is found that the flux density follows curve CD, and that when H has been reduced to zero, the flux density remaining in the iron is OD and is referred to as the *remanent flux density*.

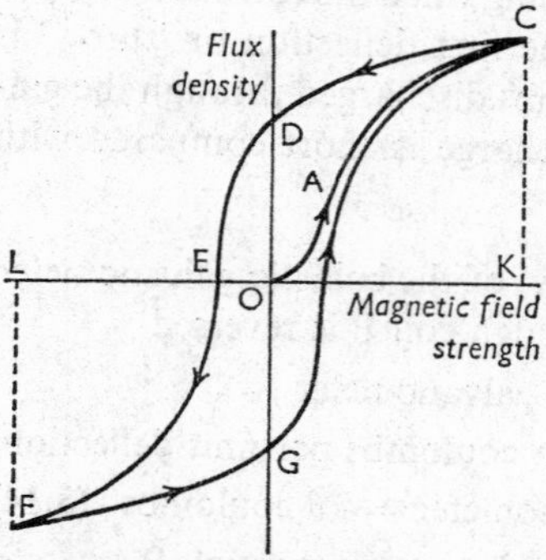

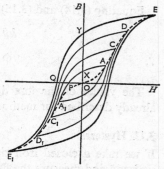

Fig. 3.15 Hysteresis loop.

If H is increased in the reverse direction, the flux density decreases, until at some value OE, the flux has been reduced to zero. The magnetic field strength OE required to wipe out the residual magnetism is termed the *coercive force*. Further increase of H causes the flux density to grow in the reverse direction as represented by curve EF. If the reversed magnetic field strength OL is adjusted to the same value as the maximum value OK in the initial direction, the final flux density LF is the same as KC.

If the magnetic field strength is varied backwards from OL to OK, the flux density follows a curve FGC similar to curve CDEF, and the closed figure CDEFGC is termed the *hysteresis loop*.

If hysteresis loops for a given iron ring are determined for different maximum values of the magnetic field strength, they are found to lie within one another, as shown in fig. 3.16. The apexes A, C, D and E of the respective loops lie on the *B–H* curve determined with increasing values of H. It will be seen that the value of the remanent flux density depends upon the value of the peak magnetization; thus, for loop A, the remanent flux density is OX, whereas for loop E, corresponding to a maximum magnetization that is approaching saturation, the remanent flux density is OY. The value of the remanent flux density obtained when the maximum magnetization reaches the saturation value of the material is termed the *remanence* of that material. Thus for the

Fig. 3.16 A family of hysteresis loops.

material having the hysteresis loops of fig. 3.16, the remanence is approximately OY.

The value of the coercive force in fig. 3.16 varies from OP for loop AA_1 to OQ for loop EE_1; and the value of the coercive force when the maximum magnetization reaches the saturation value of the material is termed the *coercivity* of that material. Thus, for fig. 3.16, the coercivity is approximately OQ. The value of the coercivity varies enormously for different materials, being about 40 000 A/m for Alnico (an alloy of iron, aluminium, nickel, cobalt and copper, used for permanent magnets) and about 3 A/m for Mumetal (an alloy of nickel, iron, copper and molybdenum).

3.12 Determination of the hysteresis loop

The circuit arrangement is shown in fig. 3.14.* With reversing switch RS on side *a* and S_2 closed, the current is adjusted by means of R_1 to the value corresponding to the maximum magnetic field strength OA (fig. 3.17) for which the loop is to be determined. RS is then reversed several times to ensure that the iron ring is in a cyclic state so that when OA is reversed, the flux density AC changes to the same value in the reverse direction. During these operations, K should be on *d*, thereby short-circuiting the flux-meter. Then with K on *c*, the deflection θ_m of F is noted when RS is reversed. In section 3.10 it was shown that the corresponding flux density B_m is given by:

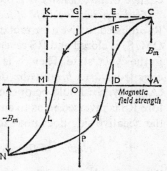

Fig. 3.17 Plotting a hysteresis loop.

$$B_m = \frac{c}{2aN_S} \times \theta_m$$

Hence points C and N of fig. 3.17 can be plotted.

From expression (3.13A), it follows that

$$change \text{ of flux density} = \frac{c}{aN_S} \times \left(\frac{\text{fluxmeter}}{\text{deflection}} \right)$$

* In actual practice, it is more convenient to insert R_2 in the connection between a pair of diagonally opposite contacts of RS; but the arrangement shown in fig. 3.14 makes it easier to follow the procedure.

and it is with the aid of this formula that we derive the remaining points required for plotting the loop.

To derive portion CFJ of loop. With RS on a, S_2 is opened and R_2 adjusted to give a current corresponding to a magnetic field strength OD (fig. 3.17). S_2 is then closed to bring the value of H back to OA, and RS is reversed several times to restore the iron to the cyclic condition. Fluxmeter deflection θ_1 is then noted when S_2 is opened.

$$\text{Corresponding change of flux density} = \frac{c}{aN_S} \times \theta_1.$$

This change of flux density is represented by EF in fig. 3.17, so that the actual flux density corresponding to magnetic field strength OD $= B_m - \text{EF} = \text{DF}$.

This test is repeated with different values of OD down to zero, the latter value being obtained by opening the circuit of R_2 so that when S_2 is opened, the current is reduced to zero and the corresponding change of flux density is represented by GJ.

To derive portion JLN of loop. To obtain point L, S_2 is opened and R_2 adjusted to give a current corresponding to magnetic field strength OM. S_2 is closed and RS reversed several times to bring the iron back to the cyclic state. Then S_2 is opened and RS simultaneously moved over from a to b, and the deflection θ_2 of the fluxmeter is noted. (Actually, S_2 should be opened the slightest fraction of a second before RS is reversed so as to ensure that the magnetic field strength on the negative side does not exceed OM.) Then:

$$\textit{change of flux density} = \text{KL} = \frac{c}{aN_S} \times \theta_2.$$

When the change of flux density exceeds B_m, as shown for KL in fig. 3.17, then:

$$\left. \begin{array}{l} \textit{actual} \text{ flux density corresponding to} \\ \text{magnetic field strength OM} \end{array} \right\} = \text{KL} - B_m = \text{ML}.$$

This test is repeated for sufficient number of negative values of the magnetic field strength to enable curve JLN to be drawn. The return portion NPC of the loop is exactly the reverse of CJN.

3.13 Current-ring theory of magnetism

It may be relevant at this point to consider why the presence of iron in a current-carrying coil increases the value of the magnetic flux and why magnetic hysteresis occurs in iron. As long ago as 1823,

André-Marie Ampère—after whom the unit of current was named—suggested that the increase in the magnetic flux might be due to electric currents circulating within the molecules of the iron. Subsequent discoveries have confirmed this suggestion, and the following brief explanation may assist in giving some idea of the current-ring theory of magnetism.

An atom consists of a nucleus of positive electricity surrounded, at distances large compared with their diameter, by electrons, which are charges of negative electricity. The electrons revolve in orbits around the nucleus, and each electron also spins around its own axis —somewhat like a gyroscope—and the magnetic characteristics of iron appear to be due mainly to this electron spin. The movement of an electron around a circular path is equivalent to a minute current flowing in a circular ring. In an iron atom, four more electrons spin round in one direction than in the reverse direction, and the axes of spin of these electrons are parallel with one another; consequently, the effect is equivalent to four current rings producing magnetic flux in a certain direction.

The iron atoms are grouped together in *domains*, each about 0·1 mm in width; and in any one domain the magnetic axes of all the atoms are parallel with one another. In an unmagnetized bar of iron, the magnetic axes of different domains are in various directions so that their magnetizing effects cancel one another. Between adjacent domains there is a region or 'wall', about 10^{-4} mm thick, within which the direction of the magnetic axes of the atoms changes gradually from that of the axes in one domain to that of the axes in the adjacent domain.

When an unmagnetized bar of iron is moved into a current-carrying solenoid, there are sudden tiny increments of the magnetic flux as the magnetic axes of the various domains are orientated so that they coincide with the direction of the m.m.f. due to the solenoid, thereby increasing the magnitude of the flux. This phenomenon is known as the *Barkhausen effect* and can be demonstrated by winding a search coil on the iron bar and connecting it through an amplifier to a loudspeaker. The sudden increments of flux due to successive orientation of the various domains, while the iron bar is being moved into the solenoid, induce e.m.f. impulses in the search coil and the effect can be heard as a rustling noise.

It follows that when a current-carrying solenoid has an iron core the magnetic flux can be regarded as consisting of two components:

(*a*) the flux produced by the solenoid without an iron core;

(*b*) the flux due to ampere-turns equivalent to the current rings formed by the spinning electrons in the orientated domains. This component reaches its maximum value when all the domains have been orientated so that their magnetic axes are in the direction of the magnetic flux. The iron is then said to be saturated.

These effects are illustrated in fig. 3.18, where graph P represents the variation of the actual flux density with magnetic field strength

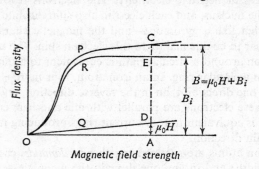

Fig. 3.18 Actual and intrinsic flux densities.

for a ferro-magnetic material. The straight line Q represents the variation of flux density with magnetic field strength if there were no iron present, thus for magnetic field strength OA,

$$AD = \mu_0 \times OA.$$

For the same magnetic field strength OA, the actual flux density B is represented by AC. Hence the flux density due to the presence of the iron $= AC - AD = AE$. This flux density is referred to as the *intrinsic magnetic flux density* or *magnetic polarization* and is represented by symbol B_i,

i.e. $$B_i = B - \mu_0 H$$

The variation of the intrinsic flux density with magnetic field strength is represented by graph R in fig. 3.18, from which it will be seen that when the magnetic field strength exceeds a certain value, the intrinsic flux density remains constant, i.e. the iron is magnetically saturated, or in other words, all the domains have been orientated so that their axes are in the direction of the magnetic field.

This alignment of the domains has a certain amount of stability, depending upon the quality of the iron and the treatment it has received during manufacture. Consequently, the iron may retain much of its magnetism after the external magnetomotive force has been removed, as already discussed in section 3.11, the remanent flux being maintained by the m.m.f. due to the electronic current-rings in the iron. A permanent magnet can therefore be regarded as an electromagnet, the relatively large flux being due to the high value of the *inherent m.m.f.* (i.e. coercive force × length of magnet) retained by the steel after it has been magnetized.

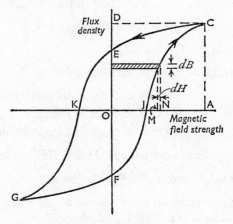

Fig. 3.19 Hysteresis loss.

A disturbance in the alignment of the domains necessitates the expenditure of energy in taking a specimen of iron through a cycle of magnetization. This energy appears as heat in the specimen and is referred to as *hysteresis loss*.

3.14 Hysteresis loss

Suppose fig. 3.19 to represent the hysteresis loop obtained on an iron ring of mean circumference l metres and cross-sectional area a square metres. Let N be the number of turns on the magnetizing coil.

Let dB = increase of flux density when the magnetic field strength is increased by a very small amount MN in dt second, and i = current in amperes corresponding to OM.

i.e. $$OM = Ni/l \qquad (3.16)$$

Instantaneous e.m.f. induced in winding

$$= -a \times dB \times N/dt \text{ volts}$$

and component of applied voltage to neutralize this e.m.f.

$$= aN \times dB/dt \text{ volts}$$

∴   instantaneous power supplied to magnetic field

$$= i \times aN \times dB/dt \text{ watts}$$

and energy supplied to magnetic field in time dt second

$$= iaN \times dB \text{ joules.}$$

From (3.16) $i = l \times OM/N$

∴   energy supplied to magnetic field in time dt

$$= (l \times OM/N) \times aN \times dB \text{ joules}$$
$$= OM \times dB \times la \text{ joules}$$
$$= \text{area of shaded strip, joules/metre}^3$$

∴   energy supplied to magnetic field when H is increased from zero

to OA = area FJCDF joules/metre³

Similarly, energy returned from magnetic field when H is reduced from OA to zero = area CDEC joules/metre³

∴   net energy absorbed by magnetic field

$$= \text{area FJCEF joules/metre}^3$$

Hence, hysteresis loss for a complete cycle

$$= \text{area of loop GFCEG joules/metre}^3 \qquad (3.17)$$

If the hysteresis loop is plotted to scales of 1 cm to x amperes/ metre along the horizontal axis and 1 cm to y teslas along the vertical axis, and if A represents the area of the loop in square centimetres, then—

hysteresis loss/cycle = Axy joules/metre³

$$= A \times (NI/l) \times \Phi/a \text{ joules/metre}^3$$

where NI/l = magnetic field strength/cm of scale

and Φ/a = flux density/cm of scale

Hence, total hysteresis loss/cycle, in joules

$$= la \times A \times (NI/l) \times \Phi/a$$
$$= A \times NI \times \Phi$$
$$= \text{(area of loop in cm}^2) \times \text{(amperes/cm of scale)}$$
$$\times \text{(webers/cm of scale)} \qquad (3.18)$$

The total hysteresis loss/cycle in a specimen can therefore be determined from the area of a loop drawn by plotting the total flux, in
webers, against the total m.m.f. in amperes.

From the areas of a number of loops similar to those shown in
fig. 3.16, the hysteresis loss per cycle was found by Steinmetz* to be
proportional to $(B_{max})^{1.6}$ for a certain quality of iron, where B_{max} represents the maximum value of the flux density. It is important to
realize that this index of 1.6 is purely empirical and has no theoretical
basis. In fact, indices as high as 3.5 have been obtained, and in
general we may say that

$$\text{hysteresis loss per cubic metre per cycle} \propto (B_{max})^x$$

where $x = 1.5$–2.5, depending upon the quality of the iron and the
range of flux density over which the measurement has been made.

From the above proof, it is evident that the hysteresis loss is proportional to the volume and to the number of cycles through which
the magnetization is taken. Hence,

if $f =$ frequency of alternating magnetization in hertz (or c/s)
and $v =$ volume of iron in cubic metres,

$$\text{hysteresis loss} = kvf(B_{max})^x \text{ watts} \qquad (3.19)$$

where k is a constant for a given specimen and a given range of flux
density.

The magnitude of the hysteresis loss depends upon: (*a*) the composition of the specimen, e.g. nickel–iron alloys such as Mumetal
and Permalloy have very narrow hysteresis loops due to their low
coercive force, and consequently have very low hysteresis loss; and
(*b*) the heat treatment and the mechanical handling to which the
specimen has been subjected. For instance, if a material has been
bent or cold-worked, the hysteresis loss is increased, but the specimen
can usually be restored to its original magnetic condition by a strain-
relieving heat treatment.

* Charles Proteus Steinmetz, U.S.A. electrical engineer, 1865–1923.

Example 3.6 *The area of the hysteresis loop obtained with a certain specimen of iron was 9·3 cm². The co-ordinates were such that 1 cm = 1000 A/m and 1 cm = 0·2 T. Calculate (a) the hysteresis loss per cubic metre per cycle and (b) the hysteresis loss per cubic metre at a frequency of 50 Hz (or c/s). (c) If the maximum flux density was 1·5 T, calculate the hysteresis loss per cubic metre for a maximum density of 1·2 T and a frequency of 30 Hz, assuming the loss to be proportional to $(B_{max})^{1.8}$.*

(a) Area of hysteresis loop in *BH* units $\Big\} = 9\cdot3 \times 1000 \times 0\cdot2 = 1860$

∴ hysteresis loss per cubic metre per cycle $\Big\} = 1860$ joules.

(b) Hysteresis loss per cubic metre at 50 Hz $\Big\} = 1860 \times 50 = 93\,000$ W.

(c) From expression (3.19),
$$93\,000 = k \times 1 \times 50 \times (1\cdot5)^{1.8}$$
$$\log (1\cdot5)^{1.8} = 1\cdot8 \times 0\cdot176 = 0\cdot3168 = \log 2\cdot074$$

∴ $$k = \frac{93\,000}{50 \times 2\cdot074} = 896$$

For $B_{max} = 1\cdot2$ T and $f = 30$ Hz,

hysteresis loss per cubic metre $\Big\} = 896 \times 30 \times (1\cdot2)^{1.8} = 37\,350$ W.

Or, hysteresis loss at 1·2 T and 30 Hz $\Big\} = 93\,000 \times \frac{30}{50} \times \left(\frac{1\cdot2}{1\cdot5}\right)^{1.8} = 37\,350$ W.

3.15 Condition for minimum volume of a permanent magnet

Let M in fig. 3.20 be a steel core to be magnetized by passing a current through winding W, and let PP be soft-iron pole-shoes of negligible reluctance. Suppose curve CDE in fig. 3.21 to represent the de-magnetizing portion of the hysteresis loop for a maximum magnetic field strength OA obtained with the closed magnetic circuit of fig. 3.20 (a). It was pointed out in section 3.13 that the remanent flux density OD may be regarded as being maintained by the *inherent magneto-motive force* due to electronic currents in the magnet. The value of the inherent m.m.f. is given by the product of the coercive force OE and the length l_m of the magnet.

With the pole-shoes PP in contact, as in fig. 3.20 (*a*), let us reduce the magnetic field strength from OA to zero and then increase it to OF in the *reverse* direction, so that the flux density in the magnet is

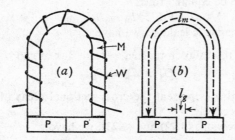

Fig. 3.20 Permanent-magnet steel core with soft iron pole-shoes.

reduced to FG (or OJ). The corresponding value of the demagnetizing magnetomotive force is (OF $\times$ l_m).

Alternatively, after the magnetic field strength has been reduced from OA to zero, with the pole-shoes in contact as in fig. 3.20 (*a*), let us pull the pole-shoes apart. The effect is to increase the reluctance of the magnetic circuit and therefore to reduce the value of the flux density. Suppose l_g to be the length of the gap required to reduce the

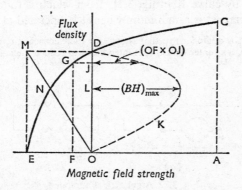

Fig. 3.21 Demagnetizing portion of hysteresis loop.

flux density in the magnet to FG. If B_g be the corresponding flux density in the gap, then

m.m.f. required for gap $= B_g l_g / \mu_0$ amperes
$= $ magnetic potential drop in gap.

As far as the reduction of the flux density in the *magnet* is concerned, the magnetic potential drop in the gap produces the same effect* as the demagnetizing m.m.f., OF $\times l_m$, and the two quantities can therefore be equated thus:

$$\text{OF} \times l_m = B_g l_g / \mu_0 \qquad (3.22)$$

Neglecting magnetic leakage, we have:

$$\text{total flux in magnet} = \text{total flux in gap}$$

i.e.
$$\text{OJ} \times a_m = B_g a_g \qquad (3.23)$$

where a_m and a_g represent the cross-sectional areas of the magnet and gap respectively.

From expressions (3.22) and (3.23), we have:

$$\text{volume of magnet} = l_m a_m = \frac{B_g l_g}{\mu_0 \times \text{OF}} \times \frac{B_g a_g}{\text{OJ}}$$

$$= \frac{B_g^2 l_g a_g}{\mu_0 (\text{OF} \times \text{OJ})} \qquad (3.24)$$

From expression (3.24), it is evident that for a given flux density in and given dimensions of the gap, the volume of the magnet is a minimum when the product (OF $\times$ OJ) is a maximum. The variation of this product for different flux densities in the *magnet core* is represented by curve K in fig. 3.21, from which it follows that the volume of magnet is a minimum when it is operated at flux density

* This effect is analogous to that shown for the electrical circuit of fig. 3.22. In fig. 3.22 (*a*), a counter e.m.f., *e*, is introduced to reduce the current to *I*,

∴
$$E - e = IR \qquad (3.20)$$

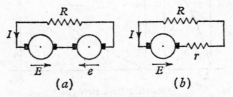

(*a*)   (*b*)

Fig. 3.22 An electrical analogue.

In fig. 3.22 (*b*), the resistance is increased by the introduction of *r* to give the same current *I*,

∴
$$E = I(R + r) \quad \text{or} \quad E - Ir = IR \qquad (3.21)$$

Comparison of (3.20) and (3.21) shows that the electric potential drop *Ir* across *r* has the same effect as the counter e.m.f. *e*,

i.e.
$$e = Ir.$$

OL. The value of the product $(BH)_{max}$ for a steel specimen is the best criterion of its suitability for use as a permanent magnet. The following table gives the magnetic data for some of the principal magnet materials:

Material	Remanence in teslas	Coercivity in A/m	Value of B for $(BH)_{max}$	Value of H for $(BH)_{max}$	$(BH)_{max}$ in J/m³
1 per cent carbon steel	0·9	4 000	0·6	$-2\,600$	1 560
6 per cent tungsten steel	1·05	5 200	0·7	$-3\,750$	2 620
35 per cent cobalt steel	0·9	20 000	0·6	$-13\,000$	7 800
Alnico (Fe–Al–Ni–Co–Cu)	0·8	40 000	0·52	$-26\,000$	13 500
Alcomax III	1·25	52 000	1·0	$-42\,000$	42 000
Ticonal G (Fe–Al–Ni–Co–Cu)	1·35	47 000	1·1	$-42\,000$	46 000

An alternative construction which gives approximate optimum working conditions is to draw a vertical line at E in fig. 3.21 and a horizontal line at D. The lines intersect at M. The straight line OM cuts the demagnetization curve at point N. The coordinates of N correspond very closely to those giving $(BH)_{max}$; in fact, if the demagnetization curve were a rectangular hyperbola, the values would be identical.

Example 3.7 *Estimate the minimum volume of* (a) *a tungsten steel magnet and* (b) *an Alcomax III magnet to maintain a flux density of* 0·5 *T across an airgap,* 4 *mm long, having an effective cross-sectional area of* 6 *cm². Assume the data given in the above table.*

(*a*) If H_1 and B_1 represent the values of the magnetic field strength and the flux density in the magnet, respectively, for $(BH)_{max}$, then for tungsten steel, $H_1 = -3750$ A/m and $B_1 = 0·7$ T.

Substituting in expression (3.22), we have:

$$3750 \times l_m = 0·5 \times 0·004/(4\pi \times 10^{-7})$$

$$\therefore \qquad l_m = 0·425 \text{ m} = 42·5 \text{ cm}.$$

Substituting in expression (3.23), we have:

$$0·7 \times a_m = 0·5 \times 6 \times 10^{-4}$$

$$\therefore \qquad a_m = 4·29 \times 10^{-4} \text{ m}^2 = 4·29 \text{ cm}^2$$

$$\therefore \quad \text{minimum volume of tungsten magnet} = 42·5 \times 4·29$$

$$= 182·5 \text{ cm}^3.$$

(*b*) For Alcomax III, $H_1 = -42\,000$ A/m and $B_1 = 1 \cdot 0$ T. From expression (3.22),

$$42\,000 \times l_m = 0 \cdot 5 \times 0 \cdot 004/(4\pi \times 10^{-7})$$

$$\therefore \qquad l_m = 0 \cdot 038 \text{ m} = 3 \cdot 8 \text{ cm.}$$

From expression (3.23),

$$1 \cdot 0 \times a_m = 0 \cdot 5 \times 6 \times 10^{-4}$$

$$\therefore \qquad a_m = 3 \times 10^{-4} \text{ m}^2 = 3 \text{ cm}^2.$$

$\therefore$ minimum volume of Alcomax III magnet

$$= 3 \cdot 8 \times 3 = 11 \cdot 4 \text{ cm}^3$$

Hence $\dfrac{\text{minimum volume of tungsten magnet}}{\text{minimum volume of Alcomax III magnet}} = \dfrac{182 \cdot 5}{11 \cdot 4} = 16$

i.e. the volume of an Alcomax III magnet is only a sixteenth of that of a tungsten magnet for the same duty.

3.16 Load line of a permanent magnet

Suppose the demagnetization curve of a magnet to be as shown in fig. 3.23, and suppose the gap between the pole-shoes to be such as to

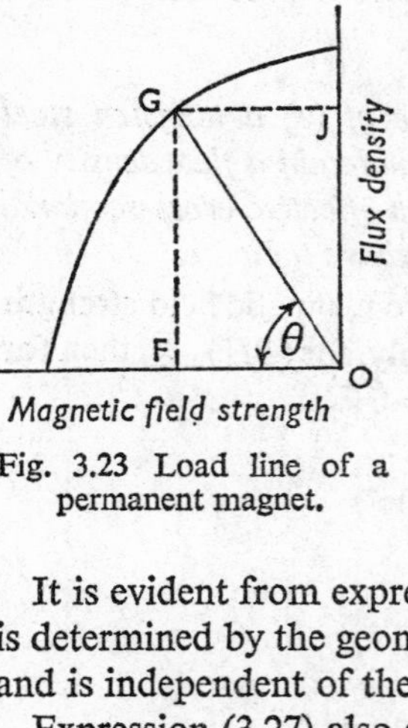

Fig. 3.23 Load line of a permanent magnet.

reduce the flux density in the magnet to OJ, as already discussed in section 3.15. Then, using the same symbols as before, we have from expression (3.22):

$$OF = B_g l_g/(\mu_0 l_m) \qquad (3.25)$$

and from expression (3.23):

$$OJ = B_g a_g/a_m \qquad (3.26)$$

Dividing (3.26) by (3.25), we have:

$$\frac{OJ}{OF} = \frac{\mu_0 l_m a_g}{l_g a_m} = \tan \theta \qquad (3.27)$$

$$= \text{slope of } \textit{load line} \text{ OG.}$$

It is evident from expression (3.27) that the slope of the load line is determined by the geometrical form of the magnet and the airgap, and is independent of the shape of the demagnetization curve.

Expression (3.27) also shows that for a given cross-sectional area of the magnet and given dimensions of the gap, the slope of the load line is directly proportional to the length, l_m, of the magnet. This means that the shorter the magnet, the smaller is the value of the flux, an effect that is attributed to *self-demagnetization* in some textbooks. This is an unfortunate term, since it is based upon the assumption of a hypothetical free pole at each end of the magnet. There is no evidence for the existence of such poles.

Example 3.8 *A Ticonal X magnet is* 30 *mm long and has a cross-sectional area of* 200 *mm². It is fitted with pole-shoes, of negligible*

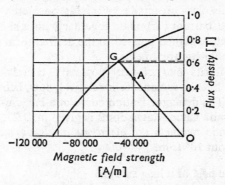

Fig. 3.24 Demagnetization curve for example 3.8.

reluctance, having an airgap, 8 *mm long, of cross-sectional area* 500 *mm². The demagnetization curve for Ticonal X is given in fig.* 3.24. *Calculate the flux density in the airgap, assuming the magnetic leakage to be negligible.*

Substituting in expression (3.27), we have:

$$\frac{OJ}{OF} = \frac{4\pi \times 10^{-7} \times 0.03 \times 500 \times 10^{-6}}{0.008 \times 200 \times 10^{-6}} = 0.118 \times 10^{-4}$$
$$= \text{slope of load line.}$$

The simplest method of locating the load line is to assume a magnetic field strength of, say, $-40\,000$ A/m; then—

$$\left.\begin{array}{l}\text{corresponding flux density}\\ \text{in magnet}\end{array}\right\} = 40\,000 \times \text{slope of load line}$$
$$= 40\,000 \times 0.118 \times 10^{-4}$$
$$= 0.472 \text{ T.}$$

With these values of $-40\,000$ A/m and 0.472 T, point A is plotted, and load line OA drawn and produced to meet the demagnetization curve at G. From fig. 3.24,

$$\text{actual value of flux density in magnet} = OJ = 0.615 \text{ T}$$

$$\therefore \qquad \text{total flux} = 0.615 \times 200 \times 10^{-6} = 123 \times 10^{-6} \text{ Wb}$$

$$\text{and} \qquad \text{flux density in gap} = \frac{123 \times 10^{-6}}{500 \times 10^{-6}} = 0.246 \text{ T}.$$

3.17 Barium-ferrite magnets

The original non-metallic permanent magnet is the natural lode-stone composed of ferrous ferrite ($FeOFe_2O_3$). Modern permanent magnets of this class are made by mixing powdered ferrous oxide with powdered barium carbonate together with a suitable bonding resin. The mixture is then pressed to the desired form and subjected to heat and magnetic treatments.

A barium-ferrite magnet may possess a coercivity as high as 240 000 A/m, but the remanence is relatively low, being of the order of 0.2–0.4 T. The demagnetization curve of a barium-ferrite magnet bearing the trade name Magnadur 2 is given in Q.27, p. 102.

These ceramic magnets are electrical insulators and have a re-sistivity of about 10^6 $\Omega\cdot$m.

3.18 Magnetic field of a long solenoid

Fig. 3.25 (a) shows a uniformly-wound solenoid, the axial length, l, of which is very large compared with the diameter. The magnetic

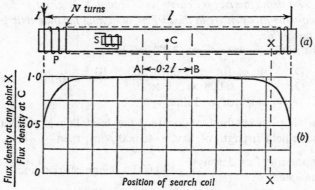

Fig. 3.25 The magnetic field of a long solenoid.

flux density at various points inside the solenoid can be determined by means of a small search-coil S connected to a ballistic galvano-meter. The deflections of the galvanometer are noted for the reversal of a given current through the solenoid winding P with the search-coil in different positions along the axis of the solenoid. The curve in fig. 3.25 (*b*) represents the result obtained with a solenoid having a length equal to 10 times the diameter. It is found that between two planes, A and B, situated 0·1*l* on each side of the centre C of such a solenoid, the variation of the flux density inside the solenoid is only about 0·1 per cent. This length of the solenoid may therefore be regarded as a part of a uniformly-wound ring (section 3.2); hence the magnetic field strength inside the central portion of a long solenoid is *I* × number of turns per metre length,

i.e. $$H = NI/l \text{ amperes/metre} \qquad (3.28)$$
where I = the magnetizing current in amperes,
N = total number of turns on the solenoid,
and l = length of solenoid in metres.

Example 3.9 *A solenoid, 1·2 m long, is uniformly wound with 800 turns. A short coil, wound with 50 turns having a mean diameter of 30 mm, is placed at the centre of the solenoid and connected to a ballistic galvanometer. The total resistance of the search-coil circuit is 2000 Ω. When a current of 5 A through the solenoid is reversed, the deflection on the galvanometer scale is 85 divisions. Calculate the ballistic constant (i.e. the no. of coulombs per unit deflection), assuming the damping of the galvanometer to be negligible and the coils to be coaxial.*

Magnetic field strength at centre of solenoid $= 5 \times 800/1·2 = 3333$ A/m

∴ flux density at centre of solenoid $= 4\pi \times 10^{-7} \times 3333$

$= 0·004\ 185$ T.

Area of search-coil $= \dfrac{\pi}{4} \times (0·03)^2 = 0·000\ 706\ 5$ m².

Flux through search-coil $= 0·004\ 185 \times 0·000\ 706\ 5$
$= 2·96 \times 10^{-6}$ Wb.

If the current is reversed in t seconds,
average e.m.f. induced in search-coil

$$= \frac{2 \times 2·96 \times 10^{-6} \times 50}{t} = \frac{296}{t} \times 10^{-6} \text{ volt}$$

and average current through galvanometer

$$= \frac{296 \times 10^{-6}}{t \times 2000} = \frac{0·148}{t} \times 10^{-6} \text{ ampere}$$

∴ quantity of electricity through galvanometer

$$= 0 \cdot 148 \times 10^{-6} \text{ coulomb}$$

and ballistic constant of galvanometer

$$= 0 \cdot 148 \times 10^{-6}/85$$
$$= 1 \cdot 74 \times 10^{-9} \text{ coulomb per division.}$$

This is one of the standard methods used for determining the ballistic constant of a galvanometer.

3.19 Magnetic energy in a non-magnetic medium

For a non-magnetic medium, the relative permeability is unity, so that the magnetic field strength and the flux density are proportional to each other and the relationship between them is represented by a straight line OD in fig. 3.26.

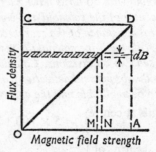

Fig. 3.26 *B/H* relationship for a non-magnetic medium.

It was shown in section 3.14 that when the flux density is increased by d*B* due to an increase MN of the magnetic field strength,

energy supplied to mag-
netic circuit   } = area of shaded strip, joules/metre³.

Hence, for a maximum flux density OC in fig. 3.26,

total energy stored in
magnetic field   } = area of triangle OCD in joules/metre³

$$= \tfrac{1}{2}\text{OA} \times \text{OC} = \tfrac{1}{2} \cdot \frac{\text{OC}}{\mu_0} \cdot \text{OC}$$

$$= \frac{\text{OC}^2}{2\mu_0} \text{ joules/metre}^3$$

Consequently, for a flux density of B teslas,

$$\text{energy stored in a non-} \atop \text{magnetic medium} \Bigg\} = \tfrac{1}{2}HB = \frac{B^2}{2\mu_0} \text{ joules/metre}^3 \qquad (3.29)$$

3.20 Magnetic pull between two iron surfaces

Consider a fixed iron core D, wound with coil C, and a movable iron core E, as shown in fig. 3.27. Suppose D and E to have the same cross-sectional area of a square metres. Also, suppose the opposite surfaces of D and E to be g metres apart and that a current I amperes through coil C produces a flux density of B teslas in the airgap. From expression (3.29),

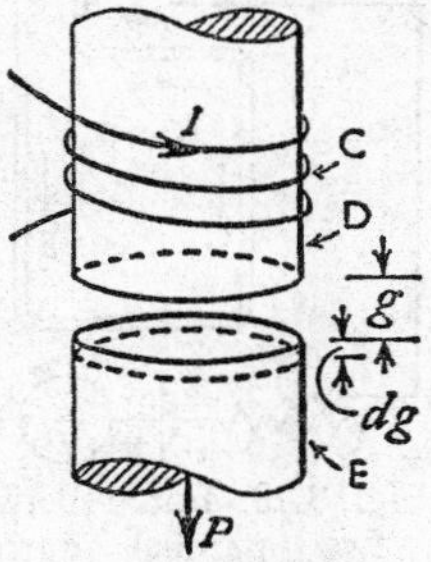

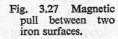

$$\text{energy stored per cubic} \atop \text{metre of airgap} \Bigg\} = B^2/2\mu_0 \text{ joules.}$$

Let P be the force of attraction in newtons between the two surfaces of D and E. Suppose iron core E to be moved a short distance dg

Fig. 3.27 Magnetic pull between two iron surfaces.

metre away from D and that the current through coil C be increased at the same time so as to keep the flux density unaltered. No e.m.f. will therefore have been induced in coil C, so that no energy will have been supplied either from the electrical circuit to the magnetic circuit or vice versa. Hence all the energy stored in the *additional* volume of airgap must have been derived from the work done when force P newtons acted through a distance dg metre, namely $P \times \text{dg}$ joules,

i.e. $\qquad\qquad P \times \text{dg} = (B^2/2\mu_0) \times a \times \text{dg}$

∴ $\qquad\qquad P = B^2 a/2\mu_0 \text{ newtons} \qquad\qquad (3.30)$

3.21 Force between two long parallel conductors carrying electric current

When two current-carrying conductors are parallel to each other, there is a force acting on each of the conductors, an effect that can be easily demonstrated by means of the apparatus referred to in section 2.7. Thus, in fig. 3.28, the rectangular loop BC has its tapered ends dipping into mercury in cups supported one directly above the other by rods P and Q, as already shown in fig. 2.11, and a current of about 10–15 amperes is passed through the loop. Part of the electrical circuit consists of a long straight rod D which can be placed alongside B, as shown in plan in fig. 3.29. When the currents in D and B are in opposite directions, as in fig. 3.29 (*a*), the two conductors repel each other and the loop (viewed from above) is deflected clock-

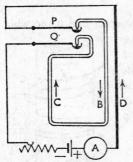

wise. On the other hand, if rod D is turned through 180° so that the currents in B and D are in the same direction, as in fig. 3.29 (*b*), the conductors attract each other and the loop turns counterclockwise.

These effects are most easily explained by first drawing the magnetic fields produced by each conductor and then combining these fields. Thus, fig. 3.30 (*a*) shows two conductors, A and B, each carrying current towards the paper. The lines of

Fig. 3.28 Force between two parallel current-carrying conductors.

magnetic flux due to current in A alone are represented by the uniformly-dotted circles in fig. 3.30 (*a*), and those due to B

alone are represented by the chain-dotted circles. It is evident that in the space between A and B the two fields tend to neutralize each other, whereas in the space outside A and B they assist each other. Hence the resultant distribution is somewhat as shown in fig. 3.30 (*b*). Since magnetic flux behaves like a stretched elastic cord, the effect is to try to move conductors A and B towards each other; in other words, there is a force of attraction between A and B.

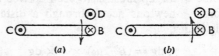

(*a*)   (*b*)

Fig. 3.29 Force between two parallel current-carrying conductors.

If the current in B is reversed, the magnetic fields due to A and B assist each other in the space between the conductors and the resultant distribution of the flux will be as shown in fig. 3.30 (*c*). The lateral pressure between the lines of flux exerts a force on the conductors tending to push them apart (section 2.3 (5)).

3.22 Magnitude of force between two parallel current-carrying conductors

Let C in fig. 3.31 represent the cross-section of a long straight conductor carrying a current *I* amperes towards the paper. The return conductor is assumed to be some distance away, so that its magnetic field may be neglected.

The magnetic flux set up by the current can be represented as concentric circles around the conductor. Let us therefore consider the flux in the cylinder of radius x metres and thickness dx metre, shown dotted in fig. 3.31. Since the outward and return conductors form

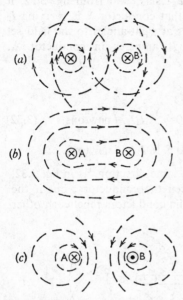

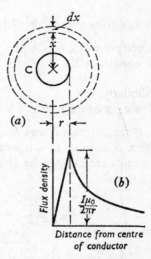

Fig. 3.30 Magnetic fields due to parallel current-carrying conductors.

Fig. 3.31 Magnetic field around a long straight conductor situated in a non-magnetic medium.

one turn, it follows that the magnetic field strength acting on this path is $I/(2\pi x)$ amperes/metre; and since the relative permeability of air and other non-magnetic mediums is unity,

$$\text{flux density at radius } x = I\mu_0/(2\pi x) \text{ teslas} \qquad (3.31)$$

Hence the flux density around a conductor is inversely proportional to the distance from the centre of the conductor, as represented by the hyperbola in fig. 3.31 (*b*). The density is a maximum at the surface of the conductor and is equal to $I\mu_0/2\pi r$ teslas, where r is the radius of the conductor in metres.

Let us next consider two parallel conductors A and B (fig. 3.32) carrying I_1 and I_2 amperes respectively and let the distance between their centres be d metres. From expression (3.31), it follows that the flux density due to I_1 alone at a distance of d metres from the centre of A is $I_1\mu_0/2\pi d$ teslas; and from fig. 3·32, it is seen that conductor B is carrying I_2 amperes at right-angles to the field set up by I_1 amperes in conductor A.

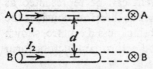

Fig. 3.32 Two parallel current-carrying conductors situated in a non-magnetic medium.

Hence from expression (2.1),

$$\left.\begin{array}{c}\text{force per metre length of}\\ \text{conductor B}\end{array}\right\} = \frac{I_1\mu_0}{2\pi d} \cdot I_2$$

$$= 2 \times 10^{-7} \times I_1 I_2/d \text{ newtons} \qquad (3.32)$$

Similarly,
force per metre length of A $= 2 \times 10^{-7} \times I_1 I_2/d$ newtons.

If the currents I_1 and I_2 are in the same direction, as in fig. 3.32, there is a force of attraction between the conductors; but if the currents are in opposite directions, the conductors repel each other

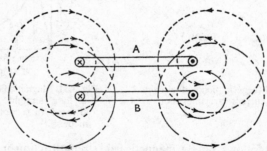

Fig. 3.33 Magnetic fields due to currents in A and B separately.

(section 3.21). If $I_1 = I_2 = 1$ A and if $d = 1$ m, it follows from expression (3.32) that the force per metre length of conductor is 2×10^{-7} N. This value forms the basis for the definition of the ampere, given in section 1.6, namely *that current which, if maintained in two straight parallel conductors of infinite length, of negligible circular cross-section, and placed 1 metre apart in a vacuum, would produce between these conductors a force of 2 × 10⁻⁷ newton per metre of length.*

3.23 Force between coils carrying electric current

Suppose A and B in fig. 3.33 to represent the cross-section of two coils carrying currents in the directions shown by the dots and crosses. Let us first consider the distribution of the magnetic fields due to the coils acting independently. Thus, current through A alone gives the

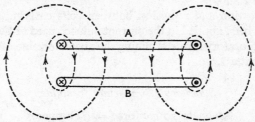

Fig. 3.34 Resultant magnetic field when currents in A and B are in the same direction.

flux distribution represented by the uniformly-dotted lines in fig. 3.33, while current in the same direction through B alone gives the distribution indicated by the chain-dotted lines. It will be seen that in the space between the coils the two fields oppose each other, while on the outside they are in the same direction. Consequently, the combined

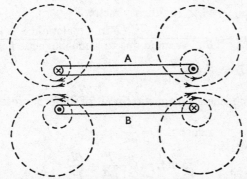

Fig. 3.35 Resultant magnetic field when currents in A and B are in opposite directions.

effect is to give the distribution shown in fig. 3.34. Since magnetic flux acts as if it is in tension, it tends to move coils A and B towards each other.

On the other hand, if the current through B is reversed, the direction of the arrowheads on the chain-dotted lines in fig. 3.33 is reversed. Consequently, the magnetic fields of A and B are in the same direction in the space between the coils and in opposition outside the coils, so that the resultant distribution becomes that shown in fig. 3.35. The magnetic flux between the coils exerts a lateral pressure, thereby producing a force of repulsion between coils A and B (see section 2.3.5).

This attraction and repulsion between coils carrying an electric current has been applied in the current balance used at the National Physical Laboratory for determining the absolute value of an electric current (section 1.6).

Summary of important formulae

$$\text{Magnetomotive force} = NI \text{ amperes}$$

$$\text{Magnetic field strength} = H = NI/l \text{ amperes/metre} \tag{3.1}$$

$$\text{Magnetic flux} = \Phi = \frac{\text{m.m.f.}}{\text{reluctance}} \text{ webers}$$

$$\text{Flux density} = B = \Phi/a \text{ teslas}$$

$$B/H = \mu = \text{absolute permeability} \tag{3.4}$$

$$= \mu_0\mu_r$$

$$\mu_0 = 4\pi \times 10^{-7} \text{ henry/metre} \tag{3.2}$$

$$\mu_r = \frac{B \text{ in a material}}{B \text{ in vacuum due to same magnetic field strength}}.$$

$$H = \frac{B}{\mu} = \frac{B}{\mu_0\mu_r}$$

For a homogeneous magnetic circuit of uniform cross-section,

$$\text{reluctance} = \frac{l}{\mu a} \tag{3.7}$$

and

$$\text{m.m.f.} = \Phi \cdot \frac{l}{\mu a} = \frac{Bl}{\mu}$$

For a composite magnetic circuit,

$$\Phi = \frac{NI}{\dfrac{l_1}{\mu_1 a_1} + \dfrac{l_2}{\mu_2 a_2}} \tag{3.10}$$

$$\text{Leakage factor} = \frac{\text{total flux}}{\text{useful flux}} \tag{3.11}$$

$$\left.\begin{array}{l}\text{Hysteresis loss/cubic}\\\text{metre/cycle}\end{array}\right\} = \left\{\begin{array}{l}\text{area of hysteresis loop, in}\\\text{joules}\end{array}\right. \tag{3.17}$$

$$\text{Hysteresis loss} = kvf(B_{\max})^x \tag{3.19}$$

where x usually lies between 1·5 and 2·5.

Permanent magnet has minimum volume when it is operated at the point where the product (BH) for the demagnetizing portion of the hysteresis loop is a maximum (3.24)

Magnetic field strength inside central portion of long solenoid

$$= I \times \text{turns/metre} \tag{3.28}$$

$$\left.\begin{array}{l}\text{Magnetic energy stored in}\\\text{non-magnetic medium}\end{array}\right\} = \frac{B^2}{2\mu_0} \text{ joules/metre}^3 \tag{3.29}$$

$$\left.\begin{array}{l}\text{Magnetic pull between two}\\\text{iron surfaces}\end{array}\right\} = \frac{B^2 a}{2\mu_0} \text{ newtons} \tag{3.30}$$

$$\left.\begin{array}{l}\text{Flux density around wire}\\\text{carrying current}\end{array}\right\} = \frac{I\mu_0}{2\pi x} \text{ teslas} \tag{3.31}$$

$$\left.\begin{array}{l}\text{Force between two conduc-}\\\text{tors carrying current}\end{array}\right\} = \frac{2 \times 10^{-7} I_1 I_2}{d} \text{ newtons/metre} \tag{3.32}$$

EXAMPLES 3

Data of B/H, when not given in question, should be taken from fig. 3.3.

1. A mild steel ring has a mean diameter of 160 mm and a cross-sectional area of 300 mm². Calculate: (*a*) the m.m.f. to produce a flux of 400 μWb, and (*b*) the corresponding values of the reluctance of the ring and of the relative permeability.

2. An iron magnetic circuit has a uniform cross-sectional area of 5 cm² and a length of 25 cm. A coil of 120 turns is wound uniformly over the magnetic circuit. When the current in the coil is 1·5 A, the total flux is 0·3 mWb; when the current is 5 A, the total flux is 0·6 mWb. For each value of current, calculate: (*a*) the magnetic field strength; (*b*) the relative permeability of the iron. (U.E.I., O.1)

3. A mild steel ring has a mean circumference of 500 mm and a uniform cross-sectional area of 300 mm². Calculate the m.m.f. required to produce a flux of 500 μWb.

   An airgap, 1 mm in length, is now cut in the ring. Determine the flux produced if the m.m.f. remains constant. Assume the relative permeability of the mild steel to remain constant at 1200. (S.A.N.C., O.1)

4. An iron ring has a mean diameter of 15 cm, a cross-section of 20 cm²
   and a radial airgap of 0·5 mm cut in it. The ring is uniformly wound
   with 1500 turns of insulated wire and a magnetizing current of 1 A
   produces a flux of 1 mWb in the airgap. Neglecting the effect of magnetic leakage and fringing, calculate: (a) the reluctance of the magnetic
   circuit; (b) the relative permeability of the iron.   (U.E.I., O.1)

5. (a) An iron ring, having a mean circumference of 750 mm and a cross-
   sectional area of 500 mm², is wound with a magnetizing coil of 120
   turns. Using the following data, calculate the current required to set up
   a magnetic flux of 630 microwebers in the ring.

Flux density (teslas)	0·9	1·1	1·2	1·3
Amperes/metre	260	450	600	820

   (b) The airgap in a magnetic circuit is 1·1 mm long and 2000 mm² in
   cross-section. Calculate: (i) the reluctance of the airgap, and (ii) the
   m.m.f. to send a flux of 700 microwebers across the airgap.
                                                             (U.L.C.I., O.1)

6. A magnetic circuit consists of a cast steel yoke which has a cross-
   sectional area of 200 mm² and a mean length of 120 mm. There are two
   airgaps, each 0·2 mm long. Calculate: (a) the m.m.f. required to produce a flux of 0·05 mWb in the airgaps; (b) the value of the relative
   permeability of cast steel at this flux density.

   The magnetization curve for cast steel is given by the following:

B (teslas)	0·1	0·2	0·3	0·4
H (amperes/metre)	170	300	380	460

                                                             (W.J.E.C., O.1)

7. An electromagnet has a magnetic circuit that can be regarded as comprising three parts in series, viz.: (a) length of 80 mm and cross-sectional area 60 mm²; (b) a length of 70 mm and cross-sectional area
   80 mm²; (c) an airgap of length 0·5 mm and cross-sectional area 60
   mm².

   Parts (a) and (b) are of a material having magnetic characteristics
   given by the following table:

H (amperes/metre)	100	210	340	500	800	1500
B (teslas)	0·2	0·4	0·6	0·8	1·0	1·2

   Determine the current necessary in a coil of 4000 turns wound on
   part (b) to produce in the airgap a flux density of 0·7 tesla. Magnetic
   leakage may be neglected.   (E.M.E.U., O.1)

8. A certain magnetic circuit may be regarded as consisting of three parts,
   A, B and C in series, each one of which has a uniform cross-sectional
   area. Part A has a length of 300 mm and a cross-sectional area of
   450 mm². Part B has a length of 120 mm and a cross-sectional area of
   300 mm². Part C is an airgap 1 mm in length and of cross-sectional
   area 350 mm². Neglecting magnetic leakage and fringing, determine
   the m.m.f. necessary to produce a flux of $3·5 \times 10^{-4}$ Wb in the airgap.

The magnetic characteristic for parts A and B is given by:

H (A/m)	400	560	800	1280	1800
B (T)	0·7	0·85	1·0	1·15	1·25

(N.C.T.E.C., O.2)

9. A magnetic circuit made of wrought iron is arranged as in fig. A. The centre limb has a cross-sectional area of 800 mm² and each of the side limbs has a cross-sectional area of 500 mm². Calculate the m.m.f. required to produce a flux of 1 mWb in the centre limb, assuming the magnetic leakage to be negligible.

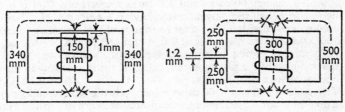

Fig. A. Fig. B.

10. A magnetic core made of Stalloy has the dimensions shown in fig. B. There is an airgap 1·2 mm long in one side limb and a coil of 400 turns is wound on the centre limb. The cross-sectional area of the centre limb is 1600 mm² and that of each side limb is 1000 mm². Calculate the exciting current required to produce a flux of 1000 μWb in the airgap. Neglect any magnetic leakage and fringing.

11. An electromagnet with its armature has an iron length of 400 mm and a cross-sectional area of 500 mm². There is a total airgap of 1·8 mm. Assuming a leakage factor of 1·2, calculate the m.m.f. required to produce a flux of 400 μWb in the armature. Points on the B/H curve are as follows:

Flux density (T)	0·8	1·0	1·2
Magnetic field strength (A/m)	800	1000	1600

12. Define the *ampere* in terms of M.K.S. units.

A coil P of 300 turns is uniformly wound on an iron ring having a cross-sectional area of 500 mm² and a mean diameter of 400 mm. Another coil S of 30 turns wound on the ring is connected to a fluxmeter having a constant of 200 microweber-turns per division. When a current of 1·5 A through P is reversed, the fluxmeter deflection is 96 divisions. Calculate: (*a*) the flux density in the ring, and (*b*) the magnetic field strength. (E.M.E.U., O.1)

13. The magnetization curve of a ring specimen of iron is determined by means of a fluxmeter. The specimen, which is uniformly wound with a magnetizing winding of 1000 turns, has a mean length of 65 cm and a cross-sectional area of 7·5 cm². A search coil of 2 turns is connected to the fluxmeter, which has a constant of 0·1 mWb-turn per scale division.

When the magnetizing current is reversed, the fluxmeter deflection is noted, the following table being the readings which are obtained:

Magnetizing current (amperes)	0·2	0·4	0·6	0·8	1·0
Fluxmeter deflection (divisions)	17·8	31·2	38	42·3	44·7

Explain the basis of the method and calculate the values of the magnetic field strength and flux density for each set of readings.

(W.J.E.C., O.2)

14. A ring specimen of mild steel has a cross-sectional area of 6 cm² and a mean circumference of 30 cm. It is uniformly wound with two coils, A and B, having 90 turns and 300 turns respectively. Coil B is connected to a ballistic galvanometer having a constant of $1·1 \times 10^{-8}$ coulomb per division. The total resistance of this secondary circuit is 200 000 Ω. When a current of 2·0 A through coil A is reversed, the galvanometer gives a maximum deflection of 200 divisions. Neglecting the damping of the galvanometer, calculate the flux density in the iron and its relative permeability. (E.M.E.U., O.1)

15. Two coils of 2000 and 100 turns respectively are wound uniformly on a non-magnetic toroid having a mean circumference of 1 m and a cross-sectional area of 500 mm². The 100-turn coil is connected to a ballistic galvanometer, the total resistance of this circuit being 5·1 kΩ. When a current of 2·5 A in the 2000-turn coil is reversed, the galvanometer is deflected through 100 divisions. Neglecting damping, calculate the ballistic constant for the instrument. (W.J.E.C., O.2)

16. Explain with the aid of diagrams and brief notes how a hysteresis loop may be obtained for an iron specimen using a fluxmeter.

A hysteresis loop is plotted against a horizontal axis which scales 1000 A/m = 1 cm and a vertical axis which scales 1 T = 5 cm. If the area of the loop is 9 cm² and the overall height is 14 cm, calculate: (*a*) the hysteresis loss in joules per cubic metre per cycle; (*b*) the maximum flux density; (*c*) the hysteresis loss in watts per kilogram, assuming the density of the material to be 7800 kg/m³.

(E.M.E.U., O.2)

17. (*a*) Explain briefly the occurrence of hysteresis loss and eddy-current loss in a transformer core and indicate how these losses may be reduced.

(*b*) The area of a hysteresis loop for a certain magnetic specimen is 160 cm² when drawn to the following scales: 1 cm represents 25 A/m and 1 cm represents 0·1 T. Calculate the hysteresis loss in watts/ kilogram when the specimen is carrying an alternating flux at 50 Hz, the maximum value of which corresponds to the maximum value attained on the hysteresis loop. The density of the material is 7600 kg/m³. (W.J.E.C., O.2)

18. Explain how the energy losses in a sample of iron subjected to an alternating magnetic field depend on the frequency and the flux density. What particular property of the material can be used as a measure of the magnitude of each type of loss?

The area of a hysteresis loop plotted for a sample of iron is 67·1 cm², the maximum flux density being 1·06 T. The scales of *B* and *H* are such that 1 cm = 0·12 T and 1 cm = 7·07 A/m. Find the loss due to hysteresis if 750 g of this iron were subjected to an alternating magnetic field of maximum flux density 1·06 T at a frequency of 60 Hz. The density of the iron is 7700 kg/m³.   (U.E.I., O.2)

19. The hysteresis loop for a certain iron ring is drawn to the following scales: 1 cm to 300 A and 1 cm to 100 μWb. The area of the loop is 37 cm². Calculate the hysteresis loss per cycle.

20. In a certain transformer, the hysteresis loss is 300 W when the maximum flux density is 0·9 T and the frequency 50 Hz. What would be the hysteresis loss if the maximum flux density were increased to 1·1 T and the frequency reduced to 40 Hz? Assume the hysteresis loss over this range to be proportional to $(B_{max})^{1·7}$.

21. The demagnetization curve for a certain magnet steel is represented by the following data:

Magnetic field strength (A/cm)	0	−144	−240	−304	−360	−400	−430
Flux density (T)	0·6	0·5	0·4	0·3	0·2	0·1	0

Derive a graph showing the variation of (*BH*) with the flux density and estimate the values of *B* and *H* at which the product (*BH*) is a maximum.

22. Name *three* materials which are suitable for permanent magnets. Explain the characteristics of these materials which render them suitable for this purpose.

The demagnetization curve for a certain permanent-magnet steel is as follows:

B in T	0	0·1	0·2	0·3	0·4	0·5	0·6	0·64
H in A/cm	−488	−464	−432	−392	−336	−240	−96	0

Determine the minimum dimensions of a magnet made from this material to maintain a flux density of 0·5 T across an airgap the dimensions of which are: length 1 cm, effective cross-sectional area 4 cm².   (U.L.C.I., O.2)

23. (*a*) Derive the condition which must be observed to ensure that a permanent magnet producing a given airgap field shall have minimum volume.

(*b*) An airgap of 300 mm² cross-section and 2·5 mm length is to be supplied with a flux density of 0·12 T. The demagnetization curve for the magnetic material used is as follows:

B (T)	0	0·25	0·5	0·75	1·0	1·25	1·40
H (A/m)	−4630	−4370	−4000	−3500	−2800	−1550	0

Calculate the dimensions of the magnet using the minimum volume of material. Neglect magnetic leakage.   (W.J.E.C., O.2)

24. Show that provided reasonable simplifying assumptions are made, the minimum volume of a given permanent-magnet material needed to establish a specified flux in a specified airgap is obtained when the value of $(B) \times (-H)$ for the material is a maximum.

A magnet of minimum volume is to be made from a material having a demagnetization curve given by:

H in A/mm	0	−12	−20	−25	−28	−30	−32
B in T	0·7	0·6	0·5	0·4	0·3	0·2	0

It is to produce a flux density of 0·4 T in an air gap of length 3 mm and cross-sectional area 500 mm². Magnetic leakage can be neglected, and pole pieces of negligible reluctance assumed. Determine the approximate length and cross-section of magnet material.

(App. El., L.U.)

25. The demagnetization curve of the steel of a permanent magnet is given by:

Magnetic field strength (A/m)	−1000	−2000	−3000	−4000
Flux density (T)	0·65	0·56	0·37	0·12

Find the length and cross-section of the magnet of minimum volume that would provide a flux density of 0·1 T in an airgap 3 mm long and of cross-section 10 cm². Assume a leakage flux between the poles of the magnet of 50 per cent of the useful flux. (S.A.N.C., O.2)

Note. If k = leakage factor = total flux in magnet/total flux in gap, total flux in magnet = $kB_g a_g$

$$\therefore \qquad a_m = kB_g a_g / OJ$$

where OJ = flux density in magnet corresponding to $(BH)_{max}$.

26. A certain magnet is 50 mm long and has a cross-sectional area of 300 mm². It is fitted with pole-shoes of negligible reluctance, arranged to give an airgap 6 mm in length and 600 mm² in cross-sectional area. If the demagnetization curve is represented by the data given in Q.21, determine the flux density in the airgap. Neglect magnetic leakage and fringing.

27. The demagnetization curve for the barium-ferrite magnet, Magnadur 2, is as follows:

H (A/mm)	0	−40	−80	−120	−160	−170	−175
B (T)	0·39	0·345	0·295	0·24	0·18	0·15	0

The magnet has a length of 35 mm and a cross-sectional area of 600 mm², and is fitted with pole pieces of negligible reluctance. Between the pole pieces there is an airgap 12 mm in length and 800 mm² in cross-sectional area. Neglecting magnetic leakage and fringing, determine the flux density in the gap.

28. A solenoid M, 1·2 m long and 80 mm diameter, is uniformly wound with 750 turns. A search coil N, 25 mm in diameter and wound with 30 turns, is mounted co-axially midway along the solenoid. Calculate the average e.m.f. induced in N when a current of 5 A through M is reversed in 0·2 s.

29. Each of the two airgaps of a moving-coil instrument is 2·5 mm long and has a cross-sectional area of 600 mm². If the flux density is 0·08 T, calculate the total energy stored in the magnetic field of the airgaps.

30. Given that the energy stored per unit volume of a magnetic field is $B^2/(2\mu_0)$, derive an expression for the pull between two magnetized surfaces.

   A lifting magnet of inverted U-shape is formed out of an iron bar 60 cm long and 10 cm² in cross-sectional area. Exciting coils of 750 turns each are wound on the two side limbs and are connected in series. There is an effective air gap of 0·1 mm between the load and the magnet at each pole. Assuming that the reluctance of the load is negligible and ignoring fringing at the airgaps and magnetic leakage, calculate the current required in the winding to lift a mass of 200 kg. Use the following magnetic data for the iron:

B in teslas	1·56	1·57	1·58
H in amperes/metre	2800	3000	3200

   (U.L.C.I., O.2)

31. A steel ring having a mean diameter of 350 mm and a cross-sectional area of 240 mm² is broken by a parallel-side airgap of length 12 mm. Short pole pieces of negligible reluctance extend the effective cross-sectional area of the airgap to 1200 mm². Taking the relative permeability of the steel as 700 and neglecting leakage, determine the current necessary in 300 turns of wire wound on the ring to produce a flux density of 0·25 T in the gap.

   Evaluate also the tractive force between the poles.

   (App. El., L.U.)

32. Each of the two pole faces of a lifting magnet has an area of 150 cm², and this may also be taken as the cross-sectional area of the 40-cm-long flux path in the magnet. Determine the m.m.f. needed on the magnet if it is to lift a 900-kg iron block separated by 0·5 mm from the pole faces. Assume the magnetic leakage factor to be 1·2, neglect fringing of the gap flux and the reluctance of the flux path in the iron block, and take the magnetization curve of the magnet material to be given by:

H (A/m)	400	600	800	1200	1600
B (T)	0·81	0·98	1·10	1·24	1·35

   (App. El., L.U.)

33. A long straight conductor, situated in air, is carrying a current of 500 A, the return conductor being far removed. Calculate the magnetic field strength and the flux density at a radius of 80 mm.

34. (*a*) The flux density in air at a point 40 mm from the centre of a long straight conductor A is 0·03 T. Assuming that the return conductor is a considerable distance away, calculate the current in A.

(*b*) In a certain magnetic circuit, having a length of 500 mm and a cross-sectional area of 300 mm², an m.m.f. of 200 A produces a flux of 400 µWb. Calculate (i) the reluctance of the magnetic circuit and (ii) the relative permeability of the core.

35. Two long parallel conductors, spaced 40 mm between centres, are each carrying a current of 5000 A. Calculate the force in newtons per metre length of conductor.

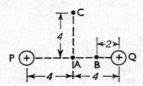

Fig. C.

36. Two long parallel conductors P and Q, situated in air and spaced 8 cm between centres, carry currents of 600 A in opposite directions. Calculate the values of the magnetic field strength and of the flux density at points A, B and C in fig. C, where the dimensions are given in centimetres.

Calculate also the values of the same quantities at the same points if P and Q are each carrying 600 A in the same direction.

CHAPTER 4

Inductance in a D.C. Circuit

4.1 Inductive and non-inductive circuits

Let us consider what happens when a coil C (fig. 4.1) and a resistor R, connected in parallel, are switched across a battery B. C consists of a large number of turns wound on an iron core D (or it may be the field winding of a generator or motor), and R may be an element of an electric heater connected in series with a *centre-zero* ammeter A_2.

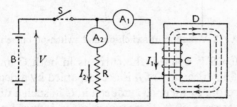

Fig. 4.1 Inductive and non-inductive circuits.

When switch S is closed, it is found that the current I_2 through R increases instantly to its final value, whereas the current i_1* through C takes an appreciable time to grow—as indicated in fig. 4.2. The final value, I_1, is equal to $\dfrac{\text{battery voltage V}}{\text{resistance of coil C}}$. In fig. 4.2, I_2 has been shown a little larger than I_1, but this is of no importance.

When S is opened, current through C decreases comparatively slowly, but the current through R instantly reverses its direction and becomes the same current as i_1; in other words, the current of C is circulating round R.

Let us now consider the reason for the difference in the behaviour of the currents in C and R.

The growth of current in C is accompanied by an increase of flux—shown dotted—in the iron core D. But it has been pointed out in section 2.10 that any change in the flux linked with a coil is ac-

---

* A lower case letter is used to represent the instantaneous value of a varying quantity.

companied by an e.m.f. induced in that coil, the direction of which—according to Lenz's Law—is always such as to oppose the change responsible for inducing the e.m.f., namely the growth of current in C. In other words, the induced e.m.f. is acting in opposition to the current and, therefore, to the applied voltage. In circuit R, the flux is so small that its induced e.m.f. is negligible.

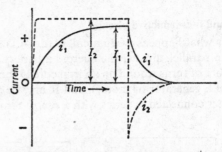

Fig. 4.2 Variation of switch-on and switch-off currents.

When switch S is opened, the currents in both C and R tend to decrease; but any decrease of i_1 is accompanied by a decrease of flux in D and therefore by an e.m.f. induced in C in such a direction as to oppose the decrease of i_1. Consequently the induced e.m.f. is now acting in the same direction as the current. But it is evident from

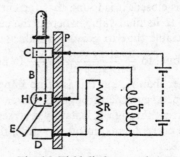

Fig. 4.3 Field-discharge switch.

fig. 4.1 that after S has been opened, the only return path for C's current is that via R; hence the reason why i_1 and i_2 are now one and the same current.

If the experiment is repeated without R, it is found that the growth of i_1 is unaffected; but when S is opened there is considerable arcing at the switch due to the maintenance of the current across the gap by the e.m.f. induced in C. The more quickly S is opened, the more rapidly does the flux in D collapse and the greater is the e.m.f. induced in C. This is the reason why it is dangerous to break quickly the full excitation of an electromagnet such as the field winding of a d.c. machine. One method of avoiding this risk is to connect a *discharge resistor* R permanently in parallel with the winding, as in fig. 4.1; but this involves a waste of energy in R while the winding is in circuit. This waste of energy can be avoided by the use of a special switch such as that shown in fig. 4.3, where blade B, pivoted at H, carries an extension E. Hinge H and jaws C and D are fixed to an insulating panel P. When the switch is being opened, E makes contact with D before B breaks contact with C, thereby connecting R and F momentarily in parallel across the source. After B has broken contact with C, resistor R and winding F form a closed circuit and the current decreases relatively slowly so that there is no risk of a dangerously high e.m.f. being induced in F. The resistance of the discharge resistor is usually about the same as that of the field winding.

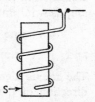

Fig. 4.4 Non-inductive resistor.

Any circuit in which a change of current is accompanied by a change of flux, and therefore by an induced e.m.f., is said to be *inductive* or to possess *self inductance* or merely *inductance*. It is impossible to have a perfectly *non-inductive* circuit, i.e. a circuit in which no flux is set up by a current; but for most purposes a circuit which is not in the form of a coil may be regarded as being practically non-inductive—even the open helix of an electric fire is almost non-inductive. In cases where the inductance has to be reduced to the smallest possible value—for instance, in resistance boxes—the wire is bent back on itself, as shown in fig. 4.4, so that the magnetizing effect of one conductor is neutralized by that of the adjacent conductor. The wire can then be coiled round an insulator S without increasing the inductance.

4.2 Unit of inductance

The unit of inductance is termed the *henry*, in commemoration of a famous American physicist, Joseph Henry (1797–1878), who, quite independently, discovered electromagnetic induction within a year after it had been discovered in this country by Michael Faraday in 1831. *A circuit has an inductance of* 1 *henry* (*or* 1 H) *if an e.m.f. of* 1 *volt is induced in the circuit when the current varies uniformly at the rate of* 1 *ampere per second.* If either the inductance or the rate of change of current be doubled, the induced e.m.f. is doubled. Hence if a circuit has an inductance of L henrys and if the current *increases* from i_1 to i_2 amperes in t seconds:

$$\left.\begin{array}{r}\text{average rate of change}\\ \text{of current}\end{array}\right\} = \frac{i_2 - i_1}{t}\ \text{amperes/second,}$$

$$\left.\begin{array}{r}\text{and average induced}\\ \text{e.m.f.}\end{array}\right\} = -L \times \text{rate of change of current}$$

$$= -L \times \frac{i_2 - i_1}{t}\ \text{volts} \tag{4.1}$$

Considering instantaneous values, if $di = $ increase of current, in amperes, in time dt second,

$$\text{rate of change of current} = di/dt\ \text{amperes/second}$$

and   e.m.f. induced in circuit $= -L \cdot di/dt$ volts   (4.2)

The minus sign in expressions (4.1) and (4.2) indicates that the direction of the induced e.m.f. is opposite to that of the current increase. It also indicates that energy is being *absorbed* from the electric circuit and stored as magnetic energy in the coil.

When the current is decreasing, the induced e.m.f. tends to prevent the decrease of the current, and its direction is therefore the same as that of the current. Energy is then being *supplied* from the magnetic circuit to the electric circuit, so that the coil is acting as a generator[*] of electrical energy.

Example 4.1 *If the current through a coil having an inductance of* 0·5 H *is reduced from* 5 A *to* 2 A *in* 0·05 s, *calculate the mean value of the e.m.f. induced in the coil.*

Average rate of change of current

$$= (2 - 5)/0 \cdot 05 = -60\ \text{amperes/second}$$

From (4.1) average e.m.f. induced in coil

$$= -0 \cdot 5 \times (-60) = 30\ \text{V.}$$

[*] In section 8.1 it is shown that in a generator the current and the generated e.m.f. are in the same direction, so that electric power is being *supplied* by the generator; whereas in an electric motor, the direction of the generated e.m.f. is opposite to that of the current, so that electric power is *absorbed* by the motor.

The direction of the induced e.m.f. is the same as that of the current, opposing its decrease.

4.3 Inductance in terms of flux-linkages per ampere

Suppose a current of I amperes through a coil of N turns to produce a flux of Φ webers, and suppose the reluctance of the magnetic circuit to remain constant so that the flux is proportional to the current. Also, suppose the inductance of the coil to be L henrys.

If the current is increased from zero to I amperes in t seconds,

average rate of change of current $= I/t$ amperes/second,

$\therefore$ average e.m.f. induced in coil $= -LI/t$ volts   (4.3)

In section 2.12, it was explained that the value of the e.m.f., in volts, induced in a coil is equal to the rate of change of flux-linkages per second. Hence, when the flux increases from zero to Φ webers in t seconds,

average rate of change of flux $= \Phi/t$ webers/second

and   average e.m.f. induced in coil $= -N\Phi/t$ volts   (4.4)

Equating expressions (4.3) and (4.4), we have:

$$-LI/t = -N\Phi/t$$

$\therefore$ $L = N\Phi/I$ henrys   (4.5)

$$= \text{flux-linkages/ampere.}$$

Considering instantaneous values,

if $d\Phi =$ increase of flux, in webers, due to an increase di ampere in dt second,

rate of change of flux $= d\Phi/dt$ webers/second

and   induced e.m.f. $= -N \cdot d\Phi/dt$ volts   (4.6)

Equating (4.2) and (4.6), we have:

$$-L \cdot di/dt = -N \cdot d\Phi/dt$$

$\therefore$ $L = N \cdot d\Phi/di$ (4.7)

$$= \frac{\text{change of flux-linkages}}{\text{change of current}}$$

For a coil having a magnetic circuit of constant reluctance, the flux is proportional to the current; consequently, $d\Phi/di$ is equal to the flux/ampere, so that

$$L = \text{flux-linkages/ampere}$$
$$= N\Phi/I \text{ henrys} \qquad (4.5)$$

This expression gives us an alternative method of defining the unit of inductance, namely: *a coil possesses an inductance of* 1 *henry if a current of* 1 *ampere through the coil produces a flux-linkage of* 1 *weber-turn.*

Example 4.2 *A coil of* 300 *turns, wound on a core of non-magnetic material, has an inductance of* 10 mH. *Calculate* (a) *the flux produced by a current of* 5 A *and* (b) *the average value of the e.m.f. induced when a current of* 5 A *is reversed in* 8 *milliseconds.*

(a) From expression (4.5),

$$10 \times 10^{-3} = 300 \times \Phi/5$$

$$\therefore \qquad \Phi = 0.167 \times 10^{-3} \text{ Wb} = 167 \text{ } \mu\text{Wb}.$$

(b) When a current of 5 A is reversed, the flux decreases from 167 μWb to zero and then increases to 167 μWb in the reverse direction,

$$\therefore \qquad \text{change of flux} = -334 \text{ } \mu\text{Wb}$$

and   average rate of change of flux $= \dfrac{-334 \times 10^{-6}}{8 \times 10^{-3}}$

$$= -0.041 \text{ } 75 \text{ Wb/s},$$

$$\therefore \qquad \text{average e.m.f. induced in coil} = -(-0.041\ 75) \times 300$$

$$= 12.5 \text{ V}.$$

Alternatively, since the current changes from 5 to −5 A,

average rate of change of current $= -\dfrac{5 \times 2}{8 \times 10^{-3}}$

$$= -1250 \text{ A/s}.$$

Hence, from expression (4.2),

average e.m.f. induced in coil $= -0.01 \times (-1250)$

$$= 12.5 \text{ V}.$$

The sign is positive because the e.m.f. is acting in the direction of the original current, at first trying to prevent the current decreasing to zero and then opposing its growth in the reverse direction.

4.4 Factors determining the inductance of a coil

Let us first consider a coil uniformly wound on a *non-magnetic* ring of uniform section—similar to that of fig. 3.1. From expression (3.2), it follows that the flux density, in teslas, in such a ring is $4\pi \times 10^{-7} \times$

the amperes/metre. Consequently, if l be the length of the magnetic circuit in metres and a its cross-sectional area in square metres, then for a coil of N turns with a current I amperes:

$$\text{magnetic field strength} = IN/l$$

and
$$\text{total flux} = \Phi = Ba = \mu_0 Ha$$
$$= 4\pi \times 10^{-7} \times (IN/l)a$$

Substituting for Φ in expression (4.5) we have:

$$\text{inductance} = L = 4\pi \times 10^{-7} \times aN^2/l \text{ henrys} \qquad (4.8)$$

Hence the inductance is proportional to the square of the number of turns and to the cross-sectional area, and is inversely proportional to the length of the magnetic circuit.

If the coil is wound on a closed iron circuit, such as a ring, the

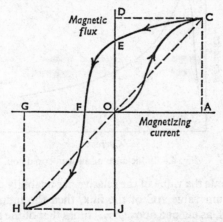

Fig. 4.5 Variation of magnetic flux with magnetizing current for a closed iron circuit.

problem of defining the inductance of such a coil becomes involved due to the fact that the variation of flux is no longer proportional to the variation of current. Suppose the relationship between the magnetic flux and the magnetizing current to be as shown in fig. 4.5; then if the iron has initially no residual magnetism, an increase of current from zero to OA causes the flux to increase from zero to AC, but when the current is subsequently reduced to zero, the decrease of flux is only DE. If the current is then increased to OG in the reverse direction, the change of flux is EJ. Consequently, we can have an infinite number of inductance values, depending upon the particular variation of current that we happen to consider.

Since we are usually concerned with the effect of inductance in an a.c. circuit, where the current varies from a maximum in one direction to the same maximum in the reverse direction, it is convenient to consider the value of the inductance as being the ratio of the change of flux-linkages to the change of current when the latter is reversed. Thus, for the case shown in fig. 4.5:

$$\text{inductance of coil} = \frac{DJ}{AG} \times \text{number of turns.}$$

This value of inductance is the same as if the flux varied linearly along the dotted line COH in fig. 4.5.

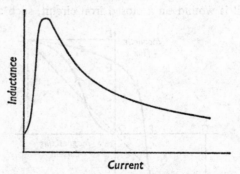

Fig. 4.6 Inductance of an iron-cored coil.

If μ_r represents the value of the relative permeability corresponding to the maximum value AC of the flux, then the inductance of the iron-cored coil, as defined above, is μ_r times that of the same coil with a non-magnetic core. Hence, from expression (4.8), we have:

inductance of iron-cored coil (for reversal of flux)

$$= 4\pi \times 10^{-7} \times \frac{aN^2}{l} \times \mu_r \text{ henrys.*} \qquad (4.9)$$

* From expression (4.9), L (henrys) $= \mu_0\mu_r \times a(\text{metres}^2) \times N^2/l(\text{metres})$

$\therefore$ absolute permeability $= \mu_0\mu_r = \dfrac{L(\text{henrys}) \times l(\text{metres})}{N^2 \times a(\text{metres}^2)}$

$$= \frac{Ll}{N^2a} \text{ henrys/metre,}$$

hence the units of absolute permeability are *henrys/metre* (or H/m); e.g. $\mu_0 = 4\pi \times 10^{-7}$ H/m.

The variations of relative permeability with magnetic field strength for various qualities of iron are shown in fig. 3.4; hence it follows from expression (4.9) that as the value of an alternating current through a coil having a closed iron circuit is increased, the value of the inductance increases to a maximum and then decreases, as shown in fig. 4.6. It will now be evident that when the value of the inductance of such a coil is stated, it is also necessary to specify the current variation for which that value has been determined.

Example 4.3 *A ring of 'Stalloy' stampings having a mean circumference of 400 mm and a cross-sectional area of 500 mm^2 is wound with 200 turns. Calculate the inductance of the coil corresponding to a reversal of a magnetizing current of* (a) 1 A, (b) 10 A.

(*a*)
$$H = \frac{1 \text{ [A]} \times 200 \text{ [turns]}}{0\cdot4 \text{ [m]}} = 500 \text{ A/m}$$

From fig. 3.3:

corresponding flux density = 1·22 T

∴   total flux = 1·22 [T] × 0·0005 [m^2] = 0·000 61 Wb.

From expression (4.5),

inductance = (0·000 61 × 200)/1·0 = 0·122 H.

(*b*)
$$H = \frac{10 \text{ [A]} \times 200 \text{ [turns]}}{0\cdot4 \text{ [m]}} = 5000 \text{ A/m.}$$

From fig. 3.3:

corresponding flux density = 1·58 T

∴   total flux = 1·58 × 0·0005 = 0·000 79 Wb

and   inductance = (0·000 79 × 200)/10 = 0·0158 H.

Example 4.4 *If a coil of 200 turns be wound on a non-magnetic core having the same dimensions as the 'Stalloy' ring for example 4.3, calculate its inductance.*

From expression (4.8) we have:

$$\text{inductance} = \frac{(4\pi \times 10^{-7}) \text{ [H/m]} \times 0\cdot0005 \text{ [m}^2\text{]} \times (200)^2 \text{ [turns}^2\text{]}}{0\cdot4 \text{ [m]}}$$

$$= 0\cdot000\ 062\ 8 \text{ H}$$

$$= 62\cdot8 \ \mu\text{H}.$$

A comparison of this inductance with the values obtained in example 4.3 for a coil of the same dimensions shows why an iron core is used when a large inductance is required.

4.5 Iron-cored inductor in a d.c. circuit

An inductor (i.e. a piece of apparatus used primarily because it possesses inductance) is frequently used in the output circuit of a rectifier to smooth out any variation (or ripple) in the direct current (section 19.8). If the inductor were made with a closed iron circuit, the relationship between the flux and the magnetizing current would be represented by curve OBD in fig. 4.7. It will be seen that if the current increases from OA to OC, the flux increases from AB to CD. If this increase takes place in t seconds, then:

average induced e.m.f.

$$= -\text{number of turns} \times \text{rate of change of flux}$$
$$= -N \times (CD - AB)/t \text{ volts} \qquad (4.10)$$

Let L_t be the *incremental* inductance of the coil over this range of flux variation, i.e. the effective value of the inductance when the flux is not proportional to the magnetizing current and varies over a relatively small range, then

$$\text{average induced e.m.f.} = -L_t \times (OC - OA)/t \text{ volts.} \quad (4.11)$$

Equating (4.10) and (4.11), we have:

$$L_t \times (OC - OA)/t = N \times (CD - AB)/t$$

$$\therefore \qquad L_t = N \times \frac{CD - AB}{OC - OA} \qquad (4.12)$$

$$= N \times \text{average slope of curve BD.}$$

From fig. 4.7 it is evident that the slope is very small when the iron is saturated. This effect is accentuated by hysteresis; thus if the current is reduced from OC to OA, the flux decreases from CD only to AE, so that the effective inductance is still further reduced.

If a short radial airgap were made in the iron ring, the flux produced by current OA would be reduced to some value AF. For the reduced flux density in the iron, the total m.m.f. required for the iron and the gap is approximately proportional to the flux; and for the same increase of current, AC, the increase of flux $= CG - AF$. As $(CG - AF)$ may be much greater than $(CD - AB)$, we have the curious result that the effective inductance of an iron-cored coil in a d.c. circuit may be increased by the introduction of an airgap.

An alternative method of increasing the flux-linkages/ampere and maintaining this ratio practically constant is to make the core of

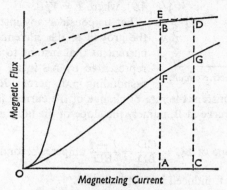

Fig. 4.7 Effect of inserting an airgap in an iron core.

compressed magnetic dust, such as small particles of iron or nickel–iron alloy, bound by shellac. This type of coil is used for 'loading' telephone lines, i.e. for inserting additional inductance at intervals along a telephone line to improve its transmission characteristics.

Example 4.5 *A laminated iron ring is wound with* 200 *turns. When the magnetizing current varies between* 5 *and* 7 *A, the magnetic flux varies between* 760 *and* 800 *μWb. Calculate the incremental inductance of the coil over this range of current variation.*

From expression (4.12) we have:

$$L_i = 200 \times \frac{(800 - 760) \times 10^{-6}}{(7 - 5)} = 0.004 \text{ H.}$$

4.6 Graphical derivation of curve of current growth in an inductive circuit

In section 4.1 the growth of current in an inductive circuit was discussed qualitatively; we shall now consider how to derive the curve showing the growth of current in a circuit of known resistance and inductance (assumed constant).

When dealing with an inductive circuit it is convenient to separate the effects of inductance and resistance by representing the inductance

L as that of an inductor or coil having no resistance and the resistance *R* as that of a resistor having no inductance, as shown in fig. 4.8. It is

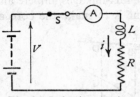

evident from the latter that the current ultimately reaches a steady value *I* (fig. 4.9), where $I = V/R$.

Let us consider any instant A during the growth of the current. Suppose the current at that instant to be *i* amperes, represented by AB in fig. 4.9. The corresponding p.d. across *R* is *Ri* volts.

Fig. 4.8 Inductive circuit.

Also at that instant the rate of change of the current is given by the slope of the curve at B, namely the slope of the tangent to the curve at B.

But the slope of BC $= \dfrac{\text{CD}}{\text{BD}} = \dfrac{I - i}{\text{BD}}$ amperes/second.

Hence e.m.f. induced in *L* at instant A

$$= -L \times \text{rate of change of current}$$
$$= -L \times (I - i)/\text{BD volts}.$$

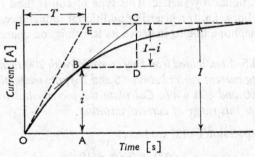

Fig. 4.9 Growth of current in an inductive circuit.

The total applied voltage *V* is absorbed partly in providing the voltage drop across *R* and partly in neutralizing the e.m.f. induced in *L*:

i.e. $$V = Ri + L \times (I - i)/\text{BD}.$$

Substituting *RI* for *V*, we have:

$$RI = Ri + L \times (I - i)/\text{BD}$$
$$\therefore \quad R(I - i) = L \times (I - i)/\text{BD}$$
hence $$\text{BD} = L/R.$$

In words, this expression means that the rate of growth of current at any instant is such that if the current continued increasing at that rate, it would reach its maximum value of I amperes in L/R seconds. Hence this period is termed the *time constant* of the circuit and is usually represented by the symbol T:

i.e.   time constant $= T = L/R$ seconds.   (4.13)

Immediately after switch S is closed, the rate of growth of the current is given by the slope of the tangent OE drawn to the curve at the origin; and if the current continued growing at this rate, it would attain its final value in time FE $= T$ seconds.

From expression (4.13) it follows that the greater the inductance and the smaller the resistance, the larger is the time constant and the longer it takes for the current to reach its final value. Also this relationship can be used to derive the curve representing the growth of current in an inductive circuit, as illustrated by the following example.

Example 4.6 *A coil having a resistance of 4 Ω and a constant inductance of 2 H is switched across a 20-V d.c. supply. Derive the curve representing the growth of the current.*

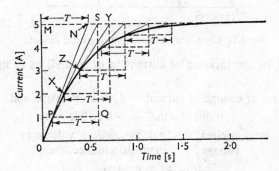

Fig. 4.10 Graph for example 4.6.

From (4.13), time constant $= T = 2/4 = 0.5$ s.

Final value of current $= I = 20/4 = 5$ A.

With the horizontal and vertical axes suitably scaled, as in fig. 4.10, draw a horizontal dotted line at the level of 5 A. Along this line mark off a period MN $= T = 0.5$ s, and join ON.

Take any point P relatively near the origin and draw a horizontal dotted line PQ = T = 0·5 s, and at Q draw a vertical dotted line QS. Join PS.

Repeat the operation from a point X on PS, Z on XY, etc.

A curve touching OP, PX, XZ, etc., represents the growth of the current. The greater the number of points used in the construction, the more accurate is the curve.

4.7 Mathematical derivation of curve of current growth in an inductive circuit

Let us again consider the circuit shown in fig. 4.8, and suppose i amperes to be the current t seconds after the switch is closed, and di

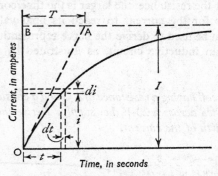

Fig. 4.11 Growth of current in an inductive circuit.

ampere to be the increase of current in dt second, as in fig. 4.11. Then:

$$\text{rate of change of current} = \mathrm{d}i/\mathrm{d}t \text{ amperes/second}$$

and

$$\text{induced e.m.f.} = -L \,.\, \mathrm{d}i/\mathrm{d}t \text{ volts.}$$

Since

$$\left.\begin{array}{c}\text{total applied} \\ \text{voltage}\end{array}\right\} = \left\{\begin{array}{l}\text{p.d. across } R + \text{voltage to} \\ \text{neutralize induced e.m.f.}\end{array}\right.$$

$\therefore$

$$V = Ri + L \,.\, \mathrm{d}i/\mathrm{d}t \qquad (4.14)$$

so that

$$V - Ri = L \,.\, \mathrm{d}i/\mathrm{d}t$$

and

$$\frac{V}{R} - i = \frac{L}{R} \,.\, \frac{\mathrm{d}i}{\mathrm{d}t}.$$

But V/R = final value of current = (say) I,

$\therefore$

$$\frac{R}{L}\,\mathrm{d}t = \frac{\mathrm{d}i}{I - i}$$

Integrating both sides, we have:

$$\frac{Rt}{L} = -\ln(I - i)^* + A$$

where $A =$ the constant of integration.

At the instant of closing the switch, $t = 0$ and $i = 0$, so that $A = \ln I$,

$$\therefore \qquad \frac{Rt}{L} = -\ln(I - i) + \ln I$$

$$= \ln\frac{I}{I - i} \qquad (4.15)$$

Hence, $\qquad (I - i)/I = e^{-Rt/L}$

$$\therefore \qquad i = I(1 - e^{-Rt/L}) \qquad (4.16)$$

This exponential relationship is often referred to as the Helmholtz equation.

Immediately after the switch is closed, the rate of change of the current is given by the slope of tangent OA drawn to the curve at the origin. If the current continued growing at this initial rate, it would attain its final value, I amperes, in T seconds, the *time constant* of the circuit (section 4.6). From Fig. 4.11, it is seen that:

initial rate of growth of current $= I/T$ amperes/second.

At the instant of closing the switch, $i = 0$; hence from expression (4.14),

$$V = L \times \text{initial rate of change of current},$$

$$\therefore \qquad \text{initial rate of change of current} = V/L$$

Hence, $\qquad I/T = V/L$

so that $\qquad T = LI/V = L/R$ seconds $\qquad (4.17)$

Substituting for R/L in (4.16), we have:

$$i = I(1 - e^{-t/T}) \qquad (4.18)$$

For $t = T$, $i = I(1 - 0.368) = 0.632I$,

hence the time constant is the time required for the current to attain 63.2 per cent of its final value.

---

* British Standards Institution recommends the use of ln and log in preference to $\log_e$ and $\log_{10}$ respectively.

*

Example 4.7 *A coil having a resistance of* 4 Ω *and a constant induc-tance of* 2 *H is switched across a* 20-V *d.c. supply. Calculate* (a) *the time constant,* (b) *the final value of the current and* (c) *the value of the current* 1 *second after the switch is closed.*

(a) Time constant $= L/R = 2/4 = 0{\cdot}5$ s.
(b) Final value of current $= V/R = 20/4 = 5$ A.
(c) Substituting $t = 1$, $T = 0{\cdot}5$ s and $I = 5$ A in (4.18),

$$i = 5(1 - e^{-1/0{\cdot}5}) = 5(1 - e^{-2})$$

From mathematical tables, $e^{-2} = 0{\cdot}1353$,

∴ $\qquad\qquad\qquad i = 5(1 - 0{\cdot}1353) = 4{\cdot}323$ A.

If such mathematical tables are not available, this section of the question can be solved by using expression (4.15), thus:

$$Rt/L = \ln I/(I - i)$$
$$= 2{\cdot}303 \log I/(I - i)$$

∴ $\qquad\qquad \log I/(I - i) = \dfrac{4 \times 1}{2 \times 2{\cdot}303} = 0{\cdot}8684$

and $\qquad\qquad I/(I - i) = 7{\cdot}386$
so that $\qquad\qquad i = 4{\cdot}323$ A.

4.8 Mathematical derivation of curve of current decay in an inductive circuit

Fig. 4.12 represents a coil, of inductance L henrys and resistance R

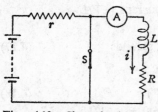

Fig. 4.12 Short-circuited in-ductive circuit.

ohms, in series with a resistor r across a battery. The function of r is to prevent the battery current becoming excessive when switch S is closed. Suppose the steady current through the coil to be I amperes when S is open. Also, suppose i amperes in fig. 4.13 to be the current through the coil t seconds after S is closed. Since the external applied voltage V of equation (4.14) is zero as far as the coil is concerned, we have:

$$0 = Ri + L \cdot di/dt$$
∴ $\qquad\qquad Ri = -L \cdot di/dt \qquad\qquad\qquad (4.19)$

In this expression, di is numerically negative since it represents a decrease of current.

Hence, $$(R/L)dt = -di/i$$

Integrating both sides, we have:

$$(R/L)t = -\ln i + A$$

where $A =$ the constant of integration.

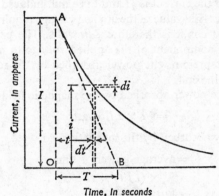

Fig. 4.13 Decay of current in an inductive circuit.

At the instant of closing switch S, $t = 0$ and $i = I$, so that

$$0 = -\ln I + A$$

$\therefore \qquad (R/L)t = \ln I - \ln i = \ln I/i$

Hence, $\qquad I/i = e^{Rt/L}$

and $\qquad i = Ie^{-Rt/L}$ \hfill (4.20)

Immediately after S is closed, the rate of decay of the current is given by the slope of the tangent AB in fig. 4.13, and

initial rate of change of current $= -I/T$ amperes/second.

Also, from equation (4.19), since initial value of i is I,

initial rate of change of current $= -RI/L$ amperes/second.

Hence, $\qquad RI/L = I/T$

so that $\qquad T = L/R = $ *time constant* of circuit,

namely, the value already deduced in section 4.6.

The curve representing the decay of the current can be derived graphically by a procedure similar to that used in section 4.6 for constructing the curve representing the current growth in an inductive circuit.

4.9 Energy stored in an inductor

If the current in a coil having a *constant* inductance of L henrys grows at a uniform rate from zero to I amperes in t seconds, the average value of the current is $\frac{1}{2}I$ and the e.m.f. induced in the coil is $-(L \times I/t)$ volts. The value of the applied voltage required to neutralize this induced e.m.f. is therefore LI/t volts. The product of the current and the component of the applied voltage to neutralize the induced e.m.f. represents the power absorbed by the magnetic field associated with the coil.

Hence average power absorbed by the magnetic field

$$= \tfrac{1}{2}I \times LI/t \text{ watts}$$

and total energy absorbed by the magnetic field

$$= \text{average power} \times \text{time}$$
$$= \tfrac{1}{2}I \times (LI/t) \times t$$
$$= \tfrac{1}{2}LI^2 \text{ joules} \qquad (4.21)$$

Let us now consider the general case of a current increasing at a uniform or a non-uniform rate in a coil having a *constant* inductance L henrys. If the current increases by di ampere in dt second,

$$\text{induced e.m.f.} = -L \,.\, di/dt \text{ volts,}$$

and if i is the value of the current at that instant,

$$\left.\begin{array}{l}\text{energy absorbed by the magnetic} \\ \text{field during time } dt \text{ second}\end{array}\right\} = iL \,.\, (di/dt) \,.\, dt$$
$$= Li \,.\, di \text{ joules.}$$

Hence total energy absorbed by the magnetic field when the current increases from 0 to I amperes

$$= L\int_0^I i \,.\, di = L \times \tfrac{1}{2}\Big[i^2\Big]_0^I$$
$$= \tfrac{1}{2}LI^2 \text{ joules} \qquad (4.21)$$

From expression (4.9), $L = N^2\mu a/l$ for a homogeneous magnetic circuit of uniform cross-sectional area.

∴ energy/cubic metre $= \frac{1}{2}I^2N^2\mu/l^2 = \frac{1}{2}\mu H^2$
$$= \frac{1}{2}HB = \frac{1}{2}B^2/(\mu_r\mu_0) \text{ joules} \quad (4.22)$$

This expression has been derived on the assumption that μ_r remains constant. When the coil is wound on a closed-iron core, the variation of μ_r renders this expression inapplicable and the energy has to be determined graphically as already explained in section 3.14. For non-magnetic materials, $\mu_r = 1$ and the energy stored per cubic metre is $\frac{1}{2}B^2/\mu_0$ joules, namely the expression already derived in section 3.19.

When an inductive circuit is opened, the current has to die away and the magnetic energy has to be dissipated. If there is no resistor in parallel with the circuit, the energy is largely dissipated as heat in the arc at the switch. With a parallel resistor, as described in section 4.1, the energy is dissipated as heat generated by the decreasing current in the total resistance of the circuit in which that current is flowing.

Example 4.8 *A coil has a resistance of 5 Ω and an inductance of 1·2 H. The current through the coil is increased uniformly from zero to 10 A in 0·2 s, maintained constant for 0·1 s and then reduced uniformly to zero in 0·3 s. Plot graphs representing the variation with time of* (a) *the current,* (b) *the induced e.m.f.,* (c) *the p.d.s across the resistance and the inductance,* (d) *the resultant applied voltage and* (e) *the power to and from the magnetic field. Assume the coil to be wound on a non-metallic core.*

The variation of current is represented by graph A in fig. 4.14; and since the p.d. across the resistance is proportional to the current, this p.d. increases from zero to (10 A × 5 Ω), namely 50 V, in 0·2 s, remains constant at 50 V for 0·1 s and then decreases to zero in 0·3 s, as represented by graph B.

During the first 0·2 s, the current is increasing at the rate of 10/0·2, namely 50 A/s,

∴ corresponding induced e.m.f. $= -50 \times 1\cdot2 = -60$ V.

During the following 0·1 s, the induced e.m.f. is zero; and during the last 0·3 s, the current is decreasing at the rate of −10/0·3, namely −33·3 A/s,

∴ corresponding induced e.m.f. $= -(-33\cdot3 \times 1\cdot2) = 40$ V.

The variation of the induced e.m.f. is represented by the uniformly-dotted graph C in fig. 4.14. Since the p.d. applied across the inductance has to neutralize the induced e.m.f., its variation is represented by the chain-dotted graph D which is exactly similar to graph C except that the signs are reversed.

The resultant voltage applied to the coil is obtained by adding graphs B and D; thus the resultant voltage increases uniformly from

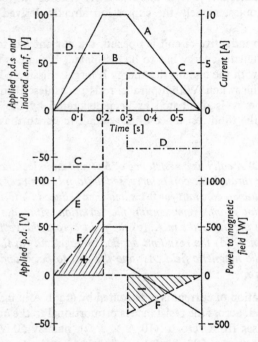

Fig. 4.14 Graphs for example 4.8.

60 to 110 V during the first 0·2 s, remains constant at 50 V for the next 0·1 s and then changes uniformly from 10 to −40 V during the last 0·3 s, as shown by graph E in fig. 4.14.

The power supplied to the magnetic field increases uniformly from zero to (10 A × 60 V), namely 600 W, during the first 0·2 s. It is zero during the next 0·1 s. Immediately the current begins to decrease, energy is being returned from the magnetic field to the electric circuit, and the power decreases uniformly from (−40 V × 10 A), namely −400 W, to zero as represented by graph F.

The positive shaded area enclosed by graph F represents the energy ($= \frac{1}{2} \times 600 \times 0\cdot2 = 60$ J) absorbed by the magnetic field during the first 0·2 s; and the negative shaded area represents the energy ($= \frac{1}{2} \times 400 \times 0\cdot3 = 60$ J) returned from the magnetic field to the electric circuit during the last 0·3 s. The two areas are obviously equal in magnitude, i.e. all the energy supplied to the magnetic field is returned to the electric circuit.

4.10 Mutual Inductance

If two coils A and C are placed relative to each other as in fig. 4.15, then when S is closed, some of the flux produced by the current in A

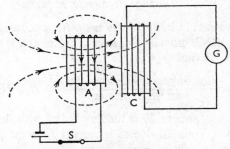

Fig. 4.15 Mutual inductance.

becomes linked with C and the e.m.f. induced in C circulates a momentary current through galvanometer G. Similarly when S is opened the collapse of the flux induces an e.m.f. in the reverse direction in C. Since a change of current in one coil is accompanied by a change of flux linked with the other coil and therefore by an e.m.f. induced in the latter, the two coils are said to have *mutual inductance*.

The unit of mutual inductance is the same as for self inductance, namely the *henry*; and *two coils have a mutual inductance of* 1 *henry if an e.m.f. of* 1 *volt is induced in one coil when the current in the other coil varies uniformly at the rate of* 1 *ampere per second.*

If two circuits possess a mutual inductance of M henrys and if the current in one circuit—termed the *primary* circuit—increases by di ampere in dt second,

$$\text{e.m.f. induced in } secondary \text{ circuit} = -M \, . \, di/dt \text{ volts} \quad (4.23)$$

The minus sign indicates that the induced e.m.f. tends to circulate a current in the secondary circuit in such a direction as to oppose the increase of flux due to the increase of current in the primary circuit.

If $d\Phi$ weber is the increase of flux linked with the secondary circuit due to the increase of di ampere in the primary,

$$\left.\begin{array}{r}\text{e.m.f. induced in secondary}\\ \text{circuit}\end{array}\right\} = -N_2 \cdot d\Phi/dt \text{ volts} \quad (4.24)$$

where N_2 = number of secondary turns. From expressions (4.23) and (4.24),

$$M \cdot di/dt = N_2 \cdot d\Phi/dt$$

$$\therefore \qquad M = N_2 \cdot d\Phi/di$$

$$= \frac{\text{change of flux-linkages with secondary}}{\text{change of current in primary}}. \quad (4.25)$$

If the relative permeability of the magnetic circuit remains constant, the ratio $d\Phi/di$ must also remain constant and is equal to the flux per ampere, so that

$$M = \frac{\text{flux-linkages with secondary}}{\text{current in primary}} = N_2\Phi_2/I_1 \quad (4.26)$$

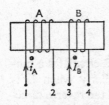

Fig. 4.16 Mutual inductance.

where Φ_2 is the flux linked with the secondary circuit due to a current I_1 in the primary circuit.

The mutual inductance between two circuits, A and B, is precisely the same whether we assume A to be the primary and B the secondary, or vice versa; for instance, if the two coils are wound on a non-metallic cylinder, as in fig. 4.16, then, from expression (4.21),

$$\left.\begin{array}{r}\text{energy in the magnetic field due to current } I_A\\ \text{in coil A alone}\end{array}\right\} = \tfrac{1}{2}L_A I_A{}^2 \text{ joules}$$

and

$$\left.\begin{array}{r}\text{energy in the magnetic field due to current } I_B\\ \text{in coil B alone}\end{array}\right\} = \tfrac{1}{2}L_B I_B{}^2 \text{ joules.}$$

Suppose the current in B to be maintained steady at I_B amperes in the direction shown in fig. 4.16, and the current in A to be increased by di ampere in dt second, then—

$$\text{e.m.f. induced in B} = -M_{12} \cdot di/dt \text{ volts}$$

where M_{12} = mutual inductance when A is primary.

If the direction of i_A is that indicated by the arrowhead in fig. 4.16, then, by Lenz's Law, the direction of the e.m.f. induced in B is anti-clockwise when the coil is viewed from the righthand end; i.e. the induced e.m.f. is in opposition to I_B and the p.d. across terminals 3 and 4 has to be increased by M_{12} . di/dt volts to maintain I_B constant. Hence the *additional* electrical energy absorbed by coil B in time dt

$$= I_B M_{12}(di/dt) \times dt = I_B M_{12} \, . \, di \text{ joules.}$$

Since I_B remains constant, the I^2R loss in B is unaffected, and there is no e.m.f. induced in coil A apart from that due to the increase of i_A; therefore this additional energy supplied to coil B is absorbed by the magnetic field. Hence, when the current in A has increased to I_A,

$$\left. \begin{array}{r} \text{total energy in magnetic} \\ \text{field} \end{array} \right\} = \tfrac{1}{2}L_A I_A{}^2 + \tfrac{1}{2}L_B I_B{}^2 + \int_0^{I_A} I_B M_{12} \, . \, di$$

$$= \tfrac{1}{2}L_A I_A{}^2 + \tfrac{1}{2}L_B I_B{}^2 + M_{12} I_A I_B \text{ joules}$$

If the direction of either I_A or I_B were reversed, the direction of the e.m.f. induced in B, while the current in A was increasing, would be the same as that of I_B, and coil B would then be acting as a generator. By the time the current in A would have reached its steady value I_A, the energy withdrawn from the magnetic field and generated in coil B would be $M_{12} I_A I_B$ joules, and final energy in magnetic field would be—

$$\tfrac{1}{2}L_A I_A{}^2 + \tfrac{1}{2}L_B I_B{}^2 - M_{12} I_A I_B \text{ joules.}$$

Hence, in general, total energy in magnetic field

$$= \tfrac{1}{2}L_A I_A{}^2 + \tfrac{1}{2}L_B I_B{}^2 \pm M_{12} I_A I_B \text{ joules} \qquad (4.27)$$

the sign being positive when the ampere-turns due to I_A and I_B are additive, and negative when they are in opposition.

If M_{21} were the mutual inductance with coil B as primary, it could be shown by a similar procedure that the total energy in the magnetic field

$$= \tfrac{1}{2}L_A I_A{}^2 + \tfrac{1}{2}L_B I_B{}^2 \pm M_{21} I_A I_B \text{ joules.}$$

Since the final conditions are identical in the two cases, the energies must be the same,

$$\therefore \qquad\qquad M_{12} I_A I_B = M_{21} I_A I_B$$

or $\qquad\qquad\qquad M_{12} = M_{21} = \text{(say) } M,$

i.e. the mutual inductance between two circuits is the same whichever circuit is taken as the primary.

When the two coils are shown on a common core, as in fig. 4.16, it is obvious that the magnetomotive forces due to i_A and I_B are additive when the directions of the currents are as indicated by the arrowheads. If, however, the coils are drawn as in fig. 4.17, it is impossible to state whether the magnetomotive forces due to currents I_A and I_B are additive or in opposition; and it is to remove this ambiguity that the dot notation has been adopted. Thus, in figs. 4.16 and 4.17, dots are inserted at ends 1 and 3 of the coils to indicate that when currents *enter both* coils (or *leave both* coils) at these ends,

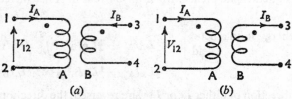

(a)   (b)

Fig. 4.17 Application of the dot notation.

as in fig. 4.17 (*a*), the magnetomotive forces of the coils are additive, and the mutual inductance is then said to be *positive*. But if I_A *enters* coil A at the dotted end and I_B *leaves* coil B at the dotted end, as in fig. 4.17 (*b*), the m.m.f.s of the coils are in opposition and the mutual inductance is then said to be *negative*.

The following illustrates the application of the dot notation but anticipates the explanation of Complex Notation given in Chapter 12.

If I_A and I_B represent the complex currents in coils A and B respectively,

V_{12} represents the complex voltage applied across terminals 1–2, the voltage being assumed positive when terminal 1 is positive with respect to terminal 2,

R_A and L_A represent the resistance and self inductance respectively of coil A,

M represents the mutual inductance of coils A and B,

and ω represents the angular frequency ($= 2\pi f$),

then if the currents are assumed to have their positive directions as indicated by the arrowheads in fig. 4.17 (*a*),

$$V_{12} = I_A(R_A + j\omega L_A) + j\omega M I_B.$$

But if the currents have their positive directions as indicated in fig. 4.17 (*b*),

$$V_{12} = I_A(R_A + j\omega L_A) - j\omega M I_B.$$

A further application of the dot notation is given at the end of section 4.12.

4.11 Relationship between mutual inductance and self inductances of two coils: coupling coefficient

Suppose a ring of non-magnetic material to be wound *uniformly* with two coils, A and B, the turns of one coil being as close as possible to those of the other coil, so that the whole of the flux produced by current in one coil is linked with all the turns of the other coil. If coil A has N_1 turns and B has N_2 turns, and if the reluctance of the magnetic circuit is S amperes/weber, then, from expression (4.5), the self-inductances of A and B are

$$L_1 = N_1\Phi_1/I_1 = N_1^2\Phi_1/(I_1N_1) = N_1^2/S \tag{4.28}$$

and $\qquad L_2 = N_2\Phi_2/I_2 = N_2^2/S \tag{4.29}$

where Φ_1 and Φ_2 are the magnetic fluxes due to I_1 in coil A and I_2 in coil B respectively, and

$$S = I_1N_1/\Phi_1 = I_2N_2/\Phi_2.$$

Since the whole of flux Φ_1 due to I_1 is linked with coil B, it follows from expression (4.26) that

$$M = N_2\Phi_1/I_1 = N_1N_2\Phi_1/(I_1N_1)$$
$$= N_1N_2/S \tag{4.30}$$

Hence, from (4.28), (4.29) and (4.30),

$$L_1L_2 = N_1^2N_2^2/S^2 = M^2$$

so that $\qquad M = \sqrt{(L_1L_2)} \tag{4.31}$

We have assumed that

(*a*) the reluctance remains constant, and

(*b*) the magnetic leakage is zero, i.e. that all the flux produced by one coil is linked with the other coil.

The first assumption means that expression (4.31) is strictly correct only when the magnetic circuit is of non-magnetic material. It is, however, approximately correct for an iron core if the latter has one or more gaps of air or other non-magnetic material, since the reluctance of such a magnetic circuit is approximately constant.

When there is magnetic leakage, i.e. when all the flux due to current in one coil is not linked with the other coil,

$$M = k\sqrt{(L_1 L_2)} \qquad (4.32)$$

where k is termed the *coupling coefficient*. 'Coupling coefficient' is a term much used in radio work to denote the degree of coupling between two coils; thus, if the two coils are close together, most of the flux produced by current in one coil passes through the other and the coils are said to be *tightly* coupled. If the coils are well apart, only a small fraction of the flux is linked with the secondary, and the coils are said to be *loosely* coupled.

Example 4.9 *Calculate the mutual inductance between the solenoid and the search coil of Example 3.9. If the self inductance of the solenoid is 2 mH and that of the search coil is 25 μH, find the coupling coefficient between the two coils.*

For a long solenoid (section 3.18),

$$\text{flux density at centre} = \frac{IN\mu_0}{l} = \frac{800I \times 4\pi \times 10^{-7}}{1 \cdot 2}$$

$$= 0 \cdot 000\ 838I \text{ tesla (or Wb/m}^2)$$

$$\therefore \quad \text{flux through search coil} = 0 \cdot 000\ 838I \times \pi \times (0 \cdot 015)^2$$

$$= 0 \cdot 592I \times 10^{-6} \text{ Wb.}$$

Since there is no iron in the solenoid, the change of flux is proportional to the change of current; hence, from (4.26),

$$\text{mutual inductance} = \frac{50 \times 0 \cdot 592I \times 10^{-6}}{I} = 29 \cdot 6 \times 10^{-6} \text{ H}$$

$$= 29 \cdot 6 \text{ μH.}$$

In expression (4.32), the self and mutual inductances must be expressed in the same units,

$$\therefore \quad \text{coupling coefficient} = \frac{29 \cdot 6}{\sqrt{(2000 \times 25)}} = 0 \cdot 132.$$

4.12 Inductance of inductively-coupled coils connected in series

Fig. 4.18 (*a*) shows two coils A and B wound coaxially on an insulating cylinder, with terminals 2 and 3 joined together. It will be evident that the fluxes produced by a current *i* through the two coils are in the same direction, and the coils are said to be cumulatively coupled. Suppose A and B to have self inductances L_A and L_B henrys respectively and a mutual inductance M henrys, and suppose the arrowheads to represent the positive direction of the current. If the current

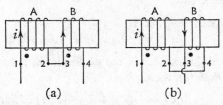

(a)   (b)

Fig. 4.18 Cumulative and differential coupling of two coils connected in series.

increases by d*i* ampere in d*t* second,

$$\left.\begin{array}{l}\text{e.m.f. induced in A due to} \\ \text{its self inductance}\end{array}\right\} = -L_A \cdot di/dt \text{ volts}$$

and
$$\left.\begin{array}{l}\text{e.m.f. induced in B due to} \\ \text{its self inductance}\end{array}\right\} = -L_B \cdot di/dt \text{ volts.}$$

Also,
$$\left.\begin{array}{l}\text{e.m.f. induced in A due to} \\ \text{increase of current in B}\end{array}\right\} = -M \cdot di/dt \text{ volts}$$

and
$$\left.\begin{array}{l}\text{e.m.f. induced in B due to} \\ \text{increase of current in A}\end{array}\right\} = -M \cdot di/dt \text{ volts.}$$

The minus signs signify that the direction of all the e.m.f.s is opposite to that of the current. Hence—

total e.m.f. induced in A and B $= -(L_A + L_B + 2M) \cdot di/dt$.

If the windings between terminals 1 and 4 be regarded as a single circuit having a self inductance L_1 henrys, then for the same increase d*i* ampere in d*t* second,

e.m.f. induced in the whole circuit $= -L_1 \cdot di/dt$ volts.

But the e.m.f. induced in the whole circuit is obviously the same as the sum of the e.m.f.s induced in A and B, i.e.

$$-L_1 \cdot di/dt = -(L_A + L_B + 2M) \cdot di/dt$$
$$\therefore \qquad L_1 = L_A + L_B + 2M \qquad (4.33)$$

Let us next reverse the direction of the current in B relative to that in A by joining together terminals 2 and 4, as in fig. 4.18 (*b*). With this differential coupling, the e.m.f., $M \cdot di/dt$, induced in coil A due to an increase di ampere in dt second in coil B, is in the same direction as the current and is therefore in opposition to the e.m.f. induced in coil A due to its self inductance. Similarly, the e.m.f. induced in B by mutual inductance is in opposition to that induced by the self inductance of B. Hence,

total e.m.f. induced in A and B
$$= -L_A \cdot di/dt - L_B \cdot di/dt + 2M \cdot di/dt$$

If L_2 be the self inductance of the whole circuit between terminals 1 and 3 in fig. 4.18 (*b*), then

$$-L_2 \cdot di/dt = -(L_A + L_B - 2M) \cdot di/dt$$
$$\therefore \qquad L_2 = L_A + L_B - 2M \qquad (4.34)$$

Hence the total inductance of inductively-coupled circuits

$$= L_A + L_B \pm 2M \qquad (4.35)$$

The positive sign applies when the coils are cumulatively coupled, the mutual inductance being then regarded as positive; and the negative sign applies when they are differentially coupled.

From expressions (4.33) and (4.34), we have

$$M = (L_1 - L_2)/4 \qquad (4.36)$$

i.e. the mutual inductance between two inductively-coupled coils is a quarter of the difference between the total self inductance of the circuit when the coils are cumulatively coupled and that when they are differentially coupled.

The above result can be derived by using complex notation (Chapter 12) together with the dot notation described in section 4.10. For simplicity, let us assume the resistance of the coils to be negligible.

If I_a represents the complex current in fig. 4.18 (*a*), where the current is assumed positive when it enters *each* coil at the dotted end, and M is therefore positive,

$$\text{p.d. across coil A} = I_a(j\omega L_A + j\omega M)$$
and $\qquad\qquad$ „ $\quad$ „ $\quad$ „ $\quad$ B $= I_a(j\omega L_B + j\omega M)$

$\therefore$ p.d. across terminals 1–4, fig. 4.18 (*a*)

$$= V_{14} = I_a(j\omega L_A + j\omega L_B + 2j\omega M) = j\omega L_1 I_a$$
$$\therefore \qquad L_1 = L_A + L_B + 2M.$$

In fig. 4.18 (*b*), the current is assumed positive when it *enters* coil A at the dotted end and *leaves* coil B at the dotted end; consequently M is negative.

Hence, if I_b represents the complex current in this case,

$$\text{p.d. across coil A} = I_b(j\omega L_A - j\omega M)$$

and $\qquad\qquad\quad\text{,, \quad ,, \quad ,, } B = I_b(j\omega L_B - j\omega M)$

$\therefore$ p.d. across terminals 1–3, fig. 4.18 (*b*)

$$= V_{13} = I_b(j\omega L_A + j\omega L_B - 2j\omega M) = j\omega L_2 I_b$$

$\therefore \qquad\qquad L_2 = L_A + L_B - 2M.$

Hence $\qquad M = (L_1 - L_2)/4$ \hfill (4.36)

Summary of important formulae

$$\text{Induced e.m.f.} = -L \cdot di/dt \text{ volts} \qquad (4.2)$$

For a magnetic circuit having constant relative permeability,

$$L = \text{flux-linkages/ampere} \qquad (4.5)$$

For a homogeneous magnetic circuit of uniform section and constant relative permeability,

$$L = N^2 \cdot \mu_0 \mu_r a/l \text{ henrys} \qquad (4.9)$$

For R and L, in series, connected across a d.c. supply,

$$\text{instantaneous current} = i = I(1 - e^{-Rt/L}) \qquad (4.16)$$
$$= I(1 - e^{-t/T}) \qquad (4.18)$$

and $\qquad\qquad \text{time constant} = T = L/R \text{ seconds} \qquad (4.17)$

For decay of current in inductive circuit,

$$\text{instantaneous current} = i = Ie^{-Rt/L} \qquad (4.20)$$

Magnetic energy stored in inductor

$$= \tfrac{1}{2}LI^2 \text{ joules} \qquad (4.21)$$

For a magnetic circuit having constant relative permeability,

$$M = \frac{\text{flux-linkages with secondary}}{\text{current in primary}} \qquad (4.26)$$

Energy stored in magnetic field of coils A and B having

$$\text{mutual inductance} = \tfrac{1}{2}L_A I_A^2 + \tfrac{1}{2}L_B I_B^2 \pm M I_A I_B \qquad (4.27)$$

$$\text{Coupling coefficient} = \frac{M}{\sqrt{(L_1 L_2)}} \qquad (4.32)$$

For inductively-coupled coils connected in series,

$$\text{total inductance} = L_A + L_B \pm 2M \qquad (4.35)$$

EXAMPLES 4

1. A 1500-turn coil surrounds a magnetic circuit which has a reluctance of 6×10^6 A/Wb. What is the inductance of the coil?
2. Calculate the inductance of a circuit in which 30 V are induced when the current varies at the rate of 200 A/s.
3. At what rate is the current varying in a circuit having an inductance of 50 mH when the induced e.m.f. is 8 V?
4. What is the value of the e.m.f. induced in a circuit having an inductance of 700 μH when the current varies at a rate of 5000 A/s?
5. A certain coil is wound with 50 turns and a current of 8 A produces a flux of 200 μWb. Calculate (a) the inductance of the coil corresponding to a reversal of the current, (b) the average e.m.f. induced when the current is reversed in 0·2 s.
6. A toroidal coil of 100 turns is wound uniformly on a non-magnetic ring of mean diameter 150 mm. The cross-sectional area of the ring is 706 mm². Estimate: (a) the magnetic field strength at the inner and outer edges of the ring when the current is 2 A; (b) the current required to produce a flux of 0·5 μWb; (c) the self inductance of the coil.

    If the ring had a small radial airgap cut in it, state, giving reasons, what alterations there would be in the answers to (a), (b) and (c).
    (S.A.N.C., O.1).
7. A coil consists of two similar sections wound on a common core. Each section has an inductance of 0·06 H. Calculate the inductance of the coil when the sections are connected (a) in series, (b) in parallel.
8. An iron rod, 1 cm diameter and 50 cm long, is formed into a closed ring and uniformly wound with 400 turns of wire. A direct current of 0·5 A is passed through the winding and produces a flux density of 0·75 T. If all the flux links with every turn of the winding, calculate: (a) the relative permeability of the iron; (b) the inductance of the coil; (c) the average value of the e.m.f. induced when the interruption of the current causes the flux in the iron to decay to 20 per cent of its original value in 0·01 s. (U.L.C.I., O.1)
9. Explain, with the aid of diagrams, the terms *self inductance* and *mutual inductance*. In what unit are they measured? Define this unit.

    Calculate the inductance of a ring-shaped coil having a mean diameter of 200 mm wound on a wooden core of diameter 20 mm. The winding is evenly wound and contains 500 turns.

If the wooden core is replaced by an iron core which has a relative permeability of 600 when the current is 5 A, calculate the new value of inductance. (U.L.C.I., O.1)

10. Name and define the unit of *self-inductance.*

A large electromagnet is wound with 1000 turns. A current of 2 A in this winding produces a flux through the coil of 0·03 Wb. Calculate the inductance of the electromagnet.

If the current in the coil is reduced from 2 A to zero in 0·1 s, what average e.m.f. will be induced in the coil? (N.C.T.E.C., O.1)

11. State Faraday's and Lenz's laws.

The field winding of a 4-pole separately-excited d.c. generator consists of 4 coils connected in series, each coil being wound with 1200 turns. If a current of 2 A produces a magnetic flux of 400 μWb, calculate: (*a*) the inductance of the field circuit; (*b*) the average value and direction of the induced e.m.f. if the field switch is opened at such a speed that the flux falls to the residual value of 20 μWb in 0·01 s. (U.E.I., O.1)

12. Explain what is meant by the self inductance of a coil and define the practical unit in which it is expressed.

A flux of 0·5 mWb is produced in a coil of 900 turns wound on a wooden ring by a current of 3 A. Calculate: (*a*) the inductance of the coil; (*b*) the average e.m.f. induced in the coil when a current of 5 A is switched off, assuming the current to fall to zero in 1 ms; (*c*) the mutual inductance between the coils, if a second coil of 600 turns was uniformly wound over the first coil. (U.E.I., O.1)

13. Define the *ampere* in terms of M.K.S. units.

An iron ring, having a mean circumference of 250 mm and a cross-sectional area of 400 mm², is wound with a coil of 70 turns. From the following data calculate the current required to set up a magnetic flux of 510 μWb.

B (teslas)	1·0	1·2	1·4
H (amperes/metre)	350	600	1250

Calculate also: (*a*) the inductance of the coil at this current; (*b*) the self-induced e.m.f. if this current is switched off in 0·005 s. (E.M.E.U., O.1)

14. Explain the meaning of *self inductance* and define the unit in which it is measured.

A coil consists of 750 turns and a current of 10 A in the coil gives rise to a magnetic flux of 1200 μWb. Calculate the inductance of the coil, and determine the average e.m.f. induced in the coil when this current is reversed in 0·01 s. (W.J.E.C., O.1)

15. Explain what is meant by the self inductance of an electric circuit and define the practical unit of self inductance.

A non-magnetic ring having a mean diameter of 300 mm and a cross-sectional area of 500 mm² is uniformly wound with a coil of 200 turns. Calculate from first principles the inductance of the winding. (App. El., L.U.)

16. Two coils, A and B, have self inductances of 120 μH and 300 μH respectively. When a current of 3 A through coil Q is reversed, the deflection on a fluxmeter connected across B is 600 μWb-turns. Calculate: (a) the mutual inductance between the coils, (b) the average e.m.f. induced in coil B if the flux is reversed in 0·1 s and (c) the coupling coefficient.

17. An iron ring having a mean diameter of 20 cm and cross-section of 10 cm² has a winding of 500 turns upon it. The ring is sawn through at one point, so as to provide an airgap in the magnetic circuit. How long should this gap be, if it is desired that a current of 4 A in the winding should produce a flux density of 1·0 T in the gap? State the assumptions made in your calculation.

What is the inductance of the winding when a current of 4 A is flowing through it?

The permeability of free space is $4\pi \times 10^{-7}$ H/m and the data for the *B-H* curve of the iron are given below:

H(A/m)	190	254	360	525	1020	1530	2230
B(T)	0·6	0·8	1·0	1·2	1·4	1·5	1·6

(App. El., L.U.)

18. A certain circuit has a resistance of 10 Ω and a constant inductance of 3 H. The current through this circuit is increased uniformly from 0 to 5 A in 0·6 s, maintained constant at 4 A for 0·1 s and then reduced uniformly to zero in 0·3 s. Draw to scale graphs representing the variation of (a) the current, (b) the induced e.m.f. and (c) the resultant applied voltage.

19. A coil having a resistance of 2 Ω and an inductance of 0·5 H has a current passed through it which varies in the following manner: (a) a uniform change from zero to 50 A in 1 s; (b) constant at 50 A for 1 s; (c) a uniform change from 50 A to zero in 2 s. Plot the current graph to a time base. Tabulate the potential difference applied to the coil during each of the above periods and plot the graph of potential difference to a time base. (U.L.C.I., O.2)

20. A coil wound with 500 turns has a resistance of 2 Ω. It is found that a current of 3 A produces a flux of 500 μWb. Calculate: (a) the inductance and the time constant of the coil: (b) the average e.m.f. induced in the coil when the flux is reversed in 0·3 s.

If the coil is switched across a 10-V d.c. supply, derive graphically a curve showing the growth of the current, assuming the inductance to remain constant.

21. Explain the term *time constant* in connection with an inductive circuit.

A coil having a resistance of 25 Ω and an inductance of 2·5 H is connected across a 50-V d.c. supply. Determine graphically: (a) the initial rate of growth of the current; (b) the value of the current after 0·15 s; and (c) the time required for the current to grow to 1·8 A.

(E.M.E.U., O.1)

22. The field winding of a d.c. machine has an inductance of 10 H and

takes a final current of 2 A when connected to a 200-V d.c. supply. Calculate: (i) the initial rate of growth of current; (ii) the time constant; and (iii) the current when the rate of growth is 5 A/s.

(W.J.E.C., O.2)

23. A 200-V d.c. supply is suddenly switched across a relay coil which has a time constant of 3 ms. If the current in the coil reaches 0·2 A after 3 ms, determine the final steady value of the current and the resistance and inductance of the coil.

Calculate the energy stored in the magnetic field when the current has reached its final steady value.

24. A coil of inductance 4 H and resistance 80 Ω is in parallel with a 200-Ω resistor of negligible inductance across a 200-V d.c. supply. The switch connecting these to the supply is then opened, the coil and resistor remaining connected together. State, in each case for an instant immediately before and for one immediately after the opening of the switch: (*a*) the current through the resistor; (*b*) the current through the coil; (*c*) the e.m.f. induced in the coil; and (*d*) the voltage across the coil.

Give rough sketch graphs, with explanatory notes, to show how these four quantities vary with time. Include intervals both before and after the opening of the switch, and mark on the graphs an approximate time scale. (App. El., L.U.)

25. A circuit consists of a 200-Ω non-reactive resistor in parallel with a coil of 4-H inductance and 100-Ω resistance. If this circuit is switched across a 100-V d.c. supply for a period of 0·06 s and then switched off, calculate the current in the coil 0·012 s after the instant of switching off. What is the maximum p.d. across the coil?

26. Define the units of: (*a*) magnetic flux, and (*b*) inductance.

Obtain an expression for the induced e.m.f. and for the stored energy of a circuit, in terms of its inductance, assuming a steady rise of current from zero to its final value and ignoring saturation.

A coil, of inductance 5 H and resistance 100 Ω, carries a steady current of 2 A. Calculate the initial rate of fall of current in the coil after a short-circuiting switch connected across its terminals has been suddenly closed. What was the energy stored in the coil, and in what form was it dissipated? (S.A.N.C., O.2)

27. If two coils have a mutual inductance of 400 μH, calculate the e.m.f. induced in one coil when the current in the other coil varies at a rate of 30 000 A/s.

28. If an e.m.f. of 5 V is induced in a coil when the current in an adjacent coil varies at a rate of 80 A/s, what is the value of the mutual inductance of the two coils?

29. If the mutual inductance between two coils is 0·2 H, calculate the e.m.f. induced in one coil when the current in the other coil is increased at a uniform rate from 0·5 to 3 A in 0·05 s.

30. If the toroid of Question 6 has a second winding of 80 turns wound over the first winding of 100 turns, calculate the mutual inductance.

31. When a current of 2 A through a coil P is reversed, a deflection of

36 divisions is obtained on a fluxmeter connected to a coil Q. If the
fluxmeter constant is 150 microweber-turns/division, what is the value
of the mutual inductance of coils P and Q?

32. Explain the meaning of the terms *self inductance* and *mutual inductance*
and define the unit by which each is measured.

A long solenoid, wound with 1000 turns, has an inductance of 120
mH and carries a current of 5 A. A search coil of 25 turns is arranged
so that it is linked by the whole of the magnetic flux. A ballistic galvanometer is connected to the search coil and the combined resistance of
the search coil and galvanometer is 200 Ω. Calculate, from first principles, the quantity of electricity which flows through the galvanometer
when the current in the solenoid is reversed.   (U.L.C.I., O.2)

33. Define the unit of mutual inductance. A cylinder, 50 mm in diameter
and 1 m long, is uniformly wound with 3000 turns in a single layer.
A second layer of 100 turns of much finer wire is wound over the first
one, near its centre. Calculate the mutual inductance between the two
coils. Derive any formula used.   (App. El., L.U.)

34. A solenoid P, 1 m long and 100 mm in diameter, is uniformly wound
with 600 turns. A search-coil Q, 30 mm in diameter and wound with 20
turns, is mounted co-axially midway along the solenoid. If Q is connected to a ballistic galvanometer, calculate the quantity of electricity
through the galvanometer when a current of 6 A through the solenoid
is reversed. The resistance of the secondary circuit is 0·1 MΩ. Find,
also, the mutual inductance between the two coils.

35. When a current of 2 A through a coil P is reversed, a deflection of 43
divisions is obtained on a fluxmeter connected to a coil Q. If the fluxmeter constant is 150 microweber-turns/division, find the mutual
inductance of coils P and Q. If the self inductances of P and Q are
5 mH and 3 mH respectively, calculate the coupling coefficient.

36. Two coils, A and B, have self inductances of 20 mH and 10 mH
respectively and a mutual inductance of 5 mH. If the currents through
A and B are 0·5 A and 2 A respectively, calculate: (*a*) the two possible
values of the energy stored in the magnetic field; and (*b*) the coupling
coefficient.

37. Two similar coils have a coupling coefficient of 0·25. When they are
connected in series cumulatively, the total inductance is 80 mH.
Calculate: (*a*) the self inductance of each coil; (*b*) the total inductance
when the coils are connected in series differentially; and (*c*) the total
magnetic energy due to a current of 2 A when the coils are connected
in series (i) cumulatively and (ii) differentially.

38. Two coils, with terminals AB and CD respectively, are inductively
coupled. The inductance measured between terminals AB is 380 µH
and that between terminals CD is 640 µH. With B joined to C, the
inductance measured between terminals AD is 1600 µH. Calculate:
(*a*) the mutual inductance of the coils; and (*b*) the inductance between
terminals AC when B is connected to D.

39. For two coils, A and B, in proximity to each other, the flux linkages
with B per unit current in A are equal to the linkages with A produced

by unit current flowing in B. Show that the mutual inductance between the coils is given by:

$$M = k\sqrt{L_aL_b} \text{ henrys,}$$

where $k \not> 1$ and L_a and L_b are the respective self inductances.

When two identical coupled coils are connected in series, the inductance of the combination is found to be 80 mH. When the connections to one of the coils are reversed, a similar measurement indicates 20 mH. Find the coupling coefficient between the two coils.

(S.A.N.C., O.2)

CHAPTER 5

Electrostatics

5.1 Electrification by Friction

Two glass rods rubbed with silk repel each other; similarly, two ebonite rods rubbed with fur repel each other. But the electrified glass and ebonite rods attract each other. The glass and ebonite therefore appear to be charged with different kinds of electricity; and these experiments show that bodies charged with the same kind of electricity repel, while bodies charged with opposite kinds of electricity attract one another.

It was about 1750 that Benjamin Franklin—an American—suggested that electricity was some form of fluid which passed from one body to the other when they were rubbed together, and that in the case of glass rubbed with silk, the electric fluid passed from the silk into the glass so that the glass contained a 'plus' or 'positive' amount of electricity. On the other hand, when ebonite was rubbed with fur, the electric fluid passed from the ebonite to the fur, leaving the ebonite with a 'minus' or 'negative' amount of electricity. Franklin's one-fluid theory has long since been discarded, but his convention still remains: thus, the glass is said to be positively charged and the ebonite negatively charged.

5.2 Structure of the atom

Every material is made up of one or more elements, an element being a substance composed entirely of atoms of the same kind; for instance, water is a combination of the elements hydrogen and oxygen, whereas common salt is a combination of the elements sodium and chlorine. The atoms of different elements differ in their structure, and this accounts for different elements possessing different characteristics.

Every atom consists of a relatively massive core or nucleus carrying a positive charge, around which *electrons* move in orbits at distances that are great compared with the size of the nucleus. Each *electron* has a mass of $9 \cdot 11 \times 10^{-31}$ kg and a *negative* charge, $-e$, equal to

140

$1 \cdot 602 \times 10^{-19}$ C. The nucleus of every atom except that of hydrogen consists of *protons* and *neutrons*. Each *proton* carries a *positive* charge, **e**, equal in magnitude to that of an electron and its mass is $1 \cdot 673 \times 10^{-27}$ kg, namely 1836 times that of an electron. A *neutron*, on the other hand, carries *no* resultant charge and its mass is approximately the same as that of a proton. Under normal conditions, an atom is neutral, i.e., the total negative charge on its electrons is equal to the total positive charge on the protons.

The atom possessing the simplest structure is that of hydrogen—it

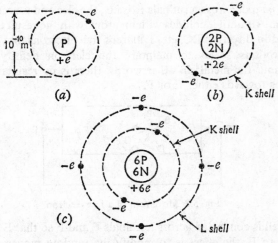

Fig. 5.1 Hydrogen, helium and carbon atoms.

consists merely of a nucleus of one proton together with a single electron which may be thought of as revolving in an orbit, of about 10^{-10} m diameter, around the proton, as in fig. 5.1 (*a*).

Fig. 5.1 (*b*) shows the arrangement of a helium atom. In this case, the nucleus consists of two protons and two neutrons, with two electrons orbiting in what is termed the K *shell*. The nucleus of a carbon atom has six protons and six neutrons and therefore carries a positive charge 6e. This nucleus is surrounded by six orbital electrons, each carrying a negative charge of −e, two electrons being in the K shell and four in the L shell, and their relative positions may be imagined to be as shown in fig. 5.1 (*c*).

The farther away an electron is from the nucleus, the smaller is the force of attraction between that electron and the positive charge on the nucleus; consequently the easier it is to detach such an electron from the atom. When atoms are packed tightly together, as in a metal, each outer electron experiences a small force of attraction towards neighbouring nuclei, with the result that such an electron is no longer bound to any individual atom, but can move at random within the metal. These electrons are termed *free* or *conduction* electrons and only a slight external influence is required to cause them to drift in a desired direction.

The full lines AB, BC, CD, etc., in fig. 5.2 represent paths of the random movement of one of the free electrons in a *metal* rod when there is *no* p.d. across terminals EF; i.e., the electron is accelerated in direction AB until it collides with an atom, with the result that it may rebound in direction BC, etc. Different free electrons move in different directions so as to maintain the electron density constant throughout the metal; in other words, there is no resultant drift of electrons towards either E or F.

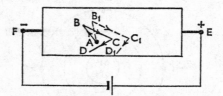

Fig. 5.2 Movement of a free electron.

If a cell is connected across terminals E and F so that E is positive relative to F, the effect is to modify the random movement of the electron as shown by the dotted lines AB_1, B_1C_1, and C_1D_1 in fig. 5.2, i.e., there is superimposed on the random movement a drift of the electron towards the positive terminal E; and the number of electrons reaching terminal E from the rod is the same as that entering the rod from terminal F. It is this drift of the electrons that constitutes the electric current in the circuit.

An atom which has lost or gained one or more free electrons is referred to as an *ion*; thus, for an atom which has lost one or more electrons, its negative charge is less than its positive charge, and such an atom is therefore termed a *positive ion*.

5.3 Movement of electrons in a conductor

In section 5.2 it was explained how the drift of electrons in a desired direction can be produced by introducing a source of electro-

motive force into the circuit. In fig. 5.3, DE represents an enlarged
view of a metal rod forming part of a closed circuit which includes a
battery B. The circles with crosses represent positive ions, namely
atoms which have lost one or more of their outermost electrons.
These ions are locked in the structure of the metal and are therefore
unable to move. The small black circles represent the free electrons
moving from left to right. This procession or drift of the electrons
takes place round the whole cir-
cuit, including battery B; i.e., the
number of electrons emerging per
second from the negative terminal
of B is exactly the same as that
entering the positive terminal per
second. It follows that an electric
current in a metal conductor consists of a movement of electrons
from a point at the *lower* potential

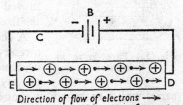

Direction of flow of electrons ⟶
⟵ Conventional direction of current

Fig. 5.3 Movement of electrons in a conductor.

to a point at the *higher* potential, namely in the opposite direction
to that taken as the conventional direction of the current. The latter
was universally adopted long before the discovery of the electron,
and so we continue to say that an electric current flows from a
point at the *higher* potential to that at the *lower* potential.

Since each electron carries a negative charge of $1 \cdot 602 \times 10^{-19}$
coulomb, it follows that when the current in a circuit is 1 ampere (or 1
coulomb/second), the number of electrons passing any given point
must be such that:

$$1 \cdot 602 \times 10^{-19} \times \text{no. of electrons/second} = 1 \text{ coulomb/second}$$
$$\therefore \qquad \text{no. of electrons/second} = 6 \cdot 24 \times 10^{18}$$

i.e., when the current in a circuit is 1 ampere, electrons are passing
any given point of the circuit at the rate of $6 \cdot 24 \times 10^{18}$ per second.

5.4 Capacitor

Two metal plates, separated by an insulator, constitute a *capacitor**
or *condenser*, namely an arrangement which has the capacity of
storing electricity as an excess of electrons on one plate and a
deficiency on the other.

The most common type of capacitor used in practice consists of

---

* Recommended by the British Standards Institution, but the term 'condenser'
is still widely used.

two strips of metal foil, represented by full lines in fig. 5.4, separated
by strips of waxed paper, shown dotted, these strips being wound
spirally, forming—in effect—two very large surfaces near to each
other. The whole assembly is thoroughly soaked in hot paraffin wax.
In radio receivers, some of the capacitors consist of two sets of metal

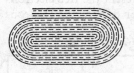

vanes, one of which is fixed and the other
set is so arranged that the vanes can be
moved into and out of the space between
the fixed vanes without touching the latter.

Fig. 5.4 Paper-insulated
capacitor.

A charged capacitor may be regarded as
a reservoir of electricity and its action can
be demonstrated by connecting a capacitor
of, say, 20 microfarads (section 5.8) in
series with a resistor R, a centre-zero microammeter A and a two-
way switch S, as in fig. 5.5. An electrostatic voltmeter ES (section
22.9) is connected across C. If R has a resistance of, say, 1 megohm,
it is found that when S is closed on *a*, the deflection on A rises
immediately to its maximum value and then falls off to zero, as
indicated by curve A in fig. 5.6. At the same time, the p.d. across C
grows in the manner shown by curve M. When S is moved over
to *b*, the current again rises immediately to the same maximum but
in the reverse direction, and then falls off as shown by curve B.
Curve N shows the corresponding variation
of p.d. across C.

If the experiment is repeated with a resistance of, say, 2 megohms, it is found that
the initial current, both on charging and
on discharging, is halved, but it takes
about twice as long to fall off, as shown
by the dotted curves D and E. Curves P
and Q represent the corresponding variation of the p.d. across C during charge and
discharge respectively.

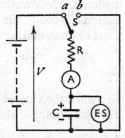

The shaded area between curve A and the
horizontal axis in fig. 5.6 represents the

Fig.   5.5   Capacitor
charged and discharged
through a resistor.

product of the average charging current (in amperes) and the time
(in seconds), namely the quantity of electricity (in coulombs) required to charge the capacitor to a p.d. of *V* volts. Similarly the
shaded area enclosed by curve B represents the same quantity of
electricity obtainable during discharge.

5.5 Hydraulic analogy

The operation of charging and discharging a capacitor may be more easily understood if we consider the hydraulic analogy given in fig.

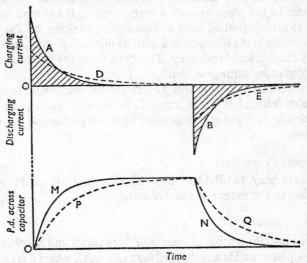

Fig. 5.6 Charging and discharging currents and p.d.s.

5.7, where P represents a piston operated by a rod R, and D is a rubber diaphragm stretched across a cylindrical chamber C. The cylinders are connected by pipes E, E and are filled with water.

When no force is being exerted on P, the diaphragm is flat, as shown dotted, and the piston is in position A. If P is pushed towards the left, water is withdrawn from G and forced into F and the diaphragm is in consequence distended, as shown by the full line. The greater the force applied to P, the greater is the amount of water displaced. But the rate at which this displacement takes place depends upon the resistance offered by pipes E, E; thus the smaller the cross-sectional area of the pipes, the longer is the time required for the steady state to be reached. The force applied to P is analogous to the e.m.f. of the battery, the quantity of water displaced corresponds to the charge, the rate at which the water passes

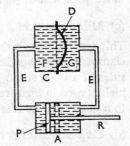

Fig. 5.7 Hydraulic analogy of a capacitor.

any point in the pipes corresponds to the current, and the cylinder C with its elastic diaphragm is the analogue of the capacitor.

When the force exerted on P is removed, the distended diaphragm forces water out of F back into G; and if the frictional resistance of the water in the pipes exceeds a certain value, it is found that the piston is merely pushed back to its original position A. The strain energy stored in the diaphragm due to its distension is converted into heat by the frictional resistance. The effect is similar to the discharge of the capacitor through a resistor.

No water can pass from F to G through the diaphragm so long as it remains intact; but if it is strained excessively it bursts, just as the insulation in a capacitor is punctured when the p.d. across it becomes excessive.

5.6 Types of capacitors

Capacitors may be divided into the following six main groups according to the nature of the dielectric:

(a) *Air capacitors*

This type usually consists of one set of fixed plates and another set of movable plates, and is mainly used for radio work where it is required to vary the capacitance.

(b) *Paper capacitors*

The electrodes consist of metal foils interleaved with paper impregnated with wax or oil and rolled into a compact form (fig. 5.4).

(c) *Mica capacitors*

This type consists either of alternate layers of mica and metal foil clamped tightly together or of thin films of silver sputtered on the two sides of a mica sheet. Owing to its relatively high cost, this type is mainly used in high-frequency circuits when it is necessary to reduce to a minimum the loss in the dielectric.

(d) *Ceramic capacitors*

The electrodes consist of metallic coatings (usually silver) on the opposite faces of a thin disc or plate of ceramic material such as the hydrous silicate of magnesia or talc—somewhat similar to that used by tailors for marking cloth. This type of capacitor is mainly used in high-frequency circuits subject to wide variation of temperature.

(e) *Polycarbonate capacitors*

Polycarbonate is a recent development in the field of plastic insulating materials. A film of polycarbonate can be produced in thicknesses down to 2 micrometres ($=2 \times 10^{-6}$ m). It is metallized with aluminium and wound to form the capacitor elements.

Polycarbonate has a relative permittivity (section 5.14) of about 2·8; it possesses high resistivity and low dielectric loss. The power factor of a polycarbonate capacitor is about 0·001 at 1 kHz and about 0·005 at 1 MHz, and the maximum operating temperature is about 125–150°C.

(f) *Electrolytic capacitors*

The type most commonly used consists of two aluminium foils, one with an oxide film and one without, the foils being interleaved with a material such as paper saturated with a suitable electrolyte, for example, ammonium borate. The aluminium oxide film is formed on the one foil by passing it through an electrolytic bath of which the foil forms the positive electrode. The finished unit is assembled in a container—usually of aluminium—and hermetically sealed. The oxide film acts as the dielectric; and as its thickness in a capacitor suitable for a working voltage of 100 V is only about 0·15 μm, a very large capacitance is obtainable in a relatively small volume.

The main disadvantages of this type of capacitor are: (*a*) the insulation resistance is comparatively low and (*b*) it is only suitable for circuits where the voltage applied to the capacitor never reverses its direction. Electrolytic capacitors are mainly used where very large capacitances are required, e.g. for reducing the ripple in the voltage wave obtained from a rectifier (section 19.6).

Solid types of electrolytic capacitors have been developed to avoid some of the disadvantages of the wet electrolytic type. In one arrangement, the wet electrolyte is replaced by manganese dioxide. In another arrangement, the anode is a cylinder of pressed sintered tantalum powder coated with an oxide layer which forms the dielectric. This oxide has a conducting coat of manganese dioxide which acts as an electron conductor and replaces the ionic conduction of the liquid electrolyte in the wet type. A layer of graphite forms the connection with a silver or copper cathode and the whole is enclosed in a hermetically-sealed steel can.

5.7 Relationship between the charge and the applied voltage

The method described in section 5.4 for determining the charge on each plate of a capacitor is instructive, but is unsuitable for accurate measurement. A much better method is to discharge the capacitor through a ballistic galvanometer (section 3.10), since the deflection of the latter is proportional to the charge.

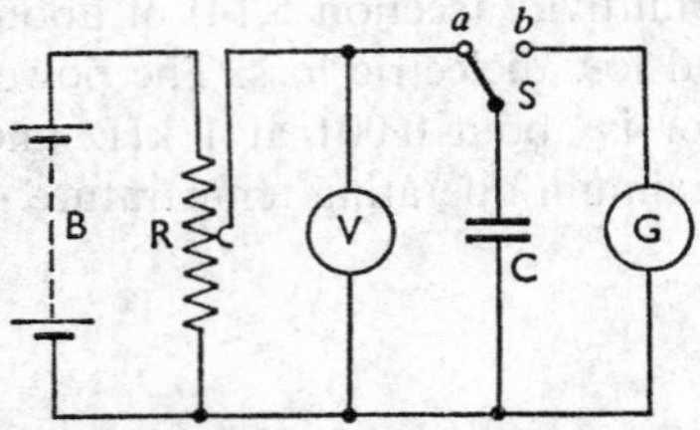

Let us charge a capacitor C (fig. 5.8) to various voltages by means of a slider on a resistor R connected across a battery B, S being on *a*; and for each voltage, note the deflection of G when C is discharged through it by moving S over to *b*. Thus, if θ is the first deflection or 'throw' observed when the capacitor, charged to a p.d. of V volts, is discharged through G, and if k is the ballistic constant of G in coulombs per unit of first deflection, then:

Fig. 5.8 Measurement of charge by ballistic galvanometer.

$$\text{discharge through G} = Q = k\theta \text{ coulombs.}$$

It is found that for a given capacitor,

$$\frac{\text{charge on C [coulombs]}}{\text{p.d. across C [volts]}} = \text{a constant} \qquad (5.1)$$

5.8 Capacitance

The property of a capacitor to store an electric charge when its plates are at different potentials is referred to as its *capacitance*.

The unit of capacitance is termed the *farad* (symbol F)—a curtailment of 'Faraday'—and may be defined as *the capacitance of a capacitor between the plates of which there appears a difference of potential of 1 volt when it is charged by 1 coulomb of electricity*.

It follows from expression (5.1) and from the definition of the farad that:

$$\frac{\text{charge [coulombs]}}{\text{applied p.d. [volts]}} = \text{capacitance [farads]}$$

or in symbols, $\qquad\qquad Q/V = C$

$\therefore \qquad\qquad\qquad\qquad Q = CV \qquad\qquad (5.2)$

In practice, the farad is found to be inconveniently large and the capacitance is usually expressed in *microfarads* (μF) or in *picofarads* (pF),

where 1 microfarad = 10^{-6} farad
and 1 picofarad = 10^{-12} farad.

Example 5.1 *A capacitor having a capacitance of* 80 μF *is connected across a 500-V d.c. supply. Calculate the charge.*

From (5.2), charge = (80×10^{-6}) [farad] $\times$ 500 [volts]
= 0·04 coulomb.

5.9 Capacitors in parallel

Suppose two capacitors, having capacitances C_1 and C_2 farads respectively, to be connected in parallel (fig. 5.9) across a p.d. of V volts. The charge on C_1 is Q_1 coulombs and that on C_2 is Q_2 coulombs, where:

$$Q_1 = C_1V \text{ and } Q_2 = C_2V.$$

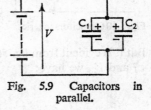

Fig. 5.9 Capacitors in parallel.

If we were to replace C_1 and C_2 by a single capacitor of such capacitance C farads that the same total charge of $(Q_1 + Q_2)$ coulombs would be produced by the same p.d., then $Q_1 + Q_2 = CV$.

Substituting for Q_1 and Q_2, we have:

$$C_1V + C_2V = CV$$
$$\therefore \qquad\qquad C = C_1 + C_2 \qquad\qquad (5.3)$$

Hence *the resultant capacitance of capacitors in parallel is the arithmetic sum of their respective capacitances.*

5.10 Capacitors in series

Suppose C_1 and C_2 in fig. 5.10 to be two capacitors connected in series with suitable centre-zero ammeters A_1 and A_2, a resistor R and a two-way switch S. When S is put over to a, A_1 and A_2 are found to indicate exactly the same charging current, each reading decreasing simultaneously from a maximum to zero, as already shown in fig. 5.6. Similarly, when S is put over to b, A_1 and A_2 indicate similar discharges. It follows that during charge the displacement of electrons from the positive plate of C_1 to the negative plate of C_2 is exactly the same as that from the upper plate (fig. 5.10) of C_2 to the lower plate of C_1. In other words, the displacement of Q coulombs of electricity is the same in every part of the circuit, and the charge on each capacitor is therefore Q coulombs.

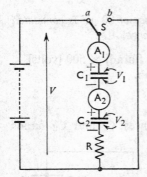

If V_1 and V_2 be the corresponding p.d.s across C_1 and C_2 respectively, then from (5.2):

$$Q = C_1V_1 = C_2V_2 \qquad (5.4)$$

so that $\quad V_1 = \dfrac{Q}{C_1},$ and $V_2 = \dfrac{Q}{C_2}.$

If we were to replace C_1 and C_2 by a single capacitor of capacitance C farads such that it would have the same charge Q coulombs with the same p.d. of V volts, then:

Fig. 5.10 Capacitors in series.

$$Q = CV, \text{ or } V = Q/C.$$

But it is evident from fig. 5.10 that $V = V_1 + V_2$. Substituting for V, V_1 and V_2, we have:

$$\frac{Q}{C} = \frac{Q}{C_1} + \frac{Q}{C_2}$$

$$\therefore \qquad \frac{1}{C} = \frac{1}{C_1} + \frac{1}{C_2} \qquad (5.5)$$

Hence *the reciprocal of the resultant capacitance of capacitors connected in series is the sum of the reciprocals of their respective capacitances.*

5.11 Distribution of voltage across capacitors in series

From expression (5.4),

$$V_2/V_1 = C_1/C_2 \qquad (5.6)$$

But $\qquad\qquad\qquad V_1 + V_2 = V$

$$\therefore \qquad\qquad\qquad V_2 = V - V_1.$$

Substituting for V_2 in (5.6), we have:

$$(V - V_1)/V_1 = C_1/C_2$$

$$\therefore \qquad V_1 = V \times \frac{C_2}{C_1 + C_2} \qquad (5.7)$$

and $\qquad\qquad V_2 = V \times \dfrac{C_1}{C_1 + C_2} \qquad (5.8)$

Example 5.2 *Three capacitors have capacitances of* 2 μF, 4 μF *and* 8 μF *respectively. Find the total capacitance when they are connected* (a) *in parallel*, (b) *in series*.

(*a*) From (5.3):

$$\text{total capacitance} = 2 + 4 + 8 = 14 \text{ μF.}$$

(*b*) If *C* be the resultant capacitance in microfarads when the capacitors are in series, then from (5.5):

$$\frac{1}{C} = \frac{1}{2} + \frac{1}{4} + \frac{1}{8} = 0\cdot5 + 0\cdot25 + 0\cdot125 = 0\cdot875$$

$$\therefore \qquad C = 1\cdot143 \text{ μF.}$$

Example 5.3 *If two capacitors having capacitances of* 6 μF *and* 10 μF *respectively are connected in series across a* 200-*V supply, find* (a) *the p.d. across each capacitor*, (b) *the charge on each capacitor*.

(*a*) Let V_1 and V_2 be the p.d.s across the 6-μF and 10-μF capacitors respectively; then, from expression (5.7),

$$V_1 = 200 \times \frac{10}{6 + 10} = 125 \text{ V}$$

and $\qquad V_2 = 200 - 125 = 75$ V.

(*b*) Charge on each capacitor

$$= \text{charge on C}_1$$
$$= 6 \times 10^{-6} \times 125 = 0\cdot000\ 75 \text{ C.}$$

5.12 Relationship between the capacitance and the dimensions of a capacitor

It follows from expression (5.3) that if two similar capacitors are connected in parallel, the capacitance is double that of one capacitor. But the effect of connecting two similar capacitors in parallel is merely to double the area of each plate. In general, we may therefore say that the capacitance of a capacitor is proportional to the area of the plates.

On the other hand, if two similar capacitors are connected in series, it follows from expression (5.5) that the capacitance is halved. We have, however, doubled the thickness of the insulation between the plates that are connected to the supply. Hence we may say in general that the capacitance of a capacitor is inversely proportional to the distance between the plates; and the above relationships may be summarized thus:

$$\text{capacitance} \propto \frac{\text{area of plates}}{\text{distance between plates}}.$$

5.13 Electric field strength and electric flux density

Let us consider a capacitor consisting of metal plates, M and N, in a glass enclosure G, shown chain-dotted in fig. 5.11, from which all the air has been removed. Let a be the area in square metres of one side of each plate and let d be the distance in metres between the plates.

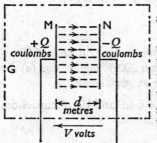

Let Q be the charge in coulombs due to a p.d. of V volts between the plates.

Suppose the dimensions of the plates to be so large compared with the distance between them that we may assume negligible fringing of the electric flux, i.e. all the electric flux may be assumed to pass straight across from M to N, as shown by the dotted lines in fig. 5.11.

The *electric field strength* (the term

Fig. 5.11 A parallel-plate capacitor.

electric force is now obsolete) in the region between the two plates M and N

is the *potential drop per unit length* or *potential gradient*, namely V/d volts/metre; and the *direction* of the electric field strength at any point is the direction of the mechanical force on a positive charge situated at that point, namely from the positively-charged plate M towards the negatively-charged plate N in fig. 5.11. The symbol for electric field strength is E^*,

i.e. $E = V/d$ volts/metre   (5.9)

In SI, *one* unit of electric flux is assumed to emanate from a positive charge of 1 coulomb and to enter a negative charge of 1 coulomb. Hence, if the charge on plates M and N is Q coulombs, electric flux between M and N $= \Psi = Q$ coulombs and

electric flux density $= D = Q/a$ coulombs/metre^2 (5.10)

where $a =$ area of dielectric, in square metres, at right-angles to direction of electric flux.

---

* Bold type is used to enable the symbols E and D for electric field strength and electric flux density respectively to be easily distinguished from the letters E and D used to represent other quantities.

From expressions (5.9) and (5.10),

$$\frac{\text{electric flux density}}{\text{electric field strength}} = \frac{D}{E} = \frac{Q}{a} \div \frac{V}{d} = \frac{Q}{V} \times \frac{d}{a} = \frac{Cd}{a}.$$

In electromagnetism, the ratio of the magnetic flux density in a vacuum to the magnetic field strength is termed the *permeability of free space* or *magnetic constant* and is represented by μ_0. Similarly, in electrostatics, the ratio of the electric flux density in a vacuum to the electric field strength is termed the *permittivity of free space* or *electric constant* and is represented by ϵ_0.

Hence, $\epsilon_0 = Cd/a$

or $C = \epsilon_0 a/d$ (5.11)

The effect of filling the space between M and N (fig. 5.11) with air at atmospheric pressure is to increase the capacitance by 0·06 per cent compared with the value when the space is completely evacuated; hence for all practical purposes, expression (5.11) can be applied to capacitors having air dielectric.

The value of ϵ_0 can be determined experimentally by charging a capacitor, of known dimensions and with vacuum dielectric, to a p.d. of V volts and then discharging it through a ballistic galvanometer having a known ballistic constant k coulombs/unit deflection. If the deflection is θ divisions,

$$Q = CV = k\theta$$

$$\therefore \qquad \epsilon_0 = C \cdot \frac{d}{a} = \frac{k\theta}{V} \cdot \frac{d}{a}.$$

From carefully conducted tests it has been found that the value of ϵ_0 is $8 \cdot 85 \times 10^{-12}$ F/m (see footnote on p. 154).

Hence the capacitance of a parallel-plate capacitor with vacuum or air dielectric is given by:

$$C = \frac{(8 \cdot 85 \times 10^{-12}) \text{ [F/m]} \times a \text{ [m}^2\text{]}}{d \text{ [m]}} \text{ farads} \qquad (5.12)$$

It may be mentioned at this point that there is a definite relationship between μ_0, ϵ_0 and the velocity of light and other electromagnetic waves; thus,

$$\frac{1}{\mu_0\epsilon_0} = \frac{1}{4\pi \times 10^{-7} \times 8 \cdot 85 \times 10^{-12}} \simeq 8 \cdot 99 \times 10^{16} \simeq (2 \cdot 998 \times 10^8)^2$$

But the velocity of light $= 2 \cdot 998 \times 10^8$ metres/second

$$\therefore \qquad \left. \begin{array}{l} \text{velocity of light} \\ \text{in metres/second} \end{array} \right\} = \frac{1}{\sqrt{(\mu_0\epsilon_0)}}. \qquad (5.13)$$

This relationship was discovered by Prof. Clerk Maxwell in 1865 and enabled him to predict the existence of radio waves about twenty years before their effect was demonstrated experimentally by Prof. H. Hertz.

5.14 Relative permittivity

If the experiment described in section 5.13 is performed with a sheet of glass filling the space between plates M and N, it is found that the value of the capacitance is greatly increased; and the ratio of the capacitance of a capacitor having a certain material as dielectric to the capacitance of that capacitor with vacuum (or air) dielectric is termed the *relative permittivity* of that material and is represented by the symbol ϵ_r. (The term *dielectric constant* is obsolete.) Values of the relative permittivity of some of the most important insulating materials are given in the following table:

Material	Relative permittivity
Air	1·0006
Paper (dry)	2–2·5
Bakelite	4·5–5·5
Glass	5–10
Rubber	2–3·5
Mica	3–7
Porcelain	6–7

From expression (5.11), it follows that if the space between the metal plates of the capacitor in fig. 5.11 is filled with a dielectric having a relative permittivity ϵ_r,

$$\text{capacitance} = C = \frac{\epsilon_0 \epsilon_r a}{d} \text{ farads} \tag{5.14}$$

$$= \frac{(8\cdot85 \times 10^{-12})[\text{F/m}] \times \epsilon_r \times a\,[\text{m}^2]}{d\,[\text{m}]} \text{ farads}$$

$$\text{and charge due to a p.d. of } V \text{ volts} \Big\} = Q = CV$$

$$= \frac{\epsilon_0 \epsilon_r a V}{d} \text{ coulombs,}$$

$$\therefore \frac{\text{electric flux density}}{\text{electric field strength}} = \frac{D}{E} = \frac{Q}{a} \div \frac{V}{d} = \frac{Qd}{Va} = \epsilon_0 \epsilon_r$$

$$= \epsilon = \text{absolute permittivity*} \tag{5.15}$$

* From expression (5.15),

$$\text{absolute permittivity} = \epsilon_0 \epsilon_r = \frac{C(\text{farads}) \times d(\text{metres})}{a(\text{metres}^2)}$$

$$= Cd/a \text{ farads/metre,}$$

hence the units of absolute permittivity are *farads/metre* (or F/m); e.g. $\epsilon_0 = 8\cdot85 \times 10^{-12}$ F/m.

This expression is similar in form to expression (3.4) deduced for the magnetic circuit, namely:

$$\frac{\text{magnetic flux density}}{\text{magnetic field strength}} = \frac{B}{H} = \mu_0\mu_r.$$

5.15 Capacitance of a multi-plate capacitor

Suppose a capacitor to be made up of n parallel plates, alternate plates being connected together as in fig. 5.12.

Let a = area of *one* side of each plate in square metres,

$\quad d$ = thickness of dielectric in metres

and $\quad \epsilon_r$ = relative permittivity of the dielectric.

Fig. 5.12 shows a capacitor with seven plates, four being connected to A and three to B. It will be seen that each side of the three plates connected to B is in contact with the dielectric, whereas only one side of each of the outer plates is in contact with it. Consequently, the useful surface area of each set of plates is $6a$ square metres.

Fig. 5.12 Multi-plate capacitor.

For n plates, the useful area of each set is $(n-1)a$ square metres;

$$\therefore \quad \text{capacitance} = \frac{\epsilon_0\epsilon_r(n-1)a}{d} \text{ farads}$$

$$= \frac{8.85 \times 10^{-12}\epsilon_r(n-1)a}{d} \text{ farads} \quad (5.16)$$

Example 5.4 *A capacitor is made with 7 metal plates connected as in fig. 5.12 and separated by sheets of mica having a thickness of 0·3 mm and a relative permittivity of 6. The area of one side of each plate is 500 cm². Calculate the capacitance in microfarads.*

Using expression (5.16), we have $n = 7$, $a = 0.05$ m², $d = 0.0003$ m and $\epsilon_r = 6$.

$$\therefore \quad C = \frac{8.85 \times 10^{-12} \times 6 \times 6 \times 0.05}{0.0003} = 0.0531 \times 10^{-6} \text{ F}$$

$$= 0.0531 \text{ μF}.$$

Example 5.5 *A p.d. of 400 V is maintained across the terminals of the capacitor of example 5.4. Calculate* (a) *the charge,* (b) *the electric* *field strength or potential gradient and* (c) *the electric flux density in the dielectric.*

(a) Charge $= Q = CV = 0.0531 \ [\mu\text{F}] \times 400 \ [\text{V}]$
$= 21.24 \ \mu\text{C}.$

(b) Electric field strength or potential gradient
$= V/d = 400 \ [\text{V}]/0.0003 \ [\text{m}] = 1 \ 333 \ 000 \ \text{V/m}$
$= 1333 \ \text{kV/m}.$

(c) Electric flux density
$= Q/a = 21.24 \ [\mu\text{C}]/(0.05 \times 6) \ [\text{m}^2]$
$= 70.8 \ \mu\text{C/m}^2.$

5.16 Capacitance of and potential gradient in a parallel-plate capacitor with composite dielectric

Suppose the space between metal plates M and N to be filled by dielectrics A and B of thickness d_1 and d_2 metres respectively, as shown in fig. 5.13 (a).

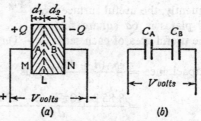

Fig. 5.13 Parallel-plate capacitor with two dielectrics.

Let $Q =$ charge in coulombs due to p.d. of V volts
and $a =$ area of each dielectric in square metres,
then $D = Q/a =$ electric flux density, in coulombs/metre², in A and B.

Let E_1 and $E_2 =$ electric field strengths in A and B respectively; then if the relative permittivities of A and B are ϵ_1 and ϵ_2 respectively,

$$\text{electric field strength in A} = E_1 = \frac{D}{\epsilon_1\epsilon_0} = \frac{Q}{\epsilon_1\epsilon_0 a}$$

and $$\text{electric field strength in B} = E_2 = \frac{D}{\epsilon_2\epsilon_0} = \frac{Q}{\epsilon_2\epsilon_0 a}.$$

Hence, $$\frac{E_1}{E_2} = \frac{\epsilon_2}{\epsilon_1} \qquad\qquad (5.17)$$

i.e. for dielectrics having the same cross-sectional area in series, the electric field strengths (or potential gradients) are inversely proportional to their relative permittivities.

Potential drop in a dielectric = electric field strength × thickness

∴   p.d. between plate M and the boundary surface L $\left.\begin{array}{c} \\ \end{array}\right\}$ = $E_1 d_1$
          between A and B

Hence all points on surface L are at the same potential, i.e. L is an *equipotential surface* and is at right-angles to the direction of the electric field strength. It follows that if a very thin metal foil were inserted between A and B, it would not alter the electric field in the dielectrics. Hence the latter may be regarded as equivalent to two capacitances, C_A and C_B, connected in series as in fig. 5.13 (b),

where $C_A = \dfrac{\epsilon_1 \epsilon_0 a}{d_1}$ and $C_B = \dfrac{\epsilon_2 \epsilon_0 a}{d_2}$

and   total capacitance between plates M and N $= \dfrac{C_A C_B}{C_A + C_B}$.

Example 5.6 *A capacitor consists of two metal plates, each 400 mm × 400 mm, spaced 6 mm apart. The space between the metal plates is filled with a glass plate 5 mm thick and a layer of paper 1 mm thick. The relative permittivities of the glass and paper are 8 and 2 respectively. Calculate (a) the capacitance, neglecting any fringing flux, and (b) the potential gradient in each dielectric in kilovolts/millimetre due to a p.d. of 10 kV between the metal plates.*

(a) Fig. 5.14 (a) shows a cross-section (not to scale) of the capacitor; and in fig. 5.14 (b), C_p represents the capacitance of the paper layer between M and the equipotential surface L and C_g represents that of the glass between L and N. From expression (5.14) we have:

$$C_p = \frac{8 \cdot 85 \times 10^{-12} \times 2 \times 0 \cdot 4 \times 0 \cdot 4}{0 \cdot 001} = 2 \cdot 83 \times 10^{-9} \text{ F}$$

and $C_g = \dfrac{8 \cdot 85 \times 10^{-12} \times 8 \times 0 \cdot 4 \times 0 \cdot 4}{0 \cdot 005} = 2 \cdot 265 \times 10^{-9} \text{ F.}$

If C is the resultant capacitance between M and N,

$$\frac{1}{C} = \frac{10^9}{2 \cdot 83} + \frac{10^9}{2 \cdot 265} = 0 \cdot 7955 \times 10^9$$

∴ $C = 1 \cdot 257 \times 10^{-9} \text{ F} = 0 \cdot 001 \ 257 \ \mu\text{F.}$

(*b*) Since C_p and C_g are in series across 10 kV, it follows from expression (5.7) that the p.d., V_p, across the paper is given by:

$$V_p = \frac{10 \times 2\cdot 265}{2\cdot 83 + 2\cdot 265} = 4\cdot 45 \text{ kV}$$

and $$V_g = 10 - 4\cdot 45 = 5\cdot 55 \text{ kV}.$$

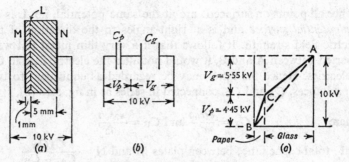

Fig. 5.14 Diagrams for example 5.6.

These voltages are represented graphically in fig. 5.14 (*c*).

$$\left.\begin{array}{l}\text{Potential gradient in the}\\ \text{paper dielectric}\end{array}\right\} = 4\cdot 45/1 = 4\cdot 45 \text{ kV/mm}$$

and $$\left.\begin{array}{l}\text{potential gradient in the}\\ \text{glass dielectric}\end{array}\right\} = 5\cdot 55/5 = 1\cdot 11 \text{ kV/mm}.$$

These potential gradients are represented by the slopes of AC and CB for the glass and paper respectively in fig. 5.14 (*c*). Had the dielectric between plates M and N been homogeneous, the potential gradient would have been 10/6 = 1·67 kV/mm, as represented by the slope of the dotted line AB in fig. 5.14 (*c*). It will therefore be seen that the effect of using a composite dielectric of two materials having different relative permittivities is to increase the potential gradient in the material having the lower relative permittivity. This effect has very important applications in high-voltage work.

5.17 Comparison of electrostatic and electromagnetic terms

It may be helpful to compare the terms and symbols used in electrostatics with the corresponding terms and symbols used in electromagnetism:

Electrostatics		Electromagnetism	
Term	Symbol	Term	Symbol
Electric flux	Ψ	Magnetic flux	Φ
Electric flux density	D	Magnetic flux density	B
Electric field strength	E	Magnetic field strength	H
Electromotive force	E	Magnetomotive force	F
Electric potential difference	V	Magnetic potential difference	—
Permittivity of free space	ϵ_0	Permeability of free space	μ_0
Relative permittivity	ϵ_r	Relative permeability	μ_r
Absolute permittivity		Absolute permeability	

$$= \frac{\text{electric flux density}}{\text{electric field strength}} \qquad = \frac{\text{magnetic flux density}}{\text{magnetic field strength}}$$

i.e. $\epsilon_0\epsilon_r = \epsilon = D/E$ i.e. $\mu_0\mu_r = \mu = B/H$

5.18 Force on an isolated charge in an electric field

Suppose L in fig. 5.15 to be a very small metal sphere carrying a positive charge of q coulombs situated in the electric field between plates M and N separated by an air dielectric, M being positive relative to N. Since like charges repel and unlike charges attract each other, there is a force acting on L urging it towards N.

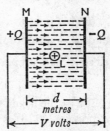

Fig. 5.15 An isolated charge in an electric field.

The movement of a positive charge from M to N is equivalent to a momentary current flowing from a point M to a point N of a wire, where the potential of M is V volts above that of N. In such a case, energy absorbed, in joules

= p.d. in volts × current in amperes × time in seconds

= p.d. in volts × charge in coulombs.

Hence, when a positive charge of q coulombs moves from M to N in fig. 5.15,

$$\text{energy absorbed} = Vq \text{ joules.}$$

Since this energy is due to the force on the charge acting through a distance d metres,

$$\begin{pmatrix}\text{force, in newtons, on} \\ \text{charge } q\end{pmatrix} \times \begin{pmatrix}\text{distance, in metres,} \\ \text{between M and N}\end{pmatrix} = Vq \text{ joules}$$

$$\therefore \qquad \text{force on charge} = qV/d = qE \text{ newtons}$$
$$= E \text{ newtons/coulomb} \qquad (5.18)$$

where E is the electric field strength in the dielectric.

If L in fig. 5.15 were carrying a negative charge, the force on L would urge it towards the positive plate M.

5.19 Deflection of an electron moving through a uniform electric field

Suppose an electron to have a velocity of v metres/second at right-angles to an electric field between two parallel plates A and B, as shown in fig. 5.16.

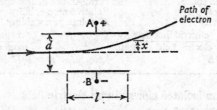

Fig. 5.16 Deflection of an electron by an electric field.

If V = p.d., in volts, between plates A and B
and d = distance, in metres, between the plates,

$$\text{electric field strength (or potential gradient)} \atop \text{between plates} \Big\} = V/d \text{ volts/metre.}$$

If e = the negative charge, in coulombs, on an electron, then from expression (5.18),

$$\text{force on electron} = eV/d \text{ newtons.}$$

If plate A is positive relative to plate B, this force deflects the electron towards plate A, as shown in fig. 5.16.

If m = mass of electron in kilograms,

transverse force on electron = mass × transverse acceleration

i.e. eV/d = m × transverse acceleration,

∴   transverse acceleration = $\dfrac{e}{m} \times \dfrac{V}{d}$ metres/second².

If the axial length of the electric field is l metres, the time taken by an electron to traverse the electric field is l/v seconds,

and $\text{final transverse velocity of} \atop \text{electron} \Big\} = \Big(\text{transverse} \atop \text{acceleration} \Big) \times \text{time}$

$$= \frac{e}{m} \times \frac{V}{d} \times \frac{l}{v} \text{ metres/second}$$

∴ deflection of electron during its movement in the electric field

= x = average transverse velocity × time

$$= \frac{1}{2} \times \frac{e}{m} \times \frac{V}{d} \times \left(\frac{l}{v}\right)^2 \text{ metres} \qquad (5.19)$$

Since $e = 1.6 \times 10^{-19}$ coulomb and $m = 9.1 \times 10^{-31}$ kilogram,

∴ $e/m = 1.76 \times 10^{11}$ C/kg.

Substituting for e/m in expression (5.19), we have:

deflection of electron = $0.88 \times 10^{11} \times (V/d) \times (l/v)^2$ metres   (5.20)

Example 5.7 *An electron has a velocity of 10^7 m/s at right-angles to the electric field between the deflecting plates of a cathode-ray tube (section 19.25). The plates are 8 mm apart and 20 mm long, and the p.d. between the plates is 50 V. Calculate the distance through which the electron is deflected during its movement through the electric field.*

From the data given in the question, $V = 50$ V, $d = 0.008$ m, $l = 0.02$ m and $v = 10^7$ m/s. Substituting in expression (5.20), we have:

$$\text{deflection} = 0.88 \times 10^{11} \times (50/0.008) \times (0.02 \times 10^{-7})^2$$
$$= 0.0022 \text{ m} = 2.2 \text{ mm}.$$

5.20 Movement of a free electron in an electric field

Suppose M and N in fig. 5.17 to be two metal plates, d metres apart, in an evacuated glass vessel G, and suppose a p.d. of V volts to be maintained between the plates. If an electron were released from the negative plate N, it follows from section 5.18 that the work done in moving the electron from N to M is eV joules, where e is the negative charge, in coulombs, on the electron.

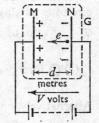

Fig. 5.17 Movement of a free electron in an electric field.

Suppose the electron to have zero initial velocity and a final velocity of v metres/second; then the final kinetic energy of the electron, immediately before its impact with plate M, is $\frac{1}{2}mv^2$ joules, where m is the mass of the electron in kilograms. Since there are no gas molecules in the space between M and N, there can be no loss of energy by collision of the electron with such molecules and all the work done on the electron is converted into kinetic energy,

$$\therefore \qquad \tfrac{1}{2}mv^2 = eV$$

and $\qquad\qquad v = \sqrt{(2Ve/m)}$ metres/second $\qquad\qquad$ (5.21)

This kinetic energy is converted into heat at the moment of impact of the electron with plate M.

Since e is $1\cdot6 \times 10^{-19}$ coulombs, an electron that is accelerated by a p.d. of 100 volts acquires kinetic energy of $1\cdot6 \times 10^{-17}$ joule. A more convenient unit for such small amounts of energy is the *electronvolt* (symbol, eV), namely the work done when an electron is moved through a p.d. of 1 volt; hence,

$$1 \text{ electronvolt} = 1\cdot6 \times 10^{-19} \text{ joule} \qquad\qquad (5.22)$$

It follows that when an electron is moved through a p.d. of 100 V,

$$\text{work done} = 100 \text{ eV}.$$

Example 5.8 *The p.d. between the anode and cathode of a vacuum diode (section 19.1) is 150 V and the current through the diode is 15 mA. Calculate (a) the maximum velocity acquired by the electrons, assuming their initial velocity to be zero, (b) the number of electrons passing per second from cathode to anode and (c) the energy absorbed in 5 minutes in (i) joules, (ii) electronvolts. Assume electron charge = $1\cdot6 \times 10^{-19}$ C and electron mass = $9\cdot11 \times 10^{-31}$ kg.*

(*a*) From expression (5.21),

$$v = \sqrt{(2 \times 150 \times 1\cdot6 \times 10^{-19})/(9\cdot11 \times 10^{-31})}$$
$$= 7\cdot27 \times 10^6 \text{ m/s.}$$

(*b*) Since a current of 1 A in a circuit corresponds to $6\cdot24 \times 10^{18}$ electrons passing a given point per second (see section 5.3),

$$\therefore \qquad \text{no. of electrons/second} = 6\cdot24 \times 10^{18} \times 0\cdot015$$
$$= 9\cdot36 \times 10^{16}$$

(*c*) (i) Energy absorbed $= VIt = 150 \times 0\cdot015 \times 5 \times 60$
$$= 675 \text{ J.}$$

Alternatively, energy absorbed/electron

$$= \tfrac{1}{2}mv^2 = \tfrac{1}{2} \times 9\cdot11 \times 10^{-31} \times (7\cdot27)^2 \times 10^{12}$$
$$= 2\cdot4 \times 10^{-17} \text{ J.}$$

$\therefore$ total energy absorbed in $\Big\}$ = (energy/electron) × no. of electrons
5 minutes

$$= 2\cdot4 \times 10^{-17} \times 9\cdot36 \times 10^{16} \times 300$$
$$= 675 \text{ J.}$$

(ii) Since 1 electronvolt $= 1\cdot6 \times 10^{-19}$ joule,

∴   energy absorbed $= 675/(1\cdot6 \times 10^{-19})$

$$= 4\cdot21 \times 10^{21} \text{ eV.}$$

Alternatively, energy absorbed/electron $= 150$ eV

∴ energy absorbed in 1 second $= 150 \times 9\cdot36 \times 10^{16}$

$$= 1\cdot404 \times 10^{19} \text{ eV}$$

and

energy absorbed in 5 minutes $= 1\cdot404 \times 10^{19} \times 300$

$$= 4\cdot21 \times 10^{21} \text{ eV.}$$

5.21 Charging and discharging currents of a capacitor

Suppose C in fig. 5.18 to represent a capacitor of, say, 30 μF connected in series with a centre-zero micro-
ammeter A across a slider S and one end
of a resistor R. A battery B is connected
across R. If S is moved at a uniform
speed along R, the p.d. applied to C,
indicated by voltmeter V, increases uniformly from 0 to V volts, as shown by
line OD in fig. 5.19.

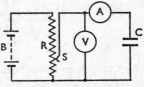

Fig. 5.18 Charging and discharging of a capacitor.

If C is the capacitance in farads and
if the p.d. across C increases uniformly from 0 to V volts in t_1
seconds,

charging current $= i_1 = Q \dfrac{\text{[coulombs or ampere seconds]}}{t_1 \text{ [seconds]}}$

$$= CV/t_1 \text{ amperes,}$$

i.e. charging current in amperes $\Big\} = \Big\{$ rate of change of charge in coulombs/ second

$$= C \text{ [farads]} \times \text{rate of change of p.d. in volts/second.}$$

Since the p.d. across C increases at a uniform rate, the charging
current, i_1, remains constant and is represented by the dotted line
LM in fig. 5.19.

Suppose the p.d. across C to be maintained constant at V volts
during the next t_2 seconds. Since the rate of change of p.d. is now
zero, the current (apart from a slight leakage current) is zero and is
represented by the dotted line NP. If the p.d. across C is then reduced
to zero at a uniform rate by moving slider S backwards, the micro-

ammeter indicates a current i_3 flowing in the reverse direction, represented by the dotted line QT in fig. 5.19. If t_3 is the time in seconds for the p.d. to be reduced from V volts to zero,

then $\qquad\qquad Q = -i_3 t_3$ coulombs

∴ $\qquad\qquad\quad i_3 = -Q/t_3 = -C \times V/t_3$ amperes

i.e. discharge current in amperes $\Big\} = \Big\{$ rate of change of charge in coulombs/ second

$\qquad\qquad\qquad\quad = C$ [farads] × rate of change of p.d. in volts/second.

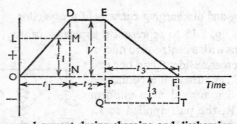

Fig. 5.19 Voltage and current during charging and discharging of a capacitor.

Since $Q = i_1 t_1 = -i_3 t_3$ (assuming negligible leakage current through C),

∴ $\qquad\qquad$ areas of rectangles OLMN and PQTF are equal.

In practice, it is seldom possible to vary the p.d. across a capacitor at a constant rate, so let us consider the general case of the p.d. across a capacitor of C farads being increased by dv volt in dt second. If the corresponding increase of charge is dq coulomb,

$$dq = C \cdot dv.$$

If the charging current at that instant is i amperes,

$$dq = i \cdot dt$$

∴ $\qquad\qquad i \cdot dt = C \cdot dv$

and $\qquad\qquad i = C \cdot dv/dt$

$\qquad\qquad\qquad = C \times$ rate of change of p.d. $\qquad$ (5.23)

If the capacitor is being discharged and if the p.d. falls by dv volt in dt second, the discharge current is given by:

$$i = C \cdot dv/dt \qquad (5.24)$$

Since dv is now negative, the current is also negative.

5.22 Graphical derivation of curve of voltage across a capacitor connected in series with a resistor across a d.c. supply

In section 5.4 we derived the curves of the voltage across a capacitor during charging and discharging from the readings on an electrostatic voltmeter connected across the capacitor. We will now consider how these curves can be derived graphically from the values of the capacitance, the resistance and the applied voltage. At the instant when S is closed on *a* (fig. 5.5), there is no p.d. across C; consequently the whole of the voltage is applied across R and the initial value of the charging current $= I = V/R$.

The growth of the p.d. across C is represented by the curve in fig. 5.20. Suppose v to be the p.d. across C and i to be the charging

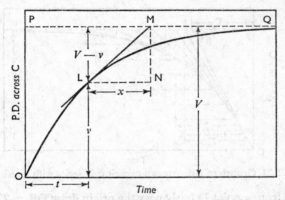

Fig. 5.20 Growth of p.d. across a capacitor in series with a resistor.

current t seconds after S is put over to *a*. The corresponding p.d. across R $= V - v$, where V is the terminal voltage of the battery.

Hence $$iR = V - v$$
and $$i = (V - v)/R \qquad (5.25)$$

If this current remained *constant* until the capacitor was fully charged, and if the time taken was x seconds:

corresponding quantity of electricity $= ix = \dfrac{V - v}{R} \times x$ coulombs.

With a constant charging current, the p.d. across C would have increased uniformly up to V volts, as represented by the tangent LM drawn to the curve at L.

But the charge added to the capacitor also

$$= \text{increase of p.d.} \times C$$
$$= (V - v) \times C$$

Hence $\dfrac{V - v}{R} \times x = C(V - v)$

and

$$x = CR = \text{the time constant, } T, \text{ of the circuit} \qquad (5.26)$$

The construction of the curve representing the growth of the p.d. across a capacitor is therefore similar to that described in section 4.6 for the growth of current in an inductive circuit. Thus, OA in fig. 5.21 represents the battery voltage V, and AB the time constant T. Join

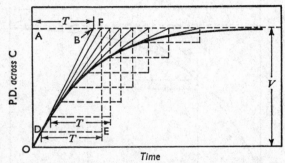

Fig. 5.21 Growth of p.d. across a capacitor in series with a resistor.

OB, and from a point D fairly near the origin draw DE $= T$ seconds and draw EF perpendicularly. Join DF, etc. Draw a curve such that OB, DF, etc., are tangents to it.

From expression (5.25) it is evident that the instantaneous value of the charging current is proportional to $(V - v)$, namely the vertical distance between the curve and the horizontal line PQ in fig. 5.20. Hence the shape of the curve representing the charging current is the inverse of that of the p.d. across the capacitor and is the same for both charging and discharging currents (assuming the resistance to be the same), and its construction is illustrated by the following example.

Example 5.9 *A 20-μF capacitor is charged to a p.d. of 400 V and then discharged through a 100 000-Ω resistor. Derive a curve representing the discharge current.*

From (5.26):

$$\text{time constant} = 100\,000\ [\Omega] \times \frac{20}{1\,000\,000}\ [\text{F}] = 2\,\text{s}.$$

$$\left.\begin{array}{r}\text{Initial value of discharge}\\ \text{current}\end{array}\right\} = \frac{V}{R} = \frac{400}{100\,000} = 0\cdot004\,\text{A} = 4\,\text{mA}.$$

Hence draw OA in fig. 5.22 to represent 4 mA and OB to represent 2 seconds. Join AB. From a point C corresponding to, say, 3·5 mA, draw CD equal to 2 seconds and DE vertically. Join CE. Repeat the

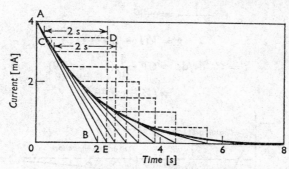

Fig. 5.22 Discharge current, example 5.9.

construction at intervals of, say, 0·5 mA and draw a curve to which AB, CE, etc., are tangents. This curve represents the variation of discharge current with time.

5.23 Mathematical derivation of curves of voltage and current when a capacitor is connected in series with a resistor across a d.c. supply

Suppose the p.d. across capacitor C in fig. 5.5, t seconds after S is switched over to a, to be v volts, and the corresponding charging current to be i amperes, as indicated in fig. 5.23. Also, suppose the p.d. to increase from v to $(v + \mathrm{d}v)$ volts in $\mathrm{d}t$ second; then, from expression (5.23),

$$i = C \cdot \mathrm{d}v/\mathrm{d}t$$

and corresponding p.d. across $\mathrm{R} = Ri = RC \cdot \mathrm{d}v/\mathrm{d}t.$

But
$$V = \text{p.d. across C} + \text{p.d. across R}$$
$$= v + RC \cdot \mathrm{d}v/\mathrm{d}t \tag{5.27}$$

$$\therefore \qquad V - v = RC \cdot \mathrm{d}v/\mathrm{d}t$$

so that
$$\frac{\mathrm{d}t}{RC} = \frac{\mathrm{d}v}{V - v}.$$

Integrating both sides, we have:

$$t/RC = -\ln(V - v) + A$$

where A = the constant of integration.

When $t = 0$, $v = 0$,

$$\therefore \qquad A = \ln V$$

so that

$$\frac{t}{RC} = \ln \frac{V}{V - v}$$

$$\therefore \qquad \frac{V}{V - v} = e^{t/RC}$$

and

$$v = V(1 - e^{-t/RC}) \qquad (5.28)$$

Also,

$$i = C \cdot dv/dt = CV \cdot \frac{d}{dt}(1 - e^{-t/RC})$$

$$= (V/R)e^{-t/RC} \qquad (5.29)$$

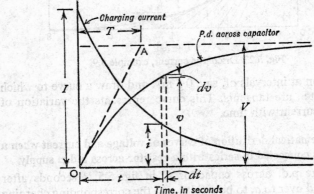

Fig. 5.23 Variation of current and p.d. during charging.

At the instant of switching on, $t = 0$ and $e^{-0} = 1$,

$$\therefore \qquad \text{initial value of current} = V/R = \text{(say) } I.$$

This result is really obvious from the fact that at the instant of switching on there is no charge on C and therefore no p.d. across it. Consequently the whole of the applied voltage must momentarily be absorbed by R.

Substituting for V/R in expression (5.29), we have:

$$\text{instantaneous charging current} = i = Ie^{-t/RC} \qquad (5.30)$$

If the p.d. across the capacitor continued increasing at the initial rate, it would be represented by OA, the tangent drawn to the initial part of the curve. If T be the *time constant* in seconds, namely the time required for the p.d. across C to increase from zero to its final value if it continued increasing at its initial rate, then:

$$\text{initial rate of increase of p.d.} = V/T \text{ volts/second} \quad (5.31)$$

But it follows from (5.27) that at the instant of closing the switch on a, $v = 0$, then:

$$V = RC \cdot dv/dt$$

$\therefore$ initial rate of change of p.d. $= dv/dt = V/RC$ (5.32)

Equating (5.31) and (5.32), we have:

$$V/T = V/RC$$

$\therefore$ $T = RC$ seconds   (5.33)

Hence we can rewrite (5.28) and (5.30) thus:

$$v = V(1 - e^{-t/T}) \quad (5.34)$$

and $i = Ie^{-t/T}$ (5.35)

Comparison of expressions (4.18) and (5.34) shows that the shape of the voltage growth across a capacitor is similar to that of the current growth in an inductive circuit.

5.24 Discharge of a capacitor through a resistor

Having charged capacitor C in fig. 5.5 to a p.d. of V volts, let us now move switch S over to b and thereby discharge the capacitor through R. The pointer of microammeter A is immediately deflected to a maximum value in the negative direction, and then the readings on both the microammeter and the voltmeter ES (fig. 5.5) decrease to zero as indicated in fig. 5.24.

Suppose the p.d. across C to be v volts t seconds after S has been moved to b, and the corresponding current to be i amperes, as in fig. 5.24, then:

$$i = -v/R \quad (5.36)$$

The negative sign indicates that the direction of the discharge current is the reverse of that of the charging current.

Suppose the p.d. across C to change by dv volt in dt second,

$\therefore$ $i = C \cdot dv/dt$ (5.37)

Since dv is now negative, i must also be negative, as already noted. Equating (5.36) and (5.37), we have:

$$-v/R = C \cdot dv/dt$$

so that

$$\frac{dt}{RC} = -\frac{dv}{v}$$

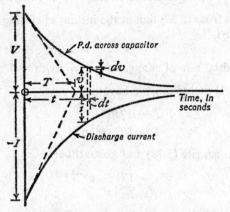

Fig. 5.24 Variation of current and p.d. during discharge.

Integrating both sides, we have:

$$t/RC = -\ln v + A.$$

When $t = 0$, $v = V$, so that $A = \ln V$.

Hence $\qquad t/RC = \ln V/v$

so that $\qquad V/v = e^{t/RC}$

and $\qquad v = Ve^{-t/RC} = Ve^{-t/T}$ $\qquad$ (5.38)

Also, $\qquad i = -v/R = -\dfrac{V}{R}e^{-t/RC} = -Ie^{-t/T}$ $\qquad$ (5.39)

where I = initial value of the discharge current = V/R.

Example 5.10 *An 8-μF capacitor is connected in series with a 0·5-MΩ resistor across a 200-V d.c. supply. Calculate: (a) the time constant; (b) the initial charging current; (c) the time taken for the p.d. across the capacitor to grow to 160 V; and (d) the current and the p.d. across the capacitor 4 s after it is connected to the supply.*

(a) From (5.33), time constant $= 0.5 \times 10^6 \times 8 \times 10^{-6}$
$$= 4 \text{ s.}$$

(b) Initial charging current $= \dfrac{V}{R} = \dfrac{200}{0.5 \times 10^6} \text{ A}$
$$= 400 \text{ } \mu\text{A.}$$

(c) From (5.34), $160 = 200(1 - e^{-t/4})$

$\therefore \qquad\qquad e^{-t/4} = 0.2.$

From mathematical tables, $t/4 = 1.61$

$\therefore \qquad\qquad t = 6.44 \text{ s.}$

Or alternatively, $e^{t/4} = 1/0.2 = 5.$

$\therefore \qquad\quad (t/4) \log e = \log 5.$

But $e = 2.718$, $\therefore$ $t = \dfrac{4 \times 0.699}{0.4343} = 6.44 \text{ s.}$

(d) From (5.34) $v = 200(1 - e^{-4/4}) = 200(1 - 0.368)$
$$= 200 \times 0.632 = 126.4 \text{ V.}$$

It will be seen that the time constant can be defined as the time required for the p.d. across the capacitor to grow from zero to 63·2 per cent of its final value.

From (5.35),

$$\text{corresponding current} = i = 400 \text{ . } e^{-1} = 400 \times 0.368$$
$$= 147 \text{ } \mu\text{A.}$$

5.25 Displacement current in a dielectric

In the preceding sections we have considered the charging current as being the movement of electrons in the conductors connecting the source to the plates of the capacitor; e.g. when switch S is put over to. *a* in fig. 5.10, electrons flow from the positive plate of C_1 via the battery to the negative plate of C_2, and the same number of electrons flow from the upper plate of C_2 to the lower plate of C_1. This current is referred to as *conduction* current.

Let us consider the capacitor of fig. 5.11, in which the metal plates M and N are in an evacuated glass enclosure G. There are no electrons in the space between the plates and therefore there cannot be any movement of electrons in this space when the capacitor is being charged. We know, however, that an electric field is being set up and that energy is being stored in the space between the plates; in other words, the space between the plates of a charged capacitor is in a state of electrostatic strain.

We do not know the exact nature of this strain (any more than we know the nature of the strain in a magnetic field), but Prof. Clerk Maxwell, in 1865, introduced the concept that any *change* in the electric flux in any region is equivalent to an electric current in that region, and he called this electric current a *displacement* current, to distinguish it from the *conduction* current referred to above.

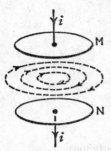

This displacement current produces a magnetic field exactly as if it had been a conduction current. For instance, when a capacitor having circular parallel metal plates M and N (fig. 5.25) is being charged by a current *i* flowing in the direction shown, a magnetic field is created in the space between the plates, as indicated by the concentric dotted lines. The plane of these concentric circles is parallel to the plates.

Fig. 5.25 Magnetic field due to displacement current.

This magnetic field disappears as soon as the displacement current ceases, i.e. as soon as the charge on the capacitor ceases to increase. When the capacitor is discharged, the magnetic field reappears in the reverse direction and again disappears when the discharge ceases. In other words, *the magnetic field is set* * *up only when the electric field is undergoing a change of intensity*. Hence, when a capacitor is being charged or discharged, we can say that the current is *continuous* around the *whole* circuit, being in the form of *conduction current* in the wires and *displacement current* in the dielectric of the capacitor. This means that we can apply Kirchhoff's First Law to plate M of the capacitor of fig. 5.25 by saying that the conduction current entering the plate is equal to the displacement current leaving that plate.

5.26 Energy stored in a charged capacitor

Suppose the p.d. across a capacitor of capacitance C farads to be increased from v to $(v + \mathrm{d}v)$ volts in $\mathrm{d}t$ second. From (5.23), the charging current, i amperes, is given by:

* Conversely, an electric field appears whenever a magnetic field changes in intensity, and it is this reciprocity between electric and magnetic fields that forms the basis of electromagnetic theory dealing with the radiation of electromagnetic waves.

$$i = C \cdot \mathrm{d}v/\mathrm{d}t.$$

Instantaneous value of power to capacitor $= iv$ watts

$$= vC \cdot \mathrm{d}v/\mathrm{d}t \text{ watts,}$$

and $\qquad$ energy supplied to capacitor during $\left.\begin{array}{l} \\ \text{interval } \mathrm{d}t \end{array}\right\} = vC \cdot \dfrac{\mathrm{d}v}{\mathrm{d}t} \cdot \mathrm{d}t$

$$= Cv \cdot \mathrm{d}v \text{ joules.}$$

Hence total energy supplied to capacitor when p.d. is increased from 0 to V volts

$$= \int_0^V Cv \cdot \mathrm{d}v = \tfrac{1}{2}C\Big[v^2\Big]_0^V$$

$$= \tfrac{1}{2}CV^2 \text{ joules} \tag{5.40}$$

For a capacitor with dielectric of thickness d metres and area a square metres,

$$\text{energy per cubic metre} = \frac{1}{2} \cdot \frac{CV^2}{ad} = \frac{1}{2} \cdot \frac{\epsilon a}{d} \cdot \frac{V^2}{ad}$$

$$= \tfrac{1}{2}\epsilon(V/d)^2 = \tfrac{1}{2}\epsilon E^2$$

$$= \tfrac{1}{2}DE = \tfrac{1}{2}D^2/\epsilon \text{ joules} \tag{5.41}$$

These expressions are similar to expressions (4.22) for the energy stored per cubic metre of a magnetic field.

Example 5.11 *A 50-μF capacitor is charged from a 200-V supply. After being disconnected it is immediately connected in parallel with a 30-μF capacitor. Find: (a) the p.d. across the combination, and (b) the electrostatic energies before and after the capacitors are connected in parallel. The 30-μF capacitor is initially uncharged.*

(*a*) From (5.2), charge $= (50 \times 10^{-6})$ [F] $\times 200$ [V]

$$= 0.01 \text{ C.}$$

When the capacitors are connected in parallel, the total capacitance is 80 μF, and the charge of 0·01 coulomb is divided between the two capacitors:

$\therefore \qquad\qquad 0.01$ [C] $= (80 \times 10^{-6})$ [F] $\times$ p.d.

$\therefore \qquad$ p.d. across capacitors $= 125$ V.

(*b*) From (5.40) it follows that when the 50-μF capacitor is charged to a p.d. of 200 V:

electrostatic energy $= \tfrac{1}{2} \times (50 \times 10^{-6})$ [F] $\times (200)^2$ [V²]

$$= 1 \text{ J.}$$

With the capacitors in parallel:

total electrostatic energy $= \frac{1}{2} \times 80 \times 10^{-6} \times (125)^2 = 0\cdot 625$ J.

It is of interest to note that there is a reduction in the energy stored in the capacitors. This loss appears as heat in the resistance of the circuit by the current responsible for equalizing the p.d.s, in the spark that may occur when the capacitors are connected in parallel, and in electromagnetic radiation if the discharge is oscillatory.

5.27 Force of attraction between oppositely-charged plates

Let us consider two parallel plates M and N (fig. 5.26) immersed in a homogeneous fluid, such as air or oil, having an absolute permittivity ϵ. Suppose the area of the dielectric to be a square metres and the distance between M and N to be x metres. If the p.d. between the plates is V volts, then from (5.41),

energy per cubic metre of dielectric $= \frac{1}{2}\epsilon(V/x)^2$ joules.

Suppose plate M to be fixed and N to be movable, and let P be the force of attraction, in newtons, between the plates. Let us next

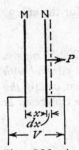

disconnect the charged capacitor from the supply and then pull plate N outwards through a distance dx metre. If the insulation of the capacitor is perfect, the charge on the plates remains constant. This means that the electric flux density and therefore the potential gradient in the dielectric must remain unaltered, the constancy of the potential gradient being due to the p.d. between plates M and N increasing in proportion to the distance between them. It follows from expression (5.41) that the energy per cubic metre of the dielectric remains constant. Consequently, all the energy in the *additional* volume of the dielectric must be derived from the work done when the force P

Fig. 5.26 At-traction be-tween charged parallel plates.

newtons acts through distance dx metre, namely $P \cdot \mathrm{d}x$ joules,

i.e. $\qquad P \cdot \mathrm{d}x = \frac{1}{2}\epsilon(V/x)^2 \cdot a \cdot \mathrm{d}x$

$\therefore \qquad P = \frac{1}{2}\epsilon a(V/x)^2$ newtons $\qquad\qquad (5.42)$

$\qquad\qquad = \frac{1}{2}\epsilon a \times$ (potential gradient in volts/metre)^2.

Example 5.12 *Two parallel metal discs, each 100 mm in diameter, are spaced 1 mm apart, the dielectric being air. Calculate the force, in* *newtons, on each disc when the p.d. between them is 1 kV.*

Area of one side of each plate $= 0\cdot7854 \times (0\cdot1)^2$
$$= 0\cdot007\ 854\ m^2.$$
Potential gradient $= 1000\ [V]/0\cdot001\ [m]$
$$= 10^6\ V/m.$$

From expression (5.42),

force $= \frac{1}{2} \times (8\cdot85 \times 10^{-12})\ [F/m] \times 0\cdot007\ 854\ [m^2] \times (10^6)^2\ [V/m]^2$
$$= 0\cdot0348\ N.$$

5.28 Dielectric strength

If the p.d. between the opposite sides of a sheet of solid insulating material is increased beyond a certain value, the material breaks down. Usually this results in a tiny hole or puncture through the dielectric so that the latter is then useless as an insulator.

The potential gradient necessary to cause breakdown of an insulating medium is termed its *dielectric strength* and is usually expressed in kilovolts/millimetre. The value of the dielectric strength of a given material decreases with increase of thickness, and the following table gives the approximate dielectric strengths of some of the most important materials:

Material	Thickness (mm)	Dielectric strength (kV/mm)
Air (at normal pressure and temperature)	0·2	5·75
	0·6	4·92
	1	4·36
	6	3·27
	10	2·98
Mica	0·01	200
	0·1	115
	1·0	61
Glass (density 2·5)	1	28·5
	5	18·3
Ebonite	1	50
Paraffin-waxed paper	0·1	40–60

Summary of important formulae

Q [coulombs] $= C$ [farads] $\times V$ [volts]   (5.2)

1 microfarad $= 10^{-6}$ farad

1 picofarad $= 10^{-12}$ farad

For capacitors in parallel,

$$C = C_1 + C_2 + \ldots \tag{5.3}$$

For capacitors in series,

$$\frac{1}{C} = \frac{1}{C_1} + \frac{1}{C_2} + \ldots \tag{5.5}$$

For C_1 and C_2 in series,

$$V_1 = V \cdot \frac{C_2}{C_1 + C_2} \tag{5.7}$$

Electric field strength in dielectric $= E = V/d \qquad (5.9)$

Electric flux density $= D = Q/a \qquad (5.10)$

Absolute permittivity $= D/E = \epsilon = \epsilon_0\epsilon_r \qquad (5.15)$

Permittivity of free space $= \epsilon_0 = 8 \cdot 85 \times 10^{-12}$ F/m

$$\frac{1}{\sqrt{(\mu_0\epsilon_0)}} = 2 \cdot 998 \times 10^8 \text{ m/s}$$

$=$ velocity of electromagnetic waves in free space.

Relative permittivity of a material

$$= \frac{\text{capacitance of capacitor with that material as dielectric}}{\text{capacitance of same capacitor with vacuum dielectric}}$$

Capacitance of parallel-plate capacitor with n plates$\Big\} = \dfrac{\epsilon(n-1)a}{d} \quad (5.16)$

For two dielectrics, A and B, of same area, in series,

$$\frac{\text{electric field strength or potential gradient in A}}{\text{electric field strength or potential gradient in B}}$$
$$= \frac{\text{relative permittivity of B}}{\text{relative permittivity of A}} \quad (5.17)$$

Force on isolated charge in electric field$\Big\} = E$ newtons/coulomb $\quad (5.18)$

Deflection of electron moving across electric field

$$= \frac{1}{2} \times \frac{e}{m} \times \frac{V}{d} \times \left(\frac{l}{v}\right)^2 \text{ metres} \tag{5.19}$$

Final velocity of free electron in electric field

$$= \sqrt{(2Ve/m)} \tag{5.21}$$

1 electronvolt $= 1 \cdot 6 \times 10^{-19}$ joule $\qquad (5.22)$

Charging current of capacitor $= \mathrm{d}q/\mathrm{d}t = C \cdot \mathrm{d}v/\mathrm{d}t$ (5.23)

For R and C in series across d.c. supply,

$$v = V(1 - \mathrm{e}^{-t/RC}) \tag{5.28}$$

and $\qquad\qquad i = I\mathrm{e}^{-t/RC}$ (5.30)

$$\text{Time constant} = T = RC \tag{5.33}$$

For C discharged through R,

$$v = V\mathrm{e}^{-t/RC} \tag{5.38}$$

and $\qquad\qquad i = -I\mathrm{e}^{-t/RC}$ (5.39)

$$\text{Energy stored in capacitor} = \tfrac{1}{2}CV^2 \text{ joules} \tag{5.40}$$

$$\left.\begin{array}{l}\text{Energy per cubic metre of}\\ \qquad\qquad\text{dielectric}\end{array}\right\} = \tfrac{1}{2}\epsilon E^2 = \tfrac{1}{2}DE$$

$$= \tfrac{1}{2}D^2/\epsilon \text{ joules} \tag{5.41}$$

$$\left.\begin{array}{l}\text{Electrostatic attraction between}\\ \qquad\qquad\text{parallel plates}\end{array}\right\} = \tfrac{1}{2}\epsilon a(V/x)^2 \text{ newtons} \tag{5.42}$$

EXAMPLES 5

1. A 20-μF capacitor is charged at a constant current of 5 μA for 10 minutes. Calculate the final p.d. across the capacitor and the corresponding charge in coulombs.
2. Three capacitors have capacitances of 10 μF, 15 μF and 20 μF respectively. Calculate the total capacitance when they are connected (*a*) in parallel, (*b*) in series.
3. A 9-μF capacitor is connected in series with two capacitors, 4 μF and 2 μF respectively, which are connected in parallel. Determine the capacitance of the combination.

   If a p.d. of 20 V is maintained across the combination, determine the charge on the 9-μF capacitor and the energy stored in the 4-μF capacitor.   (U.E.I., O.2)
4. Two capacitors, having capacitances of 10 μF and 15 μF respectively, are connected in series across a 200-V d.c. supply. Calculate: (*a*) the charge on each capacitor; (*b*) the p.d. across each capacitor. Also find the capacitance of a single capacitor that would be equivalent to these two capacitors in series.
5. Three capacitors of 2, 3 and 6 μF respectively are connected in series across a 500-V d.c. supply. Calculate: (*a*) the charge on each capacitor; (*b*) the p.d. across each capacitor; and (*c*) the energy stored in the 6-μF capacitor.
6. A certain capacitor has a capacitance of 3 μF. A capacitance of 2·5 μF is required by combining this capacitance with another. Calculate the

capacitance of the second capacitor and state how it must be connected to the first.

7. A capacitor A is connected in series with two capacitors B and C connected in parallel. If the capacitances of A, B and C are 4, 3 and 6 μF respectively, calculate the equivalent capacitance of the combination.

If a p.d. of 20 V is maintained across the whole circuit, calculate the charge on the 3-μF capacitor.

8. Three capacitors, A, B and C, are connected in series across a 200-V d.c. supply. The p.d.s across the capacitors are 40, 70 and 90 V respectively. If the capacitance of A is 8 μF, what are the capacitances of B and C?

9. Two capacitors, A and B, are connected in series across a 200-V d.c. supply. The p.d. across A is 120 V. This p.d. is increased to 140 V when a 3-μF capacitor is connected in parallel with B. Calculate the capacitances of A and B.

10. Show from first principles that the total capacitance of two capacitors having capacitances C_1 and C_2 respectively, connected in parallel, is $C_1 + C_2$.

A circuit consists of two capacitors A and B in parallel connected in series with another capacitor C. The capacitances of A, B and C are 6 μF, 10 μF and 16 μF respectively. When the circuit is connected across a 400-V d.c. supply, calculate: (i) the potential difference across each capacitor; (ii) the charge on each capacitor.   (U.E.I., O.1)

11. On what factors does the capacitance of a parallel-plate capacitor depend?

Derive an expression for the resultant capacitance when two capacitors are connected in series.

Two capacitors, A and B, having capacitances of 20 μF and 30 μF respectively, are connected in series to a 600-V d.c. supply. Determine the p.d. across each capacitor.

If a third capacitor C is connected in parallel with A and it is then found that the p.d. across B is 400 V, calculate the value of C and the energy stored in it.   (N.C.T.E.C., O.2)

12. Derive an expression for the energy stored in a capacitor of C farads when charged to a potential difference of V volts.

A capacitor of 4 μF capacitance is charged to a p.d. of 400 V and then connected in parallel with an uncharged capacitor of 2 μF capacitance. Calculate the p.d. across the parallel capacitors and the energy stored in the capacitors before and after being connected in parallel. Explain the difference.   (E.M.E.U., O.1)

13. Derive expressions for the equivalent capacitance of a number of capacitors: (*a*) in series; (*b*) in parallel.

Two capacitors of 4 μF and 6 μF capacitance respectively are connected in series across a p.d. of 250 V. Calculate the p.d. across each capacitor and the charge on each.

The capacitors are disconnected from the supply p.d. and reconnected in parallel with each other, with terminals of similar polarity being joined together. Calculate the new p.d. and charge for each capacitor.

What would have happened if, in making the parallel connection, the connections of one of the capacitors had been reversed?

(U.L.C.I., O.2)

14. Show that the total capacitance of two capacitors having capacitances C_1 and C_2 connected in series is $C_1C_2/(C_1 + C_2)$.

A 5-μF capacitor is charged to a potential difference of 100 V and then connected in parallel with an uncharged 3-μF capacitor. Calculate the potential difference across the parallel capacitors.

(U.E.I., O.1)

15. Find an expression for the energy stored in a capacitor of capacitance C farads charged to a p.d. of V volts.

A 3-μF capacitor is charged to a p.d. of 200 V and then connected in parallel with an uncharged 2-μF capacitor. Calculate the p.d. across the parallel capacitors and the energy stored in the capacitors before and after being connected in parallel. Account for the difference.

(App. El., L.U.)

16. Explain the terms *electric field strength* and *permittivity*.

Two square metal plates, each of size 200 mm, are completely immersed in insulating oil of relative permittivity 5 and spaced 3 mm apart. A p.d. of 600 V is maintained between the plates. Calculate: (*a*) the capacitance of the capacitor; (*b*) the charge stored on the plates; (*c*) the electric field strength in the dielectric; (*d*) the electric flux density. (U.L.C.I., O.1)

17. A capacitor consists of two metal plates, each having an area of 900 cm², spaced 3 mm apart. The whole of the space between the plates is filled with a dielectric having a relative permittivity of 6. A p.d. of 500 V is maintained between the two plates. Calculate: (*a*) the capacitance; (*b*) the charge; (*c*) the electric field strength; (*d*) the electric flux density. (E.M.E.U., O.1)

18. Describe with the aid of a diagram what happens when a battery is connected across a simple capacitor comprising two metal plates separated by a dielectric.

A capacitor consists of two metal plates, each having an area of 600 cm², separated by a dielectric 4 mm thick which has a relative permittivity of 5. When the capacitor is connected to a 400-V d.c. supply, calculate: (i) the capacitance; (ii) the charge; (iii) the electric field strength, (iv) the electric flux density. (U.E.I., O.1)

19. Define: (*a*) the farad; (*b*) the relative permittivity.

A capacitor consists of two square metal plates of side 200 mm, separated by an air space 2 mm wide. The capacitor is charged to a p.d. of 200 V and a sheet of glass having a relative permittivity of 6 is placed between the metal plates immediately they are disconnected from the supply. Calculate: (*a*) the capacitance with air dielectric; (*b*) the capacitance with glass dielectric; (*c*) the p.d. across the capacitor after the glass plate has been inserted; (*d*) the charge on the capacitor.

(U.E.I., O.1)

20. What factors affect the capacitance that exists between two parallel metal plates insulated from each other?

A capacitor consists of two similar, square, aluminium plates, each 100 mm × 100 mm, mounted parallel and opposite each other. Calculate the capacitance when the distance between the plates is 1 mm and the dielectric is mica of relative permittivity 7·0.

If the plates are connected to a circuit which provides a constant current of 2 μA, how long will it take the potential difference of the plates to change by 100 V, and what will be the increase in the charge?

(S.A.N.C., O.1)

21. What are the factors which determine the capacitance of a parallel-plate capacitor? Mention how a variation in each of these factors will influence the value of capacitance.

Calculate the capacitance in microfarads of a capacitor having 11 parallel plates separated by mica sheets 0·2 mm thick. The area of one side of each plate is 1000 mm² and the relative permittivity of mica is 5.

(W.J.E.C., O.1)

22. A parallel-plate capacitor has a capacitance of 300 pF. It has 9 plates, each 40 mm × 30 mm, separated by mica having a relative permittivity of 5. Calculate the thickness of the mica.

23. A capacitor consists of two parallel metal plates, each of area 2000 cm² and 5 mm apart. The space between the plates is filled with a layer of paper 2 mm thick and a sheet of glass 3 mm thick. The relative permittivities of the paper and glass are 2 and 8 respectively. A potential difference of 5 kV is applied between the plates. Calculate: (a) the capacitance of the capacitor; (b) the potential gradient in each dielectric; (c) the total energy stored in the capacitor.

(N.C.T.E.C., O.2)

24. Obtain from first principles an expression for the capacitance of a single-dielectric, parallel-plate capacitor in terms of the plate area, the distance between plates and the permittivity of the dielectric.

A sheet of mica, 1 mm thick and of relative permittivity 6, is interposed between two parallel brass plates 3 mm apart. The remainder of the space between the plates is occupied by air. Calculate the area of each plate if the capacitance between them is 0·001 μF. Assuming that air can withstand a potential gradient of 3 MV/m, show that a p.d. of 5 kV between the plates will not cause a flashover.

(S.A.N.C., O.2)

25. Explain what is meant by electric field strength in a dielectric and state the factors upon which it depends.

Two parallel metal plates of large area are spaced at a distance of 10 mm from each other in air, and a p.d. of 5000 V is maintained between them. If a sheet of glass, 5 mm thick and having a relative permittivity of 6, is introduced between the plates, what will be the maximum electric field strength and where will it occur?

(App. El., L.U.)

26. Two capacitors of capacitance 0·2 μF and 0·05 μF are charged to voltages of 100 V and 300 V respectively. The capacitors are then

connected in parallel by joining terminals of corresponding polarity together. Calculate:

(a) the charge on each capacitor before being connected in parallel,
(b) the energy stored on each capacitor before being connected in parallel,
(c) the charge on the combined capacitors,
(d) the p.d. between the terminals of the combination,
(e) the energy stored in the combination.

27. A capacitor consists of two metal plates, each 200 mm × 200 mm, spaced 1 mm apart, the dielectric being air. The capacitor is charged to a p.d. of 100 V and then discharged through a ballistic galvanometer having a ballistic constant of 0·0011 microcoulomb per scale division. The amplitude of the first deflection is 32 divisions. Calculate the value of the absolute permittivity of air.

Calculate also the electric field strength and the electric flux density in the air dielectric when the terminal p.d. is 100 V.

28. When the capacitor of Q. 27 is immersed in oil, charged to a p.d. of 30 V and then discharged through the same galvanometer, the first deflection is 27 divisions. Calculate: (a) the relative permittivity of the oil, (b) the electric field strength and the electric flux density in the oil when the terminal p.d. is 30 V and (c) the energy stored in the capacitor.

29. A capacitor is charged to a p.d. of 50 V and then discharged through a ballistic galvanometer having a constant of $1·6 \times 10^{-10}$ coulomb/division. If the first deflection of the galvanometer is 124 divisions, what is the value of the capacitance? If the capacitor consists of two metal plates, each 300 mm × 300 mm, spaced 2 mm apart in air, calculate the absolute permittivity of air.

30. A 0·1-μF capacitor is charged to a p.d. of 5 V. Calculate: (a) the charge in microcoulombs; (b) the number of electrons displaced.

31. When the current in a wire is 600 A, what is the number of electrons per second passing a given point of the wire?

32. If the number of electrons passing per second through an ammeter is 7×10^{16}, what should be the ammeter reading?

33. If there are 3×10^{15} electrons passing per second between two metal surfaces and if the p.d. between the surfaces is 200 V, calculate: (a) the energy absorbed in 20 minutes (i) in joules (ii) in electron-volts, and (b) the final velocity of the electrons assuming that their initial velocity is zero.

34. In a thermionic valve, the current between the filament and the anode is 3 mA and the p.d. is 120 V. Calculate: (a) the number of electrons per second passing from the filament to the anode; (b) the energy absorbed in 5 minutes in (i) electronvolts and (ii) joules; and (c) the final velocity of the electrons, assuming their initial velocity to be zero.

35. An electron is situated between two parallel plates and just outside the

negative plate. The plates are spaced 5 mm apart in an evacuated bulb and a p.d. of 400 V is maintained between them. Calculate the force on the electron.

If the electron, starting from rest, travels to the positive plate, calculate: (*a*) the kinetic energy of the electron immediately before impact with that plate, and (*b*) the time taken.

36. Explain what is meant by the terms, *electric field strength, relative permittivity* and *equipotential surface*.

An electron has a velocity of $1 \cdot 5 \times 10^7$ m/s at right-angles to the uniform electric field between two parallel deflecting plates of a cathode-ray tube. If the plates are 25 mm long and spaced 9 mm apart and the p.d. between the plates is 75 V, calculate how far the electron is deflected sideways during its movement through the electric field. Assume electronic charge to be $1 \cdot 6 \times 10^{-19}$ C and electronic mass to be $9 \cdot 1 \times 10^{-31}$ kg. **(App. El., L.U.)**

37. Define the *electronvolt*.

An electron, travelling at a velocity of 10^7 m/s, enters a transverse electric field at right-angles. If this field is 2 cm wide, determine: (*a*) the field strength necessary to deflect the path of the electron through 30°; (*b*) the resultant velocity of the electron on leaving the transverse field.

What would the angular deflection be if the initial velocity was doubled?

Derive any expression used. **(W.J.E.C., O.2)**

38. Derive an expression for the velocity acquired by a particle of mass *m* and chargé *e*, which starts from rest and passes through a potential difference *V* in an electric field.

Two parallel plates situated 2 cm apart in vacuum have a potential difference of 100 V between them. If the electron leaves the cathode with negligible velocity, determine: (*a*) the velocity of the electron when it strikes the anode; (*b*) the transit time.

If the anode dissipation is 0·16 W, how many electrons per second pass between cathode and anode? **(W.J.E.C., O.2)**

39. Define the units of *electric field strength, electric flux density* and *permittivity*.

An electron of charge $1 \cdot 6 \times 10^{-19}$ C is at rest in a vacuum between two parallel plates spaced 20 mm apart. If the potential difference between these plates is 500 V, calculate the force on the electron.
(N.C.T.E.C., O.2)

40. A 20-μF capacitor is charged and discharged thus:

Steady charging current of 0·02 A from 0 to 0·5 second.
    ,,   ,,   ,,   0·01 A from 0·5 to 1·0 second.
Zero current   from 1·0 to 1·5 seconds.
Steady discharging current of 0·01 A from 1·5 to 2·0 seconds.
    ,,   ,,   ,,   0·005 A from 2·0 to 4·0 seconds.
Draw graphs to scale showing how the current and the capacitor voltage vary with time.

41. Define the *time constant* of a circuit that includes a resistor and capacitor connected in series.

    A 100-μF capacitor is connected in series with an 8000-Ω resistor. Determine the time constant of the circuit. If the combination is connected suddenly to a 100-V d.c. supply, find: (*a*) the initial rate of rise of p.d. across the capacitor; (*b*) the initial charging current; (*c*) the ultimate charge in the capacitor; and (*d*) the ultimate energy stored in the capacitor.   (W.J.E.C., O.2)

42. A 10-μF capacitor connected in series with a 50-kΩ resistor is switched across a 50-V d.c. supply. Derive graphically curves showing how the charging current and the p.d. across the capacitor vary with time.

43. A 2-μF capacitor is joined in series with a 2-MΩ resistor to a d.c. supply of 100 V. Draw a current–time graph and explain what happens in the period after the circuit is made, if the capacitor is initially uncharged.

    Calculate the current flowing and the energy stored in the capacitor at the end of an interval of 4 seconds from the start.

                                                                (N.C.T.E.C., O.2)

44. Derive an expression for the current flowing at any instant after the application of a constant voltage V to a circuit having a capacitance C in series with a resistance R.

    Determine, for the case in which $C = 0.01$ μF, $R = 100\,000$ Ω and $V = 1000$ V, the voltage to which the capacitor has been charged when the charging current has decreased to 90 per cent of its initial value, and the time taken for the current to decrease to 90 per cent of its initial value.   (App. El., L.U.)

45. Derive an expression for the stored electrostatic energy of a charged capacitor.

    ·A 10-μF capacitor in series with a 10-kΩ resistor is connected across a 500-V d.c. supply. The fully charged capacitor is disconnected from the supply and discharged by connecting a 1000-Ω resistor across its terminals. Calculate: (*a*) the initial value of the charging current; (*b*) the initial value of the discharge current; and (*c*) the amount of heat, in joules, dissipated in the 1000-Ω resistor.

                                                                (App. El., L.U.)

46. A 20-μF capacitor is found to have an insulation resistance of 50 MΩ, measured between the terminals. If this capacitor is charged off a d.c. supply of 230 V, find the time required after disconnection from the supply for the p.d. across the capacitor to fall to 60 V. Prove any formula used.   (App. El., L.U.)

47. A circuit consisting of a 6-μF capacitor, an electrostatic voltmeter and a resistor in parallel, is connected across a 140-V d.c. supply. It is then disconnected and the reading on the voltmeter falls to 70 V in 127 s. When the test is performed without the resistor, the time taken for the same fall in voltage is 183 s. Calculate the resistance of the resistor.

48. A constant d.c. voltage of V volts is applied across two plane parallel electrodes. Derive expressions for the electric field strength and flux density in the field between the electrodes and the charge on the

*

electrodes. Hence, or otherwise, derive an expression for the capacitance of a parallel-plate capacitor.

An electronic flash tube requires an energy input of 8·5 joules which is obtained from a capacitor charged from a 2000-V d.c. source. The capacitor is to consist of two parallel plates, 11 cm in width, separated by a dielectric of thickness 0·1 mm and relative permittivity 5·5. Calculate the necessary length of each capacitor plate.

(U.E.I., O.2)

49. An electrostatic device consists of two parallel conducting plates, each of area 1000 cm². When the plates are 1 cm apart in air, the attractive force between them is 0·1 newton. Calculate the potential difference between the plates. Find also the energy stored in the system.

If the device is used in a container filled with a gas of relative permittivity 4, what effect does this have on the force between the plates? (U.L.C.I., O.2)

50. The energy stored in a certain capacitor when connected across a 400-V d.c. supply is 0·3 joule. Calculate: (*a*) the capacitance, and (*b*) the charge on the capacitor.

51. A variable capacitor having a capacitance of 800 pF is charged to a p.d. of 100 V. The plates of the capacitor are then separated until the capacitance is reduced to 200 pF. What is the change of p.d. across the capacitor? Also, what is the energy stored in the capacitor when its capacitance is: (*a*) 800 pF; (*b*) 200 pF? How has the increase of energy been supplied?

52. A 200-pF capacitor is charged to a p.d. of 50 V. The dielectric has a cross-sectional area of 300 cm² and a relative permittivity of 2·5. Calculate the energy stored per cubic metre of the dielectric.

53. A parallel-plate capacitor, with the plates 2 cm apart, is immersed in oil having a relative permittivity of 3. The plates are charged to a p.d. of 25 kV. Calculate the force between the plates in newtons per square metre of plate area and the energy stored in joules per cubic metre of the dielectric.

54. A capacitor consists of two metal plates, each 600 mm × 500 mm, spaced 1 mm apart. The space between the metal plates is occupied by a dielectric having a relative permittivity of 6, and a p.d. of 3 kV is maintained between the plates. Calculate: (*a*) the capacitance in pico-farads; (*b*) the electric field strength and the electric flux density in the dielectric; and (*c*) the force of attraction, in newtons, between the plates.

CHAPTER 6

Direct-Current Machines

6.1 General arrangement of a d.c. machine

Fig. 6.1 shows the general arrangement of a four-pole d.c. generator or motor. The fixed part consists of four iron cores C, referred to as *pole cores*, attached to an iron or steel ring R, called the *yoke*. The pole cores are usually made of steel plates riveted together and bolted to the yoke, which may be of cast steel or fabricated rolled steel. Each pole core has pole tips, partly to support the field

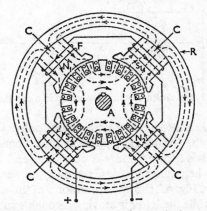

Fig. 6.1 General arrangement of a 4-pole d.c. machine.

winding and partly to increase the cross-sectional area and thus reduce the reluctance of the airgap. Each pole core carries a winding F so connected as to excite the poles alternately N and S.

The armature core A consists of iron laminations, about 0·4–0·6 mm thick, insulated from one another and assembled on the shaft in the case of small machines and on a cast-iron spider in the case of large machines. The purpose of laminating the core is to reduce the eddy-current loss (section 9.1). Slots are stamped on the periphery of the laminations, partly to accommodate and provide mechanical security to the armature winding and partly to give a shorter

185

186 Electrical Technology

airgap for the magnetic flux to cross between the pole face and the armature 'teeth'. In fig. 6.1, each slot has two circular conductors,* insulated from each other.

The dotted lines in fig. 6.1 represent the distribution of the *useful* magnetic flux, namely that flux which passes into the armature core and is therefore cut by the armature conductors when the armature revolves. It will be seen from fig. 6.1 that the magnetic flux which emerges from N_1 divides, half going towards S_1 and half towards S_2. Similarly, the flux emerging from N_2 divides equally between S_1 and S_2.

Suppose the armature to revolve clockwise, as shown by the curved arrow in fig. 6.1. Applying Fleming's Righthand Rule (section 2.10), we find that the e.m.f. generated in the conductors is towards the paper in those moving under the N poles and outwards

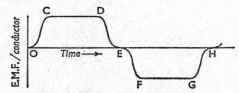

Fig. 6.2 Waveform of e.m.f. generated in a conductor.

from the paper in those moving under the S poles. If the airgap is of uniform length, the e.m.f. generated in a conductor remains constant while it is moving under a pole face, and then decreases rapidly to zero when the conductor is midway between the pole tips of adjacent poles. Fig. 6.2 shows the variation of the e.m.f. generated in a conductor while the latter is moving through two pole pitches, a *pole pitch* being the distance between the centres of adjacent poles. Thus, at instant O, the conductor is midway between the pole tips of, say, S_2 and N_1, and CD represents the e.m.f. generated while the conductor is moving under the pole face of N_1, the e.m.f. being assumed positive when its direction is towards the paper in fig. 6.1. At instant E, the conductor is midway between the pole tips of N_1 and S_1; and portion EFGH represents the variation of the e.m.f.

* The term *conductor*, when applied to armature windings, refers to the active portion of the winding, namely that part which cuts the flux, thereby generating an e.m.f.; for example, if an armature has 40 slots and if each slot contains 8 wires, the armature is said to have 8 conductors per slot and a total of 320 conductors.

while the conductor is moving through the next pole pitch. The variation of e.m.f. during interval OH in fig. 6.2 is repeated indefinitely, so long as the speed is maintained constant.

A d.c. generator, however, has to give a voltage that remains constant in direction and in magnitude, and it is therefore necessary to use a *commutator* to enable a steady or direct voltage to be obtained from the alternating e.m.f. generated in the rotating conductors.

Fig. 6.3 shows a longitudinal or axial section and an end elevation of half of a relatively small commutator. It consists of a large num-

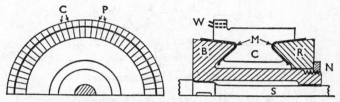

Fig. 6.3 Commutator of a d.c. machine.

ber of wedge-shaped copper segments or bars C, assembled side by side to form a ring, the segments being insulated from one another by thin mica sheets P. The segments are shaped as shown so that they can be clamped securely between a V-ring B, which is part of a cast-iron bush or sleeve, and another V-ring R which is tightened and kept in place by a nut N. The bush is keyed to shaft S.

The copper segments are insulated from the V-rings by collars of micanite M, namely thin mica flakes cemented together with shellac varnish and moulded to the exact shape of the rings. These collars project well beyond the segments so as to reduce surface leakage of current from the commutator to the shaft. At the end adjacent to the winding, each segment has a milled slot to accommodate two armature wires W which are soldered to the segment.

6.2 Ring-wound armature

The action of the commutator is most easily understood if we consider the earliest form of armature winding, namely the ring-wound armature shown in fig. 6.4. In this diagram, C represents a core built of sheet-iron rings or laminations, insulated from one another. The core is wound with eight coils, each consisting of two turns; and the two ends of any one coil are connected to adjacent segments of the commutator. P and Q represent two carbon brushes, namely two blocks of specially treated carbon, bearing on the commutator. Actually these brushes are pressed by springs against the outer surface of the segments, but to avoid confusion in fig. 6.4 they are shown on the inside. The term 'brush' is a relic of the time when current

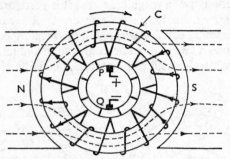

Fig. 6.4 Ring-wound armature.

was collected by a bundle of copper wires arranged somewhat like a brush.

The dotted lines in fig. 6.4 represent the distribution of the magnetic flux; and it will be seen that this magnetic flux is cut only by that part of the winding that lies on the external surface of the core. It is also evident from fig. 6.4 that between brushes Q and P, there are two paths in parallel through the armature winding and that all the conductors are divided equally between these two paths. Furthermore, as the armature rotates, this state of affairs remains unaltered.

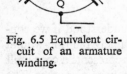

Fig. 6.5 Equivalent circuit of an armature winding.

Suppose the armature to be driven clockwise. Then, by applying the Righthand Rule of section 2.10, we find that the direction of the e.m.f.s generated in conductors under the N pole is towards the paper, and that of the e.m.f.s in the conductors under the S pole is outwards from the paper, as indicated by the arrowheads. The result is that in each of the two *parallel* paths, the same number of conductors is generating e.m.f. acting from Q towards P. The effect is similar to that obtained if equal numbers of cells (in series) were connected in two parallel circuits, as shown in fig. 6.5. In this case it is clear that P is the positive and Q the negative terminal, and that the total e.m.f. is equal to the sum of the e.m.f.s of the cells connected in series between Q and P. Also, the current in each path is a half of the total current flowing outwards at P and returning at Q. Similarly, in the armature winding, the total e.m.f. between brushes P and Q depends upon the e.m.f. per conductor and the number of conductors in series per path between Q and P; and the current in each conductor is half the total current at each brush.

It follows from fig. 6.4 that the conductors generating e.m.f. are those which are moving opposite a pole, and that in each path the number of conductors simultaneously generating e.m.f. remains constant from instant to instant and is unaffected by the rotation of the armature. Hence, the p.d. between P and Q must also remain practically constant from instant to instant—a result made possible by the commutator.

The ring winding of fig. 6.4 has the disadvantages: (*a*) it is very expensive to wind, since each turn has to be taken round the core by hand; (*b*) only a small portion of each turn is effective in cutting magnetic flux. Hence, modern armatures are 'drum' wound, as shown in fig. 6.6.

6.3 Double-layer drum windings

Let us consider a four-pole armature with, say, 11 slots, as in fig. 6.6. In order that all the coils may be similar in shape and therefore

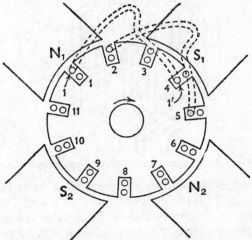

Fig. 6.6 Arrangement of a double-layer winding.

may be wound to the correct shape before being assembled on the core, they have to be made such that if side 1 of a coil occupies the outer half of one slot, the other side 1′ occupies the inner half of another slot. This necessitates a kink in the end-connections in order that the coils may overlap one another as they are being assembled.

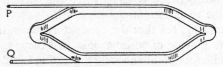

Fig. 6.7 An armature coil.

Fig. 6.7 shows the shape of the end-connections of a single coil consisting of a number of turns, and fig. 6.8 (*b*) shows how three coils, 1–1′, 2–2′ and 3–3′, are arranged in the slots so that their end-connections overlap one another, the end elevation of the end-connections of coils 1–1′ and 3–3′ being as shown in fig. 6.8 (*a*). The end-connection of coil 2–2′ has been omitted from fig. 6.8 (*a*) to

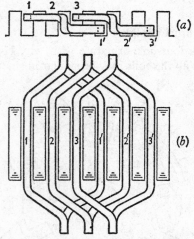

Fig. 6.8 Arrangement of overlap of end-connections.

enable the shape of the other end-connections to be shown more clearly. In fig. 6.7, the two ends of the coil are brought out to P and Q; and as far as the connections to the commutator segments are concerned, the number of turns on each coil is of no consequence.

From fig. 6.6 it is evident that if the e.m.f.s generated in conductors

1 and 1′ are to assist each other, 1′ must be moving under a S pole when 1 is moving under a N pole; thus, by applying the Righthand Rule (section 2.10) to fig. 6.6 and assuming the armature to be rotated clockwise, we find that the direction of the e.m.f. generated in conductor 1 is towards the paper whereas that generated in conductor 1′ is outwards from the paper. Hence, the distance between coil sides 1 and 1′ must be approximately a pole pitch. With 11 slots it is impossible to make the distance between 1 and 1′ exactly a pole pitch, and in fig. 6.6 one side of coil 1–1′ is shown in slot 1 and the other side is in slot 4. The coil is then said to have a *coil span* of 4 − 1, namely 3. In practice, the coil span must be a whole number and is approximately equal to $\dfrac{\text{total number of slots}}{\text{total number of poles}}$.

In the example shown in fig. 6.6, a very small number of slots has for simplicity been chosen. In actual machines the number of slots

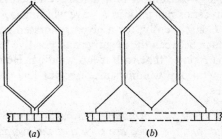

(a)   (b)

Fig. 6.9 (a) Coil of a lap winding; (b) coil of a wave winding.

per pole usually lies between 10 and 15 and the coil span is slightly less than the value given by the above expression.

Let us now return to the consideration of the 11-slot armature. The 11 coils are assembled in the slots with a coil span of 3, and we are now faced with the problem of connecting to the commutator segments the 22 ends that are projecting from the winding.

Apart from a few special windings, armature windings can be divided into two groups, depending upon the manner in which the wires are joined to the commutator, namely:

    (a) lap windings,

    (b) wave windings.

In lap windings the two ends of any one coil are taken to adjacent segments as in fig. 6.9 (a), where a coil of two turns is shown; whereas in wave windings the two ends of each coil are bent in opposite directions and taken to segments some distance apart, as in fig. 6.9 (*b*).*

A lap winding has as many paths in parallel between the negative and positive brushes as there are of poles; for instance, with an 8-pole lap winding, the armature conductors form eight parallel paths between the negative and positive brushes. A wave winding, on the other hand, has only two paths in parallel, irrespective of the number of poles. Hence, if a machine has *p* pairs of poles,

no. of parallel paths with a lap winding $= 2p$

and   „ „   „   „   „ „ wave winding $= 2$.

For a given cross-sectional area of armature conductor and a given current density in the conductor, it follows that the total current from a lap winding is *p* times that from a wave winding. On the other hand, for a given number of armature conductors, the number of conductors in series per path in a wave winding is *p* times that in a lap winding. Consequently, for a given generated e.m.f. per conductor, the voltage between the negative and positive brushes with a wave winding is *p* times that with a lap winding. Hence it may be said that, in general, lap windings are used for low-voltage, heavy-current machines.

Example 6.1 *An eight-pole armature is wound with* 480 *conductors. The magnetic flux and the speed are such that the average e.m.f. generated in each conductor is* 2·2 *V, and each conductor is capable of carrying a full-load current of* 100 *A. Calculate the terminal voltage on no load, the output current on full load and the total power generated on full load when the armature is* (a) *lap-connected and* (b) *wave-connected.*

(*a*) With the armature lap-connected,

no. of parallel paths in the⎫ $=$ number of poles
armature winding⎭ $= 8$

∴   no. of conductors/path $= 480/8 = 60$.

Terminal voltage on no load $=$ (e.m.f./conductor) $\times$
(no. of conductors/path)
$= 2·2 \times 60 = 132$ V.

---

* For fuller treatment of lap and wave windings, see *Principles of Electricity*, 2nd edition, by A. Morley and E. Hughes.

Output current on full load = (full-load current per conductor)
$$\times \text{ (no. of parallel paths)}$$
$$= 100 \times 8 = 800 \text{ A.}$$

Total power generated on full load $\Big\}$ = output current × generated e.m.f.

$$= 800 \times 132 = 105\ 600 \text{ W}$$
$$= 105 \cdot 6 \text{ kW.}$$

(*b*) With the armature wave-connected,

no. of parallel paths = 2

$\therefore$ no. of conductors/path = 480/2 = 240.

Terminal voltage on no load = $2 \cdot 2 \times 240 = 528$ V.

Output current on full load = $100 \times 2 = 200$ A.

Total power generated on full load $\Big\}$ = $200 \times 528 = 105\ 600$ W

$$= 105 \cdot 6 \text{ kW.}$$

It will be seen from the above example that the total power generated by a given machine is the same whether the armature winding is lap- or wave-connected.

6.4 Calculation of e.m.f. generated in an armature winding

When an armature is rotated through one revolution, each conductor cuts the magnetic flux emanating from all the N poles and also that entering all the S poles. Consequently,

if Φ = useful flux per pole, in webers, entering or leaving the armature

p = number of *pairs* of poles

and N = speed in revolutions/minute,

time of 1 revolution = $60/N$ seconds

and time taken by a conductor to move one pole pitch

$$= \frac{60}{N} \cdot \frac{1}{2p} \text{ seconds}$$

$\therefore$ average rate at which conductor cuts the flux

$$= \Phi \div \left(\frac{60}{N} \cdot \frac{1}{2p}\right) = \frac{2\Phi Np}{60} \text{ webers/second}$$

and average e.m.f. generated in each conductor

$$= \frac{2\Phi Np}{60} \text{ volts.}$$

If Z = total number of armature conductors,
$\quad c$ = number of parallel paths through winding between positive and negative brushes
$\quad$ = 2 for a wave winding
and $\quad$ = $2p$ for a lap winding,
$\therefore \quad Z/c$ = number of conductors in series in each path.

The brushes are assumed to be in contact with segments connected to conductors in which no e.m.f. is being generated, and the e.m.f. generated in each conductor, while it is moving between positions of zero e.m.f., varies as shown by curve OCDE in fig. 6.2. The number of conductors in series in each of the parallel paths between the brushes remains practically constant; hence

total e.m.f. between brushes = average e.m.f. per conductor $\times$
no. of conductors in series per path

$$= \frac{2\Phi Np}{60} \times \frac{Z}{c}$$

i.e. $$E = 2\frac{Z}{c} \times \frac{Np}{60} \times \Phi \text{ volts.} \tag{6.1}$$

Example 6.2 *A four-pole wave-connected armature has 51 slots with 12 conductors per slot and is driven at 900 rev/min. If the useful flux per pole is 25 mWb, calculate the value of the generated e.m.f.*

Total number of conductors = Z = 51 $\times$ 12 = 612; c = 2; p = 2; N = 900 rev/min; Φ = 0·025 weber.
Using expression (6.1), we have:

$$E = 2 \times \frac{612}{2} \times \frac{900 \times 2}{60} \times 0\cdot025$$

$$= 459 \text{ volts.}$$

Example 6.3 *An eight-pole lap-connected armature, driven at 350 rev/min, is required to generate 260 V. The useful flux per pole is about 0·05 Wb. If the armature has 120 slots, calculate a suitable number of conductors per slot.*

For an eight-pole lap winding, $c = 8$.

Hence,
$$260 = 2 \times \frac{Z}{8} \times \frac{350 \times 4}{60} \times 0\cdot05$$

$\therefore$ $\qquad\qquad\qquad Z = 890$ (approximately)

and number of conductors per slot $= 890/120 = 7\cdot4$ (approx.).

This value must be an even number; hence 8 conductors per slot would be suitable.

Since this arrangement involves a total of $8 \times 120 = 960$ conductors, and since a flux of $0\cdot05$ weber per pole with 890 conductors gave 260 V, then with 960 conductors, the same e.m.f. is generated with a flux of $0\cdot05 \times (890/960) = 0\cdot0464$ Wb/pole.

6.5 Armature reaction

By armature reaction is meant the effect of armature ampere-turns upon the value and the distribution of the magnetic flux entering and leaving the armature core.

Let us, for simplicity, consider a two-pole machine having an

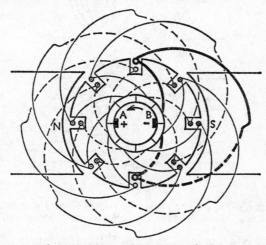

Fig. 6.10 A two-pole armature winding.

armature with eight slots and two conductors per slot, as shown in fig. 6.10. The curved lines between the conductors and the commutator segments represent the front end-connections of the armature winding and those on the outside of the armature represent the back end-connections. The armature winding—like all modern d.c.

windings—is of the double-layer type, the end-connections of the outer layer being represented by full lines and those of the inner layer by dotted lines.

Brushes A and B are placed so that they are making contact with conductors which are moving midway between the poles and have

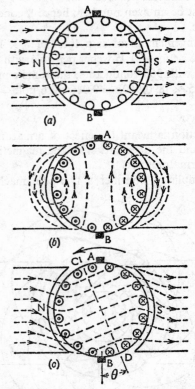

Fig. 6.11 Flux distribution due to (*a*) field current alone, (*b*) armature current alone, (*c*) field and armature currents of a d.c. generator.

therefore no e.m.f. induced in them. If the armature is driven anticlockwise, the direction of the e.m.f.s generated in the various conductors is represented by the dots and crosses. By following the directions of these e.m.f.s it is found that brush A is positive and B negative. In diagrams where the end-connections are omitted, it is usual to show the brushes midway between the poles, as in fig. 6.11.

In general, an armature has ten to fifteen slots per pole, so that the conductors are more uniformly distributed around the armature core than is suggested by fig. 6.10; and for simplicity we may omit the slots and consider the conductors uniformly distributed as in fig. 6.11 (*a*). The latter shows the distribution of flux when there is no armature current, the flux in the gap being practically radial and uniformly distributed.

Fig. 6.11 (*b*) shows the distribution of the flux set up by current flowing through the armature winding in the direction that it will actually flow when the machine is loaded as a *generator*. It will be seen that at the centre of the armature core and in the pole shoes the direction of this flux is at right-angles to that due to the field winding; hence the reason why the flux due to the armature current is termed *cross flux*.

Fig. 6.11 (*c*) shows the resultant distribution of the flux due to the combination of the fluxes in fig. 6.11 (*a*) and (*b*); thus over the leading* halves of the pole-faces the cross flux is in opposition to the main flux, thereby reducing the flux density, whereas over the trailing halves the two fluxes are in the same direction, so that the flux density is strengthened. Apart from the effect of magnetic saturation, the increase of flux over one half of the pole face is the same as the decrease over the other half, and the total flux per pole remains practically unaltered. Hence, in a generator, the effect of armature reaction is to twist or distort the flux in the direction of rotation.

One important consequence of this distortion of the flux is that the magnetic neutral axis is shifted through an angle θ from AB to CD; in other words, with the machine on no load and the flux distribution of fig. 6.11 (*a*), conductors are moving parallel to the magnetic flux and therefore generating no e.m.f. when they are passing axis AB. When the machine is loaded as a generator and the flux distorted as in fig. 6.11 (*c*), conductors are moving parallel to the flux and generating no e.m.f. when they are passing axis CD.

An alternative and in some respects a better method of representing the effect of armature current is to draw a developed diagram of the armature conductors and poles, as in fig. 6.12 (*a*). The direction of the current in the conductors is indicated by the dots and crosses.

In an actual armature, the two conductors forming one turn are situated approximately a pole pitch apart, as in fig. 6.10; but *as far*

---

* The pole tip which is first met during revolution by a point on the armature or stator surface is known as the *leading pole tip* and the other as the *trailing pole tip*.

as the magnetic effect of the currents in the armature conductors is concerned, the end connections could be arranged as shown by the dotted lines in fig. 6.12 (*a*). From the latter, it will be seen that the conductors situated between the vertical axes CC_1 and DD_1 act as if they formed concentric coils producing a magnetomotive force having its maximum value along axis AA_1. Similarly, the currents in the conductors to the left of axis CC_1 and to the right of DD_1 produce a magnetomotive force that is a maximum along axis BB_1.

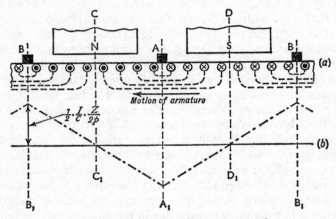

Fig. 6.12 Distribution of armature m.m.f.

Since the conductors are assumed to be distributed uniformly around the armature periphery, the distribution of the m.m.f. is represented by the chain-dotted line* in fig. 6.12 (*b*). These lines pass through zero at points C_1 and D_1 midway between the brushes.

If $I = total$ armature current, in amperes,

        $Z =$ number of armature conductors,

        $c =$ number of parallel paths

and $p =$ number of pairs of poles,

$$\text{current/conductor} = I/c$$

and $\text{conductors/pole} = Z/(2p)$

$\therefore$ $\text{ampere-conductors/pole} = \dfrac{I}{c} \cdot \dfrac{Z}{2p}.$

---

* The m.m.f. is assumed to be positive when the direction of the flux produced by it is from the airgap into the armature core (fig. 6.13).

Since two armature conductors constitute one turn,

$$\text{ampere-turns/pole} = \frac{1}{2} \cdot \frac{I}{c} \cdot \frac{Z}{2p} \qquad (6.2)$$

This expression represents the armature m.m.f. at each brush axis.

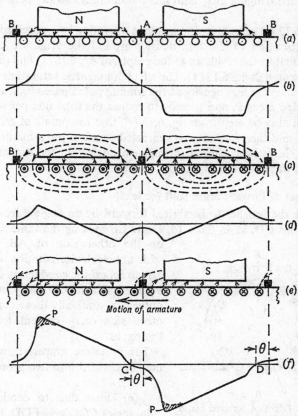

Fig. 6.13 Distribution of main flux, cross flux and resultant flux.

The effect of the armature ampere-turns upon the distribution of the magnetic flux is represented in fig. 6.13. The dotted lines in fig. 6.13 (*a*) represent the distribution of the magnetic flux in the airgap on *no load*. The corresponding variation of the flux density over the periphery of the armature is represented by the ordinates of fig. 6.13 (*b*). Fig. 6.13 (*c*) and (*d*) represent the cross flux due to the armature ampere-turns alone, the armature current being assumed in the direction in which it flows when the machine is loaded as a generator. It will be seen that the flux density in the gap increases from zero at the centre of the pole face to a maximum at the pole tips and then decreases rapidly owing to the increasing length of the path of the fringing flux, until it is a minimum midway between the poles.

Fig. 6.13 (*e*) represents the machine operating as a generator, and the distribution of the flux density around the armature core is approximately the resultant of the graphs of fig. 6.13 (*b*) and (*d*), and is represented in fig. 6.13 (*f*). The effect of magnetic saturation would be to reduce the flux density at the trailing pole tips, as indicated by the shaded areas P, and thereby to reduce the total flux per pole.

It will also be seen from fig. 6.13 (*f*) that the points of zero flux density, and therefore of zero generated e.m.f. in the armature conductors, have been shifted through an angle θ in the direction of rotation to points C and D.

6.6 Effect of forward brush shift (or lead)

Suppose the brushes to be shifted forward by an angle θ from axis AB to axis CD, as in fig. 6.14. Draw EF making the same angle θ

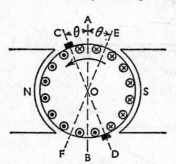

Fig. 6.14 Effect of forward brush shift.

on the other side of AB. Then with anticlockwise rotation of the armature, *all* the conductors on the lefthand side of CD carry current outwards and all those on the other side carry current towards the paper.

The armature ampere-turns can now be divided into two groups:

(*a*) Those due to conductors in angles COE and FOD, shown separately in fig. 6.15 (*a*). From the latter, it is obvious that these conductors are carrying current in such a direction as to try to set up a flux in opposition to that produced by the field winding, and their effect is to reduce the flux through the armature. Hence the ampere-turns due to these conductors are referred to as *demagnetizing* or *back* ampere-turns.

(*b*) Those due to conductors in angles COF and EOD and shown separately in fig. 6.15 (*b*). The ampere-turns due to the current in these conductors are responsible for the distortion of the flux and are therefore termed the *distorting* or *cross* ampere-turns.

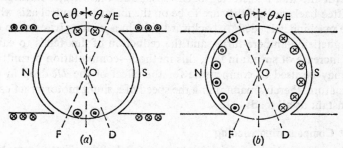

Fig. 6.15 Demagnetizing and distorting armature ampere-turns.

6.7 Calculation of demagnetizing and distorting ampere-turns

Let θ = brush shift in electrical degrees (i.e. the span corresponding to 2 pole-pitches is regarded as 360 electrical degrees)

= $p \times$ brush shift in angular degrees.

From expression (6.2),

$$\text{ampere-turns/pole} = \frac{1}{2} \cdot \frac{I}{c} \cdot \frac{Z}{2p}$$

From fig. 6.15 (*a*) it is seen that for every 360 electrical degrees there are 4θ electrical degrees containing conductors that are responsible for the demagnetizing ampere-turns.

$$\therefore \quad \begin{array}{c} \text{demagnetizing ampere-} \\ \text{turns/pole} \end{array} \Bigg\} = \begin{pmatrix} \text{total ampere-} \\ \text{turns/pole} \end{pmatrix} \times \frac{4\theta}{360}$$

$$= \frac{1}{2} \cdot \frac{I}{c} \cdot \frac{Z}{2p} \cdot \frac{4\theta}{360}. \qquad (6.3)$$

Since all the armature ampere-turns which do not produce demagnetization are responsible for distortion of the flux,

$\therefore$ distorting ampere-turns/pole

$$= \begin{pmatrix} \text{total ampere-} \\ \text{turns/pole} \end{pmatrix} - \begin{pmatrix} \text{demagnetizing ampere-} \\ \text{turns/pole} \end{pmatrix}$$

$$= \frac{1}{2} \cdot \frac{I}{c} \cdot \frac{Z}{2p} \Big(1 - \frac{4\theta}{360} \Big) \qquad (6.4)$$

6.8 Armature reaction in a d.c. motor

The direction of the armature current in a d.c. motor is *opposite* to that of the generated e.m.f., whereas in a generator the current is in the *same* direction as the generated e.m.f. It follows that in a d.c. motor the flux is distorted backwards; and the brushes have to be shifted backwards if they are to be on the magnetic neutral axis when the machine is loaded. A backward shift in a motor gives rise to de-magnetizing ampere-turns, and the reduction of flux tends to cause an increase of speed; in fact, this method—commutation permitting—may be used to compensate for the effect of the IR drop in the armature, thereby maintaining the speed of a shunt motor practically constant at all loads.

6.9 Compensating winding

It has already been explained in section 6.5 that the effect of armature reaction is to alter the distribution of the flux density under the pole

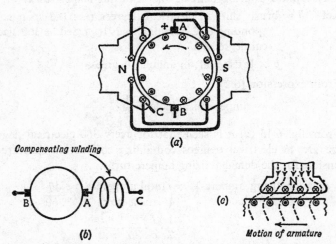

Fig. 6.16 Arrangement and connection of a compensating winding.

face, as shown in fig. 6.13. If the load fluctuates rapidly over a wide range, the sudden shifting backwards and forwards of the flux may induce a sufficiently high e.m.f.* in an armature coil to produce an arc across the top of the mica sheet separating the commutator

* This e.m.f. is a maximum in coils that are approximately midway between the brushes.

segments to which the coil is connected. Such an arc may then extend to adjacent segments until it ultimately results in a flashover between the brushes. Hence, machines subject to sudden fluctuation of load may have to be fitted with a *compensating winding*. The latter is placed in slots in the pole shoes and connected in series with the armature winding in such a way that the ampere-turns of the compensating winding are equal and opposite to those due to the armature conductors that are opposite the pole face, as shown in fig. 6.16 (*a*) and (*b*).

The combined effect of currents in the armature and compensating windings of a d.c. generator is to skew the flux in the airgap as indicated in fig. 6.16 (*c*), but to leave the density of the flux unaltered. The pull exerted on the armature teeth by the skewed flux has a tangential component tending to oppose the rotation of the core. In the case of a motor, the flux is skewed in the opposite direction, thereby exerting on the teeth a pull having a tangential component responsible for the rotation of the armature.

From expression (6.2),

$$\text{armature ampere-turns/pole} = \frac{1}{2} \cdot \frac{I}{c} \cdot \frac{Z}{2p}.$$

Hence, for a uniform flux density under the pole face,

$$\left.\begin{array}{r}\text{ampere-turns/pole for com-}\\ \text{pensating winding}\end{array}\right\} = \frac{1}{2} \cdot \frac{I}{c} \cdot \frac{Z}{2p} \cdot \frac{\text{pole arc}}{\text{pole pitch}} \qquad (6.5)$$

$$\simeq 0.7 \times \text{armature ampere-turns/pole}$$

Owing to its high cost, a compensating winding is fitted only to relatively large machines subject to sudden fluctuations of load, e.g. rolling-mill motors.

6.10 Commutation

The e.m.f. generated in a conductor of a d.c. armature is an alternating e.m.f. and the current in a conductor is in one direction when the conductor is moving under a N-pole and in the reverse direction when it is moving under a S-pole. This reversal of current in a coil has to take place while the two commutator segments to which the coil is connected are being short-circuited by a brush, and the process is termed *commutation*. The duration of this short circuit is usually about $\frac{1}{500}$ second. The reversal of, say, 100 A in an inductive circuit in such a short time is likely to present difficulty and may cause considerable sparking at the brushes.

For simplicity in considering the variation of current in the short-circuited coil we can represent the coils and the commutator segments as in fig. 6.17, where the two ends of any one coil are connected. to adjacent segments, as in a lap winding.

If the current per conductor is I and if the armature is moving

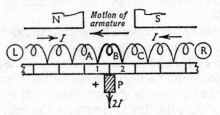

Fig. 6.17 Portion of armature winding.

from right to left, then—assuming the brush to be positive—coil C is carrying current from right to left (R to L), whereas coil A is carrying current from L to R. We shall therefore examine the variation of current in coil B which is connected to segments 1 and 2.

The current in coil B remains at its full value from R to L until segment 2 begins to make contact with brush P, as in fig. 6.18 (*a*). As the area of contact with segment 2 increases, current i_1 flowing to the brush via segment 2 increases and current $(I - i_1)$ through coil

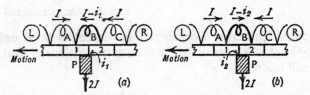

Fig. 6.18 Coil B near the beginning and the end of commutation.

B decreases. If the current distribution between segments 1 and 2 were determined by the areas of contact only, the current through coil B would decrease linearly, as shown by line M in fig. 6.19. It follows that when the brush is making equal areas of contact with segments 1 and 2, current through B would be zero; and further movement of the armature would cause the current through B to grow in the reverse direction.

Fig. 6.18 (*b*) represents the position of the coils near the end of the period of short circuit. The current from segment 1 to P is then i_2 and that flowing from left to right through B is $(I - i_2)$. The short circuit is ended when segment 1 breaks contact with P, and the current through coil B should by that instant have attained its full value from L to R. Under these conditions there should be no spark-

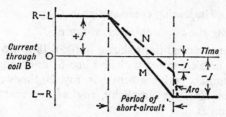

Fig. 6.19 Variation of current in the short-circuited coil.

ing at the brush, and this linear variation of the current in the short-circuited coil is referred to as *straight line* or *linear commutation*.

It was explained in section 6.5 that the armature current gives rise to a magnetic field; thus, fig. 6.11 (*b*) shows the flux set up by the armature current alone. From the direction of this cross-flux and assuming anticlockwise rotation, we can deduce that an e.m.f. is generated towards the paper in a conductor moving in the vicinity of brush A, namely in the direction in which current was flowing in the conductor before the latter was short-circuited by the brush. The same conclusion may be derived from a consideration of the resultant distri-bution of flux given in fig. 6.13 (*e*), where a conductor moving in the region of brush A is generating an e.m.f. in the same direction as that generated when the conductor was moving under the preceding main pole. This generated e.m.f.—often referred to as the *react-ance voltage*—is responsible for delaying the reversal of the current in the short-circuited coils as shown by curve N in fig. 6.19. The result is that when segment 1 is due to break contact with the brush, as in fig. 6.20, the current through coil B has grown to some value i (fig. 6.19); and the remainder, namely $(I - i)$, has to pass between segment 1 and the brush in the form of an arc. This arc is

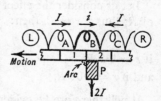

Fig. 6.20 Arcing when segment leaves brush.

rapidly drawn out and the current through B grows quickly from i to I, as shown in fig. 6.19.

It is this reactance voltage that is mainly responsible for sparking at the brushes of d.c. machines, and most methods of reducing sparking are directed towards the reduction or neutralization of the reactance voltage.

6.11 Methods of improving commutation

(a) *By increasing the brush-contact resistance*

Let us consider the conditions shown in fig. 6.18 (*b*), where segment 1 is about to break contact with the brush; and suppose that the resistance of contact with segment 1 is, say, 20 times that with segment 2. Let us also, for simplicity, neglect any reactance voltage generated in coil B.

If r = resistance of contact between segment 2 and brush,

then $20r$ = resistance of contact between segment 1 and brush.

Let R = resistance of coil B.
From fig. 6.18 (*b*) it will be seen that:

p.d. between 1 and P = p.d. across B + p.d. between 2 and P

i.e. $i_2 \times 20r = (I - i_2)R + (2I - i_2)r$

$$\therefore \qquad\qquad i_2 = I \cdot \frac{R + 2r}{R + 21r}.$$

Let us consider the effect of varying the value of r relative to that of R; thus if $r = R$, then:

$$i_2 = 0 \cdot 136I.$$

If $r = 0 \cdot 1\ R$, $i_2 = 0 \cdot 39I$,

and if $r = 0 \cdot 01\ R$, $i_2 = 0 \cdot 841I$.

It will therefore be evident that if the brush contact resistance is made very low, for example by using copper gauze brushes, a large current is still flowing from the leading segment to the brush even when the area of contact has decreased to a very small value. This means that the current density becomes very high and an arc is easily formed as the segment leaves the brush. By the use of carbon brushes the contact resistance is considerably increased and commutation greatly improved. Various grades of carbon brushes possessing different contact resistances are manufactured; and the most suitable grade for a particular machine is very largely a matter of experiment.

(b) *By shifting the brushes forward in a generator and backward in a motor*

If the brushes of a generator were moved forward to the magnetic neutral zones C and D in fig. 6.13 (*f*), the short-circuited coils would not be generating any e.m.f. and the reversal of the current in a short-circuited coil would then be determined by the relative resistances of that coil and of the areas of contact between the brush and the segments, as already explained for method (*a*). In general, however, it is

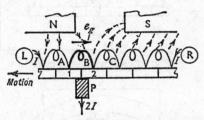

Fig. 6.21 Brush ahead of magnetic neutral axis.

desirable to move the brushes a little further forward than the magnetic neutral zone in order that the short-circuited coil may be cutting the fringing flux of the main pole immediately in front, as shown in fig. 6.21. The e.m.f., e_g, generated in B is then of assistance in hastening or forcing the reversal of current in that coil.

The disadvantage of this method is that, for best commutation, the brushes have to be shifted for every variation of load. Also, the larger the forward shift of the brushes of a generator or the backward shift for a motor, the greater are the demagnetizing ampere-turns.

(c) *By using commutating poles*

Instead of moving the brushes of a generator forward beyond the magnetic neutral zone so as to generate in the short-circuited coil an e.m.f. which assists the reversal of the current, we can obtain the same effect by inserting auxiliary poles between the main poles. These auxiliary poles are termed *commutating poles*, *compoles* or *interpoles*, and their polarity must be the same as that of the main pole immediately ahead, as in fig. 6.22. The brushes are fixed on the geometric neutral axis, and the exciting winding of the commutating poles is connected in series with the armature winding in order that the flux set up in these poles may be proportional to the current to be commutated.

The e.m.f., e_g, generated in that part of the short-circuited coil that

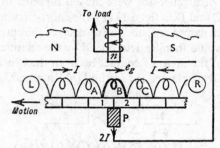

Fig. 6.22 Commutating pole.

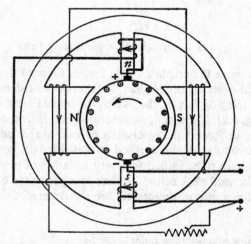

Fig. 6.23 Connections of a shunt generator with compoles.

is opposite the face of the compole must be sufficient to neutralize the e.m.f. generated by the cross-flux in the remainder of the coil and to provide the surplus e.m.f. necessary to hasten or force the reversal of the current. Since the value of e_g should be proportional to the armature current and since the flux in the compoles is proportional to the armature current, it follows that if the value of e_g is satisfactory on full load, it is automatically satisfactory at all loads between no load and full load.

Fig. 6.23 shows the connections of the compole and shunt windings for a two-pole generator driven anticlockwise. It will be seen that the compole and armature ampere-turns are in opposition to one another; consequently the number of ampere-turns on the compoles must be sufficient to neutralize the armature ampere-turns and to provide the extra ampere-turns required to send the necessary flux through the magnetic circuit of the compoles. In practice, the number of compole ampere-turns per pole is about 1·2–1·3 times the number of armature ampere-turns per pole.

Summary of important formulae

$$E = \frac{2Z}{c} \cdot \frac{Np}{60} \cdot \Phi \text{ volts.} \tag{6.1}$$

where $c = 2$ for wave winding

   $= 2p$ for lap winding

and $p = $ no. of *pairs* of poles.

$$\text{Armature ampere-turns/pole} = \frac{1}{2} \cdot \frac{I}{c} \cdot \frac{Z}{2p} \tag{6.2}$$

$$\text{Demagnetizing ampere-turns/pole} = \frac{1}{2} \cdot \frac{I}{c} \cdot \frac{Z}{2p} \cdot \frac{4\theta}{360} \tag{6.3}$$

where $\theta = $ brush lead in *electrical* degrees.

$$\text{Distorting ampere-turns/pole} = \frac{1}{2} \cdot \frac{I}{c} \cdot \frac{Z}{2p}\left(1 - \frac{4\theta}{360}\right) \tag{6.4}$$

$$\left.\begin{matrix}\text{Ampere-turns/pole for compensat-}\\ \text{ing winding}\end{matrix}\right\} = \frac{1}{2} \cdot \frac{I}{c} \cdot \frac{Z}{2p} \cdot \frac{\text{arc}}{\text{pitch}} \tag{6.5}$$

EXAMPLES 6

1. A six-pole armature is wound with 498 conductors. The flux and the speed are such that the average e.m.f. generated in each conductor is 2 V. The current in each conductor is 120 A. Find the total current and the generated e.m.f. of the armature if the winding is connected (*a*) wave, (*b*) lap. Also find the total power generated in each case.
2. A four-pole armature is wound with 564 conductors and driven at 800 rev/min, the flux per pole being 20 mWb. The current in each conductor is 60 A. Calculate the total current, the e.m.f. and the electrical power generated in the armature if the conductors are connected (*a*) wave, (*b*) lap.

3. An eight-pole lap-connected armature has 96 slots with 6 conductors per slot and is driven at 500 rev/min. The useful flux per pole is 0·09 Wb. Calculate the generated e.m.f.

4. A four-pole armature has 624 lap-connected conductors and is driven at 1200 rev/min. Calculate the useful flux per pole required to generate an e.m.f. of 250 V.

5. A six-pole armature has 410 wave-connected conductors. The useful flux per pole is 0·025 Wb. Find the speed at which the armature must be driven if the generated e.m.f. is to be 485 V.

6. The wave-connected armature of a four-pole d.c. generator is required to generate an e.m.f. of 520 V when driven at 660 rev/min. Calculate the flux per pole required if the armature has 144 slots with 2 coil slides per slot, each coil consisting of 3 turns.

7. The armature of a four-pole d.c. generator has 47 slots, each containing 6 conductors. The armature winding is wave-connected, and the flux per pole is 25 mWb. At what speed must the machine be driven to generate an e.m.f. of 250 V?

8. Develop from first principles an expression for the e.m.f. of a d.c. generator.

   Calculate the e.m.f. developed in the armature of a two-pole d.c. generator, whose armature has 280 conductors and is revolving at 1000 rev/min. The flux per pole is 0·03 Wb. (N.C.T.E.C., O.1)

9. Draw a neat labelled diagram of the cross-section of a four-pole d.c. shunt-connected generator. What are the essential functions of the field coils, armature, commutator and brushes?

   The e.m.f. generated by a four-pole d.c. generator is 400 V when the armature is driven at 1000 rev/min. Calculate the flux per pole if the wave-wound armature has 39 slots with 16 conductors per slot.

   (U.E.I., O.1)

10. Explain the function of the commutator in a d.c. machine.

    A six-pole d.c. generator has a lap-connected armature with 480 conductors. The resistance of the armature circuit is 0·02 Ω. With an output current of 500 A from the armature, the terminal voltage is 230 V when the machine is driven at 900 rev/min. Calculate the useful flux per pole and derive the expression employed. (App. El., L.U.)

11. A 300-kW, 500-V, eight-pole d.c. generator has 768 armature conductors, lap-connected. Calculate the number of demagnetizing and cross ampere-turns/pole when the brushes are given a lead of 5 electrical degrees from the geometric neutral. Neglect the effect of the shunt current.

12. Write a short essay describing the effects of (*a*) armature reaction and (*b*) poor commutation on the performance of a d.c. machine. Indicate in your answer how the effects of armature reaction may be reduced and how commutation may be improved. (E.M.E.U., O.2)

13. (*a*) Explain how armature reaction occurs in a d.c. generator and the effect it has on the flux distribution of the machine.

    (*b*) Explain the difficulties of commutation in a d.c. generator. What methods are used to overcome these difficulties? (W.J.E.C., O.2)

14. A four-pole motor has a wave-connected armature with 888 conductors. The brushes are displaced backwards through 5 angular degrees from the geometrical neutral. If the total armature current is 90 A, calculate: (*a*) the cross and the back ampere-turns/pole, and (*b*) the additional field current to neutralize this demagnetization, if the field winding has 1200 turns/pole.

15. Calculate the number of turns/pole required for the commutating poles of the d.c. generator referred to in Q. 11, assuming the compole ampere-turns/pole to be about 1·3 times the armature ampere-turns/pole and the brushes to be in the geometric neutral.

16. An eight-pole generator has a lap-connected armature with 640 conductors. The ratio of pole arc/pole pitch is 0·7. Calculate the ampere-turns/pole of a compensating winding to give uniform airgap density when the total armature current is 900 A.

17. Define the temperature coefficient of resistance.

A four-pole machine has a wave-wound armature with 576 conductors. Each conductor has a cross-sectional area of 5 mm² and a mean length of 800 mm. Assuming the resistivity for copper to be 0·0173 μΩ·m at 20° C and the temperature coefficient of resistance to be 0·004/°C at 0° C, calculate the resistance of the armature winding at its working temperature of 50° C. (U.E.I., O.1)

CHAPTER 7

Direct-Current Generators

7.1 Armature and field connections

The general arrangement of the brush and field connections of a four-pole machine is shown in fig. 7.1. The four brushes B make contact with the commutator. The positive brushes are connected to the positive terminal A and the negative brushes to the negative terminal A_1. From fig. 6.10 it will be seen that the brushes are situated

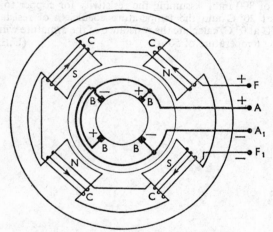

Fig. 7.1 Armature and field connections.

approximately in line with the centres of the poles. This position enables them to make contact with conductors in which little or no e.m.f. is being generated since these conductors are then moving between the poles.

The four exciting or field coils C are usually joined in series and the ends are brought out to terminals F and F_1. These coils must be so connected as to produce N and S poles alternately. The arrow-heads in fig. 7.1 indicate the direction of the field current when F is positive.

In general, we may divide the methods used for connecting the field and armature windings into the following groups:

(a) *Separately-excited generators*—the field winding being connected to a source of supply other than the armature of its own machine.

(b) *Self-excited generators*, which may be subdivided into:

(i) *Shunt-wound generators*—the field winding being connected across the armature terminals.

(ii) *Series-wound generators*—the field winding being connected in series with the armature winding.

(iii) *Compound-wound generators*—a combination of shunt and series windings.

Before we discuss the above systems in greater detail, let us consider the relationship between the magnetic flux and the exciting ampere-turns of a generator on no load. From fig. 6.1 it will be seen that the ampere-turns of one field coil have to maintain the flux through one airgap, a pole core, part of the yoke, one set of armature teeth and part of the armature core. The number of ampere-turns required for the *airgap* is directly proportional to the flux and is represented by the straight line OA in fig. 7.2. For low values of the flux, the number of ampere-turns required to send the flux through the *iron* portion of the magnetic circuit is very small; but when the flux exceeds a certain value, some parts—especially the teeth—begin to get saturated and the number of ampere-turns increases far more rapidly than the flux, as shown by curve B. Hence, if DE represents the number of ampere-turns/pole required to maintain flux OD across the airgap and if DF represents the number of ampere-turns/pole to send this flux through the iron portion of the magnetic circuit, then:

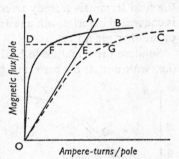

Fig. 7.2 Magnetization curve of a machine.

total ampere-turns/pole to produce flux OD = DE + DF = DG.

By repeating this procedure for various values of the flux, we can derive the *magnetization curve* C representing the relationship between the useful magnetic flux per pole and the total ampere-turns/pole.

7.2 Separately-excited generator

Fig. 7.3 shows a simple method of representing the armature and field windings, A and A_1 being the armature terminals and F and F_1

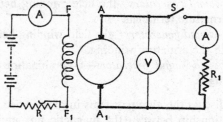

Fig. 7.3 Separately-excited generator.

the field terminals already referred to in fig. 7.1. The field winding is connected in series with a resistor R and an ammeter A to a battery or another generator.

Suppose the armature to be driven at a *constant* speed on no load (i.e. with switch S open) and the exciting current to be increased from zero up to the maximum permissible value and then reduced to zero, the generated e.m.f. being noted for various values of the magnetizing current. It is found that the relationship between the two quantities is represented by curves P and Q in fig. 7.4, P being for increasing values of the excitation and Q for decreasing values. The difference between the curves is due to hysteresis (section 3.11), and OR represents the e.m.f. generated by the residual magnetism in the poles. If the test is repeated, the e.m.f. curve is found to follow the dotted line and then merge into curve P. These curves are termed the *internal* or *open-circuit characteristics* of the machine.

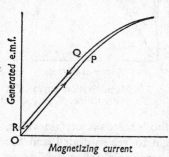

Fig. 7.4 Open-circuit characteristic.

From expression (6.1), it is evident that for a given speed, the e.m.f. generated in a given machine is proportional to the flux per

$e \propto \phi/\text{pole}$.

$I \propto NI/\text{pole}$

pole; and for a given field winding, the number of ampere-turns/pole is proportional to the magnetizing current. Hence the shape of the open-circuit characteristic P in fig. 7.4 is the same as that of the magnetization curve C in fig. 7.2; and curves P and Q in fig. 7.4 indicate how the flux varies with increasing and decreasing values of the magnetizing current.

Let us next consider the effect of load upon the terminal voltage. This can be determined experimentally by closing switch S (fig. 7.3), varying the resistance of resistor R_1 and noting the terminal voltage for each load current. Curve M in fig. 7.5 is typical of the relationship for such a machine. The decrease in the terminal voltage with increase of load is mainly due to the resistance drop in the armature circuit; thus, if the load current is 100 A and the resistance of the armature circuit is $0\cdot08\Omega$, the voltage drop in the armature circuit is $100 \times 0\cdot08 = 8$ V. Consequently, if the generated e.m.f. be 235 V, the terminal p.d. $= 235 - 8 = 227$ V. In general, if $E =$ generated e.m.f., $I_a =$ armature current, $R_a =$ resistance of armature circuit and $V =$ terminal p.d., then:

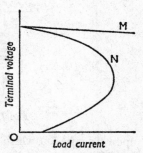

Fig. 7.5 Load characteristics of separately-excited and shunt generators.

$$V = E - I_a R_a \quad \cdot \qquad (7.1)$$

Another factor that may cause a decrease in the terminal voltage is the decrease of flux and therefore of the generated e.m.f. due to (i) the demagnetizing ampere-turns of the armature if the brushes are given a forward shift (section 6.6) and (ii) magnetic saturation in the armature teeth due to distortion of the flux (section 6.5).

The curve representing the variation of terminal voltage with load current is termed the *load* or *external characteristic* of the generator.

The separately-excited generator has the disadvantage of requiring a separate source of direct current and is therefore employed only in special cases, for instance when a wide range of terminal voltage is required (see section 8.8).

7.3 Shunt-wound generator

The field winding is connected in series with a field regulating resistor R across the armature terminals as shown in fig. 7.6, and is therefore in parallel or 'shunt' with the load. The power absorbed by the shunt circuit is limited to about 2–3 per cent of the rated output of the generator by winding the field coils with a large number of turns of comparatively thin wire.

A shunt generator will excite only if the poles have some residual magnetism and the resistance of the shunt circuit is less than some critical value, the actual value depending upon the machine and

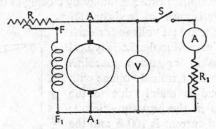

Fig. 7.6 Shunt-wound generator.

upon the speed at which the armature is driven. Suppose curve P in fig. 7.7 to represent the open-circuit characteristic of a shunt generator with *increasing* excitation; then, for a shunt current OA, the e.m.f. is AB and

$$\text{corresponding resistance of shunt circuit} = \frac{\text{terminal voltage}}{\text{shunt current}}$$

$$= AB/OA = \tan BOA$$

$$= \text{slope of OB.}$$

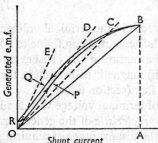

Fig. 7.7 Variation of e.m.f. with shunt circuit resistance.

When the machine is running at its rated speed with the shunt circuit disconnected, the residual magnetism generates an e.m.f. OR. Immediately the shunt circuit is closed, this e.m.f., OR, circulates a current OC (fig. 7.8) through the shunt winding; but the magnetic flux produced by this current generates an e.m.f. CD, which in turn circulates a larger shunt current, etc., until the e.m.f. reaches the value AB. During this continuous process of building up the excitation, the e.m.f. generated by a given shunt current is greater than that required to send that current through the resistance of the shunt circuit. For instance, when the shunt current is OF (fig. 7.8), the generated e.m.f. is FH, but the voltage required to send this current through the resistance of the shunt circuit is FG. The difference between FH and FG, namely GH, is the voltage required to neutralize the e.m.f. induced in the field winding* due to the growth of the field current; thus if R_f and L are the resistance and inductance respectively of the shunt circuit, then for a field current I_f amperes increasing at the rate of di/dt amperes/second,

$$FH = FG + GH$$

i.e.   generated e.m.f. $= E = I_f R_f + L \cdot di/dt$

∴ $di/dt = (E - I_f R_f)/L$

               $= HG/L$ in fig. 7.8

Fig. 7.7 shows the effect of varying the resistance of the shunt circuit; thus, when the shunt circuit is broken, the resistance is infinite and the e.m.f. OR is due to the residual magnetism. When the shunt circuit is closed and its resistance reduced to tan EOA, the e.m.f. is only slightly larger than OR; but when it is reduced to tan DOA, the value of the e.m.f. is extremely sensitive to variation of the resistance—a very slight reduction of the shunt resistance is accompanied by a relatively large increase of e.m.f. Since tan DOA is the value of the shunt circuit resistance

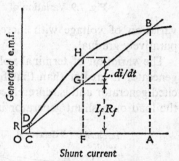

Fig. 7.8 Growth of the excitation of a shunt generator.

at which the excitation increases rapidly, this value is known as the *critical resistance of the shunt circuit*.

By drawing lines such as OE, OD, OC and OB corresponding to various values of the shunt circuit resistance, we can derive curves P and Q of fig. 7.9. Portion *ab* of curve P shows that the e.m.f. increases very slowly with decrease of resistance so long as the latter is greater than the critical value; but when the resistance is about the critical value, the e.m.f. is very unstable—a slight decrease of

* The resistance and inductance of the armature winding are being neglected, since they are usually very small compared with the corresponding values of the shunt circuit.

resistance or a slight increase of speed is accompanied by a large increase of e.m.f.

It will be seen from curve Q in fig. 7.9 that hysteresis causes the

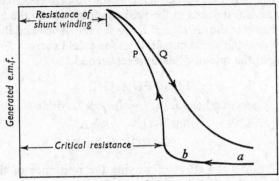

Fig. 7.9 Variation of e.m.f. with shunt resistance.

variation of voltage with increasing shunt resistance to be comparatively gradual.

The variation of terminal voltage with load current for a shunt generator is greater than that for the corresponding separately-excited generator and is represented by curve N in fig. 7.5. Thus, when the load on a shunt generator is increased, the decrease in the ter-

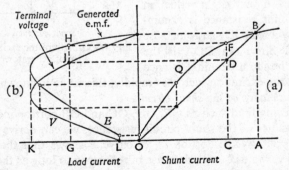

Fig. 7.10 Derivation of the load characteristic of a shunt generator.

minal voltage, due partly to the increased $I_a R_a$ drop in the armature winding and partly to armature reaction, is accompanied by a decrease in the shunt current. Consequently, the flux and the generated e.m.f. are reduced, thereby causing a further reduction in the terminal voltage.

The load characteristic of a shunt generator can be derived from the open-circuit characteristic by the construction shown in fig. 7.10, on the assumption that the effect of armature reaction (section 6.5) is negligible. Curve Q represents the variation of generated e.m.f. with *decreasing* excitation, and AB represents the terminal voltage of the generator on no load when the resistance of the shunt circuit is adjusted to tan BOA. For this shunt resistance, the relationship between the terminal voltage and the shunt current is represented by the straight line OB. Hence, when the machine is loaded so that the terminal voltage is CD, the shunt current is OC and the generated e.m.f. is CF; and from expression (7.1) we have

$$. \text{CD} = \text{CF} - I_a R_a$$

$$\therefore \qquad I_a = \frac{\text{CF} - \text{CD}}{R_a} = \frac{\text{DF}}{R_a} \qquad (7.2)$$

where I_a = armature current

and R_a = resistance of armature circuit.

Corresponding load current = armature current — shunt current

$$= \text{DF}/R_a - \text{OC}$$

$$= \text{OG in fig. 7.10 (}b\text{).}$$

It follows from expression (7.2) that the armature current is proportional to the vertical intercept between curve Q and straight line OB. Hence we can derive curves E and V in fig. 7.10 (b) representing the relationships between the load current and the generated e.m.f. and terminal voltage respectively; thus, on the vertical line at G corresponding to load current OG, points H and J are the horizontal projections of F and D respectively. It will be seen that as the resistance of the load is reduced, the load current increases to a maximum value OK and then decreases to OL when the terminals are short-circuited, OL being the current due to the e.m.f. generated by the residual magnetism.

The shunt machine is the type of d.c. generator most frequently employed, but the load current must be limited to a value that is well below the maximum value, thereby avoiding excessive variation of the terminal voltage.

7.4 Series-wound generator

Since the whole of the armature current passes through the field winding (fig. 7.11), the latter consists of comparatively few turns of thick wire or strip.

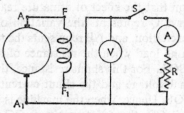

Fig. 7.11 Series-wound generator

When the machine is on open circuit, i.e. when S is open, the terminal voltage OP (fig. 7.12) is very small, being due to the residual flux in the poles. If S is closed with the load resistance R comparatively large, the machine does not excite; but as R is reduced, a value is reached when a slight reduction of R is accompanied by a relatively large increase of terminal voltage. The full line in fig. 7.12

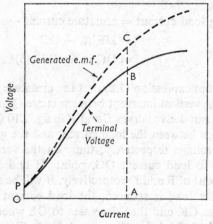

Fig. 7.12 Characteristics of a series generator.

shows the variation of terminal voltage with load current for a certain series generator driven at a constant speed; and the dotted line represents the generated e.m.f. of the same machine when driven at the same speed with the armature on open circuit and the field separately excited. Thus, for a current OA, AC is the generated e.m.f. and AB is the corresponding terminal voltage on load. The difference, BC, is mainly due to the *IR* drop in the field and armature windings.

It is obvious from fig. 7.12 that a series-wound generator is quite unsuitable when the voltage is to be maintained constant or even approximately constant over a wide range of load current.

7.5 Compound-wound generator

In fig. 7.13, C represents the shunt coils and S the series coils, these windings being usually connected* so that their ampere-turns assist

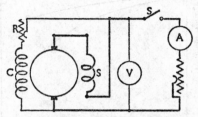

Fig. 7.13 Compound-wound generator.

one another. Consequently the larger the load, the greater are the flux and the generated e.m.f.

If curve S in fig. 7.14 represents the load characteristic with shunt winding alone, then by the addition of a small series winding the fall of terminal voltage with increase of load is reduced as indicated by curve P. Such a machine is said to be *under-compounded*. By increasing the number of series turns we can arrange for the machine to maintain its terminal voltage (curve Q) practically constant between no load and full load, in which case the machine is said to be *level-compounded*. If the number of series turns be increased still further, the terminal voltage increases with increase of load—as represented by curve T. The machine is then said to be *over-compounded*.

An over-compounded generator may be used to compensate for the voltage drop in a feeder system (section 1.18). For instance, if the

---

* In practice, it is of little consequence whether the shunt winding is connected 'long-shunt' as in fig. 7.13 or 'short-shunt', i.e. directly across the armature terminals, since the shunt current is very small compared with the full-load current and the number of series turns is very small compared with the number of shunt turns.

voltage at the far end of a feeder is to be maintained constant at, say, 460 V and if the total resistance of the feeder is 0·2 Ω, the voltage drop due to a current of 100 A is 20 V. Consequently the voltage at the transmitting end must be (460 + 20) = 480 V. When there is no load on the feeder, the voltage at the transmitting end is the same as that at the far end, namely 460 V. Hence a generator, over-compounded to give 460 V on no load and 480 V with 100 A, maintains

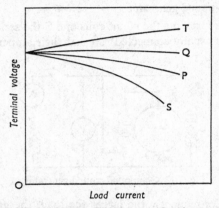

Fig. 7.14 Load characteristics of shunt-wound and compound-wound generators.

the voltage at the far end of the feeder approximately constant over this range of load.

Example 7.1 *A shunt generator is to be converted into a level-compounded generator by the addition of a series field winding. From a test on the machine with shunt excitation only, it is found that the shunt current is 4·1 A to give 440 V on no load and 5·8 A to give the same voltage when the machine is supplying its full load of 200 A. The shunt winding has 1200 turns per pole. Find the number of series turns required per pole.*

Ampere-turns/pole required on no load

$$= 4·1 \times 1200 = 4920 \text{ At.}$$

Ampere-turns/pole required on full load

$$= 5·8 \times 1200 = 6960 \text{ At.}$$

Hence ampere-turns/pole to be provided by the series winding

$$= 6960 - 4920 = 2040 \text{ At,}$$

∴ number of series turns per pole,

$$= 2040/200 = 10.$$

The shunt current is so small compared with the full-load current that it is of little consequence whether it is included in or omitted from the above calculation.

EXAMPLES 7

1. Sketch neat graphs to show the internal or open-circuit characteristics of a separately-excited d.c. generator. Why is a field regulator necessary with such a machine?

   A four-pole d.c. generator gives 410 V on open circuit when driven at 900 rev/min. Calculate the flux per pole if the wave-connected armature winding has 39 slots with 16 conductors per slot.

   (U.E.I.)

2. A six-pole d.c. generator having a lap-connected armature winding is required to give a terminal voltage of 240 V when supplying an armature current of 400 A. The armature has 84 slots and is driven at 700 rev/min. The resistance of the armature circuit is 0·03 Ω and the useful flux per pole is about 0·03 Wb. Calculate the number of conductors per slot and the actual value of the useful flux per pole.

3. Explain why the terminal voltage of a d.c. shunt-excited generator falls as the current supplied by the machine is increased.

   The resistance of the field circuit of a shunt-excited d.c. generator is 200 Ω. When the output of the generator is 100 kW, the terminal voltage is 500 V and the generated e.m.f. 525 V. Calculate: (*a*) the armature resistance, and (*b*) the value of the generated e.m.f. when the output is 60 kW, if the terminal voltage is then 520 V.

   (E.M.E.U.)

4. A short-shunt compound generator has armature, shunt-field and series-field resistances of 0·8 Ω, 45 Ω and 0·6 Ω respectively and supplies a load of 5 kW at 250 V. Calculate the e.m.f. generated in the armature.

5. The following table gives the open-circuit voltages for different field currents of a shunt generator driven at a constant speed:

Terminal voltage (V)	120	240	334	400	444	470
Field current (A)	0·5	1·0	1·5	2·0	2·5	3·0

Plot a graph showing the variation of generated e.m.f. with exciting current and from this graph derive the value of the generated e.m.f. when the shunt circuit has a resistance of (*a*) 160 Ω, (*b*) 210 Ω and (*c*) 300 Ω. Also find the value of the critical resistance of the shunt circuit.

6. The following table gives open-circuit voltages for different values of field current for a d.c. generator:

Open-circuit voltage (V)	120	240	334	400	444	470
Field current (A)	0·5	1·0	1·5	2·0	2·5	3·0

Express these values graphically and determine therefrom the generated e.m.f. when the shunt field circuit has a resistance of 200 Ω. Find also the critical resistance of the shunt field circuit. The effects of armature resistance and brush volt drop may be neglected.

(E.M.E.U., O.1)

7. The following table relates to the open-circuit curve of a shunt generator running at 750 rev/min:

Generated voltage (V)	10	172	300	360	385	395
Field current (A)	0	1	2	3	4	5

Determine the no-load terminal voltage if the field circuit resistance is 125 Ω.

If the speed is now halved, what is the resultant terminal voltage? At the reduced speed, what value of field circuit resistance will give a no-load terminal voltage of 175 V? (N.C.T.E.C., O.2)

8. What conditions must be fulfilled for the self-excitation of a d.c. shunt generator?

The open-circuit characteristic of a shunt-excited d.c. machine, run at 1200 rev/min, is:

Terminal voltage (V)	47	85	103	114	122	127	135	141
Field current (A)	0·2	0·4	0·6	0·8	1·0	1·2	1·6	2·0

The field coil resistance is 55 Ω. Determine: (i) the value of the field-regulating resistance to enable the machine to generate 120 V on open circuit when run at 1200 rev/min; (ii) the value of the open-circuit voltage when the regulator is set to 20 Ω and the speed is reduced to 800 rev/min. (U.E.I., O.2)

9. A shunt generator on test is separately excited and driven at 1000 rev/min. The open-circuit test figures are as follows:

Field current (A)	0	0·157	0·477	0·663	0·805	1·0	1·29
Induced e.m.f. (V)	12·5	100	300	400	450	500	550

The machine is then shunt-connected and run at 1000 rev/min with a constant field-circuit resistance of 540 Ω. The armature-circuit resistance is 1 Ω. Deduce and plot the load characteristic of this machine, neglecting the effect of armature reaction. What is the approximate value of the maximum load current? (E.M.E.U.)

10. The open-circuit characteristic of a shunt generator when separately excited and running at 1000 rev/min is given by:

Open-circuit voltage	56	112	150	180	200	216	230
Field amperes	0·5	1·0	1·5	2·0	2·5	3·0	3·5

If the generator is shunt-connected and runs at 1100 rev/min with a total field resistance of 80 Ω, determine:

(*a*) the no-load e.m.f.;
(*b*) the output current when the terminal voltage is 200 V if the armature resistance is 0·1 Ω.
(*c*) the terminal voltage of the generator when giving the maximum output current.

Neglect the effect of armature reaction and of brush contact drop.
(N.C.T.E.C.)

11. The open-circuit characteristic of a certain d.c. generator, when running at 500 rev/min, is given below:

Field current (A)	2	4	6	8	10
Generated e.m.f. (V)	71	135	179	202	214

The generator is now shunt-connected and run at 600 rev/min, the resistance of the field circuit being 25 Ω. What e.m.f. will it generate?

If the generator is placed on load and the terminal p.d. becomes 230 V, what is the load current? The armature resistance is 0·2 Ω and armature reaction is to be neglected. (App. El., L.U.)

12. Calculate the number of series turns per pole required on a compound generator to enable it to maintain the voltage constant at 460 V between no load and full load of 100 kW. Without any series winding, it is found that the shunt current has to be 2 A on no load and 2·65 A on full load to maintain the voltage constant at 460 V. Number of turns per pole on the shunt winding = 2000.

If the series coils were wound with 8 turns per pole and had a total resistance of 0·01 Ω, what value of diverter resistance would be required to give level compounding?

13. A 250-V, 50-kW shunt generator has 1000 turns on each pole of its field winding. On no load a current of 3·5 A in the field winding produces a terminal voltage of 250 V, but on full load the shunt current has to be increased to 5 A for the same terminal voltage at the same speed. Calculate the number of series field turns per pole required for level compounding.

CHAPTER 8

Direct-Current Motors

8.1 A d.c. machine as generator or motor

There is no difference of construction between a d.c. generator and a d.c. motor. In fact, the only difference is that in a generator the generated e.m.f. is greater than the terminal voltage, whereas in a motor the generated e.m.f. is less than the terminal voltage. For instance, suppose a shunt generator D (fig. 8.1) to be driven by an engine and

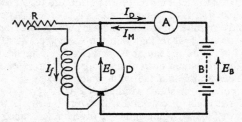

Fig. 8.1 Shunt-wound machine as generator or motor.

connected through a centre-zero ammeter A to a battery B. If the field regulator R is adjusted until the reading on A is zero, the e.m.f., E_D, generated in D is then exactly equal to the e.m.f., E_B, of the battery. If R is now reduced, the e.m.f. generated in D exceeds that of B, and the excess e.m.f. is available to circulate a current I_D through the resistance of the armature circuit, the battery and the connecting conductors. Since I_D is in the same direction as E_D, machine D is a generator of electrical energy.

Next, suppose the supply of steam or oil to the engine driving D to be cut off. The speed of the set falls, and as E_D decreases, I_D becomes less, until, when $E_D = E_B$, there is no circulating current. But E_D continues to decrease and becomes less than E_B, so that a current I_M flows in the reverse direction. Hence B is now supplying electrical energy to drive D as an electric motor.

The speed of D continues to fall until the difference between E_D and

E_B is sufficient to circulate the current necessary to maintain the rotation of D.

It will be noticed that the direction of the field current I_t is the same whether D is running as a generator or a motor.

The relationship between the current, the e.m.f., etc., for machine D may be expressed thus:

If E = e.m.f. generated in armature,

V = terminal voltage,

R_a = resistance of armature circuit,

and I_a = armature current,

then, when D is operating as a generator, it follows from expression (7.1) that:

$$E = V + I_a R_a. \tag{8.1}$$

When the machine is operating as a motor, the e.m.f., E, is less than the applied voltage V, and the direction of the current I_a is the reverse of that when the machine is acting as a generator; hence

$$E = V - I_a R_a$$

or $$V = E + I_a R_a. \tag{8.2}$$

Since the e.m.f. generated in the armature of a motor is in opposition to the applied voltage, it is sometimes referred to as a *back e.m.f.*

Example 8.1 *The armature of a d.c. machine has a resistance of 0·1 Ω and is connected to a 230-V supply. Calculate the generated e.m.f. when it is running* (a) *as a generator giving 80 A and* (b) *as a motor taking 60 A.*

(a) Voltage drop due to armature resistance = 80 × 0·1 = 8 V.
From (8.1),   generated e.m.f.= 230 + 8 = 238 V.

(b) Voltage drop due to armature resistance = 60 × 0·1 = 6 V.
From (8.2),   generated e.m.f. = 230 − 6 = 224 V.

8.2 Speed of a motor

In section 6.4 it was shown that the relationship between the generated e.m.f., speed, flux, etc., is represented by:

$$E = 2\frac{Z}{c} \cdot \frac{Np}{60} \cdot \Phi \tag{6.1}$$

For a given machine, Z, c and p are fixed; and in such a case we can write:

$$E = kN\Phi$$

where

$$k = 2\frac{Z}{c} \cdot \frac{p}{60}.$$

Substituting for E in expression (8.2) we have:

$$V = kN\Phi + I_aR_a$$

$$\therefore \qquad N = \frac{V - I_aR_a}{k\Phi}. \qquad (8.3)$$

The value of I_aR_a is usually less than 5 per cent of the terminal voltage V, so that:

$$N \simeq \frac{V}{k\Phi}. \qquad (8.4)$$

In words, this expression means that the speed of an electric motor is approximately proportional to the voltage applied to the armature and inversely proportional to the flux; and all methods of controlling the speed involve the use of either or both of these relationships.

Example 8.2 *A four-pole motor is fed at 440 V and takes an armature current of 50 A. The resistance of the armature circuit is 0·28 Ω. The armature winding is wave-connected with 888 conductors and the useful flux per pole is 0·023 Wb. Calculate the speed.*

From expression (8.2) we have:

$$440 = \text{generated e.m.f.} + 50 \times 0\cdot28$$

$$\therefore \qquad \text{generated e.m.f.} = 440 - 14 = 426 \text{ V}.$$

Substituting in the e.m.f. equation (6.1), we have:

$$426 = 2 \times \frac{888}{2} \times \frac{N \times 2}{60} \times 0\cdot023$$

$$\therefore \qquad N = 626 \text{ rev/min}.$$

Example 8.3 *A motor runs at 900 rev/min off a 460-V supply. Calculate the approximate speed when the machine is connected across a 200-V supply. Assume the new flux to be 0·7 of the original flux.*

If Φ be the original flux, then from expression (8.4):

$$900 = \frac{460}{k\Phi}$$

$$\therefore \qquad k\Phi = 0\cdot511$$

and new speed $= \dfrac{\text{new voltage}}{k \times \text{original flux} \times 0.7}$ (approximately)

$$= \frac{200}{0.511 \times 0.7} = 559 \text{ rev/min.}$$

8.3 Torque of an electric motor

If we start with equation (8.2) and multiply each term by I_a, namely the total armature current, we have:

$$VI_a = EI_a + I_a^2 R_a.$$

But VI_a represents the total electrical power supplied to the armature, and $I_a^2 R_a$ represents the loss due to the resistance of the armature circuit. The difference between these two quantities, namely EI_a, therefore represents the mechanical power developed by the armature. All of this mechanical power is not available externally, since some of it is absorbed as friction loss at the bearings and at the brushes and some is wasted as hysteresis loss (section 3.14) and in circulating eddy currents in the iron core (section 9.1).

If T is the torque,* in newton metres, exerted on the armature to develop the mechanical power just referred to, and if N is the speed in revolutions/minute, then from expression (1.7),

mechanical power developed $= 2\pi TN/60$ watts.

Hence $\qquad 2\pi TN/60 = EI_a \qquad\qquad\qquad$ (8.5)

$$= 2\frac{Z}{c} \cdot \frac{Np}{60} \cdot \Phi \cdot I_a$$

$\therefore \qquad\qquad T = 0.318 \dfrac{I_a}{c}. Zp\Phi \text{ newton metres}\dagger \quad$ (8.6)

For a given machine, Z, c and p are fixed; in which case:

$$T \propto I_a \times \Phi \qquad\qquad\qquad (8.7)$$

* In many textbooks, the value of the torque is derived from expression (2.1) This method gives the correct result; but it should be realized that with a slotted armature, flux density in the slots is extremely low, so that there is practically no force on the conductors. Practically the whole of the torque is exerted on the teeth (see section A.4 of the Appendix).

† Expression (8.6) also represents the torque to be applied to the armature of a d.c. *generator* to generate the output power of and the I^2R loss in the *armature winding*. Additional torque has to be applied to supply the iron loss in the armature core and the friction and windage losses of the generator.

Or, in words, the torque of a given d.c. motor is proportional to the product of the armature current and the flux per pole.

Example 8.4 *A d.c. motor takes an armature current of* 110 *A at* 480 *V. The resistance of the armature circuit is* 0·2 Ω. *The machine has* 6 *poles and the armature is lap-connected with* 864 *conductors. The flux per pole is* 0·05 *Wb. Calculate* (a) *the speed and* (b) *the gross torque developed by the armature.*

(a)   Generated e.m.f. $= 480 - (110 \times 0\cdot2)$

$$= 458 \text{ V.}$$

Since the armature winding is lap-connected, $c = 6$.
Substituting in expression (6.1), we have:

$$458 = 2 \times \frac{864}{6} \times \frac{N \times 3}{60} \times 0\cdot05$$

$\therefore$ $N = 636$ rev/min.

(b) Mechanical power developed by armature $= 110 \times 458$

$$= 50\,380 \text{ W.}$$

Substituting in expression (8.5) we have:

$$2\pi T \times (636/60) = 50\,380$$

$\therefore$ $T = 756$ N·m.

Alternatively, using expression (8.6) we have:

$$T = 0\cdot318 \times (110/6) \times 864 \times 3 \times 0\cdot05$$
$$= 756 \text{ N·m.}$$

Example 8.5 *The torque required to drive a d.c. generator at* 15 *rev/s is* 2 *kN·m. The iron, friction and windage losses in the machine are* 8 *kW. Calculate the power generated in the armature winding.*

Driving torque $= 2$ kN·m $= 2000$ N·m.

From expression (1.7), power required to drive the generator

$$= 2\pi \times 2000 \text{ [N·m]} \times 15 \text{ [rev/s]}$$
$$= 188\,400 \text{ W} = 188\cdot4 \text{ kW.}$$

Since iron, friction and windage losses = 8 kW

∴   power generated in armature winding = 188·4 − 8
$$= 180\cdot4 \text{ kW.}$$

8.4 Starting resistor

If the armature of example 8.2 were stationary and then switched directly across a 440-V supply, there would be no generated e.m.f. and the current would tend to grow to 440/0·28 = 1572 A. Such a current, in addition to subjecting the armature to a severe mechanical shock, would blow the fuses, thereby disconnecting the supply from the motor. It is therefore necessary (except with very

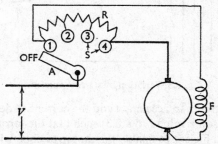

Fig. 8.2 Shunt-wound motor with starter.

small motors) to connect a variable resistor in series with the armature, the resistance being reduced as the armature accelerates. Such an arrangement is termed a *starter*. If the starting current in the above example is to be limited to, say, 80 A, the total resistance of the starter and armature must be 440/80 = 5·5 Ω, so that the resistance of the starter alone must be (5·5 − 0·28) = 5·22 Ω.

Fig. 8.2 shows a starting resistor R subdivided between four contact-studs S and connected to a shunt-wound motor. One end of the shunt winding is joined to stud 1; consequently when arm A is moved from 'Off' to that stud, the full voltage is applied to the shunt winding and the whole of R is in series with the armature. The armature current instantly grows to a value I_1 (fig. 8.3) where:

$$I_1 = \frac{\text{supply voltage } V}{\text{resistance of (armature + starter)}}. \tag{8.8}$$

Since the torque is proportional to (armature current × flux), it follows that the maximum torque is immediately available to accelerate the armature.

As the armature accelerates, its e.m.f. grows and the armature current decreases as indicated by curve *ab*. When the current has fallen to some pre-arranged value I_2, arm A is moved over to stud 2, thereby cutting out sufficient resistance to allow the current to rise once more to I_1. This operation is repeated until A is on stud 4 and the whole of the starting resistor is cut out of the armature circuit.

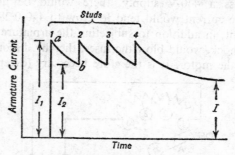

Fig. 8.3 Variation of starting current.

The motor continues to accelerate and the current to decrease until it settles down at some value I (fig. 8.3) such that the torque due to this current is just sufficient to enable the motor to cope with its load.

It is evident from fig. 8.2 that when A is on stud 4 the whole of R is in the field circuit. The effect upon the field current, however, is almost negligible. Thus, for the example considered at the beginning

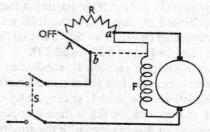

Fig. 8.4 Wrong methods of connecting the shunt winding.

of this section, the shunt current of such a machine would not exceed 2 A, so that the resistance of shunt winding F would be at least 440/2, namely 220 Ω. Consequently the addition of 5·22 Ω means a decrease of only 2·4 per cent in the field current.

If the field winding were connected to *a* as in fig. 8.4, then at the instant of starting, the voltage across the field winding would be very small, namely that across the armature winding. In the above example, for instance, this p.d. is only $80 \times 0.28 = 22.4$ V. Consequently the torque would be extremely small and the machine would probably refuse to start. On the other hand, if the field winding were connected to *b*, as shown dotted in fig. 8.4, it would be directly across the supply; but if the motor were stopped by moving the starter arm back to 'Off', the field would still remain excited. If such a field current were switched off by opening switch S quickly, the sudden collapse of the flux would induce a very high e.m.f. in F and might result in a breakdown of the insulation.

With the connections shown in fig. 8.2, it is obvious that the armature winding, the starting resistor and the field winding form a closed circuit. Consequently when A returns to 'Off', the kinetic energy of the machine and its load maintains rotation for an appreciable period, and during most of that time the machine continues to excite as a shunt generator. The field current therefore decreases comparatively slowly and there is no risk of an excessive e.m.f. being induced.

8.5 Protective devices on starters

It is very desirable to provide the starter with protective devices to enable the starter arm to return to 'Off',

(*a*) when the supply fails, thus preventing the armature being directly across the mains when this voltage is restored, and

(*b*) when the motor becomes overloaded or develops a fault causing the machine to take an excessive current.

Fig. 8.5 shows a starter fitted with such features. The no-volt release NVR consists of a coil wound on a U-shaped iron core. The starter arm carries an iron plate B, so that when it is in the 'On' position, as shown, the core of NVR is magnetized by the field current and holds B against the tendency of a spiral spring G to return A to 'Off'. Should the supply fail, the motor stops, NVR is de-energized and the residual magnetism in it is not sufficient to hold A against the counterclockwise torque of G.

If a connection C be made between the core of NVR and the supply side of the coil, the shunt current flows from A via B and C, and the starting resistor is thereby short circuited.

The coil of the *overload release* OLR is wound on an iron core and connected in series with the motor. An iron plate L, pivoted at one end, carries an extension which, when lifted, connects together two pins p, p. When the current taken by the motor exceeds a certain value, the magnetic pull on L is sufficient to lift it, thereby short circuiting the coil of NVR and releasing the starter arm A. The critical value of the overload current is dependent upon the length of airgap between L and the core of OLR and is controlled by a screw adjustment S.

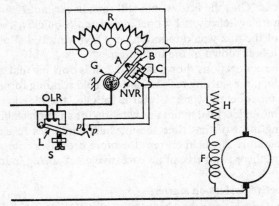

Fig. 8.5 Starter with no-volt and overload releases.

8.6 Speed characteristics of electric motors

With very few exceptions, d.c. motors are shunt-, series- or compound-wound. The connections of a shunt motor have already been given in figs. 8.2 and 8.5; and figs. 8.6 and 8.7 show the connections for series and compound motors respectively, the starter in each case being shown in the 'On' position. (Students should always make a practice of including starters in diagrams of motor connections.) In compound motors, the series and shunt windings almost invariably assist each other, as indicated in fig. 8.7.

The speed characteristic of a motor usually represents the variation of speed with input current or input power, and its shape can be easily derived from expression (8.3), namely:

$$N = \frac{V - I_a R_a}{k\Phi}.$$

In shunt motors, the flux Φ is only slightly affected by the armature current and the value of $I_a R_a$ at full load rarely exceeds 5 per cent of V, so that the variation of speed with input current may be represented by curve A in fig. 8.8. Hence shunt motors are suitable where the speed has to remain approximately constant over a wide range of load.

In series motors, the flux increases at first in proportion to the current and then less rapidly owing to magnetic saturation (fig. 7.2).

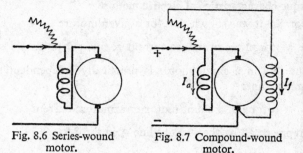

Fig. 8.6 Series-wound motor.   Fig. 8.7 Compound-wound motor.

Also R_a in the above expression now includes the resistance of the field winding. Hence the speed is roughly inversely proportional to the current, as indicated by curve B in fig. 8.8. It will be seen that if the load falls to a very small value, the speed may become danger-

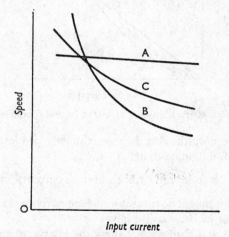

Fig. 8.8 Speed characteristics of shunt, series and compound motors.

ously high. A series motor should therefore not be employed when there is any such risk; for instance, it should never be belt-coupled to its load except in very small machines such as vacuum cleaners.

Since the compound motor has a combination of shunt and series excitations, its characteristic (curve C in fig. 8.8) is intermediate between those of the shunt and series motors, the exact shape depending upon the values of the shunt and series ampere-turns.

8.7 Torque characteristics of electric motors

In section 8.3 it was shown that for a given motor:

$$\text{torque} \propto \text{armature current} \times \text{flux per pole.}$$

Since the flux in a shunt motor is practically independent of the armature current:

$$\therefore \qquad \text{torque of a shunt motor} \propto \text{armature current}$$

and is represented by the straight line A in fig. 8.9.

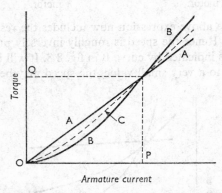

Fig. 8.9 Torque characteristics of shunt, series and compound motors.

In a series motor the flux is approximately proportional to the current up to full load, so that:

$$\text{torque of a series motor} \propto (\text{armature current})^2, \text{ approx.}$$

Above full load, magnetic saturation becomes more marked and the torque does not increase so rapidly.

Curves A, B and C in fig. 8.9 show the relative shapes of torque curves for shunt, series and compound motors having the same full-load torque OQ with the same full-load armature current OP, the exact shape of curve C depending upon the relative value of the shunt and series ampere-turns at full load.

From fig. 8.9 it is evident that for a given current below the full-load value the shunt motor exerts the largest torque; but for a given current above that value the series motor exerts the largest torque.

The maximum permissible current at starting is usually about 1·5 times the full-load current. Consequently where a large starting torque is required, such as for hoists, cranes, electric trains, etc., the series motor is the most suitable machine.

Example 8.6 *A series motor runs at* 600 *rev/min when taking* 110 *A from a* 230-*V supply. The resistance of the armature circuit is* 0·12 Ω *and that of the series winding is* 0·03 Ω. *Calculate the speed when the current has fallen to* 50 *A, assuming the useful flux per pole for* 110 *A to be* 0·024 *Wb and that for* 50 *A to be* 0·0155 *Wb.*

Total resistance of armature and series windings

$$= 0·12 + 0·03 = 0·15 \ \Omega,$$

∴   e.m.f. generated when current is 110 A

$$= 230 - 110 \times 0·15 = 213·5 \ V.$$

In section 8.2 it was shown that for a given machine:

generated e.m.f. = a constant (say k) × speed × flux.

Hence with 110 A, $213·5 = k \times 600 \times 0·024$

∴ $k = 14·82.$

With 50 A,   generated e.m.f. $= 230 - 50 \times 0·15 = 222·5 \ V.$

But the new e.m.f. generated $= k \times$ new speed × new flux

∴ $222·5 = 14·82 \times$ new speed $\times 0·0155$

∴   speed for 50 A $= 969$ rev/min.

8.8 Speed control of d.c. motors

It has already been explained in section 8.2 that the speed of a d.c. motor can be altered by varying either the flux or the armature voltage or both; and the methods most commonly employed are:

(*a*) A variable resistor, termed a *field regulator*, in series with the shunt winding—only applicable to shunt and compound motors. Such a field regulator is indicated by H in fig. 8.5. When the resistance is increased, the field current, the flux and the generated e.m.f. are reduced. Consequently more current flows through the armature and the increased torque enables the armature to accelerate until the generated e.m.f. is again nearly equal to the applied voltage (see example 8.7).

With this method it is possible to increase the speed to three or four times that at full excitation, but it is not possible to reduce the speed below that value. Also, with any given setting of the regulator, the speed remains approximately constant between no load and full load.

(*b*) A resistor, termed a *controller*, in series with the armature. The electrical connections for a controller are exactly the same as for a starter, the only difference being that in a controller the resistor elements are designed to carry the armature current indefinitely, whereas in a starter they can only do so for a comparatively short time without getting excessively hot.

For a given armature current, the larger the controller resistance in circuit, the smaller is the p.d. across the armature and the lower, in consequence, is the speed.

This system has several disadvantages: (i) the relatively high cost of the controller; (ii) much of the input energy may be dissipated in the controller and the overall efficiency of the motor considerably reduced thereby; (iii) the speed may vary greatly with variation of load due to the change in the p.d. across the controller causing a corresponding change in the p.d. across the motor; thus, if the supply voltage is 240 V, and if the current decreases so that the p.d. across the controller falls from, say, 100 to 40 V, then the p.d. across the motor increases from 140 V to 200 V.

The principal advantage of the system is that speeds from zero upwards are easily obtainable, and the method is chiefly used for controlling the speed of cranes, hoists, trains, etc., where the motors are frequently started and stopped and where efficiency is of secondary importance.

(*c*) Exciting the field winding off a constant-voltage system and supplying the armature from a separate generator, as shown in fig. 8.10. M represents the motor whose speed is to be controlled; M_1G is a motor-generator set consisting of a shunt motor coupled to a separately-excited generator. The voltage applied to the armature of M can be varied between zero and a maximum by means of R. If provision is made for reversing the excitation of G, the speed of M can then be varied from a maximum in one direction to the same maximum in the reverse direction.

This method is often referred to as the *Ward–Leonard* system and is used for controlling the speed of motors driving colliery winders, rolling mills, etc.

(*d*) When an a.c. supply is available, the voltage applied to the armature can be controlled by thyristors, the operation of which is explained in section 20.17. Briefly, the thyristor is a solid-state rectifier

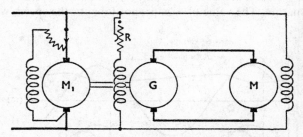

Fig. 8.10 Ward–Leonard system of speed control.

which is normally non-conducting in the forward and reverse directions. It is provided with an extra electrode, termed the *gate*, so arranged that when a pulse of current is introduced into the gate circuit, the thyristor is 'fired', i.e. it conducts in the forward direction. Once it is fired, the thyristor continues to conduct until the current falls below the holding value.

Fig. 8.11 shows a simple arrangement for controlling a d.c. motor from a single-phase supply. Field winding F is separately excited via bridge-connected rectifiers J (section 19.10) and armature A is

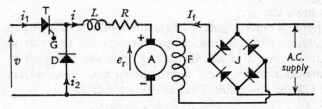

Fig. 8.11. Thyristor system of speed control.

supplied via thyristor T. Gate G of the thyristor is connected to a firing circuit which supplies a current pulse once every cycle. The arrangement of the firing circuit is not shown as it is too involved for inclusion in this diagram. In fig. 8.11, R and L represent the resistance and inductance respectively of the armature winding and an external inductor that may be inserted to increase the inductance of the circuit. A diode D is connected across the armature and the inductor.

The sine wave in fig. 8.12(a) represents the supply voltage and the wavy line MN represents the e.m.f., e_r, generated by the rotation of the armature. The value of e_r is proportional to the speed, which in

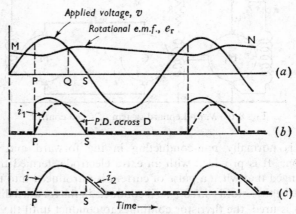

Fig. 8.12. Waveforms for fig. 8.11.

turn varies with the armature current. Thus, when the current exceeds the average value, the armature accelerates, and then decelerates when the current falls below the average value. The speed fluctuation indicated in fig. 8.12(a) is exaggerated to illustrate the effect more clearly.

Suppose a current pulse to be applied to gate G at instant P in fig. 8.12(a). The resulting current through the thyristor and the armature grows at a rate depending upon the difference between the applied voltage, v, and the rotational e.m.f., e_r, and upon the ratio L/R for the circuit. Thus at any instant:

$$v = e_r + iR + L \cdot di/dt$$

$$\therefore \qquad i = (v - e_r - L \cdot di/dt)/R \qquad (8.9)$$

While the thyristor is conducting, the p.d. across the armature and inductor, and therefore across diode D, varies as shown by the full-line waveform in fig. 8.12(b).

At instant Q, $e_r = v$, so that $i = -(L/R) \cdot di/dt$. The current is now decreasing so that di/dt is negative; hence the current is still positive and is therefore continuing to exert a *driving* torque on the armature.

At instant S, the supply voltage is reversing its direction. A reverse current from the supply flows through D, thereby making the cathode of the thyristor *positive* relative to its anode. Consequently, the current i_1 through the thyristor falls to zero, as indicated by the dotted line in fig. 8.12(*b*), so that the thyristor reverts to its nonconducting state.

Current is now confined to the closed circuit formed by the armature and diode D. Since the p.d. across D is practically zero, equation (8.9) can now be written:

$$i = -(e_r + L \cdot \mathrm{d}i/\mathrm{d}t)/R.$$

Current i is now decreasing at sufficiently high rate for the value of $L \cdot \mathrm{d}i/\mathrm{d}t$ to exceed e_r. Since the value of di/dt is negative, the direction of the current is unaltered. Hence the current still continues to exert a driving torque on the armature, the energy supplied to the load being recovered partly from that stored in the inductance of the circuit while the current was growing and partly from that stored as kinetic energy in the motor and load during acceleration.

The dotted waveform in fig. 8.12(*c*) represents current i_2 through diode D. The latter is often referred to as a free-wheeling diode since it carries current when the thyristor *ceases* to conduct.

The later the instant of firing the thyristor, the smaller is the average voltage applied to the armature and the lower the speed in order that the motor may take the same average current from the supply to enable it to maintain the *same* load torque. Thus the speed of the motor can be controlled over a wide range.

An increase of motor load causes the speed to fall, thereby allowing a larger current pulse to flow during the conducting period. The fluctuation of current can be reduced by (*a*) using two thyristors to give full-wave rectification when the supply is single-phase and (*b*) using three or six thyristors when the supply is three-phase.

An important application of the thyristor is the speed control of series motors in battery-driven vehicles. The principle of operation is that pulses of the battery voltage are applied to the motor, and the *average* value of the voltage across the motor is controlled by varying the ratio of the 'on' and 'off' durations of the pulses. Thus, if the 'on' period is t_1 and the 'off' period is t_2, average motor voltage = battery voltage $\times t_1/(t_1 + t_2)$.

Fig. 8.13 show the essential features of this method of speed control. The series motor M is connected in series with thyristor T

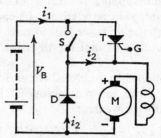

Fig. 8.13. Chopper speed control of a series motor.

across the battery. A free-wheeling diode D is connected in parallel with the motor, and a switch S is used to short-circuit the thyristor at the end of each 'on' period.

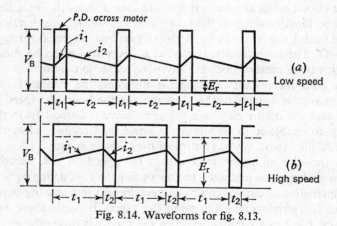

Fig. 8.14. Waveforms for fig. 8.13.

With S open, the thyristor is fired by a current pulse applied to gate G, causing a current i_1 to flow through T and M, as shown in fig. 8.14. After an interval t_1, switch S is closed for sufficient time to allow the thyristor current to fall to zero, thereby enabling the thyristor to revert to its non-conducting state.

After S is opened, the current through M decreases at a rate such that the e.m.f. induced in the inductance of the field and armature windings exceeds the rotational e.m.f. by an amount sufficient to circulate a current i_2 around the closed circuit formed by the motor and diode D. After an interval t_2, the operation is repeated, as indicated in fig. 8.14.

For a low speed, the ratio t_1/t_2 is small, as shown in fig. 8.14(a). A higher speed is obtained by increasing t_1 and reducing t_2, as in fig. 8.14(b). The dotted horizontal lines in fig. 8.14 represent the average values of the rotational e.m.f. E_r generated in the armature. The average value of the current is determined by the torque requirement, exactly as described in earlier sections.

The battery is supplying power only during the 'on' intervals t_1. During the 'off' intervals t_2, current i_2 is exerting a *driving* torque on the armature, the energy supplied to the mechanical load during these intervals being derived from the magnetic fields of the series and armature windings and from the kinetic energy of the motor and load.

The frequency of the pulses may be as high as 3000 per second, and the arrangement used to control the firing of the thyristor is referred to as a *chopper circuit*, but the details of this circuit are too complex for inclusion here.

The importance of speed control by thyristors will be realized from the fact that of all the thyristors used in the United Kingdom during 1969, about 60 per cent were for battery-driven vehicles and about 25 per cent for d.c. motor control from mains supply.

Example 8.7 *A shunt motor is running at* 626 *rev/min* (*example* 8.2) *when taking an armature current of* 50 *A from a* 440-*V supply. The armature circuit has a resistance of* 0·28 Ω. *If the flux is suddenly reduced by* 5 *per cent, find:* (a) *the maximum value to which the current increases momentarily and the ratio of the corresponding torque to the initial torque;* (b) *the ultimate steady value of the armature current, assuming the torque due to the load to remain unaltered.*

(*a*) From example 8.2:

initial e.m.f. generated $= 400 - 50 \times 0·28 = 426$ V.

Immediately after the flux is reduced 5 per cent, i.e. before the speed has begun to increase:

new e.m.f. generated $= 426 \times 0·95 = 404·7$ V

∴   corresponding voltage drop due to armature resistance

$$= 440 - 404 \cdot 7 = 35 \cdot 3 \text{ V}$$

and corresponding armature current $= 35 \cdot 3/0 \cdot 28 = 126$ A.

From expression (8.7):

torque of a given machine α armature current $\times$ flux

$$\therefore \quad \frac{\text{new torque}}{\text{initial torque}} = \frac{\text{new current}}{\text{initial current}} \times \frac{\text{new flux}}{\text{initial flux}}$$

$$= \frac{126}{50} \times 0 \cdot 95 = 2 \cdot 394.$$

Hence the sudden reduction of 5 per cent in the flux is accompanied by more than a twofold increase of torque; this is the reason why the motor accelerates.

(*b*) After the speed and current have attained steady values, the torque will have decreased to the original value, so that:

new current $\times$ new flux $=$ original current $\times$ original flux

$$\therefore \quad \text{new armature current} = 50 \times \frac{1}{0 \cdot 95} = 52 \cdot 6 \text{ A}.$$

Example 8.8 *A shunt motor runs on no load at* 700 *rev/min off a* 440-*V supply. The resistance of the shunt circuit is* 240 Ω, *Calculate the additional resistance required in the shunt circuit to raise the no-load speed to* 1000 *rev/min. The following table gives the relationship between the flux and the shunt current:*

Shunt current (A)	0·5	0·75	1·0	1·25	1·5	1·75	2·0
Flux/pole (mWb)	6	8	9·4	10·2	10·8	11·2	11.5

When the motor is on no load, the voltage drop due to the armature resistance is negligible.

Initial shunt current $= 440/240 = 1 \cdot 83$ A.

From a graph representing the data given in the above table, corresponding flux/pole $= 11 \cdot 3$ mWb.

Since the speed is inversely proportional to the flux,

flux/pole at 1000 rev/min $= 11 \cdot 3 \times 700/1000$

$$= 7 \cdot 91 \text{ mWb}.$$

From the graph it is found that the shunt current to produce a flux of 7·91 mWb/pole is 0·74 A,

∴   corresponding resistance of shunt circuit $= 440/0.74$
$$= 595\ \Omega,$$

and   additional resistance required$\left.\right\}$ $= 595 - 240 = 335\ \Omega.$
            in shunt circuit

Example 8.9 *A shunt motor, supplied at* 230 *V, runs at* 900 *rev/min when the armature current is* 30 *A. The resistance of the armature circuit is* 0·4 Ω*. Calculate the resistance required in series with the armature to reduce the speed to* 600 *rev/min, assuming that the armature current is then* 20 *A.*

Initial e.m.f. generated $= 230 - 30 \times 0.4 = 218$ V.

Since the excitation remains constant, the generated e.m.f. is proportional to the speed.

∴   e.m.f. generated at 600 rev/min $= 218 \times 600/900 = 145.3$ V

Hence voltage drop due to total$\left.\right\}$
      resistance of armature $= 230 - 145.3 = 84.7$ V,
          circuit

and   total resistance of armature$\left.\right\}$ $= 84.7/20 = 4.235\ \Omega$
              circuit

∴   additional resistance required in$\left.\right\}$ $= 4.235 - 0.4$
        armature circuit $= 3.835\ \Omega.$

Summary of important formulae

For a generator, $E = V + I_a R_a$ (8.1)

For a motor, $V = E + I_a R_a$ (8.2)

For a given motor, $N = \dfrac{V - I_a R_a}{k\Phi} \simeq \dfrac{V}{k\Phi}$ (8.3)

$$T = 0.318\frac{I_a}{c} \cdot Zp\Phi \text{ newton metres} \quad (8.6)$$

For a given motor, $T \propto I_a \Phi$ (8.7)

EXAMPLES 8

1. A shunt machine has armature and field resistances of 0·04 Ω and 100 Ω respectively. When connected to a 460-V d.c. supply and driven as a

generator at 600 rev/min, it delivers 50 kW. Calculate its speed when running as a motor and taking 50 kW from the same supply.

Show that the direction of rotation of the machine as a generator and as a motor under these conditions is unchanged.

2. Obtain from first principles an expression for the e.m.f, of a 2–pole d.c. machine, defining the symbols used.

A 100-kW, 500-V, 750-rev/min, d.c. shunt generator, connected to constant-voltage bus-bars, has field and armature resistances of 100 Ω and 0·1 Ω respectively. If the prime-mover fails, and the machine continues to run, taking 50 A from the bus-bars, calculate its speed. Neglect brush-drop and armature reaction effects.   (S.A.N.C., O.2)

3. Derive the e.m.f. equation for a d.c. machine.

A d.c. shunt motor has an armature resistance of 0·5 Ω and is connected to a 200-V supply. If the armature current taken by the motor is 20 A, what is the e.m.f. generated by the armature?

What is the effect of: (*a*) inserting a resistor in the field circuit; (*b*) inserting a resistor in the armature circuit if the armature current is maintained at 20 A?   (E.M.E.U., O.1)

4. Explain clearly the effect of the back e.m.f. of a shunt motor. What precautions must be taken when starting a shunt motor?

A four-pole d.c. motor is connected to a 500-V d.c. supply and takes an armature current of 80 A. The resistance of the armature circuit is 0·4Ω. The armature is wave wound with 522 conductors and the useful flux per pole is 0·025 Wb. Calculate: (*a*) the back e.m.f. of the motor; (*b*) the speed of the motor; (*c*) the torque in newton metres developed by the armature.   (U.E.I., O.1)

5. A shunt machine is running as a motor off a 500-V system, taking an armature current of 50 A. If the field current is suddenly increased so as to increase the flux by 20 per cent, calculate the value of the current that would momentarily be fed back into the mains. Neglect the shunt current and assume the resistance of the armature circuit to be 0·5 Ω.

6. A shunt motor is running off a 220-V supply taking an armature current of 15 A, the resistance of the armature circuit being 0·8 Ω. Calculate the value of the generated e.m.f.

If the flux were suddenly reduced by 10 per cent, to what value would the armature current increase momentarily?

7. A six-pole d.c. motor has a wave-connected armature with 87 slots, each containing 6 conductors. The flux per pole is 20 mWb and the armature has a resistance of 0·13 Ω. Calculate the speed when the motor is running off a 240-V supply and taking an armature current of 80 A. Calculate also the torque, in newton metres, developed by the armature.

8. A four-pole, 460-V shunt motor has its armature wave-wound with 888 conductors. The useful flux per pole is 0·02 Wb and the resistance of the armature circuit is 0·7 Ω. If the armature current is 40 A, calculate: (*a*) the speed, and (*b*) the torque in newton metres.   (U.E.I., O.1)

9. A four-pole motor has its armature lap-wound with 1040 conductors and runs at 1000 rev/min when taking an armature current of 50 A

from a 250-V d.c. supply. The resistance of the armature circuit is 0·2 Ω. Calculate: (a) the useful flux per pole; (b) the torque developed by the armature in newton metres. (U.E.I., O.1)

10. A d.c. shunt generator delivers 5 kW at 250 V when driven at 1500 rev/min. The shunt circuit resistance is 250 Ω and the armature circuit resistance is 0·4 Ω. The iron, friction and windage losses are 250 W. Determine the torque (in newton metres) required to drive the machine at the above load. (W.J.E.C., O.2)

11. Explain why a d.c. shunt-wound motor needs a starter on constant-voltage mains. A shunt-wound motor has a field resistance of 350 Ω and an armature resistance of 0·2 Ω and runs off a 250-V supply. The armature current is 55 A and the motor speed is 1000 rev/min. Assuming a straight-line magnetization curve, calculate: (a) the additional resistance required in the field circuit to increase the speed to 1100 rev/min for the same armature current, and (b) the speed with the original field current and an armature current of 100 A.
(E.M.E.U., O.1)

12. Explain the necessity for using a starter with a d.c. motor.
A 240-V d.c. shunt motor has an armature of resistance of 0·2 Ω. Calculate: (a) the value of resistance which must be introduced into the armature circuit to limit the starting current to 40 A; (b) the e.m.f. generated when the motor is running at a constant speed with this additional resistance in circuit and with an armature current of 30 A.
(N.C.T.E.C., O.1)

13. A separately-excited d.c. motor is fed from a 460-V d.c. supply. The speed is to be controlled by field weakening. When the machine was driven as a separately-excited generator at 750 rev/min, the following open-circuit curve was obtained:

Generated e.m.f. (V)	10	172	300	360	385	395
Field current (A)	0	1	2	3	4	5

Plot a curve showing the no-load speed against field current. Assume negligible losses. (N.C.T.E.C., O.2)

14. A long-shunt, compound-wound motor connected to a 250-V supply has a full-load output of 15 kW, and at this load is found to have an efficiency of 82 per cent. The armature has a resistance of 0·2 Ω and the series field winding (which is designed to change the flux/pole by 0·002 Wb on full load) has a resistance of 0·3 Ω. The shunt field, which produces a flux per pole of 0·02 Wb, has a resistance of 250 Ω. When the machine runs on no load, its speed is found to be 500 rev/min and the current taken from the supply is 2·5 A.

Calculate the speed on full load if the machine is cumulatively compounded.

The iron parts of the machine operate on the linear part of their magnetization characteristics throughout and the effect of temperature on resistance may be neglected. (E.M.E.U., O.2)

15. A series d.c. motor is fed from a 400-V d.c. supply. Calculate and plot the variation of motor speed (rad/s) against armature current, given the following information:

(*a*) When the machine is driven as a generator and separately excited, the following open-circuit curve was obtained at a speed of 100 rad/s:

Field current (A)	0	5	15	25	35	40	50
Open-circuit voltage (V)	0	87	245	335	375	385	395

(*b*) Armature resistance $= 1.6\ \Omega$; field resistance $= 0.9\ \Omega$.
Assume negligible brush voltage drop. (N.C.T.E.C., O.2)

16. A d.c. shunt motor takes an armature current of 20 A from a 230-V supply. Resistance of the armature circuit is $0.5\ \Omega$. Calculate the resistance required in series with the armature to halve the speed if: (*a*) the load torque is constant; (*b*) the load torque is proportional to the square of the speed.

17. Calculate the torque, in newton metres, developed by a d.c. motor having an armature resistance of $0.25\ \Omega$ and running at 750 rev/min when taking an armature current of 60 A from a 480-V supply.

18. A six-pole, lap-wound, 220-V, shunt-excited d.c. machine takes an armature current of 2·5 A when unloaded at 950 rev/min. When loaded, it takes an armature current of 54 A from the supply and runs at 950 rev/min. The resistance of the armature circuit is $0.18\ \Omega$ and there are 1044 armature conductors.

For the loaded condition, calculate: (i) the generated e.m.f.; (ii) the useful flux per pole; (iii) the useful torque developed by the machine in newton metres. (U.E.I., O.2)

19. A d.c. shunt motor runs at 900 rev/min from a 480-V supply when taking an armature current of 25 A. Calculate the speed at which it will run from a 240-V supply when taking an armature current of 15 A. The resistance of the armature circuit is $0.8\ \Omega$. Assume the flux per pole at 240 V to have decreased to 75 per cent of its value at 480 V.

20. A shunt motor, connected across a 440-V supply, takes an armature current of 20 A and runs at 500 rev/min. The armature circuit has a resistance of $0.6\ \Omega$. If the magnetic flux is reduced by 30 per cent and the torque developed by the armature increases by 40 per cent, what are the values of the armature current and of the speed? (App. El., L.U.)

21. A d.c. series motor connected across a 460-V supply runs at 500 rev/min when the current is 40 A. The total resistance of the armature and field circuits is $0.6\ \Omega$. Calculate the torque on the armature in newton metres. (App. El., L.U.)

22. A d.c. series motor has the following magnetization data:

Field current (amperes)	5	10	15	20	25
E.M.F. (volts) at 800 rev/min	180	356	488	548	570

The armature resistance is $1.4\ \Omega$ and the field circuit resistance $0.6\ \Omega$. Plot the speed/current curve when the motor is connected across a 480-V supply.

23. The output power of a shunt motor running off a 240-V supply is 16 kW. Calculate the value of the starter resistance necessary to limit the

starting current to 1·5 times the full-load current if the full-load efficiency of the motor is 88 per cent and the resistance of the armature circuit is 0·2 Ω. Neglect the shunt current.

Also, calculate the generated e.m.f. of the motor when the current has fallen to the full-load value, assuming that the whole of the starter resistance is still in circuit.

24. A d.c. shunt motor runs off a 230-V supply and has an armature resistance of 0·3 Ω. Calculate: (*a*) the resistance required in series with the armature to limit the armature current to 75 A at starting, and (*b*) the value of the generated e.m.f. when the armature current has fallen to 50 A with this value of resistance still in circuit.

25. Assuming that the e.m.f. generated by a d.c. machine is $E = k\Phi\omega$ volts, deduce the expression for the generated torque as $T = k\Phi I_a$ newton metres, where $\Phi =$ flux/pole in webers, $\omega =$ speed of armature in radians per second, $I_a =$ armature current in amperes and $k =$ a constant.

A d.c. shunt motor runs at 70 rad/s from a 460-V supply and takes an armature current of 100 A. Resistors of 2·52 Ω and 1·64 Ω are connected in series with the armature and in parallel with the armature respectively. The field excitation remains direct from the 460-V supply, but the generated torque is halved. Determine the new running speed. Neglect brush voltage drop. The armature resistance is 0·4 Ω.

(N.C.T.E.C., O.2)

26. A d.c. series motor, having armature and field resistances of 0·06 Ω and 0·04 Ω respectively, was tested by driving it at 2000 rev/min and measuring the open-circuit voltage across the armature terminals, the field being supplied from a separate source. One of the readings taken was: field current, 350 A; armature p.d., 1560 V. From the above information, obtain a point on the speed/current characteristic, and one on the torque/current characteristic for normal operation at 750 V and at 350 A. Take the torque due to rotational loss as 50 N·m. Assume that brush drop and field weakening due to armature reaction can be neglected. (S.A.N.C., O.2)

27. A shunt motor runs on no load at 800 rev/min off a 240-V supply with no external resistor in the field circuit. The armature current is 2 A. Calculate the resistance required in series with the shunt winding so that the motor may run at 950 rev/min when taking an armature current of 30 A. Shunt winding resistance = 160 Ω; armature resistance = 0·4 Ω. It may be assumed that the flux is proportional to the field current.

28. The resistance of the armature circuit of a 250-V shunt motor is 0·3 Ω and its full-load speed is 1000 rev/min. Calculate the resistance required in series with the armature to reduce the speed with full-load torque to 800 rev/min, the full-load armature current being 50 A. If the load torque is then halved, at what speed will the motor run? Neglect the effect of armature reaction.

29. A d.c. series motor, connected to a 440-V supply, runs at 600 rev/min when taking a current of 50 A. Calculate the value of a resistor which,

when inserted in series with the motor, will reduce the speed to 400 rev/min, the gross torque being then half its previous value. Resistance of motor = 0·2 Ω. Assume the flux to be proportional to the field current.

30. A series motor runs at 900 rev/min when taking 30 A at 230 V. The total resistance of the armature and field circuits is 0·8 Ω. Calculate the values of the additional resistance required in series with the machine to reduce the speed to 500 rev/min if the gross torque is: (*a*) constant; (*b*) proportional to the speed; (*c*) proportional to the square of the speed. Assume the magnetic circuit to be unsaturated.

CHAPTER 9

Efficiency of D.C. Machines

9.1 Losses in generators and motors

The losses in d.c. machines can be classified thus:

i. *Armature losses*

(a) I^2R loss in armature winding. The resistance of an armature winding can be measured by the voltmeter–ammeter method, as described in section 9.4. If the resistance measurement is made at room temperature, the resistance at normal working temperature should be calculated. Thus, if the resistance be R_1 at room temperature of, say, 15° C and if 50° C be the temperature *rise* of the winding after the machine has been operating on full load for 3 or 4 hours, then from expression 1.12, we have:

$$\text{resistance at } 65° \text{ C} = R_1 \times \frac{1 + (0.004\,26 \times 65)}{1 + (0.004\,26 \times 15)} = 1.2R_1.$$

(b) Iron loss in the armature core due to hysteresis and eddy currents. Hysteresis loss has been discussed in section 3.14 and is dependent upon the quality of the iron. It is proportional to the frequency and is approximately proportional to the square of the flux. The eddy-current loss is due to circulating currents set up in the iron laminations. Had the core been of solid iron, as shown in fig. 9.1 (a) for a two-pole machine, then if the armature were rotated, e.m.f.s would be generated in the core in exactly the same way as they are generated in conductors placed on the armature, and these e.m.f.s would circulate currents—known as *eddy currents*—in the core as shown dotted in fig. 9.1 (a), the rotation being assumed clockwise when the armature is viewed from the righthand side of the machine. Owing to the very low resistance of the core, these eddy currents would be considerable and would cause a large loss of power in and excessive heating of the armature.

If the core is made of laminations insulated from one another, the eddy currents are confined to their respective sheets, as shown dotted in fig. 9.1 (*b*), and the eddy-current loss is thereby reduced. Thus, if the core is split up into five laminations, the e.m.f. per lamination is only a fifth of that generated in the solid core. Also, the cross-sectional area per path is reduced to about a fifth, so that the resistance per path is roughly five times that of the solid core. Consequently the current per path is about $\frac{1}{25}$ of that in the solid core. Hence:

$$\frac{I^2R \text{ loss per lamination}}{I^2R \text{ loss in solid core}} = \left(\frac{1}{25}\right)^2 \times 5 = \frac{1}{125} \text{ (approx.)}$$

Since there are five laminations,

$$\frac{\text{total eddy-current loss in laminated core}}{\text{total eddy-current loss in solid core}} = \frac{5}{125} = \left(\frac{1}{5}\right)^2.$$

It follows that the eddy-current loss is approximately proportional to the square of the thickness of the laminations. Hence the

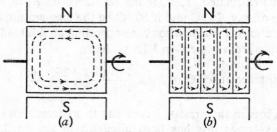

Fig. 9.1 Eddy currents.

eddy-current loss can be reduced to any desired value, but if the thickness of the laminations is made less than about 0·4 mm, the reduction in the loss does not justify the extra cost of construction. Eddy-current loss can also be reduced considerably by the use of silicon-iron alloy—usually about 4 per cent of silicon—due to the resistivity of this alloy being much higher than that of ordinary iron.

Since the e.m.f.s induced in the core are proportional to the frequency and the flux, therefore the eddy-current loss is proportional to (frequency × flux)^2.

ii. *Commutator losses*

(*a*) Loss due to the contact resistance between the brushes and the segments. This loss is dependent upon the quality of the brushes. For carbon brushes, the p.d. between a brush and the commutator, over a wide range of current, is usually about 1 volt per positive set of brushes and 1 volt per negative set, so that the total contact-resistance loss, in watts, is approximately 2 × total armature current.

(*b*) Loss due to friction between the brushes and the commutator. This loss depends upon the total brush pressure, the coefficient of friction and the peripheral speed of the commutator.

iii. *Excitation losses*

(*a*) Loss in the shunt circuit (if any) equal to the product of the shunt current and the terminal voltage. In shunt generators this loss increases a little between no load and full load, since the shunt current has to be increased to maintain the terminal voltage constant; but in level-compounded generators and in shunt and compound motors, it remains approximately constant.

(*b*) Losses in series, compole and compensating windings (if any). These losses are proportional to the square of the armature current.

iv. *Bearing friction and windage losses*

The bearing friction loss is roughly proportional to the speed, but the windage loss, namely the power absorbed in setting up circulating currents of air, is proportional to the cube of the speed. The windage loss is very small unless the machine is fitted with a cooling fan.

v. *Stray load loss*

It was shown in section 6.5 that the effect of armature reaction is to distort the flux, the flux densities at certain points of the armature being increased; consequently the iron loss is also increased. This stray loss is usually neglected as it is difficult to estimate its value. Consequently the value of the efficiency determined by the Swinburne method (section 9.4 (*b*)) is a little higher than that obtained from the input and output powers of the machine.

9.2 Efficiency of a d.c. machine

(a) *Generator*

If R_a = total resistance of armature circuit (including the brush-contact resistance and the resistance of series and compole windings, if any),

       I = output current,

       I_s = shunt current

and I_a = armature current = $I + I_s$,

then   total loss in armature circuit = $I_a^2 R_a$.

If $V =$ terminal voltage,

$$\text{loss in shunt circuit} = I_\text{s}V.$$

This includes the loss in the shunt regulating resistor.

If $C =$ sum of iron, friction and windage losses,

$$\text{total losses} = I_\text{a}^2R_\text{a} + I_\text{s}V + C$$
$$\text{Output power} = IV$$

$\therefore \qquad \text{input power} = IV + I_\text{a}^2R_\text{a} + I_\text{s}V + C$

and $\qquad \text{efficiency} = \dfrac{IV}{IV + I_\text{a}^2R_\text{a} + I_\text{s}V + C}$ \hfill (9.1)

(b) *Motor*

If $I =$ input current and $I_\text{s} =$ shunt current, then

$$\text{armature current} = I_\text{a} = I - I_\text{s}.$$

Using the symbols given above for a generator, we have:

$$\text{total losses} = I_\text{a}^2R_\text{a} + I_\text{s}V + C$$
$$\text{Input power} = IV$$

$\therefore \qquad \text{output power} = IV - I_\text{a}^2R_\text{a} - I_\text{s}V - C$

and $\qquad \text{efficiency} = \dfrac{IV - I_\text{a}^2R_\text{a} - I_\text{s}V - C}{IV}$ \hfill (9.2)

9.3 Approximate condition for maximum efficiency

Let us assume (*a*) that the shunt current is negligible compared with the armature current at load corresponding to maximum efficiency and (*b*) that the shunt, iron and friction losses are independent of the load; then from expression (9.1),

$$\text{efficiency of a generator} = \frac{IV}{IV + I^2R_\text{a} + I_\text{s}V + C}$$
$$= \frac{V}{V + IR_\text{a} + \dfrac{1}{I}(I_\text{s}V + C)} \tag{9.3}$$

This efficiency is a maximum or a minimum when the denominator of (9.3) is a minimum or a maximum respectively, namely when:

$$\frac{\text{d}}{\text{d}I}\left\{V + IR_\text{a} + \frac{1}{I}(I_\text{s}V + C)\right\} = 0$$

i.e. $\qquad R_\text{a} - \dfrac{1}{I^2}(I_\text{s}V + C) = 0$ \hfill (9.4)

$\therefore \qquad I^2R_\text{a} = I_\text{s}V + C$ \hfill (9.5)

The condition for the denominator in expression (9.3) to be a minimum, and therefore the efficiency to be a maximum, is that the lefthand side of expression (9.4), when differentiated with respect to I, should be positive; thus,

$$\frac{\mathrm{d}}{\mathrm{d}I}\left\{R_\mathrm{a} - I^{-2}(I_\mathrm{s}V + C)\right\} = 2I^{-3}(I_\mathrm{s}V + C)$$

Since this quantity is positive, it follows that expression (9.5) represents the condition for maximum efficiency; i.e. the efficiency of the generator is a maximum when the load is such that the variable loss is equal to the constant loss.

Precisely the same conclusion can be derived for a motor from expression (9.2).

9.4 Determination of efficiency

(a) *By direct measurement of input and output powers*
In the case of small machines the output power can be measured by some form of mechanical brake as that shown in fig. 9.2, where a belt (or ropes) on an air- or water-cooled pulley has one end attached

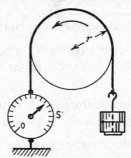

Fig. 9.2 Brake test on a motor.

via a spring balance S to the floor and has a mass of m kilograms suspended at the other end.

Weight of the suspended mass $= W$
$$\simeq 9{\cdot}81\, m \text{ newtons.}$$
If reading on spring balance $= S$ newtons,
∴   net pull due to friction $= (W - S)$ newtons.
If $r =$ effective radius of brake, in metres
and $N =$ speed of pulley, in revolutions/minute,
torque due to brake friction $= (W - S)r$ newton metres
and   output power $= 2\pi(W - S)rN/60$ watts   (9.6)

If V = supply voltage, in volts
and I = current taken by motor, in amperes,
       input power = IV watts

and   efficiency $= \dfrac{2\pi(W - S)rN}{60 \times IV} = \dfrac{2\pi(9{\cdot}81\ m - S)rN}{60 \times IV}$ (9·7)

The size of machine that can be tested by this method is limited by the difficulty of dissipating the heat generated at the brake.

(b) *By separate measurement of the losses (Swinburne's* Method)*

Let us for simplicity consider the case of a shunt generator or motor. The machine is run as a motor on no load with the supply voltage V adjusted to the rated voltage, namely the voltage stamped on the

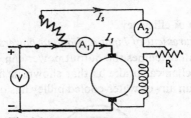

Fig. 9.3 Measurement of no-load loss.

nameplate. The speed is adjusted to its rated value by means of shunt regulator R (fig. 9.3). If I_1 and I_s be the readings on ammeters A_1 and A_2 respectively,

          input power to armature = I_1V watts

and   shunt loss† = I_sV watts

The input power to the armature supplies: (i) iron loss in core; (ii) friction loss at bearings and commutator; (iii) windage loss; and (iv) a very small I^2R loss due to the resistance of the armature circuit. This I^2R loss can be calculated, but usually it is negligible compared with the other losses (see Example 9.1).

The resistance of the armature circuit (including the compole

---

* Sir James Swinburne, F.R.S. (1858–1958), was a pioneer in the development of electric lamps, electrical machinery and synthetic resins.

† In the case of a d.c. generator, the value of I_s noted when the machine is running as a motor on no load is less than that when the machine is running as a generator at the same speed and terminal voltage. Consequently the shunt and iron losses in the no-load test are less than those in the machine when operating as a generator, and the efficiency calculated by the Swinburne method is therefore slightly higher than the true efficiency.

windings, if any) is determined by disconnecting one end of the shunt circuit, as in fig. 9.4, and noting the voltages across the armature for various values of current with the armature stationary. The resistance decreases slightly with increase of armature current, as shown in fig. 9.5. This effect is due to the brush contact resistance being approximately inversely proportional to the current.

If the windings are at room temperature when these measurements are being made, the resistance at normal working temperature should be calculated (section 9.1).

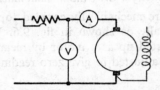

Fig. 9.4 Measurement of armature resistance.

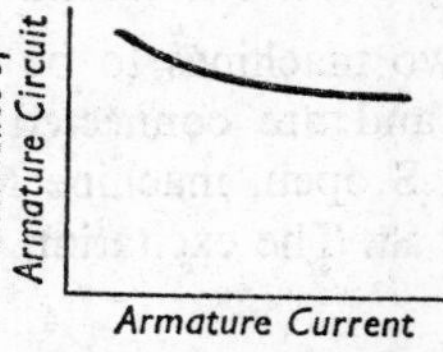

Fig. 9.5 Variation of armature resistance with current.

If I = line current at which the efficiency is to be calculated,

$$\left.\begin{array}{r}\text{corresponding armature} \\ \text{current}\end{array}\right\} = I_a = I + I_s,\ \text{as a generator,}$$

$$= I - I_s,\ \text{as a motor.}$$

If R_a = resistance of armature circuit at working temperature for current I_a (fig. 9.5),

$$\text{corresponding loss in armature circuit} = I_a{}^2 R_a.$$

Since iron, friction and windage losses = $I_1 V$

and   shunt loss = $I_s V$

∴   total losses = $I_a{}^2 R_a + (I_1 + I_s)V.$

If the machine is a generator, output power = IV,

∴   input power = $IV + I_a{}^2 R_a + (I_1 + I_s)V$

and   efficiency $= \dfrac{IV}{IV + I_a{}^2 R_a + (I_1 + I_s)V}$ (9.8)

If the machine is a motor, input power = IV,

∴   output power = $IV - I_a{}^2 R_a - (I_1 + I_s)V$

and   efficiency $= \dfrac{IV - I_a{}^2 R_a - (I_1 + I_s)V}{IV}$ (9.9)

The main advantages of the Swinburne method are: (i) the power required to test a large machine is comparatively small, and (ii) the data enable the efficiency to be calculated at any load.

The main disadvantages are: (i) no account is taken of the stray losses (section 9.1), and (ii) the test does not enable the performance of the machine on full load to be checked; for instance, it does not indicate whether commutation on full load is satisfactory and whether the temperature rise is within the specified limit.

(c) *Regenerative or Hopkinson* method for two similar shunt machines*

The two machines to be tested are mechanically coupled to each other and are connected electrically as shown in fig. 9.6. With switch S open, machine M is started up as a motor by means of starter St. The excitation of G is adjusted to give zero reading on

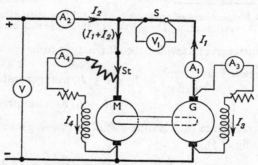

Fig. 9.6 Regenerative or Hopkinson test.

voltmeter V_1 and S is then closed. Current I_1 is adjusted to any desired value by increasing the excitation of G or reducing the excitation of M, and the corresponding readings on the ammeters are noted. If V be the supply voltage, then:

$$\text{output power of G} = I_1 V$$

and

$$\text{input power to M} = (I_1 + I_2)V.$$

An approximate value of the efficiency can be derived by assuming the efficiencies of the two machines to be equal; thus, if η be the efficiency of each machine,

---

* John Hopkinson (1849–98) was Professor of Electrical Engineering at King's College, London.

output power of M $= \eta(I_1 + I_2)V$

= input power to G

∴   output power of G $= \eta \times$ input power to G

$= \eta^2(I_1 + I_2)V.$

Hence $\eta^2(I_1 + I_2)V = I_1V$

∴ $$\eta = \sqrt{\left(\frac{I_1}{I_1 + I_2}\right)} \qquad (9.10)$$

The armature current of M may be 15–30 per cent greater than that of G so that the armature I^2R loss in M is considerably greater than that in G. On the other hand, the excitation of G is appreciably greater than that of M so that the excitation and iron losses in G are correspondingly greater than those in M. Most of these differences can be allowed for by calculating the efficiencies thus:

If R_a = resistance of the armature circuit of each machine, and if I_3 and I_4 = shunt currents of G and M respectively,

armature I^2R loss in G $= (I_1 + I_3)^2R_a$

and   armature I^2R loss in M $= (I_1 + I_2 - I_4)^2R_a.$

Also, loss in shunt circuit of G $= I_3V$

and   loss in shunt circuit of M $= I_4V.$

But   total losses in G and M = power supplied from mains

$= I_2V$

∴   iron, friction and windage losses in G and M

$= I_2V - \{(I_1 + I_3)^2R_a + (I_1 + I_2 - I_4)^2R_a + (I_3 + I_4)V\}$

= (say) C.

We shall now make the only assumption necessary with this method, namely that these losses are equally divided between the two machines.

∴   iron, friction and windage losses per machine $= \frac{1}{2}C$.

For the generator,

total losses $= (I_1 + I_3)^2R_a + I_3V + \frac{1}{2}C =$ (say) P_g

and   output power $= I_1V$

∴   efficiency of G $= \dfrac{I_1V}{I_1V + P_g}$ \qquad (9.11)

For the motor,

$$\text{total losses} = (I_1 + I_2 - I_4)^2 R_a + I_4 V + \tfrac{1}{2}C$$
$$= (\text{say}) \; P_m$$

and input power $= (I_1 + I_2)V$

$$\therefore \quad \text{efficiency of M} = \frac{(I_1 + I_2)V - P_m}{(I_1 + I_2)V} \tag{9.12}$$

The advantages of the Hopkinson method are: (i) the total power required to test two machines is small compared with the power of each machine; (ii) the machines can be tested under full-load conditions so that the commutation and the temperature rise can be checked; and (iii) the efficiency is being determined under load conditions so that the stray loss is being taken into account.

The main disadvantage is the necessity for two practically identical machines to be available.

Example 9.1 *A 100-kW, 460-V shunt generator was run as a motor on no load at its rated voltage and speed. The total current taken was 9·8 A, including a shunt current of 2·7 A. The resistance of the armature circuit (including compoles) at normal working temperature was 0·11 Ω. Calculate the efficiencies at (a) full load and (b) half load.*

(a) Output current at full load $= \dfrac{100 \times 1000}{460} = 217 \cdot 5 \; \text{A},$

$\therefore$ armature current at full load $= 217 \cdot 5 + 2 \cdot 7 = 220 \cdot 2 \; \text{A}.$

Copper loss in armature circuit at full load $\Big\} = (220)^2 \times 0 \cdot 11 = 5325 \; \text{W}.$

Loss* in shunt circuit $= 2 \cdot 7 \times 460 = 1242 \; \text{W}.$

Armature current on no load $= 9 \cdot 8 - 2 \cdot 7 = 7 \cdot 1 \; \text{A},$

$\therefore$ input power to armature on no load $\Big\} = 7 \cdot 1 \times 460 = 3265 \; \text{W}.$

But loss in armature circuit on no load $\Big\} = (7 \cdot 1)^2 \times 0 \cdot 11 = 5 \cdot 5 \; \text{W}.$

This loss is less than 0·2 per cent of the input power and could therefore have been neglected.

* See footnote on p. 256.

Hence, iron, friction and windage losses $= 3260$ W,

and total losses at full load $= 5325 + 1242 + 3260 = 9827$ W

$$= 9.83 \text{ kW.}$$

Input power at full load $= 100 + 9.83 = 109.8$ kW

and efficiency at full load $= \dfrac{100}{109.8} = 0.911$ per unit*

$$= 91.1 \text{ per cent.}$$

(b) Output current at half load $\Big\} = 108.7$ A,

$\therefore$ armature current at half load $\Big\} = 108.7 + 2.7 = 111.4$ A

and copper loss in armature circuit at half load $\Big\} = (111.4)^2 \times 0.11 = 1365$ W

$\therefore$ total losses at half load $= 1365 + 1242 + 3260 = 5867$ W

$$= 5.87 \text{ kW.}$$

Input power at half load $= 50 + 5.87 = 55.87$ kW

and efficiency at half load $= \dfrac{50}{55.87} = 0.895$ per unit

$$= 89.5 \text{ per cent.}$$

Example 9.2 *Two 1000-kW, 500-V shunt machines were tested by the Hopkinson method. When the output current of the generator was 2000 A, the input current from the supply mains (I_2 in fig. 9.6) was 380 A, and the shunt currents of the generator and motor were 22 and 17 A respectively. The resistance of the armature circuit of each machine was 0·01 Ω. Calculate the efficiency of the generator at full load by assuming (a) equal efficiencies and (b) equal iron and friction losses.*

Output current of generator on full load $\Big\} = \dfrac{1000 \times 1000}{500} = 2000$ A.

---

* Per-unit value $= \dfrac{\text{actual value (in any unit)}}{\text{basic or rated value (in same unit)}}$

Thus, if the rated voltage of a system is 200 V, a voltage of 190 V is $190/200 = 0.95$ p.u.; or if an ammeter is reading 5·2 A when the correct value is 5 A, the error is $(5.2 - 5)/5 = 0.04$ p.u. It will be obvious that the percentage value is 100 times the per-unit value.

(*a*) From expression (9.10),

$$\text{efficiency on full load} = \sqrt{\left(\frac{2000}{2000 + 380}\right)}$$
$$= 0{\cdot}9165 \text{ per unit}$$
$$= 91{\cdot}65 \text{ per cent.}$$

(*b*) Armature current of G $= 2000 + 22 = 2022$ A

$\therefore$ copper loss in armature $\left.\begin{array}{c}\\ \text{circuit of G}\end{array}\right\} = (2022)^2 \times 0{\cdot}01 = 40\ 900$ W
$$= 40{\cdot}9 \text{ kW.}$$

Loss in shunt circuit of G $= 22 \times 500 = 11\ 000$ W
$$= 11 \text{ kW.}$$

Armature current of M $= 2000 + 380 - 17 = 2363$ A

$\therefore$ copper loss in armature $\left.\begin{array}{c}\\ \text{circuit of M}\end{array}\right\} = (2363)^2 \times 0{\cdot}01 = 55\ 850$ W
$$= 55{\cdot}85 \text{ kW.}$$

Loss in shunt circuit of M $= 17 \times 500 = 8500$ W
$$= 8{\cdot}5 \text{ kW.}$$

But total losses in G and M $= 380 \times 500 = 190\ 000$ W
$$= 190 \text{ kW}$$

$\therefore$ iron, friction and wind-$\left.\begin{array}{c}\\ \text{age losses in G and M}\end{array}\right\} = 190 - (40{\cdot}9 + 11 + 55{\cdot}85 + 8{\cdot}5)$
$$= 73{\cdot}75 \text{ kW}$$

so that

iron, friction and windage $\left.\begin{array}{c}\\ \text{losses in each machine}\end{array}\right\} = 36{\cdot}9$ kW

Hence total losses in G on full $\left.\begin{array}{c}\\ \text{load}\end{array}\right\} = 40{\cdot}9 + 11 + 36{\cdot}9$
$$= 88{\cdot}8 \text{ kW.}$$

Input power to G on full load $= 1000 + 88{\cdot}8 = 1088{\cdot}8$ kW

$\therefore$ efficiency on full load $= \dfrac{1000}{1088{\cdot}8} = 0{\cdot}9185$ per unit

$$= 91{\cdot}85 \text{ per cent.}$$

Summary of important formulae

$$\text{Efficiency of generator} = \frac{IV}{IV + I_a^2 R_a + I_s V + C} \qquad (9.1)$$

where $C =$ iron, friction and windage losses.

$$\text{Efficiency of motor} = \frac{IV - I_a^2 R_a - I_s V - C}{IV} \qquad (9.2)$$

For maximum efficiency,

$$\text{variable loss} = \text{constant loss} \qquad (9.5)$$

For the brake shown in fig. 9.2,

$$\text{efficiency} = \frac{2\pi(9{\cdot}81\, m - S)rN}{60 \times IV} \qquad (9.7)$$

For Swinburne test,

$$\text{efficiency of generator} = \frac{IV}{IV + I_a^2 R_a + (I_1 + I_s)V} \qquad (9.8)$$

$$\text{efficiency of motor} = \frac{IV - I_a^2 R_a - (I_1 + I_s)V}{IV} \qquad (9.9)$$

where $I_1 = $ armature current on no load.

For Hopkinson test,

$$\text{approximate value of efficiency} = \sqrt{\left(\frac{I_1}{I_1 + I_2}\right)} \qquad (9.10)$$

where $I_1 = $ output current of generator and $I_2 = $ current from supply mains.

EXAMPLES 9

1. The armature of a certain four-pole d.c. generator has an armature eddy-current loss of 500 W when driven at a given speed and field excitation. If the speed is increased by 15 per cent and the flux is increased by 10 per cent, calculate the new value of the eddy-current loss.
2. A shunt-wound generator has a full-load output of 10 kW at a terminal voltage of 240 V. The armature and shunt circuit resistances are 0·6 Ω and 160 Ω respectively. The mechanical and iron losses total 500 W. Calculate the power, in kilowatts, required at the driving shaft at full load and the corresponding value of the efficiency. What will be the approximate output power of the generator at maximum efficiency and what will be the value of that efficiency?
3. If the d.c. machine referred to in Q. 2 is run as a motor taking 40 A from a 240-V supply, what are the values of (*a*) the output power, in kilowatts, and (*b*) the efficiency? Calculate also (*c*) the approximate value of the input power when the efficiency is a maximum, and (*d*) the value of that efficiency.
4. A d.c. shunt motor has an output of 8 kW when running at 750 rev/min off a 480-V supply. The resistance of the armature circuit is 1·2 Ω and that of the shunt circuit is 800 Ω. The efficiency at that load is 83 per

cent. Determine: (*a*) the no-load armature current; (*b*) the speed when the motor takes 12 A; and (*c*) the armature current when the gross torque is 60 N·m. Assume the flux to remain constant.

5. In a brake test on a d.c. motor, the effective load on the brake pulley was 265 N, the effective diameter of the pulley 650 mm and the speed 720 rev/min. The motor took 35 A at 220 V. Calculate the output power, in kilowatts, and the efficiency at this load.

6. In a test on a d.c. motor with the brake shown in fig. 9.2., the mass suspended at one end of the belt was 30 kg and the reading on the spring balance was 65 N. The effective diameter of the brake wheel was 400 mm and the speed was 960 rev/min. The input to the motor was 23 A at 240 V. Calculate: (*a*) the output power and (*b*) the efficiency.

7. A d.c. shunt machine has an armature resistance of 0·5 Ω and a field-circuit resistance of 750 Ω. When run under test as a motor, with no mechanical load, and with 500 V applied to the terminals, the line current was 3 A. Allowing for a drop of 2 V at the brushes, estimate the efficiency of the machine when it operates as a generator with an output of 20 kW at 500 V, the field-circuit resistance remaining unchanged. State the assumptions made.   (S.A.N.C., O.2)

8. Enumerate the various losses which take place in a shunt motor when it is running on no load.

A 230-V shunt motor, running on no load and at normal speed, takes an armature current of 2·5 A from 230-V mains. The field-circuit resistance is 230 Ω and the armature-circuit resistance is 0·3 Ω. Calculate the motor output, in kilowatts, and the efficiency when the total current taken from the mains is 35 A.

If the motor is used as a 230-V shunt generator, find the efficiency and input power for an output current of 35 A.   (E.M.E.U.)

9. In a test on a d.c. shunt generator whose full-load output is 200 kW at 250 V, the following figures were obtained:

(*a*) When running light as a motor at full speed, the line current was 36 A, the field current 12 A and the supply voltage 250 V.

(*b*) With the machine at rest, a p.d. of 6 V produced a current of 400 A through the armature circuit.

Explain how these results may be utilized to obtain the efficiency of the generator, and obtain the efficiency at full load and half load. Neglect voltage drop at the brushes.   (U.E.I.)

10. Enumerate the losses in a d.c. shunt motor and explain how each loss is affected by a change of load.

The current taken by a 460-V shunt motor when running light is 7 A. The resistance of the armature circuit is 0·15 Ω and that of the field circuit is 230 Ω. Calculate the output (in kilowatts) and the efficiency when the current taken is 130 A. Calculate also the armature current at which the efficiency is a maximum.   (App. El., L.U.)

11. In a test on two similar 440-V, 200-kW generators, the circulating current is equal to the full-load current, and in addition, 90 A are

taken from the supply. Calculate the approximate efficiency of each machine.

12. Describe the Hopkinson test for obtaining the efficiency of two similar shunt motors. The readings obtained in such a test were as follows: line voltage, 100 V; motor current, 30 A; generator current, 25 A; armature resistance of each machine, 0·25 Ω. Calculate the efficiency of each machine from these results, ignoring the field currents and assuming that their iron and mechanical losses are the same.

(U.L.C.I.)

13. Describe with the aid of a diagram the Hopkinson test for two similar 230-V shunt machines. The following figures were obtained during such a test on two similar machines: armature currents, 37 A and 30 A, field currents, 0·85 A and 0·8 A. Calculate the efficiencies of the machines and indicate the given currents on a circuit diagram. The armature resistances are each 0·33 Ω. (E.M.E.U.)

CHAPTER 10

Alternating Voltage and Current

10.1 Generation of an alternating e.m.f.

Fig. 10.1 shows a loop AB carried by a spindle DD rotated at a constant speed in an anticlockwise direction in a uniform magnetic field due to poles NS. The ends of the loop are brought out to two slip-rings C_1 and C_2, attached to but insulated from DD. Bearing on

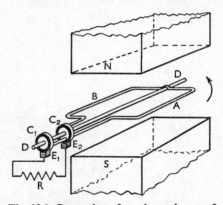

Fig. 10.1 Generation of an alternating e.m.f.

these rings are carbon brushes E_1 and E_2, which are connected to an external resistor R. When the plane of the loop is horizontal, the two sides A and B are moving parallel to the direction of the magnetic flux; therefore, no flux is being cut and no e.m.f. is being generated in the loop.

In fig. 10.2 (a), the vertical dotted lines represent lines of magnetic flux and loop AB is shown after it has rotated through an angle θ from the horizontal position, namely the position of zero e.m.f. Suppose the peripheral velocity of each side of the loop to be v metres per second; then at the instant shown in fig. 10.2, this peripheral velocity can be represented by the length of a line AL drawn at right-angles to the plane of the loop. We can resolve AL into two components AM and AN, perpendicular and parallel re-

266

spectively to the direction of the magnetic flux, as shown in fig. 10.2 (*b*).

Since $\angle MLA = 90° - \angle MAL = \angle MAO = \theta,$

∴ $AM = AL \sin \theta = v \sin \theta.$

The e.m.f. generated in A is due entirely to the component of the velocity perpendicular to the magnetic field. Hence, if *B* is the flux

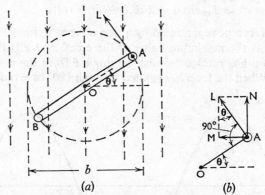

(*a*) (*b*)

Fig. 10.2 Instantaneous value of generated e.m.f.

density in teslas (or webers/metre^2) and if *l* is the length in metres of each of the parallel sides A and B of the loop, it follows from expression (2.3) that:

e.m.f. generated in one side of loop $= Blv \sin \theta$ volts

and

total e.m.f. generated in loop $= 2\,Blv \sin \theta$ volts   (10.1)

i.e. the generated e.m.f. is proportional to $\sin \theta$. When $\theta = 90°$, the plane of the loop is vertical and both sides of the loop are cutting the magnetic flux at the maximum rate, so that the generated e.m.f. is then at its maximum value E_m. From expression (10.1), it follows that when $\theta = 90°$, $E_m = 2Blv$ volts.

If b = breadth of the loop in metres

and n = speed of rotation in revolutions/second,

then $v = \pi bn$ metres/second

and $E_m = 2Bl \times \pi bn$ volts

$= 2\pi BAn$ volts

where $A = lb$ = area of loop in square metres.

If the loop is replaced by a coil of N turns in series, each turn having an area of A square metres,

maximum value of e.m.f. generated in coil

$$= E_m = 2\pi BAnN \text{ volts} \tag{10.2}$$

and instantaneous value of e.m.f. generated in coil

$$= e^* = E_m \sin\theta = 2\pi BAnN \sin\theta \text{ volts} \tag{10.3}$$

This e.m.f. can be represented by a sine wave as in fig. 10.3, where E_m represents the maximum value of the e.m.f. and e is the value after the loop has rotated through an angle θ from the position of zero e.m.f. When the loop has rotated through 180° or π radians, the

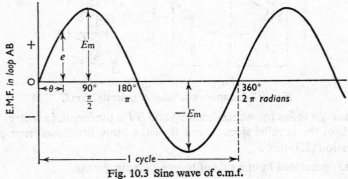

Fig. 10.3 Sine wave of e.m.f.

e.m.f. is again zero. When θ is varying between 180° and 360° (π and 2π radians), side A of the loop is moving towards the right in fig. 10.1 and is therefore cutting the magnetic flux in the opposite direction to that during the first half-revolution. Hence, if we regard the e.m.f. as positive while θ is varying between 0 and 180°, it is negative while θ is varying between 180° and 360°; i.e. when θ varies between 180° and 270°, the value of the e.m.f. increases from zero to $-E_m$ and then decreases to zero as θ varies between 270° and 360°. Subsequent revolutions of the loop merely produce a repetition of the e.m.f. wave.

Each repetition of a variable quantity, recurring at equal intervals,

* Lower-case letters are used to represent instantaneous values and upper-case letters represent definite values such as maximum, average or r.m.s. values. In a.c. circuits, capital I and V without any subscript represent r.m.s. values.

is termed a *cycle*, and the duration of one cycle is termed its *period* (or *periodic time*). The number of such cycles that occur in one second is termed the *frequency* of that quantity. The unit of frequency is the *hertz* (Hz) in memory of Heinrich Rudolf Hertz, who, in 1888, was the first to demonstrate experimentally the existence and properties of electromagnetic radiation predicted by Maxwell in 1865.

Example 10.1 *A coil of* 100 *turns is rotated at* 1500 *rev/min in a magnetic field having a uniform density of* 0·05 *T, the axis of rotation being at right angles to the direction of the flux. The mean area per turn is* 40 *cm²*. *Calculate* (a) *the frequency*, (b) *the period*, (c) *the maximum value of the generated e.m.f. and* (d) *the value of the generated e.m.f. when the coil has rotated through* 30° *from the position of zero e.m.f.*

(*a*) Since the e.m.f. generated in the coil undergoes one cycle of variation when the coil rotates through one revolution,

$$\therefore \qquad \text{frequency} = \text{no. of cycles/second}$$
$$= \text{no. of revolutions/second}$$
$$= 1500/60 = 25 \text{ Hz.}$$

(*b*) $\qquad\qquad$ Period = time of 1 cycle
$$= 1/25 = 0·04 \text{ s.}$$

(*c*) From expression (10.2),

$$E_m = 2\pi \times 0·05 \times 0·004 \times 100 \times 1500/60 = 3·14 \text{ volts.}$$

(*d*) For $\theta = 30°$, $\sin 30° = 0·5$,

$$\therefore \qquad\qquad e = 3·14 \times 0·5 = 1·57 \text{ volts.}$$

In practice, the waveform of the e.m.f. generated in a conductor of an a.c. generator (or alternator) is usually similar to that shown in fig. 6.2. In an actual machine, however, the conductors are distributed in slots around the periphery of the machine and the effect is to make the waveform of the resultant e.m.f. nearly sinusoidal (section 15.4). In most a.c. calculations, the waveforms of both the voltage and the current are assumed to be sinusoidal.

10.2 Relationship between frequency, speed and number of pole pairs
The waveform of the e.m.f. generated in an alternator undergoes one complete cycle of variation when the conductors move past a

N and a S pole; and the shape of the wave over the negative half is exactly the same as that over the positive half. This symmetry of the positive and negative half-cycles does not necessarily hold for waveforms of voltage and current in circuits incorporating rectifiers, thermionic valves or transistors.

If an alternator has p *pairs* of poles and if its speed is n revolutions/second, then

frequency $= f =$ no. of cycles/second

$\qquad\qquad = $ no. of cycles/revolution $\times$ no. of revs/second

$\qquad\qquad = pn$ hertz $\qquad\qquad\qquad\qquad\qquad\qquad$ (10.4)

Thus if a two-pole machine is to generate an e.m.f. having a frequency of 50 Hz, then from expression (10.4),

$$50 = 1 \times n$$
$$\therefore \qquad \text{speed} = 50 \text{ revolutions/second}$$
$$= 50 \times 60 = 3000 \text{ rev/min.}$$

Since it is not possible to have fewer than two poles, the highest speed at which a 50-Hz alternator can be operated is 3000 rev/min.

10.3 Average and r.m.s. values of an alternating current

Let us first consider the general case of a current the waveform of which cannot be represented by a simple mathematical expression. For instance, the wave shown in fig. 10.4 is typical of the current taken by a transformer on no load. If n equidistant midordinates, i_1, i_2, etc., are taken over either the positive or the negative half-cycle, then:

$$\left.\begin{array}{r}\textit{average} \text{ value of current over} \\ \text{half a cycle}\end{array}\right\} = I_{av} = \frac{i_1 + i_2 + \ldots + i_n}{n} \qquad (10.5)$$

Or, alternatively,

$$\text{average value of current} = \frac{\text{area enclosed over half-cycle}}{\text{length of base over half-cycle}} \qquad (10.6)$$

This method of expressing the average value is the more convenient when we come to deal with sinusoidal waves.

In a.c. work, however, the average value is of comparatively little importance. This is due to the fact that it is the power produced by the electric current that usually matters. Thus, if the current represented in fig. 10.4 (*a*) is passed through a resistor having resistance R ohms, the heating effect of i_1 is $i_1^2 R$, that of i_2 is $i_2^2 R$, etc., as shown in fig. 10.4 (*b*). The variation of the heating effect during the second half-cycle is exactly the same as that during the first half-cycle,

$$\therefore \qquad \text{average heating effect} = \frac{i_1^2 R + i_2^2 R + \ldots + i_n^2 R}{n}$$

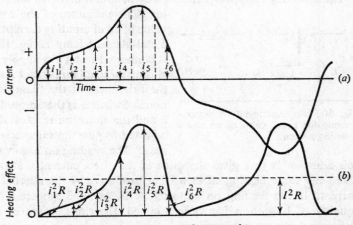

Fig. 10.4 Average and r.m.s. values.

Suppose I to be the value of *direct* current through the same resistance R to produce a heating effect equal to the average heating effect of the alternating current, then:

$$I^2 R = \frac{i_1^2 R + i_2^2 R + \ldots + i_n^2 R}{n}$$

$$\therefore \qquad I = \sqrt{\left(\frac{i_1^2 + i_2^2 + \ldots + i_n^2}{n}\right)} \qquad (10.7)$$

$$= \text{square } root \text{ of the } mean \text{ of the } squares \text{ of the current}$$

$$= \text{root-mean-square (or r.m.s.) value of the current.}$$

This quantity is also termed the *effective* value of the current. It will be seen that the *r.m.s. or effective value of an alternating current is measured in terms of the* direct *current that produces the same heating effect in the same resistance.*

Alternatively, the average heating effect may be expressed:

average heating effect over $\frac{1}{2}$-cycle

$$= \frac{\text{area enclosed by } i^2R \text{ curve over } \frac{1}{2}\text{-cycle}}{\text{length of base}} \qquad (10.8)$$

This is a more convenient expression to use when deriving the r.m.s. value of a sinusoidal current.

The following simple experiment may be found useful in illustrat-

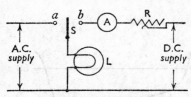

Fig. 10.5 An experiment to demonstrate the r.m.s. value of an alternating current.

ing the significance of the r.m.s. value of an alternating current. A metal-filament lamp L (fig. 10.5) is connected to an a.c. supply by closing switch S on contact a and the brightness of the filament is noted. Switch S is then moved to b and the slider on resistor R is adjusted to give the same brightness.* The reading on a moving-coil ammeter A then gives the value of the direct current that produces the same heating effect as that produced by the alternating current. If the reading on ammeter A is, say, 0·3 ampere when equality of brightness has been attained, the r.m.s. value of the alternating current is 0·3 ampere.

The r.m.s. value is always greater than the average except for a rectangular wave, in which case the heating effect remains constant so that the average and the r.m.s. values are the same.

$$\text{Form factor of a wave} = \frac{\text{r.m.s. value}}{\text{average value}} \qquad (10.9)$$

$$\text{Peak or crest factor of a wave} = \frac{\text{peak or maximum value}}{\text{r.m.s. value}} \qquad (10.10)$$

10.4 Average and r.m.s. values of sinusoidal currents and voltages

If I_m is the maximum value of a current which varies sinusoidally as shown in fig. 10.6 (*a*), the instantaneous value i is represented by:

$$i = I_m \sin \theta$$

where θ = angle in radians from instant of zero current.

---

* For more precise adjustment, an illumination photometer (section 21.7) can be placed at a convenient distance from the lamp, and R adjusted to give the same reading when S is moved over from a to b.

For a very small interval $d\theta$ radian, the area of the shaded strip is $i \cdot d\theta$ ampere radians. The use of the unit 'ampere radian' avoids converting the scale on the horizontal axis from radians to seconds.

$\therefore$ total area enclosed by the current wave over $\frac{1}{2}$-cycle

$$= \int_0^\pi i \cdot d\theta = I_\mathrm{m} \int_0^\pi \sin\theta \cdot d\theta = -I_\mathrm{m}\left[\cos\theta\right]_0^\pi = -I_\mathrm{m}\left[-1-1\right]$$

$= 2I_\mathrm{m}$ ampere-radians.

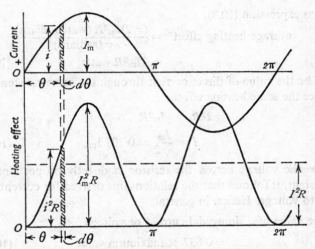

Fig. 10.6 Average and r.m.s. values of a sinusoidal current.

From expression (10.6),

$$\left.\begin{array}{r}\text{average value of current over}\\ \text{half a cycle}\end{array}\right\} = \frac{2I_\mathrm{m}\ (\text{ampere radians})}{\pi\ (\text{radians})}$$

i.e. $\qquad\qquad I_\mathrm{av} = 0.637\,I_\mathrm{m}$ amperes $\qquad\qquad$ (10.11)

If the current be passed through a resistor having resistance R ohms, instantaneous heating effect $= i^2R$ watts.

The variation of i^2R during a complete cycle is shown in fig. 10.6 (b). During interval $d\theta$ radian, heat generated is $i^2R \cdot d\theta$ watt radians and is represented by the area of the shaded strip. Hence:

heat generated during the first
half-cycle} = area enclosed by the i^2R curve

$$= \int_0^\pi i^2 R . d\theta = I_m{}^2 R \int_0^\pi \sin^2 \theta . d\theta$$

$$= \frac{I_m{}^2 R}{2} \int_0^\pi (1 - \cos 2\theta) . d\theta$$

$$= \frac{I_m{}^2 R}{2} \left[\theta - \tfrac{1}{2} \sin 2\theta \right]_0^\pi$$

$$= \frac{\pi}{2} I_m{}^2 R \text{ watt radians.}$$

From expression (10.8),

$$\text{average heating effect*} = \frac{(\pi/2) I_m{}^2 R \text{ (watt radians)}}{\pi \text{ (radians)}}$$

$$= \tfrac{1}{2} I_m{}^2 R \text{ watts} \qquad (10.12)$$

If I be the value of direct current through the same resistance to produce the same heating effect,

$$I^2 R = \tfrac{1}{2} I_m{}^2 R$$

∴ $$I = \frac{I_m}{\sqrt{2}} = 0.707 \, I_m \qquad (10.13)$$

Since the voltage across the resistor is directly proportional to the current, it follows that the relationships derived for current also apply to voltage. Hence, in general:

average value of a sinusoidal current or voltage

$$= 0.637 \times \text{maximum value} \qquad (10.14)$$

r.m.s. value of a sinusoidal current or voltage

$$= 0.707 \times \text{maximum value} \qquad (10.15)$$

From expressions (10.9) and (10.10),

$$\text{form factor of a sine wave} = \frac{0.707 \times \text{maximum value}}{0.637 \times \text{maximum value}}$$

$$= 1.11 \qquad (10.16)$$

* This result is really obvious from the fact that $\sin^2 \theta = \tfrac{1}{2} - \tfrac{1}{2} \cos 2\theta$. In words, this means that the square of a sine wave may be regarded as being made up of two components: (a) a constant quantity equal to half the maximum value of the $\sin^2 \theta$ curve, and (b) a cosine curve having twice the frequency of the $\sin \theta$ curve. From fig. 10.6 it is seen that the curve of the heating effect undergoes two cycles of change during one cycle of current. The average value of component (b) over a complete cycle is zero; hence the average heating effect is $\tfrac{1}{2} I_m{}^2 R$.

and peak or crest factor of a sine $\left.\right\}$ wave $\left.\right\}$ $= \dfrac{\text{maximum value}}{0{\cdot}707 \times \text{maximum value}}$

$$= 1{\cdot}414 \qquad (10.17)$$

Example 10.2 *A moving-coil ammeter, a thermal* ammeter and a rectifier are connected in series with a resistor across a 110-V a.c. supply. The circuit has a resistance of 50 Ω to current in one direction and an infinite resistance to current in the reverse direction. Calculate:* (a) *the readings on the ammeters, and* (b) *the form and peak factors of the current wave. Assume the supply voltage to be sinusoidal.*

(*a*) Maximum value of the voltage $= 110/0{\cdot}707 = 155{\cdot}5$ V

$\therefore$ maximum value of the current $= 155{\cdot}5/50 = 3{\cdot}11$ A.

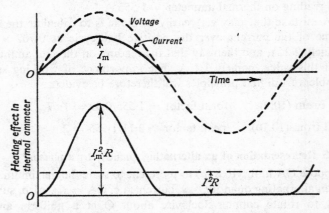

Fig. 10.7 Waveforms of voltage, current and power for example 10.2.

During the positive half-cycle the current is proportional to the voltage and is therefore sinusoidal, as shown in fig. 10.7 (*a*),

$\therefore$ average value of current over the positive $\left.\right\}$ half-cycle $\left.\right\}$ $= 0{\cdot}637 \times 3{\cdot}11$

$$= 1{\cdot}98 \text{ A.}$$

During the negative half-cycle, the current is zero. Owing, however, to the inertia of the moving system, the moving-coil ammeter reads the average value of the current over the *whole* cycle,

---

* A thermal ammeter is an instrument the operation of which depends upon the heating effect of a current (see section 22.7).

$$\therefore \qquad \text{reading on M.C. ammeter} = \frac{1 \cdot 98}{2} = 0 \cdot 99 \text{ A.}$$

The variation of the heating effect in the thermal ammeter is shown in fig. 10.7 (*b*), the maximum power being $I_m{}^2 R$, where R is the resistance of the instrument.

From expression (10.12) it is seen that the average heating effect over the positive half-cycle is $\frac{1}{2} I_m{}^2 R$; and since no heat is generated during the second half-cycle, it follows that the average heating effect over a complete cycle is $\frac{1}{4} I_m{}^2 R$.

If I be the direct current to produce the same heating effect,

$$I^2 R = \tfrac{1}{4} I_m{}^2 R$$

$$\therefore \qquad I = \tfrac{1}{2} I_m = 3 \cdot 11/2 = 1 \cdot 555 \text{ A,}$$

i.e. reading on thermal ammeter = 1·555 A.

A mistake that may very easily be made is to calculate the r.m.s. value of the current over the positive half-cycle as $0 \cdot 707 \times 3 \cdot 11$, namely 2·2 A, and then say that the reading on thermal ammeter is half this value, namely 1·1 A. The importance of working such a problem from first principles will therefore be evident.

(*b*) From (10.9),   form factor = 1·555/0·99 = 1·57

and from (10.10),   peak factor = 3·11/1·555 = 2.

10.5 Representation of an alternating quantity by a phasor*

Suppose OA in fig. 10.8 (*a*) to represent to scale the maximum value of an alternating quantity, say, current; i.e. OA = I_m. Also, suppose OA to rotate counter-clockwise about O at a uniform angular velocity. This is purely a conventional direction which has been universally adopted. An arrowhead is drawn at the outer end of the phasor, partly to indicate which end is assumed to move and partly to indicate the precise length of the phasor when two or more phasors happen to coincide.

Fig. 10.8 (*a*) shows OA when it has rotated through an angle θ from the position occupied when the current was passing through its

---

* It has been the practice to refer to a line, such as OA in fig. 10.8 (*a*), as a *vector* when it represents the magnitude and phase of a sinusoidal alternating quantity. Strictly, however, a vector should only be used to represent the magnitude and direction of a quantity *in space*, e.g. the magnitude and direction of a force in mechanics. The term *phasor* has been adopted to replace the term *vector* for representing graphically the magnitude and phase of a sinusoidal alternating current or voltage.

zero value. If AB and AC are drawn perpendicular to the horizontal and vertical axes respectively:

$$OC = AB = OA \sin \theta$$
$$= I_m \sin \theta$$
$$= i, \text{ namely the value of the current at that instant.}$$

Hence the projection of OA on the vertical axis represents to scale the instantaneous value of the current. Thus when $\theta = 90°$, the projection is OA itself; when $\theta = 180°$, the projection is zero and

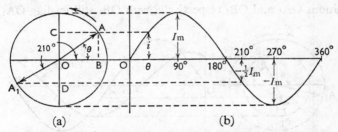

(a)   (b)

Fig. 10.8 Phasor representation of an alternating quantity.

corresponds to the current passing through zero from a positive to a negative value; when $\theta = 210°$, the phasor is in position OA_1, and the projection $= OD = \frac{1}{2}OA_1 = -\frac{1}{2}I_m$; and when $\theta = 360°$, the projection is again zero and corresponds to the current passing through zero from a negative to a positive value. It follows that OA rotates through one revolution or 2π radians in one cycle of the current wave.

If f is the frequency in hertz, then OA rotates through f revolutions or $2\pi f$ radians in 1 second. Hence the angular velocity of OA is $2\pi f$ radians/second and is denoted by the symbol ω (omega):

i.e. $\omega = 2\pi f$ radians/second.

If the time taken by OA in fig. 10.8 to rotate through an angle θ radians be t seconds, then:

$$\theta = \text{angular velocity} \times \text{time}$$
$$= \omega t = 2\pi f t \text{ radians.}$$

We can therefore express the instantaneous value of the current thus:

$$i = I_m \sin \theta = I_m \sin \omega t = I_m \sin 2\pi f t.$$

Let us next consider how two quantities such as voltage and current can be represented by a phasor diagram. Fig. 10.9 (*b*) shows the voltage leading the current by an angle ϕ. In fig. 10.9 (*a*), OA represents the maximum value of the current and OB that of the voltage. The angle between OA and OB must be the same angle ϕ as in fig. 10.9 (*b*). Consequently when OA is along the horizontal axis, the current at that instant is zero and the value of the voltage is represented by the projection of OB on the vertical axis. These values correspond to instant O in fig. 10.9 (*b*).

After the phasors have rotated through an angle θ, they occupy positions OA₁ and OB₁ respectively, with OB₁ still leading OA₁ by

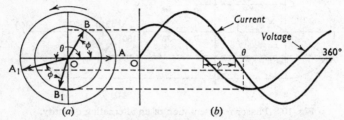

Fig. 10.9 Phasor representation of quantities differing in phase.

the same angle ϕ; and the instantaneous values of the current and voltage are again given by the projections of OA₁ and OB₁ on the vertical axis, as shown by the horizontal dotted lines.

If the instantaneous value of the current is represented by

$$i = I_\mathrm{m} \sin \theta,$$

then the instantaneous value of the voltage is represented by

$$v = V_\mathrm{m} \sin (\theta + \phi)$$

where $I_\mathrm{m} = \mathrm{OA}$ and $V_\mathrm{m} = \mathrm{OB}$ in fig. 10.9 (*a*).

The current in fig. 10.9 is said to *lag* the voltage by an angle ϕ or the voltage is said to *lead* the current by an angle ϕ. The *phase difference* ϕ between the two phasors remains constant, irrespective of their position.

10.6 Addition and subtraction of sinusoidal alternating quantities

Suppose OA and OB in fig. 10.10 to be phasors representing to scale the maximum values of, say, two alternating voltages having the same frequency but differing in phase by an angle ϕ. Complete the parallelogram OACB and draw the diagonal OC. Project OA, OB and OC on to the vertical axis. Then for the positions shown in fig. 10.10:

$$\text{instantaneous value of OA} = \text{OD}$$
$$\text{,,} \qquad \text{,,} \quad \text{,,} \quad \text{OB} = \text{OE}$$
$$\text{and} \qquad\qquad \text{,,} \qquad\qquad \text{,,} \quad \text{,,} \quad \text{OC} = \text{OF}.$$

Since AC is parallel and equal to OB, DF = OE,

$$\therefore \qquad\qquad \text{OF} = \text{OD} + \text{DF} = \text{OD} + \text{OE}$$

i.e. $\quad \left.\begin{array}{c} \text{the instantaneous} \\ \text{value of OC} \end{array}\right\} = \left\{\begin{array}{l} \text{sum of the instantaneous values of} \\ \text{OA and OB.} \end{array}\right.$

Hence OC represents the maximum value of the resultant voltage to the scale that OA and OB represent the maximum values of the

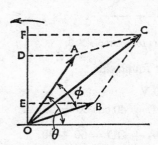

Fig. 10.10 Addition of phasors. Fig. 10.11 Subtraction of phasors.

separate voltages. OC is therefore termed the *phasor sum* of OA and OB; and it is evident that OC is less than the arithmetic sum of OA and OB except when the latter are in phase with each other. This is the reason why it is seldom correct in a.c. work to add voltages or currents together arithmetically.

If voltage OB is to be subtracted from OA, then OB is produced backwards so that OB_1 is equal and opposite to OB (fig. 10.11). The diagonal OD of the parallelogram drawn on OA and OB_1 represents the *phasor difference* of OA and OB.

Example 10.3 *The instantaneous values of two alternating voltages are represented respectively by $v_1 = 60 \sin \theta$ volts and $v_2 = 40 \sin(\theta - \pi/3)$ volts. Derive an expression for the instantaneous value of* (a) *the sum and* (b) *the difference of these voltages.*

*

(*a*) It is usual to draw the phasors in the position corresponding to $\theta = 0$,* i.e. OA in fig. 10.12 is drawn to scale along the X axis to represent 60 volts, and OB is drawn $\pi/3$ radians or 60° behind OA to represent 40 volts. The diagonal OC of the parallelogram drawn on OA and OB represents the phasor sum of OA and OB. By measurement, OC = 87 volts and angle ϕ between OC and the X axis is 23·5°, namely 0·41 radian; hence:

instantaneous sum of the two voltages $= 87 \sin (\theta - 0.41)$ volts.

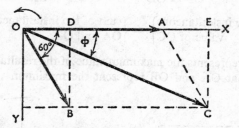

Fig. 10.12 Addition of phasors for example 10.3.

Alternatively, this expression can be found thus:

horizontal component of OA = 60 V

horizontal component of OB = OD = 40 cos 60° = 20 V.

$\therefore$ resultant horizontal component $\left. \right\}$ = OA + OD = 60 + 20

$= 80$ V = OE in fig. 10.12.

Vertical component of OA = 0

vertical component of OB = BD $= -40 \sin 60°$

$= -34.64$ V.

$\therefore$ resultant vertical component $\left. \right\}$ $= -34.64$ V = CE.

The minus sign merely indicates that the resultant vertical component is *below* the horizontal axis and that the resultant voltage must therefore lag relative to the reference phasor OA.

* The idea of a phasor rotating continuously serves to establish its physical significance, but its application in circuit analysis is simplified by *fixing* the phasor in position corresponding to $t = 0$, as in fig. 10.12, thereby eliminating the time function. Such a phasor represents the magnitude of the sinusoidal quantity and its phase relative to a reference quantity, e.g. in fig. 10.12, phasor OB lags the reference phasor OA by 60°.

Hence maximum value of resultant$\left.\begin{array}{}\\ \text{voltage}\end{array}\right\}$ = OC = $\sqrt{\{(80)^2 + (-34 \cdot 64)^2\}}$

$$= 87 \cdot 2 \text{ V.}$$

If ϕ is the phase difference between OC and OA,

$$\tan \phi = \text{EC/OE} = -34 \cdot 64/80 = -0 \cdot 433$$
$$\therefore \qquad \phi = -23 \cdot 4° = -0 \cdot 41 \text{ radian}$$

and instantaneous value of$\left.\begin{array}{}\\ \text{resultant voltage}\end{array}\right\}$ = $87 \cdot 2 \sin (\theta - 0 \cdot 41)$ volts.

(*b*) The construction for subtracting OB from OA is obvious from

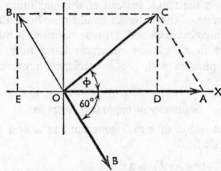

Fig. 10.13 Subtraction of phasors for example 10.3.

fig. 10.13. By measurement, OC = 53 volts and $\phi = 41° = 0 \cdot 715$ radian.

$\therefore$ instantaneous difference of the$\left.\begin{array}{}\\ \text{two voltages}\end{array}\right\}$ = $53 \sin (\theta + 0 \cdot 715)$ volts.

Alternatively, resultant horizontal$\left.\begin{array}{}\\ \text{component}\end{array}\right\}$ = OA − OE = 60 − 20

$$= 40 \text{ V} = \text{OD in fig. 10.13,}$$

and resultant vertical component = $B_1E = 34 \cdot 64$ V

$$= \text{DC in fig. 10.13.}$$

$\therefore$ maximum value of resultant$\left.\begin{array}{}\\ \text{voltage}\end{array}\right\}$ = OC = $\sqrt{\{(40)^2 + (34 \cdot 64)^2\}}$

$$= 52 \cdot 9 \text{ V}$$

and $\qquad\qquad \tan \phi = \text{DC/OD} = 34 \cdot 64/40$

$$= 0 \cdot 866$$

$\therefore \qquad\qquad\qquad \phi = 40 \cdot 9° = 0 \cdot 714$ radian

and instantaneous value of$\left.\begin{array}{}\\ \text{resultant voltage}\end{array}\right\}$ = $52 \cdot 9 \sin (\theta + 0 \cdot 714)$ volts.

10.7 Phasor diagrams drawn with r.m.s. values instead of maximum values

It is important to note that when alternating voltages and currents are represented by phasors it is assumed that their waveforms are sinusoidal. It has already been shown that for sine waves the r.m.s. or effective value is 0·707 times the maximum value. Furthermore, ammeters and voltmeters are almost invariably calibrated to read the r.m.s. values. Consequently it is much more convenient to make the length of the phasors represent r.m.s. rather than maximum values. If the phasors of fig. 10.12, for instance, were drawn to represent to scale the r.m.s. instead of the maximum values of the voltages, the shape of the diagram would remain unaltered and the phase relationships between the various quantities would remain unaffected. Hence in all phasor diagrams from now onwards, the lengths of the phasors will, for convenience, represent the r.m.s. values.

Summary of important formulae

Instantaneous value of e.m.f. generated in a coil rotating in a uniform magnetic field

$$= e = E_m \sin \theta$$
$$= 2\pi BANn \sin \theta \text{ volts} \tag{10.3}$$
$$f = np \tag{10.4}$$

For n equidistant mid-ordinates over half a cycle,

$$\text{average value} = \frac{i_1 + i_2 + \ldots + i_n}{n} \tag{10.5}$$

and r.m.s. or effective value $= \sqrt{\left(\dfrac{i_1{}^2 + i_2{}^2 + \ldots + i_n{}^2}{n}\right)} \tag{10.7}$

For sinusoidal waves,

$$\text{average value} = 0.637 \times \text{maximum value} \tag{10.11}$$
and r.m.s. or effective value $= 0.707 \times \text{maximum value} \tag{10.13}$

$$\text{Form factor} = \frac{\text{r.m.s. value}}{\text{average value}} \tag{10.9}$$
$$= 1.11 \text{ for a sine wave.}$$

$$\text{Peak or crest factor} = \frac{\text{peak or maximum value}}{\text{r.m.s. value}} \tag{10.10}$$
$$= 1.414 \text{ for a sine wave.}$$

EXAMPLES 10

Note. In questions reproduced from examination papers, the term 'vector' has been replaced by 'phasor' by kind permission of the respective institutions.

1. A coil is wound with 300 turns on a square former having sides 50 mm in length. Calculate the maximum value of the e.m.f. generated in the coil when it is rotated at 2000 rev/min in a uniform magnetic field of density 0·8 tesla. What is the frequency of this e.m.f.?
2. Explain what is meant by the terms *wave form, frequency* and *average value.*
   A square coil of side 10 cm, having 100 turns, is rotated at 1200 rev/min about an axis through the centre and parallel with two sides in a uniform magnetic field of density 0·4 T. Calculate: (*a*) the frequency; (*b*) the root-mean-square value of the induced e.m.f.; (*c*) the instantaneous value of the induced e.m.f. when the coil is at a position 40 degrees after passing its maximum induced voltage.
                                                                      (U.E.I., O.1)
3. A rectangular coil, measuring 30 cm by 20 cm and having 40 turns, is rotated about an axis coinciding with one of its longer sides at a speed of 1500 rev/min in a uniform magnetic field of flux density 0·075 T. Find, from first principles, an expression for the instantaneous e.m.f. induced in the coil, if the flux is at right-angles to the axis of rotation.
   Evaluate this e.m.f. at an instant 0·002 s after the plane of the coil has been perpendicular to the field.   (U.L.C.I., O.2)
4. Calculate the speed at which an eight-pole alternator must be driven in order that it may generate an e.m.f. having a frequency of 60 Hz.
5. An alternator driven at 375 rev/min generates an e.m.f. having a frequency of 50 Hz. Calculate the number of poles.
6. If a four-pole alternator is driven at a speed of 1800 rev/min, what is the frequency of the generated e.m.f.?
7. Define the following terms: *period, phase difference.*
   An alternating voltage had the following instantaneous values, in volts, measured at equal intervals of time over half cycle:

   0,   30,   40,   45,   55,   80,   90,   56,   0.

   Draw the waveform to scale over half a cycle and determine the average and the root-mean-square values of the voltage.   (U.E.I., O.1)
8. The following ordinates were taken during a half-cycle of a symmetrical alternating-current wave, the current varying in a linear manner between successive points:

Phase angle, in degrees	0	15	30	45	60	75	90
Current, in amperes	0	3·6	8·4	14	19·4	22·5	25
Phase angle, in degrees	105	120	135	150	165	180	
Current, in amperes	25·2	23	15·6	9·4	4·2	0	

Determine: (*a*) the mean value; (*b*) the r.m.s. value; (*c*) the form factor. (N.C.T.E.C., O.1)

9. Explain the significance of the root-mean-square value of an alternating current or voltage waveform. Define the form factor of such a waveform.

Calculate from first principles the r.m.s. value and form factor of an alternating voltage having the following values over half a cycle, both half-cycles being symmetrical about the zero axis:

Time, in milliseconds	0	1	2	3	4
Voltage, in volts	0	100	100	100	0

These voltage values are joined by straight lines. (U.L.C.I., O.1)

10. A triangular voltage wave has the following values over one-half cycle, both half-cycles being symmetrical about the zero axis:

Time (ms)	0	10	20	30	40	50	60	70	80	90	100
Voltage (V)	0	2	4	6	8	10	8	6	4	2	0

Plot half cycle of the waveform and hence determine: (*a*) the average value; (*b*) the r.m.s. value; (*c*) the form factor. (U.L.C.I., O.1)

11. Describe, and explain the action of, an ammeter suitable for measuring the r.m.s. value of a current.

An alternating current has a periodic time $2T$. The current for a time one-third of T is 50 A; for a time one-sixth of T, it is 20 A; and zero for a time equal to one-half of T. Calculate the r.m.s. and average values of this current. (E.M.E.U. O.2)

12. A triangular voltage wave has a periodic time of $\frac{3}{100}$ s. For the first $\frac{2}{100}$ s of each cycle it increases uniformly at the rate of 1000 V/s, while for the last $\frac{1}{100}$ s it falls away uniformly to zero. Find, graphically or otherwise: (*a*) its average value; (*b*) its r.m.s. value; (*c*) its form factor. (E.M.E.U., O.2)

13. Define the root-mean-square value of an alternating current. Explain why this value is more generally employed in a.c. measurements than either the average or the peak value.

Under what circumstances would it be necessary to know (*a*) the average and (*b*) the peak value of an alternating current or voltage?

Calculate the ratio of the peak values of two alternating currents which have the same r.m.s. values, when the waveform of one is sinusoidal and that of the other triangular. What effect would lack of symmetry of the triangular wave about its peak value have upon this ratio? (App. El., L.U.)

14. A voltage, 100 sin 314*t* volts, is maintained across a circuit consisting of a half-wave rectifier in series with a 50-Ω resistor. The resistance of the rectifier may be assumed to be negligible in the forward direction and infinity in the reverse direction. Calculate the average and the r.m.s. values of the current.

15. State what is meant by the root-mean-square value of an alternating current and explain why the r.m.s. value is usually more important than either the maximum or the mean value of the current.

A moving-coil ammeter and a moving-iron ammeter are connected in series with a rectifier across a 110-V (r.m.s.) a.c. supply. The total resistance of the circuit in the conducting direction is 60 Ω and that in the reverse direction may be taken as infinity. Assuming the waveform of the supply voltage to be sinusoidal, calculate from first principles the reading on each ammeter. (App. El., L.U.)

16. If the waveform of a voltage has a form factor of 1·15 and a peak factor of 1·5, and if the peak value is 4·5 kV, calculate the average and the r.m.s. values of the voltage.

17. An alternating current was measured by a d.c. milliammeter in conjunction with a full-wave rectifier. The reading on the milliammeter was 7 mA. Assuming the waveform of the alternating current to be sinusoidal, calculate (a) the r.m.s. value and (b) the maximum value of the alternating current.

18. An alternating current, when passed through a resistor immersed in water for 5 minutes, just raised the temperature of the water to boiling point. When a direct current of 4 A was passed through the same resistor under identical conditions, it took 8 minutes to boil the water. Find the r.m.s. value of the alternating current. Neglect other factors than heat given to the water.

If a rectifier type of ammeter connected in series with the resistor read 5·2 A when the alternating current was flowing, find the form factor of the alternating current.

19. Explain what is meant by the *r.m.s. value* of an alternating current.

In a certain circuit supplied from 50-Hz mains, the potential difference has a maximum value of 500 V and the current has a maximum value of 10 A. At the instant $t = 0$, the instantaneous values of the p.d. and the current are 400 V and 4 A respectively, both increasing positively. Assuming sinusoidal variation, state trigonometrical expressions for the instantaneous values of the p.d. and the current at time t.

Calculate the instantaneous values at the instant $t = 0·015$ s and find the angle of phase difference between the p.d. and the current.

Sketch the complexor (phasor) diagram. (S.A.N.C., O.1)

20. Explain with the aid of a sketch how the r.m.s. value of an alternating current is obtained.

An alternating current i is represented by:

$$i = 10 \sin 942t \text{ amperes.}$$

Determine (a) the frequency, (b) the period, (c) the time taken from $t = 0$ for the current to reach a value of 6 A for a first and second time, (d) the energy dissipated when the current flows through a 20-Ω resistor for 30 minutes. (S.A.N.C., O.1)

21. (a) Explain the term *r.m.s. value* as applied to an alternating current.

(b) An alternating current flowing through a circuit has a maximum value of 70 A, and lags the applied voltage by 60°. The maximum value of the voltage is 100 V, and both current and voltage waveforms are sinusoidal. Plot the current and voltage waveforms in their correct relationship for the positive half of the voltage. What is the value of the current when the voltage is at a positive peak?

(W.J.E.C., O.1)

22. Two sinusoidal e.m.f.s of peak values 50 V and 20 V respectively but differing in phase by 30° are induced in the same circuit. Draw the phasor diagram and find the peak and r.m.s. values of the resultant e.m.f.

23. Two impedances are connected in parallel to the supply, the first takes a current of 40 A at a lagging phase angle of 30°, and the second a current of 30 A at a leading phase angle of 45°. Draw a phasor diagram to scale to represent the supply voltage and these currents. From this diagram, by construction, determine the total current taken from the supply and its phase angle.

24. Two circuits connected in parallel take alternating currents which can be expressed trigonometrically as:

$i_1 = 13 \sin 314t$ amperes and $i_2 = 12 \sin (314t + \pi/4)$ amperes.

Sketch the waveforms of these currents to illustrate maximum values and phase relationships.

By means of a phasor diagram drawn to scale, determine the resultant of these currents, and express it in trigonometric form. Give also the r.m.s. value and the frequency of the resultant current.

(E.M.E.U., O.1)

25. The voltage drops across two components, when connected in series across an a.c. supply, are: $v_1 = 180 \sin 314t$ volts and $v_2 = 120 \sin (314t + \pi/3)$ volts respectively. Determine with the aid of a phasor diagram: (*a*) the voltage of the supply in trigonometric form; (*b*) the r.m.s. voltage of the supply; (*c*) the frequency of the supply.

(E.M.E.U., O.1)

26. Three e.m.f.s, $e_A = 50 \sin \omega t$, $e_B = 80 \sin (\omega t - \pi/6)$ and $e_C = 60 \cos \omega t$ volts, are induced in three coils connected in series so as to give the phasor sum of the three e.m.f.s. Calculate the maximum value of the resultant e.m.f. and its phase relative to e.m.f. e_A. Check the results by means of a phasor diagram drawn to scale.

If the connections to coil B were reversed, what would be the maximum value of the resultant e.m.f. and its phase relative to e_A?

(App. El., L.U.)

27. Find graphically or otherwise the resultant of the following four voltages:

$e_1 = 25 \sin \omega t$, $e_2 = 30 \sin (\omega t + \pi/6)$,

$e_3 = 30 \cos \varphi t$, $e_4 = 20 \sin (\omega t - \pi/4)$

Express the answer in a similar form.   (U.L.C.I., O.2)

28. Four e.m.f.s, $e_1 = 100 \sin \omega t$, $e_2 = 80 \sin (\omega t - \pi/6)$, $e_3 = 120 \sin (\omega t + \pi/4)$ and $e_4 = 100 \sin (\omega t - 2\pi/3)$, are induced in four coils connected in series so that the sum of the four e.m.f.s is obtained. Find graphically *or* by calculation the resultant e.m.f. and its phase difference with (*a*) e_1 and (*b*) e_2.

If the connections to the coil in which the e.m.f. e_2 is induced are reversed, find the new resultant e.m.f.   (E.M.E.U., O.2)

29. The currents in three circuits connected in parallel to a voltage source are: (i) 4 A in phase with the applied voltage; (ii) 6 A lagging the applied voltage by 30°; (iii) 2 A leading the applied voltage by 45°. Represent these currents to scale on a phasor diagram, showing their correct relative phase displacement with each other.

Determine, graphically or otherwise, the total current taken from the source, and its phase angle with respect to the supply voltage.

(U.L.C.I., O.1)

CHAPTER 11

Single-Phase Circuits

11.1 Alternating current in a circuit possessing resistance only

Consider a circuit having a resistance R ohms connected across the terminals of an alternator A, as in fig. 11.1, and suppose the alternating voltage to be represented by the sine wave of fig. 11.2. If the value of the voltage at any instant B is v volts, the value of the current* at that instant is given by:

$$i = v/R \text{ amperes.}$$

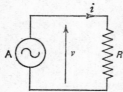

Fig. 11.1 Circuit with resistance only.

When the voltage is zero, the current is also zero; and since the current is proportional to the voltage, the waveform of the current is exactly the same as that of the voltage. Also the two quantities are *in phase* with each other; that is, they pass through their zero values at the same instant and attain their maximum values in a given direction at the same instant. Hence the current wave is as shown dotted in fig. 11.2.

If V_m and I_m be the maximum values of the voltage and current respectively, it follows that:

$$I_m = V_m/R \tag{11.1}$$

But the r.m.s. value of a sine wave is 0·707 times the maximum value, so that:

r.m.s. value of voltage $= V = 0·707\ V_m$

and   r.m.s. value of current $= I = 0·707\ I_m.$

---

* An arrow or arrowhead, as in fig. 11.1, is used to indicate the direction in which the current flows when it is regarded positive. It is immaterial which direction is chosen as positive; but once it has been decided upon for a given circuit or network, the same direction must be adhered to for all the currents and voltages involved in that circuit or network.

Substituting for I_m and V_m in (11.1) we have:

$$\frac{I}{0 \cdot 707} = \frac{V}{0 \cdot 707 R}$$

$$\therefore \qquad I = V/R \qquad (11.2)$$

Hence Ohm's Law can be applied without any modification to an a.c. circuit possessing resistance only.

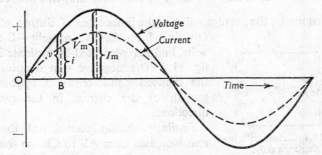

Fig. 11.2 Voltage and current waveforms for a resistive circuit.

If the instantaneous value of the applied voltage is represented by:

$$v = V_m \sin \theta,$$

then instantaneous value of current in a resistive circuit $\Big\} = i = \dfrac{V_m}{R} \sin \theta \qquad (11.3)$

The phasors representing the voltage and current in a resistive circuit are shown in fig. 11.3. The two phasors are actually coincident,

Fig. 11.3 Phasor diagram for a resistive circuit.

but are drawn slightly apart so that the identity of each may be clearly recognized. As mentioned on p. 280, it is usual to draw the phasors in the position corresponding to $\theta = 0$. Hence the phasors representing the voltage and current of expression (11.3) are drawn along the X axis.

11.2 Alternating current in a circuit possessing inductance only

Let us consider the effect of a sinusoidal current flowing through a coil having an inductance of L henrys and a negligible resistance, as in fig. 11.4. For instance, let us consider what is happening during the first quarter-cycle of fig. 11.5. This quarter-cycle has been divided into three equal intervals, OA, AC and CF seconds. During interval OA, the current increases from zero to AB; hence the average rate of change of current is AB/OA amperes/second, and is represented by ordinate JK drawn midway between O and A. From expression (4.2), the e.m.f., in volts, induced in a coil

$$= -L \times \text{rate of change of current in amperes per second};$$

consequently, the average value of the induced e.m.f. during interval OA is $-L \times$ AB/OA, namely $-L \times$ JK volts, and is represented by ordinate JQ in fig. 11.5. The negative sign denotes that the induced e.m.f. tends to oppose the growth of the current in the positive direction.

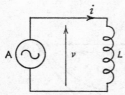

Fig. 11.4 Circuit with inductance only.

Similarly, during interval AC, the current increases from AB to CE, so that the average rate of change of current is DE/AC amperes/second, which is represented by ordinate LM in fig. 11.5; and the corresponding induced e.m.f. is $-L \times$ LM volts and is represented by LR. During the third interval CF, the average rate of change of current is GH/CF, namely NP amperes/second; and the corresponding induced e.m.f. is $-L \times$ NP volts and is represented by NS. At instant F, the current has ceased growing but has not yet begun to decrease; consequently the rate of change of current is then zero. The induced e.m.f. will therefore have decreased from a maximum at O to zero at F. Curves can now be drawn through the derived points, as shown in fig. 11.5.

During the second quarter-cycle, the current decreases, so that the rate of change of current is negative and the induced e.m.f. becomes positive, tending to prevent the current decreasing. Since the sine wave of current is symmetrical about ordinate FH, the curves representing the rate of change of current and the e.m.f. induced in the coil will be symmetrical with those derived for the first quarter-cycle. Since the rate of change of current at any instant is proportional to the slope of the current wave at that instant, it is evident that the value of the induced e.m.f. increases from zero at F to a maximum at T and then decreases to zero at U in fig. 11.5.

By using shorter intervals, for example by taking ordinates at intervals of 10° and noting the corresponding values of the ordinates with the aid of trigonometrical tables, it is possible to derive fairly accurately the shapes of the curves representing the rate of change of current and the induced e.m.f.

From fig. 11.5, it will be seen that the induced e.m.f. attains its

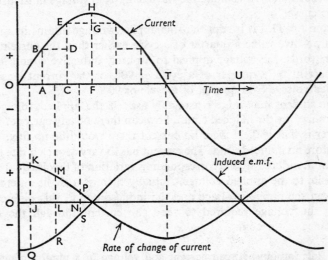

Fig. 11.5 Waveforms of current, rate of change of current and induced e.m.f.

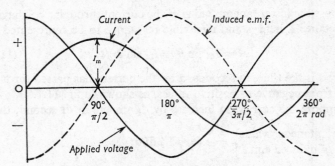

Fig. 11.6 Voltage and current waveforms for a purely inductive circuit.

maximum positive value a quarter of a cycle *after* the current has done the same thing—in fact, it goes through all its variations a quarter of a cycle after the current has gone through similar variations. Hence the induced e.m.f. is said to *lag* the current by a quarter of a cycle or the current is said to *lead* the induced e.m.f. by a quarter of a cycle.

Since the resistance of the coil is assumed negligible, we may regard the whole of the applied voltage as being absorbed in neutralizing the induced e.m.f. Hence the curve of applied voltage in fig. 11.6 can be drawn exactly equal and opposite to that of the induced e.m.f.; and since the latter is sinusoidal, the wave of applied voltage must also be a sine curve.

From fig. 11.6 it is seen that the applied voltage attains its maximum positive value a quarter of a cycle earlier than the current; in other words, the voltage applied to a purely inductive circuit leads the current by a quarter of a cycle or 90°, or the current lags the applied voltage by a quarter of a cycle or 90°.

The student may quite reasonably ask: If the applied voltage is neutralized by the induced e.m.f., how can there be any current? The answer is that if there were no current there would be no flux, and therefore no induced e.m.f. The current has to vary at such a rate that the e.m.f. induced by the corresponding variation of flux is equal and opposite to the applied voltage. Actually there is a slight difference between the applied voltage and the induced e.m.f., this difference being the voltage required to send the current through the low resistance of the coil.

11.3 Relationship between current and voltage in a purely inductive circuit

Suppose the instantaneous value of the current through a coil having inductance L henrys and negligible resistance to be represented by

$$i = I_m \sin \theta = I_m \sin 2\pi ft \qquad (11.4)$$

where t is the time, in seconds, after the current has passed through zero from negative to positive values, as shown in fig. 11.7.

Suppose the current to increase by di ampere in dt second, then

$$\left.\begin{array}{r}\text{instantaneous value} \\ \text{of induced e.m.f.}\end{array}\right\} = e = -L \cdot di/dt$$

$$= -LI_m\frac{d}{dt}(\sin 2\pi ft)$$

$$= -2\pi fLI_m \cos 2\pi ft$$

$$= 2\pi fLI_m \sin (2\pi ft - \pi/2) \qquad (11.5)$$

Since f represents the number of cycles/second, the duration of 1 cycle $= 1/f$ second. Consequently:

when $t = 0$, $\cos 2\pi ft = 1$

and induced e.m.f. $= -2\pi fLI_m$.

When $t = 1/2f$, $\cos 2\pi ft = \cos \pi = -1$,

and induced e.m.f. $= 2\pi fLI_m$.

Hence the induced e.m.f. is represented by the dotted curve in fig. 11.7, lagging the current by a quarter of a cycle.

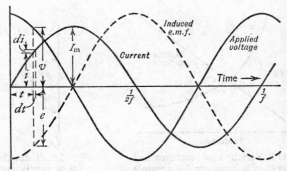

Fig. 11.7 Voltage and current waveforms for a purely inductive circuit.

Since the resistance of the circuit is assumed negligible, the whole of the applied voltage is absorbed in neutralizing the induced e.m.f.,

$\therefore$ instantaneous value of $\left.\right\}$ applied voltage $= v$

$$= 2\pi fLI_m \cos 2\pi ft$$
$$= 2\pi fLI_m \sin (2\pi ft + \pi/2) \qquad (11.6)$$

Comparison of expressions (11.4) and (11.6) shows that the applied voltage leads the current by a quarter of a cycle. Also, from expression (11.6), it follows that the maximum value V_m of the applied voltage is $2\pi fLI_m$,

i.e. $V_m = 2\pi fLI_m$, so that $V_m/I_m = 2\pi fL$.

If I and V be the r.m.s. values, then:

$$\frac{V}{I} = \frac{0 \cdot 707\, V_m}{0 \cdot 707\, I_m} = 2\pi fL$$
$$= \textit{inductive reactance}.$$

The inductive reactance is expressed in ohms and is represented by the symbol X_L.

Hence
$$I = \frac{V}{2\pi fL} = \frac{V}{X_L} \tag{11.7}$$

The inductive reactance is proportional to the frequency and the current produced by a given voltage is inversely proportional to the frequency, as shown in fig. 11.8.

The phasor diagram for a purely inductive circuit is given in fig.

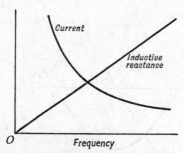

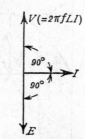

Fig. 11.8 Variation of reactance and current with frequency for a purely inductive circuit.

Fig. 11.9 Phasor diagram for a purely inductive circuit.

11.9, where E represents the r.m.s. value of the e.m.f. induced in the circuit; and V, equal and opposite to E, represents the r.m.s. value of the applied voltage.

11.4 Mechanical analogy of an inductive circuit

One of the most puzzling things to a student commencing the study of alternating currents is the behaviour of a current in an inductive circuit. For instance, why should the current in fig. 11.6 be at its maximum value when there is no applied voltage? Why should there be no current when the applied voltage is at its maximum? Why should it be possible to have a voltage applied in one direction and a current flowing in the reverse direction, as is the case during the second and fourth quarter-cycles in fig. 11.6?

It may therefore be found helpful to consider a simple mechanical analogy—the simpler the better. In mechanics, the *inertia* of a body opposes any change in the *speed* of that body. The effect of inertia is therefore analogous to that of *inductance* in opposing any change in the *current*.

Suppose we take a heavy metal cylinder C (fig. 11.10), such as a pulley or an armature, and roll it backwards and forwards on a horizontal surface between two extreme positions A and B. Let us consider the forces and the speed while C is being rolled from A to B. At first the speed is zero, but the force applied to the body is at its maximum, causing C to accelerate towards the right. This applied force is reduced—as indicated by the length of the arrows in fig. 11.10—until it is zero when C is midway between A and B. C ceases to accelerate and will therefore have attained its maximum speed from left to right.

Immediately after C has passed the mid-point, the direction of the

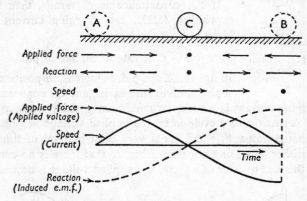

Fig. 11.10 Mechanical analogy of a purely inductive circuit.

applied force is reversed and increased until the body is brought to rest at B and then begins its return movement.

The reaction of C, on the other hand, is equal and opposite to the applied force and corresponds to the e.m.f. induced in the inductive circuit.

From an inspection of the arrows in fig. 11.10 it is seen that the speed in a given direction is a maximum a quarter of a complete oscillation after the applied force has been a maximum in the same direction, but a quarter of an oscillation before the reaction reaches its maximum in that direction. This is analogous to the current in a purely inductive circuit lagging the applied voltage by a quarter of a cycle and leading the induced e.m.f. by a quarter of a cycle. Also it is evident that when the speed is a maximum the applied force is zero, and that when the applied force is a maximum the speed is zero;

and during the second half of the movement indicated in fig. 11.10, the direction of motion is opposite to that of the applied force. These relationships correspond exactly to those found for a purely inductive circuit.

11.5 Alternating current in a circuit possessing capacitance only

Fig. 11.11 shows a capacitor C connected in series with an ammeter A across the terminals of an alternator; and the alternating voltage applied to C is represented in fig. 11.12. Suppose this voltage to be positive when it makes plate D positive relative to plate E.

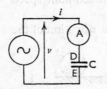

If the capacitance is C farads, then from expression (5.23), the charging current i is given by:

Fig. 11.11 Circuit with capacitance only.

$$i = C \times \text{rate of change of p.d.}$$

In fig. 11.12, the p.d. is increasing positively at the maximum rate at instant O; consequently the charging current is also at its maximum positive value at that instant. A quarter of a cycle later, the applied voltage has reached its maximum value V_m; and for a very brief interval of time the p.d. is neither increasing nor decreasing, so that there is no current. During the next quarter of a cycle, the applied voltage is decreasing.

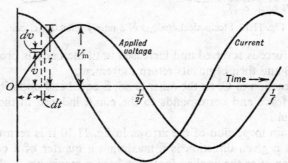

Fig. 11.12 Voltage and current waveforms for a purely capacitive circuit.

Consequently the capacitor discharges, the discharge current being in the negative direction.

When the voltage is passing through zero, the slope of the voltage curve is at its maximum, i.e. the p.d. is varying at the maximum rate; consequently the current is also a maximum at that instant.

11.6 Relationship between current and voltage in a purely capacitive circuit

Suppose the instantaneous value of the voltage* applied to a capacitor having capacitance C farads to be represented by:

$$v = V_m \sin \theta = V_m \sin 2\pi ft \qquad (11.8)$$

If the applied voltage increases by dv volt in dt second (fig. 11.12), then from (5.23),

instantaneous value
of current $= i = C \cdot dv/dt$

$$= C\frac{d}{dt}(V_m \sin 2\pi ft)$$

$$= 2\pi fCV_m \cos 2\pi ft$$

$$= 2\pi fCV_m \sin (2\pi ft + \pi/2) \quad (11.9)$$

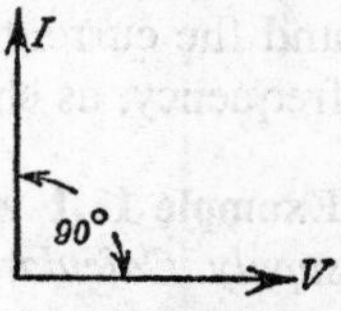

Fig. 11.13 Phasor diagram for a purely capacitive circuit.

Comparison of expressions (11.8) and (11.9) shows that the current leads the applied voltage by a quarter of a cycle, and the current and voltage can be represented by phasors as in fig. 11.13.

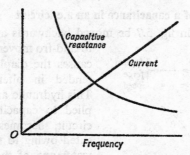

Fig. 11.14 Variation of reactance and current with frequency for a purely capacitive circuit.

From expression (11.9) it follows that the maximum value I_m of the current is $2\pi fCV_m$,

$$\therefore \qquad \frac{V_m}{I_m} = \frac{1}{2\pi fC}.$$

---

* It will be noted that we start with the voltage wave in this case, whereas with inductance we started with the current wave. The reason for this is that in the case of inductance, we derive the induced e.m.f. by differentiating the current expression; whereas with capacitance, we derive the current by differentiating the voltage expression.

Hence, if I and V be the r.m.s. values,

$$\frac{V}{I} = \frac{1}{2\pi f C} = \text{capacitive reactance} \qquad (11.10)$$

The capacitive reactance is expressed in ohms and is represented by the symbol X_C.

Hence, $I = 2\pi f C V = V/X_C$ \qquad (11.11)

The capacitive reactance is inversely proportional to the frequency, and the current produced by a given voltage is proportional to the frequency, as shown in fig. 11.14.

Example 11.1 *A 30-μF capacitor is connected across a 400-V, 50-Hz supply. Calculate* (a) *the reactance of the capacitor and* (b) *the current.*

(a) From expression (11.10):

$$\text{reactance} = \frac{1}{2 \times 3.14 \times 50 \times 30 \times 10^{-6}} = 106.2 \ \Omega.$$

(b) From expression (11.11):

$$\text{current} = 400/106.2 = 3.77 \text{ A}.$$

11.7 Analogies of a capacitance in an a.c. circuit

If the piston P in fig. 5.7 be moved backwards and forwards, the to-and-fro movement of the water causes the diaphragm to be distended in alternate directions. This hydraulic analogy, when applied to capacitance in an a.c. circuit, becomes rather complicated owing to the inertia of the water and of the piston; and as we do not want to take the effect of inertia into account at this stage, it is more convenient to consider a very light flexible strip L (fig. 11.15), such as a metre rule, having one end rigidly clamped.

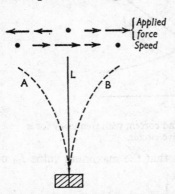

Fig. 11.15 Mechanical analogy of a capacitive circuit.

Let us apply an alternating force comparatively slowly by hand so as to oscillate L between positions A and B.

When L is in position A, the applied force is at its maximum towards the *left*. As the force is reduced, L moves towards the *right*. Immediately L has passed the centre position, the applied force has to be increased towards the right, while the speed in this direction is decreasing. These variations are indicated by the lengths of the arrows in fig. 11.15. From the latter it is seen that the speed towards the right is a maximum a quarter of a cycle before the applied force is a maximum in the same direction. The speed is therefore the analogue of the alternating current, and the applied force is that of the applied voltage. Hence capacitance in an electrical circuit is analogous to elasticity in mechanics, whereas inductance is analogous to inertia (section 11.4).

11.8 Alternating current in a circuit possessing resistance, inductance and capacitance in series

We have already considered resistive, inductive and capacitive circuits separately. An actual circuit, however, may have resistance

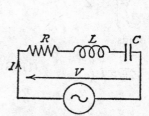

Fig. 11.16 Circuit with *R*, *L*
and *C* in series.

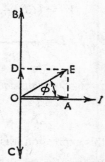

Fig. 11.17 Phasor diagram
for fig. 11.16.

and inductance,* or resistance and capacitance, or resistance, inductance and capacitance in series. Hence, if we first consider the general case of *R*, *L* and *C* in series, we can adapt the results to the other two cases by merely omitting the capacitive or the inductive reactance from the expressions derived for the general case.

* A coil possesses resistance and inductance, and its resistance can be considered as part of the total resistance of the circuit. If the coil has an iron core, the hysteresis and eddy-current losses in the core are equivalent to an increase in the effective resistance of the coil. Thus, if the iron loss in the core is 40 W when the current in the coil is 5 A, the resistor to give the same loss when connected in series with the circuit has a resistance of $40/5^2$, namely 1·6 Ω. Hence if the winding has a resistance of, say, 4 Ω, the effective resistance of the coil is (4 + 1·6), namely 5·6 Ω.

Fig. 11.16 shows a circuit having resistance R ohms, inductance L henrys and capacitance C farads in series, connected across an a.c. supply of V volts (r.m.s.) at a frequency of f hertz. Let I be the r.m.s. value of the current in amperes.

From section 11.1, the p.d. across R is RI volts in phase with the current and is represented by phasor OA in phase with OI in fig. 11.17.* From section 11.3, the p.d. across L is $2\pi fLI$, and is represented by phasor OB, leading the current by 90°; and from section 11.6, the p.d. across C is $I/(2\pi fC)$ and is represented by phasor OC lagging the current by 90°.

Since OB and OC are in direct opposition, their resultant is OD = OB − OC, OB being assumed greater than OC in fig. 11.17; and the supply voltage is the phasor sum of OA and OD, namely OE. From fig. 11.17,

$$OE^2 = OA^2 + OD^2 = OA^2 + (OB - OC)^2$$

$$\therefore \qquad V^2 = (RI)^2 + \left(2\pi fLI - \frac{I}{2\pi fC}\right)^2$$

so that

$$I = \frac{V}{\sqrt{\left\{R^2 + \left(2\pi fL - \frac{1}{2\pi fC}\right)^2\right\}}} = \frac{V}{Z}. \qquad (11.12)$$

where $\qquad Z = impedance$ of circuit in ohms

$$= \frac{V}{I} = \sqrt{\left\{R^2 + \left(2\pi fL - \frac{1}{2\pi fC}\right)^2\right\}} \qquad (11.13)$$

From this expression it is seen that:

$$\text{resultant reactance} = 2\pi fL - \frac{1}{2\pi fC}$$

$$= \text{inductive reactance—capacitive reactance.}$$

If $\qquad \phi = $ phase difference between the current and the supply voltage,

$$\tan \phi = \frac{AE}{OA} = \frac{OD}{OA} = \frac{OB - OC}{OA} = \frac{2\pi fLI - I/(2\pi fC)}{RI}$$

$$= \frac{\text{inductive reactance} - \text{capacitive reactance}}{\text{resistance}} \qquad (11.14)$$

* When drawing a phasor diagram of this type, one should always start with the phasor representing the quantity that is common to the various components of the circuit. Thus with a series circuit it is the current that is the same in all the components, so that in fig. 11.17, OI is the first phasor to be drawn. For parallel circuits it is the voltage that is common to the various components, so that the voltage phasor is then the first to be drawn.

$$\cos \phi = \frac{\text{OA}}{\text{OE}} = \frac{RI}{ZI} = \frac{\text{resistance}}{\text{impedance}} \qquad (11.15)$$

$$\text{and } \sin \phi = \frac{\text{AE}}{\text{OE}} = \frac{\text{resultant reactance}}{\text{impedance}} \qquad (11.16)$$

If the inductive reactance is greater than the capacitive reactance, $\tan \phi$ is positive and the current lags the supply voltage by an angle ϕ; if less, $\tan \phi$ is negative, signifying that the current leads the supply voltage by an angle ϕ.

If a circuit consists of a coil having resistance R ohms and inductance L henrys, such a circuit can be considered as possessing resistance and inductance in series; and from (11.13),

$$\text{impedance} = \sqrt{\{R^2 + (2\pi fL)^2\}}$$

and from (11.14) the phase angle by which the current lags the supply voltage is given by:

$$\phi = \tan^{-1} (2\pi fL/R).$$

Example 11.2 *A coil having a resistance of 12 Ω and an inductance of 0·1 H is connected across a 100-V, 50-Hz supply. Calculate:* (a) *the reactance and the impedance of the coil;* (b) *the current; and* (c) *the phase difference between the current and the applied voltage.*

When solving problems of this kind, students should first of all draw a circuit diagram (fig. 11.18) and insert all the known quantities.

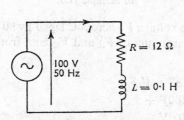

Fig. 11.18 Circuit diagram for example 11.2.

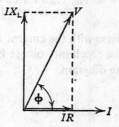

Fig. 11.19 Phasor diagram for example 11.2.

They should then proceed with the phasor diagram, fig. 11.19. It is not essential to draw the phasor diagram to exact scale, but it is helpful to draw it approximately correctly since it is then easy to make a rough check of the calculated values.

(a)　　　Reactance $= X_L = 2\pi fL$

$$= 2\pi \times 50 \times 0{\cdot}1 = 31{\cdot}4 \ \Omega.$$

Impedance $= Z = \sqrt{(R^2 + X_L{}^2)}$

$$= \sqrt{(12^2 + 31{\cdot}4^2)} = 33{\cdot}6 \ \Omega.$$

(b)　　　Current $= I = V/Z = 100/33{\cdot}6 = 2{\cdot}975$ A.

(c)　　　Tan $\phi = X/R = 31{\cdot}4/12 = 2{\cdot}617$

$\therefore$　　　　　　　　$\phi = 69°.$

Example 11.3 *A metal-filament lamp, rated at 750 W, 100 V, is to be connected in series with a capacitor across a 230-V, 60-Hz supply. Calculate (a) the capacitance required and (b) the phase angle between the current and the supply voltage.*

(a) The circuit is given in fig. 11.20, where R represents the lamp. In the phasor diagram of fig. 11.21, the voltage V_R across R is in

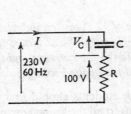

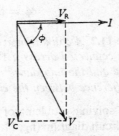

Fig. 11.20 Circuit diagram　　　Fig. 11.21 Phasor diagram
for example 11.3.　　　　　　　　for example 11.3.

phase with the current I, while the voltage V_C across C lags I by 90°. The resultant voltage V is the phasor sum of V_R and V_C, and from the diagram:

$$V^2 = V_R{}^2 + V_C{}^2$$

$\therefore$　　　　$(230)^2 = (100)^2 + V_C{}^2$

$\therefore$　　　　　　　　$V_C = 207$ V.

Rated current of lamp $= \dfrac{750 \text{ watts}}{100 \text{ volts}} = 7{\cdot}5$ A.

From (11.11)　　　　$7{\cdot}5 = 2 \times 3{\cdot}14 \times 60 \times C \times 207$

$\therefore$　　　　　　　　$C = 96 \times 10^{-6} \ \dot{\text{F}} = 96 \ \mu\text{F}.$

(b) If $\phi =$ phase angle between the current and the supply voltage,

$$\cos \phi = V_R/V \text{ (from fig. 11.21)}$$
$$= 100/230 = 0.435$$
$$\therefore \qquad \phi = 64° 12'.$$

Example 11.4 *A circuit having a resistance of* 12 Ω, *an inductance of* 0·15 *H and a capacitance of* 100 μF *in series, is connected across a* 100-*V,* 50-*Hz supply. Calculate:* (a) *the impedance;* (b) *the current;* (c) *the voltages across R, L and C;* (d) *the phase difference between the current and the supply voltage.*

The circuit diagram is the same as that of fig. 11.16.

(*a*) From (11.13),

$$Z = \sqrt{\left\{ (12)^2 + \left(2 \times 3.14 \times 50 \times 0.15 - \frac{10^6}{2 \times 3.14 \times 50 \times 100} \right)^2 \right\}}$$

$$= \sqrt{\{144 + (47.1 - 31.85)^2\}} = 19.4 \ \Omega.$$

(*b*) Current $= V/Z = 100/19.4 = 5.15$ A.

(*c*) Voltage across $R = V_R = 12 \times 5.15 = 61.8$ V,

voltage across $L = V_L = 47.1 \times 5.15 = 242.5$ V

and voltage across $C = V_C = 31.85 \times 5.15 = 164$ V.

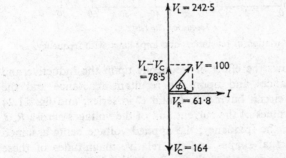

Fig. 11.22 Phasor diagram for example 11.4.

These voltages and current are represented by the respective phasors in fig. 11.22.

(*d*) Phase difference between current and supply voltage

$$= \phi = \cos^{-1} (V_R/V) = \cos^{-1} 61.8/100 = 51° 50'.$$

Or, alternatively, from (11.14),

$$\phi = \tan^{-1} (47.1 - 31.85)/12 = \tan^{-1} 1.271 = 51° 48'.$$

11.9 Circuit with R, L and C in series: Effect of frequency variation

In fig. 11.17, we assumed the capacitive reactance to be less than the inductive reactance, and in consequence the expressions for capacitive reactance in (11.12), (11.13) and (11.14) have a negative sign in front of them. This was a matter of chance in the above discussion, but it is, in general, found convenient to regard inductive reactance as positive and capacitive reactance as negative.*

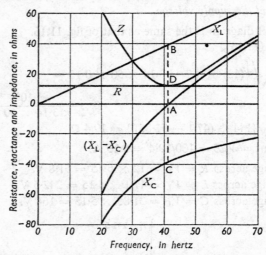

Fig. 11.23 Variation of reactances and impedance with frequency.

Fig. 11.23 shows the effect of frequency upon the inductive and capacitive reactances and upon the resultant reactance and the impedance of a circuit having R, L and C in series; and fig. 11.24 shows how the values of the current and of the voltages across R, L and C vary with the frequency, the applied voltage being assumed constant. The actual shapes and the relative magnitudes of these curves depend upon the values chosen for R, L and C; and the curves in figs. 11.23 and 11.24 have been derived for the circuit of Ex. 11.4, where $R = 12\ \Omega$, $L = 0{\cdot}15$ H, $C = 100\ \mu$F and applied voltage $= 100$ V.

It will be seen from fig. 11.23 that for frequency OA, the inductive

* These positive and negative signs are merely conventions. A 100-Ω capacitive reactance has exactly the same effect as 100-Ω inductive reactance as far as the *magnitude* of the current for a given voltage is concerned. It is only the *phase* of the current that is affected.

reactance AB and the capacitive reactance AC are equal in magnitude so that their resultant is zero. Consequently the impedance is then the same as the resistance AD of the circuit. Furthermore, as the frequency is reduced below OA or increased above OA, the impedance increases and therefore the current decreases. Also, it will be seen from fig. 11.24 that when the frequency is OA, the voltages across L and C are equal and each is much greater than the supply voltage, namely 100 V. Such a condition is referred to as *resonance*, an effect that is extremely important in radio work, partly because it provides a simple method of increasing the sensitivity of a radio

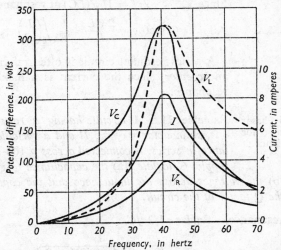

Fig. 11.24 Effect of frequency variation upon voltages across R, L and C in series.

receiver and partly because it gives selectivity, i.e. it enables a signal of given frequency to be considerably magnified so that it can be separated from signals of other frequencies.

11.10 Resonance in a circuit having R, L and C in series

From the preceding section, it is evident that for a series circuit, resonance occurs when:

$$\text{inductive reactance} - \text{capacitive reactance} = 0$$

i.e. when
$$2\pi f L = \frac{1}{2\pi f C}$$

or
$$f = \frac{1}{2\pi\sqrt{(LC)}} \tag{11.17}$$

At this frequency, $Z = R$ and $I = V/R$. If the resistance is small compared with the inductive and capacitive reactances, the p.d.s across the latter, namely $2\pi fLI$ and $I/2\pi fC$, are many times the supply voltage, as represented by the phasors in fig. 11.25.

Voltage magnification at resonance $= \dfrac{\text{voltage across } L \text{ (or } C)}{\text{supply voltage}}$

$$= \frac{2\pi fLI}{RI} = \frac{2\pi fL}{R} \qquad (11.18)$$

$$= Q \text{ factor of the circuit.}$$

Since $\qquad 2\pi f = 1/\sqrt{(LC)}$ at resonance,

$$\therefore \qquad Q \text{ factor} = \frac{2\pi fL}{R} = \frac{1}{R}\sqrt{\left(\frac{L}{C}\right)} \qquad (11.19)$$

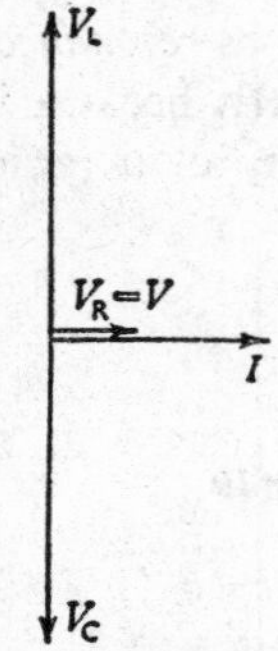

A series resonant circuit is often referred to as an *acceptor*, since the current is a maximum at resonance.

Fig. 11.25 Phasor diagram for series circuit at resonance.

Example 11.5 *A circuit, having a resistance of 4 Ω, an inductance of* 0·5 *H and a variable capacitance in series, is connected across a* 100-V, 50-Hz *supply. Calculate:* (a) *the capacitance to give resonance;* (b) *the voltages across the inductance and the capacitance; and* (c) *the Q factor of the circuit.*

(a) For resonance, $\quad 2\pi fL = 1/2\pi fC$

$$\therefore \qquad\qquad C = \frac{1}{(2 \times 3\cdot14 \times 50)^2 \times 0\cdot5}$$

$$= 20\cdot3 \times 10^{-6}\ \text{F} = 20\cdot3\ \mu\text{F}.$$

(b) At resonance, $\qquad I = V/R = 100/4 = 25$ A.

$\therefore$ p.d. across inductance $= V_L = 2 \times 3\cdot14 \times 50 \times 0\cdot5 \times 25$

$$= 3925\ \text{V}$$

and p.d. across capacitor $= V_C = 3925$ V.

Or alternatively, from (11.11),

$$V_C = \frac{25 \times 10^6}{2 \times 3\cdot14 \times 50 \times 20\cdot3} = 3925\ \text{V}.$$

(c) From (11.18),

$$Q \text{ factor} = \frac{2 \times 3\cdot14 \times 50 \times 0\cdot5}{4} = 39\cdot25.$$

The voltages and current are represented by the respective phasors, but not to scale, in fig. 11.25, and fig. 11.26 shows how the current taken by this circuit varies with frequency, the applied voltage being assumed constant at 100 V.

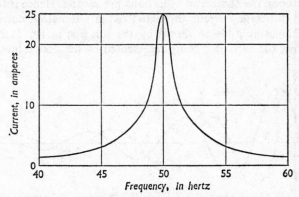

Fig. 11.26 Variation of current with frequency for circuit of example 11.5.

In this example the voltages across the inductance and the capacitance at resonance are each nearly forty times the supply voltage. A fuller explanation, however, is necessary to understand the physical significance of such a result.

11.11 Natural frequency of oscillations in a circuit possessing inductance and capacitance

Suppose C in fig. 11.27 to be a capacitor whose capacitance can be varied from, say, 1 to 20 μF, and suppose L to be an inductor whose inductance is variable between, say, 0·01 and 0·1H. A loudspeaker P is connected across a resistor R having a low resistance. A two-way switch S enables C to be charged from a battery B and discharged through L, and R is adjusted to give a

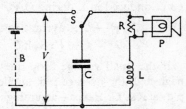

Fig. 11.27 A capacitor discharged through an inductor.

convenient volume of sound in P. It is found that each time C is discharged through L, a pizzicato note is emitted by P, similar to the sound produced by plucking the string of a violin or 'cello.

Further, the pitch of the note can be varied by varying either C or L—the larger the capacitance and the inductance, the lower the pitch. But it is well known that a musical sound requires the vibration of some medium for its production and that the pitch is dependent upon the number of vibrations per second. Hence it follows that the discharge current through P is an alternating current of diminishing amplitude as shown by the full line in fig. 11.28, the p.d. across the capacitor being represented by the dotted line.

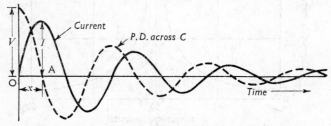

Fig. 11.28 Voltage and current waveforms for fig. 11.27.

From expression (5.40), the energy stored in C at instant O is $\frac{1}{2}CV^2$ joules. At instant A there is no p.d. across and therefore no energy in C; but the current is I amperes, so that the energy stored in L is $\frac{1}{2}LI^2$ joules (section 4.9). If we neglect the energy wasted in the resistance of the circuit during interval OA, then:

$$\tfrac{1}{2}LI^2 = \tfrac{1}{2}CV^2$$

and
$$I = V\sqrt{(C/L)} \qquad (11.20)$$

If x seconds be the duration of the quarter-cycle OA, average e.m.f. induced in L during OA

$$= -L \times \text{average rate of change of current}$$
$$= -L \times I/x \text{ volts.}$$

If the small voltage drop due to the resistance of the circuit be neglected, the average p.d. applied to L is the same as the average p.d. across C. Hence—assuming a sine wave—we have

$$\frac{LI}{x} = \frac{2}{\pi} V \text{ (section 10.4).}$$

Substituting for I the value given in (11.20):

$$\frac{LV}{x} \times \sqrt{\frac{C}{L}} = \frac{2}{\pi} \text{ V,}$$

$$\therefore \qquad x = \frac{\pi}{2}\sqrt{(LC)} \text{ seconds},$$

so that the duration of one cycle $= 4x = 2\pi\sqrt{(LC)}$ seconds

and
$$\text{frequency} = \frac{1}{2\pi\sqrt{(LC)}} \qquad (11.21)$$

This quantity is termed the *natural frequency* of the circuit and represents the frequency with which energy is oscillating backwards and forwards between the capacitor and the inductor, the energy being at one moment stored as electrostatic energy in the capacitor, and a quarter of a cycle later as magnetic energy in the inductor. Owing to loss in the resistance of the circuit, the net amount of energy available to be passed backwards and forwards between L and C gradually decreases.

11.12 Oscillation of energy at resonance

In section 11.10 it was explained that resonance occurs in an a.c. circuit when

$$f = \frac{1}{2\pi\sqrt{(LC)}} \qquad (11.17)$$

From expressions (11.17) and (11.21) it follows that the condition for resonance is that the frequency of the applied alternating voltage

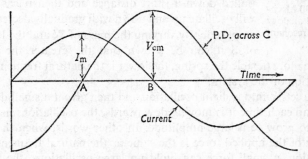

Fig. 11.29 Waveforms of current and p.d. for a capacitor.

is the same as the natural frequency of oscillation of the circuit. This condition enables a large amount of energy to be maintained in oscillation between L and C; thus, if the current and the p.d. across the capacitor in a resonant circuit be represented by the curves of fig. 11.29, the magnetic energy stored in L at instant A is $\frac{1}{2}LI_m^2$ joules, and the electrostatic energy in C at instant B is $\frac{1}{2}CV_{cm}^2$

joules, where V_{cm} represents the maximum value of the voltage across the capacitor.

Since $I_m = 2\pi f C V_{cm}$, and from (11.17), $L = \dfrac{1}{(2\pi f)^2 C}$,

$$\therefore \qquad \tfrac{1}{2}LI_m^2 = \tfrac{1}{2} \times \frac{1}{(2\pi f)^2 C} \times (2\pi f C V_{cm})^2$$

$$= \tfrac{1}{2}C V_{cm}^2,$$

i.e. the magnetic energy in L at instant A is equal to the electrostatic energy in C at instant B, and the power taken from the supply is simply that required to move this energy backwards and forwards between L and C through the resistance of the circuit.

11.13 Mechanical analogy of a resonant circuit

It was pointed out in section 11.4 that inertia in mechanics is analogous to inductance in the electric circuit, and in section 11.7 that elasticity is analogous to capacitance. A very simple mechanical analogy of an electrical circuit possessing inductance, capacitance and a very small resistance can therefore be obtained by attaching a mass W (fig. 11.30) to the lower end of a helical spring S, the upper end of which is rigidly supported. If W is pulled down a short distance and then released, it will oscillate up and down with gradually decreasing amplitude. By varying the mass of W and the length of S it can be shown that the greater the mass and the more flexible the spring, the lower is the natural frequency of oscillation of the system.

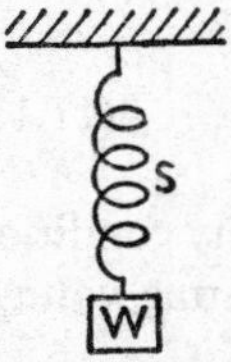

Fig. 11.30 Mechanical analogy of a resonant circuit.

If we set W into a slight oscillation and then give it a small downward tap each time it is moving downwards, the oscillations may be made to grow to a large amplitude. In other words, when the frequency of the applied force is the same as the natural frequency of oscillation, a small force can build up large oscillations, the work done by the applied force being that required to supply the losses involved in the transference of energy backwards and forwards between the kinetic and potential forms of energy.

Examples of resonance are very common; for instance, the rattling of a loose member of a vehicle at a particular speed or of a loudspeaker diaphragm when reproducing a sound of a certain pitch, and the oscillations of the pendulum of a clock and of the balance wheel of a watch due to the small impulse given regularly through the escapement mechanism from the mainspring.

11.14 Alternating current in a circuit possessing resistance, inductance and capacitance in parallel

Suppose three circuits possessing resistance R ohms, inductance L henrys and capacitance C farads respectively to be connected in parallel, as in fig. 11.31, across a supply voltage V of frequency f hertz.

Current through $R = I_R = V/R$, in phase with V,

current through $L = I_L = \dfrac{V}{2\pi fL}$, lagging V by $90°$

and current through $C = I_C = 2\pi fCV$, leading V by $90°$.

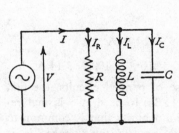

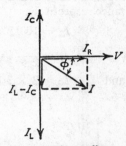

Fig. 11.31 R, L and C in parallel.

Fig. 11.32 Phasor diagram for fig. 11.31.

These currents and the supply voltage are represented by the phasors in fig. 11.32, where it is assumed that I_L is greater than I_C.

Resultant of I_L and $I_C = I_L - I_C$,

and current from supply $= I =$ phasor sum of I_R and $(I_L - I_C)$

$$= \sqrt{\{I_R{}^2 + (I_L - I_C)^2\}}$$

If $\phi =$ phase difference between the supply voltage and the resultant current,

$$\tan \phi = (I_L - I_C)/I_R.$$

If $\tan \phi$ is positive, the resultant current lags the supply voltage by an angle ϕ; if negative, the resultant current leads.

Example 11.6 *Three circuits possessing a resistance of* 50 Ω, *an inductance of* 0·15 *H, and a capacitance of* 100 μF *respectively, are*

*

connected in parallel across a 100-*V*, 50-*Hz supply. Calculate:* (a) *the current in each circuit;* (b) *the resultant current; and* (c) *the phase angle between the resultant current and the supply voltage.*

(*a*) The circuit diagram is given in fig. 11.33, where I_R, I_L and I_C

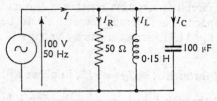

Fig. 11.33 Circuit diagram for example 11.6.

represent the currents through the resistance, inductance and capacitance respectively.

$$I_R = 100/50 = 2 \text{ A,}$$

$$I_L = \frac{100}{2 \times 3 \cdot 14 \times 50 \times 0 \cdot 15} = 2 \cdot 125 \text{ A}$$

and $\quad I_C = 2 \times 3 \cdot 14 \times 50 \times 100 \times 10^{-6} \times 100 = 3 \cdot 14$ A.

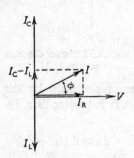

Fig. 11.34 Phasor diagram for example 11.6.

In the case of parallel circuits, the first phasor (fig. 11.34) to be drawn is that representing the quantity that is common to those circuits, namely the voltage. I_R is then drawn in phase with V, I_L lagging 90° and I_C leading 90°.

(*b*) The resultant of I_C and I_L

$$= I_C - I_L = 3 \cdot 14 - 2 \cdot 125$$
$$= 1 \cdot 015 \text{ A, leading by 90°.}$$

The current I taken from the alternator is the resultant of I_R and ($I_C - I_L$), and from fig. 11.34:

$$I^2 = I_R{}^2 + (I_C - I_L)^2 = 2^2 + (1 \cdot 015)^2 = 5 \cdot 03$$

∴ $\quad I = 2 \cdot 24$ A.

(*c*) From fig. 11.34:

$$\cos \phi = \frac{I_R}{I} = \frac{2}{2 \cdot 24} = 0 \cdot 893$$

∴ $\qquad \phi = 26° \, 45'.$

Since I_C is greater than I_L, the resultant current leads the supply voltage by 26° 45'.

11.15 Resonance in parallel circuits

If the values of L and C in fig. 11.31 were such that $I_L = I_C$, $(I_L - I_C)$ would be zero and the resultant current $I = I_R$; so that, theoretically, current could be flowing backwards and forwards round L and C without any current having to be supplied from the mains. In practice, the inductive circuit has some resistance, however small, so that the parallel inductive and capacitive circuits are actually as represented in fig. 11.35.

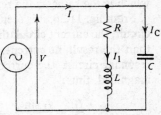

Fig. 11.35 Coil and capacitor in parallel.

Current through inductive circuit $= I_1 = \dfrac{V}{\sqrt{\{R^2 + (2\pi f L)^2\}}}$

and phase angle between $\left.\begin{array}{l} I_1 \\ \text{and } V \end{array}\right\} = \phi = \tan^{-1} \dfrac{2\pi f L}{R}$.

Since R is assumed to be very small compared with $2\pi f L$, ϕ is nearly 90°.

Current taken by capacitor $= I_C = 2\pi f C V$, leading V by 90°.

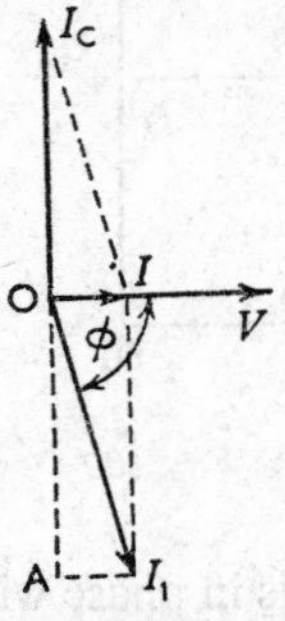

If I_1 and I_C be such that the resultant current I is in phase with the supply voltage as shown in fig. 11.36, the circuit is said to be in resonance.

From fig. 11.36, $\quad I_C = OA = I_1 \sin\phi \qquad (11.22)$

But $\quad \sin\phi = \dfrac{\text{reactance of coil}}{\text{impedance of coil}} = \dfrac{2\pi f L}{\sqrt{\{R^2 + (2\pi f L)^2\}}}$

Substituting for I_C, I_1 and $\sin\phi$ in (11.22), we have:

$$2\pi f C V = \dfrac{2\pi f L V}{R^2 + (2\pi f L)^2}.$$

so that $\quad f = \dfrac{1}{2\pi L}\sqrt{(L/C - R^2)}$

Fig. 11.36 Phasor diagram for fig. 11.35.

If R is very small compared with $2\pi f L$, as in radio circuits (see example 11.7),

$$C \simeq \dfrac{1}{(2\pi f)^2 L}$$

$$\therefore \qquad\qquad f = \frac{1}{2\pi\sqrt{(LC)}} \qquad\qquad (11.23)$$

which is the same as the resonance frequency of a series circuit.

From fig. 11.36, it is seen that when resonance occurs in a parallel circuit, the current circulating in L and C may be many times greater than the resultant current; in other words, by means of a parallel resonant circuit, the current taken from the supply can be greatly magnified, thus:

$$\frac{I_C}{I} = \frac{I_1 \sin\phi}{I} = \frac{\sin\phi}{\cos\phi} = \tan\phi = \frac{2\pi fL}{R}$$

$$= Q \text{ factor of circuit.}$$

It will be noted that the Q factor is a measure of voltage magnification in a series circuit and of current magnification in a parallel circuit.

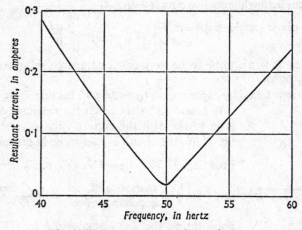

Fig. 11.37 Resonance curve for a rejector.

The resultant current in a resonant parallel circuit is in phase with the supply voltage,

$$\text{and impedance of such a} \atop \text{circuit} \Big\} = \frac{V}{I} = \frac{V}{\text{OA}\cot\phi}$$

$$= \frac{V}{I_C}\tan\phi = \frac{1}{2\pi fC} \cdot \frac{2\pi fL}{R}$$

$$= L/CR \qquad\qquad (11.24)$$

This means that a resonant parallel circuit is equivalent to a non-reactive resistor of $L/(CR)$ ohms. This quantity is termed the *dynamic impedance* of the circuit; and it is obvious that the lower the resistance of the coil, the higher is the dynamic impedance of the parallel circuit. This type of circuit, when used in radio work, is referred to as a *rejector*, since its impedance is a maximum and the resultant current a minimum at resonance.

Fig. 11.37 represents the variation of current with frequency in a parallel resonant circuit consisting of an inductor of 4 Ω resistance and 0·5 H inductance connected in parallel with a 20·3-μF capacitor across a constant voltage of 100 V, the values being the same as those of the series circuit of example 11.5. Since figs. 11.26 and 11.37 refer to circuits having the same values of R, L and C, it is of interest to note that a change of 10 per cent in the frequency from the resonance value reduces the current to about an eighth in the series circuit and increases the resultant current about eight times when the circuits are in parallel.

Example 11.7 *A tuned circuit consisting of a coil having an inductance of 200 μH and a resistance of 20 Ω in parallel with a variable capacitor is connected in series with a resistor of 8000 Ω across a 60-V supply having a frequency of 1 MHz. Calculate:* (a) *the value of C to give resonance;* (b) *the dynamic impedance and the Q factor of the tuned circuit; and* (c) *the current in each branch.*

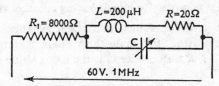

Fig. 11.38 Circuit diagram for example 11.7.

(a) 1 MHz = 10^6 Hz,

∴ reactance of $L = 2 \times 3·14 \times 10^6 \times 200 \times 10^{-6} = 1256$ Ω,

so that the resistance of the coil is very small compared with its reactance.

Substituting in expression (11.23), we have:

$$10^6 = \frac{1}{2 \times 3·14\sqrt{(200 \times 10^{-6} \times C)}}$$

∴ $$C = 126·7 \times 10^{-12} \text{ F} = 126·7 \text{ pF}.$$

(b) $\left.\begin{array}{l}\text{Dynamic} \\ \text{impedance}\end{array}\right\} = \dfrac{L}{CR} = \dfrac{200 \times 10^{-6}}{126 \cdot 7 \times 10^{-12} \times 20} = 79\,000\ \Omega$

and $\quad Q$ factor $= \dfrac{2\pi fL}{R} = \dfrac{2 \times 3 \cdot 14 \times 10^6 \times 200 \times 10^{-6}}{20}$

$$= 62 \cdot 8.$$

(c) Total equivalent resistance of circuit

$$= 79\,000 + 8000 = 87\,000\ \Omega,$$

$\therefore \qquad\qquad$ current $= 60/87\,000$ A $= 0 \cdot 69$ mA.

P.d. across tuned circuit $= 0 \cdot 69 \times 10^{-3} \times 79\,000 = 54 \cdot 5$ V

$\therefore \quad$ current through inductive branch of tuned circuit

$$= \frac{54 \cdot 5}{\sqrt{\{(20)^2 + (1256)^2\}}} = \frac{54 \cdot 5}{1256}\ \text{A}$$

$$= 43 \cdot 4\ \text{mA},$$

$\left.\begin{array}{l}\text{and current} \\ \text{through } C\end{array}\right\} = 2 \times 3 \cdot 14 \times 10^6 \times 126 \cdot 7 \times 10^{-12} \times 54 \cdot 5$ A

$$= 43 \cdot 4\ \text{mA}.$$

Actually there must be a very slight difference between the values of the currents in the two parallel branches, but it is so small as to be negligible. The results show that the current in each of the parallel circuits is about 62·8 times the resultant current taken from the supply.

11.16 Power in a non-reactive circuit having constant resistance

In section 10.3 it was explained that when an alternating current flows through a resistor of R ohms, the average heating effect over a complete cycle is I^2R watts, where I is the r.m.s. value of the current in amperes.

If V volts be the r.m.s. value of the applied voltage, then for a non-reactive circuit having constant resistance R ohms, $V = IR$:

$\therefore \qquad$ average value of the power $= I^2R = I \times IR$
$$= IV \text{ watts}.$$

Hence the power in a non-reactive circuit is given by the product of the ammeter and voltmeter readings, exactly as in a d.c. circuit.

11.17 Power in a purely inductive circuit

Consider a coil wound with such thick wire that the resistance is negligible in comparison with the inductive reactance X_L ohms. If such a coil is connected across a supply voltage V, the current is given by $I = V/X_L$ amperes. Since the resistance is very small, the heating effect and therefore the power are also very small, even though the voltage and the current be large. Such a curious conclusion—so different from anything we have experienced in d.c. circuits —requires fuller explanation if its significance is to be properly understood. Let us therefore consider fig. 11.39, which shows the applied voltage and the current for a purely inductive circuit, the current lagging the voltage by a quarter of a cycle.

The power at any instant is given by the product of the voltage and the current at that instant; thus at instant L, the applied voltage

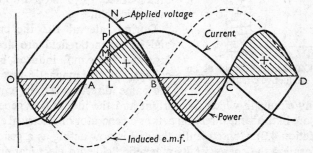

Fig. 11.39 Power curve for a purely inductive circuit.

is LN volts and the current is LM amperes, so that the power at that instant is LN × LM watts and is represented to scale by LP.

By repeating this calculation at various instants we can deduce the curve representing the variation of power over one cycle. It is seen that during interval OA the applied voltage is positive, but the current is negative, so that the power is negative; and that during interval AB, both the current and the voltage are positive, so that the power is positive.

The power curve is found to be symmetrical about the horizontal axis OD. Consequently the shaded areas marked '—' are exactly equal to those marked '+', so that the mean value of the power over the complete cycle OD is zero.

It is necessary, however, to consider the significance of the positive and negative areas if we are to understand what is really taking place. So let us consider an alternator P (fig. 11.40) connected to a coil Q whose resistance is negligible, and let us assume that the voltage and current are represented by the graphs in fig. 11.39. At instant A, there is no current and therefore no magnetic field through and around Q. During interval AB, the growth of the current is accompanied by a growth of flux as shown by the dotted lines in fig. 11.40. But the existence of a magnetic field involves some kind of a strain in the space occupied by the field and the storing up of energy in that field, as already dealt with in section 4.9. The current, and therefore the magnetic energy associated with it, reach their maximum values at instant B; and since the loss in the coil is assumed negligible, it follows that at that instant the whole of the energy

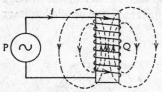

supplied to the coil during interval AB, and represented by the shaded area marked '+', is stored up in the magnetic field.

Fig. 11.40 Magnetic field of an inductive circuit.

During interval BC the current and its magnetic field are decreasing; and the e.m.f. induced by the collapse of the magnetic flux is in the same direction as the current.

But any circuit in which the current and the induced or generated e.m.f. are in the same direction acts as a generator of electrical energy (see section 4.2). Consequently the coil is now acting as a generator transforming the energy of its magnetic field into electrical energy, the latter being sent to alternator P to drive it as a motor. The energy thus returned is represented by the shaded area marked '—' in fig. 11.39; and since the positive and negative areas are equal, it follows that during alternate quarter-cycles electrical energy is being sent from the alternator to the coil, and during the other quarter-cycles the same amount of energy is sent back from the coil to the alternator. Consequently the net energy absorbed by the coil during a complete cycle is zero; in other words, the average power over a complete cycle is zero.

11.18 Power in a purely capacitive circuit

In this case, the current leads the applied voltage by a quarter of a cycle, as shown in fig. 11.41; and by multiplying the corresponding instantaneous values of the voltage and current, we can derive the curve representing the variation of power. During interval OA, the voltage and current are both positive so that the power is positive, i.e. power is being supplied from the alternator to the capacitor; and the shaded area enclosed by the power curve during interval OA represents the value of the electrostatic energy stored in the capacitor at instant A.

During interval AB, the p.d. across the capacitor decreases from its maximum value to zero and the whole of the energy stored in the capacitor at instant A is returned to the alternator; consequently the net energy absorbed during the half cycle OB is zero. Similarly, the

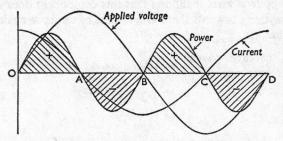

Fig. 11.41 Power curve for a purely capacitive circuit.

energy absorbed by the capacitor during interval BC is returned to the alternator during interval CD. Hence the average power over a complete cycle is zero.

11.19 Power in a circuit possessing resistance and reactance in series

Let us consider the general case of the current differing in phase from the applied voltage; thus in fig. 11.42 (*a*), the current is shown lagging the voltage by an angle ϕ.

Let instantaneous value of voltage $= v = V_m \sin \theta$

then instantaneous value of current $= i = I_m \sin (\theta - \phi)$.

At any instant, the value of the power is given by the product of the voltage and the current at that instant, i.e. instantaneous value of power $= vi$ watts.

By multiplying the corresponding instantaneous values of voltage and current, the curve representing the variation of power in fig. 11.42 (*b*) can be derived:

$$\text{i.e.} \quad \begin{aligned} \text{instantaneous} \\ \text{power} \end{aligned} \Bigg\} = vi = V_m \sin \theta \,.\, I_m \sin (\theta - \phi)$$

$$= \tfrac{1}{2} V_m I_m \{\cos \phi - \cos (2\theta - \phi)\}$$

$$= \tfrac{1}{2} V_m I_m \cos \phi - \tfrac{1}{2} V_m I_m \cos (2\theta - \phi).$$

From this expression, it is seen that the instantaneous value of the power consists of two components:

(1) $\frac{1}{2}V_m I_m \cos \phi$, which contains no reference to θ and therefore remains constant in value.

(2) $\frac{1}{2}V_m I_m \cos (2\theta - \phi)$, the term 2θ indicating that it varies at twice the supply frequency; thus in fig. 11.42 (b), it is seen that the power undergoes two cycles of variation for one cycle of the voltage wave. Furthermore, since the average value of a cosine curve over a *complete* cycle is zero, it follows that this component does not contribute anything towards the *average* value of the power taken from the alternator.

Hence,
$$\left.\begin{array}{c}\text{average power}\\ \text{over one cycle}\end{array}\right\} = \frac{1}{2}V_m I_m \cos \phi$$
$$= \frac{V_m}{\sqrt{2}} \cdot \frac{I_m}{\sqrt{2}} \cdot \cos \phi$$
$$= VI \cos \phi \qquad (11.25)$$

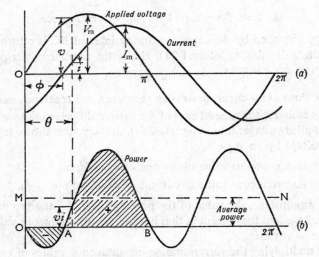

Fig. 11.42 Voltage, current and power curves.

where V and I are the r.m.s. values of the voltage and current respectively. In fig. 11.42 (b), the average power is represented by the height above the horizontal axis of the dotted line MN drawn midway between the positive and negative peaks of the power curve.

It will be noticed that during interval OA in fig. 11.42 (b), the power is negative, and the shaded negative area represents energy returned from the circuit to the alternator. The shaded positive area during interval AB represents energy supplied from the alternator to the circuit; and the difference between the two areas represents the net energy absorbed by the circuit during interval OB. The larger the phase difference between the voltage and current, the smaller is the difference between the positive and negative areas and the smaller, therefore, is the average power over the complete cycle.

11.20 Power factor

In a.c. work the product of the r.m.s. values of the applied voltage and current is VI *voltamperes*, the latter term being used to distinguish this quantity from the power, expressed in watts. The number of watts is equal to or less than the number of voltamperes and the latter has, in general, to be multiplied by a quantity termed *power factor* to give the power in watts:

i.e.   power in watts = no. of voltamperes × power factor

$$= VI \times \text{power factor} \tag{11.26}$$

or   power factor = $\left\{ \dfrac{\text{power in watts}}{\substack{\text{product of r.m.s. values of} \\ \text{voltage and current}}} \right.$ (11.27)

Comparison of expressions (11.25) and (11.26) shows that for *sinusoidal* voltage and current:

$$\text{power factor} = \cos \phi \tag{11.28}$$

From the general phasor diagram of fig. 11.17 for a *series* circuit, it follows that:

$$\cos \phi = \frac{IR}{V} = \frac{IR}{IZ} = \frac{\text{resistance}}{\text{impedance}} \tag{11.15}$$

It has become the practice to say that the power factor is *lagging* when the *current lags the supply voltage* and *leading* when the *current leads the supply voltage*. This means that the supply voltage is regarded as the reference quantity.

11.21 Active and reactive currents

If a current I lags the applied voltage V by an angle ϕ, as in fig. 11.43, it can be resolved into two components,* OA in phase with the voltage and OB lagging by 90°.

---

* If the phasor diagram of fig. 11.43 refers to a circuit possessing resistance and inductance in series, OA and OB must *not* be labelled I_R and I_L respectively. This is an error frequently made by beginners.

Since power $= IV \cos \phi = V \times \mathrm{OI} \cos \phi = V \times \mathrm{OA}$ watts, therefore OA is termed the *active* or *power* component of the current:

i.e. active component of current $= I \cos \phi$ $\qquad$ (11.29)

Power due to component $\mathrm{OB} = V \times \mathrm{OB} \cos 90° = 0$, so that OB is termed the *reactive* or *wattless* component of the current:

i.e. reactive component of current $= I \sin \phi$ $\qquad$ (11.30)

and reactive voltamperes (or vars) $= VI \sin \phi$ $\qquad$ (11.31)

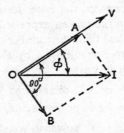

Fig. 11.43 Active and reactive components of current.

Example 11.8 *A coil having a resistance of 6 Ω and an inductance of 0·03 H is connected across a 50-V, 60-Hz supply. Calculate: (a) the current; (b) the phase angle between the current and the applied voltage; (c) the power factor; (d) the voltamperes; (e) the power; and (f) the reactive voltamperes (or vars).*

(*a*) The phasor diagram for such a circuit is given in fig. 11.19.

$$\text{Reactance of circuit} = 2\pi fL = 2 \times 3\cdot14 \times 60 \times 0\cdot03$$
$$= 11\cdot31 \ \Omega.$$

From (11.13), impedance $= \sqrt{\{6^2 + (11\cdot31)^2\}} = 12\cdot8 \ \Omega$

and $\qquad$ current $= 50/12\cdot8 = 3\cdot91$ A.

(*b*) From (11.14), $\tan \phi = \dfrac{X}{R} = \dfrac{11\cdot31}{6} = 1\cdot885.$

∴ $\qquad\qquad\qquad \phi = 62° \ 3'.$

(*c*) From (11.28), power factor $= \cos 62° \ 3' = 0\cdot469$

or, from (11.15), „ $\quad$ „ $\quad = 6/12\cdot8 = 0\cdot469.$

(*d*) Voltamperes $= 50 \times 3\cdot91 = 195\cdot5$ VA.

(*e*) Power $=$ voltamperes $\times$ power factor
$$= 195\cdot5 \times 0\cdot469 = 91\cdot7 \text{ W.}$$

Or, alternatively:

$$\text{power} = I^2R = (3 \cdot 91)^2 \times 6 = 91 \cdot 7 \text{ W}.$$

(*f*) Reactive voltamperes (or vars) = voltamperes $\times \sin \phi$
$$= 195 \cdot 5 \times \sin 62° 3'$$
$$= 172 \cdot 6 \text{ VAr}.$$

Example 11.9 *A single-phase motor operating off a 400-V, 50-Hz supply is developing 10 kW with an efficiency of 84 per cent and a power factor of 0·7 lagging. Calculate: (a) the input kilovoltamperes; (b) the active and reactive components of the current; and (c) the reactive kilovoltamperes (or kilovars).*

(*a*) $$\text{Efficiency} = \frac{\text{output power in watts}}{\text{input power in watts}}$$
$$= \frac{\text{output power in watts}}{IV \times \text{p.f.}}$$

$\therefore$ $$0 \cdot 84 = \frac{10 \times 1000}{IV \times 0 \cdot 7}$$

so that $$IV = 17\ 000 \text{ VA}.$$

$\therefore$ input kilovoltamperes = 17 kVA.

(*b*) Current taken by motor $= \dfrac{\text{input voltamperes}}{\text{voltage}}$
$$= 17\ 000/400 = 42 \cdot 5 \text{ A}$$

$\therefore$ $\left.\begin{array}{l}\text{active component} \\ \text{of current}\end{array}\right\} = 42 \cdot 5 \times 0 \cdot 7 = 29 \cdot 75 \text{ A}.$

Since $\sin \phi = \sqrt{(1 - \cos^2 \phi)} = \sqrt{\{1 - (0 \cdot 7)^2\}} = 0 \cdot 714$,

$\therefore$ $\left.\begin{array}{l}\text{reactive component of} \\ \text{current}\end{array}\right\} = 42 \cdot 5 \times 0 \cdot 714 = 30 \cdot 35 \text{ A}.$

(*c*) $\left.\begin{array}{l}\text{Reactive kilovoltamperes} \\ \text{(or kilovars)}\end{array}\right\} = 400 \times 30 \cdot 35/1000$
$$= 12 \cdot 14 \text{ kVAr}.$$

Example 11.10 *Calculate the capacitance required in parallel with the motor of example 11.9 to raise the power factor to 0·9 lagging.*

The circuit and phasor diagrams are given in figs. 11.44 and 11.45 respectively, M being the motor taking a current I_M of 42·5 A.

Current I_C taken by the capacitor must be such that when combined with I_M, the resultant current I lags the voltage by an angle ϕ, where $\cos \phi = 0.9$. From fig. 11.45,

$$\text{active component of } I_M = I_M \cos \phi_M = 42.5 \times 0.7$$
$$= 29.75 \text{ A}$$

and   active component of $I = I \cos \phi = I \times 0.9$.

These components are represented by OA in fig. 11.45,

$$\therefore \qquad\qquad I = 29.75/0.9 = 33.06 \text{ A}.$$

Reactive component of $I_M = I_M \sin \phi_M$
$$= 30.35 \text{ A (from Example 11.9)}$$

and   reactive component of $I = I \sin \phi$
$$= 33.06\sqrt{\{1 - (0.9)^2\}}$$
$$= 33.06 \times 0.436 = 14.4 \text{ A}.$$

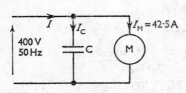

Fig. 11.44 Circuit diagram for
example 11.10.

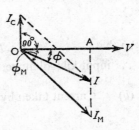

Fig. 11.45 Phasor diagram
for fig. 11.44.

From fig. 11.45 it will be seen that:

$$I_C = \text{reactive component of } I_M - \text{reactive component of } I$$
$$= 30.35 - 14.4 = 15.95 \text{ A}.$$

But $I_C = 2\pi f C V$

$\therefore \qquad 15.95 = 2 \times 3.14 \times 50 \times C \times 400$

and $C = 127 \times 10^{-6} \text{ F} = 127 \text{ }\mu\text{F}$.

From the above example it will be seen that the effect of connecting a 127-μF capacitor in parallel with the motor is to reduce the current taken from the supply from 42.5 to 33.06 A, without altering either the current or the power taken by the motor. This enables an economy to be effected in the size of the generating plant and in the cross-sectional area of conductor in the cable.

Example 11.11 *An alternator is supplying a load of* 300 *kW at a power factor of* 0·6 *lagging. If the power factor is raised to unity, how many more kilowatts can the alternator supply for the same kVA loading?*

Since the power in kW

$$= \text{number of kilovoltamperes} \times \text{power factor},$$

∴ number of kilovoltamperes

$$= 300/0·6 = 500 \text{ kVA.}$$

When the power factor is raised to unity:

number of kilowatts = number of kilovoltamperes = 500 kW.

Hence increased power supplied by alternator

$$= 500 - 300 = 200 \text{ kW.}$$

11.22 The practical importance of power factor

If an alternator is rated to give, say, 2000 A at a voltage of 400 V, it means that these are the highest current and voltage values the machine can give without the temperature exceeding a safe value. Consequently the rating of the alternator is given as 400 × 2000/ 1000 = 800 kVA. The phase difference between the voltage and the current depends upon the nature of the load and not upon the generator. Thus if the power factor of the load is unity, the 800 kVA are also 800 kW; and the engine driving the generator has to be capable of developing this power together with the losses in the generator. But if the power factor of the load is, say, 0·5, the power is only 400 kW; so that the engine is developing only about one-half of the power of which it is capable, though the alternator is supplying its rated output of 800 kVA.

Similarly, the conductors connecting the alternator to the load have to be capable of carrying 2000 A without excessive temperature rise. Consequently they can transmit 800 kW if the power factor is unity, but only 400 kW at 0·5 power factor, for the same rise of temperature.

It is therefore evident that the higher the power factor of the load, the greater is the *power* that can be generated by a given alternator and transmitted by a given conductor.

The matter may be put another way by saying that, for a *given power*, the lower the power factor, the larger must be the size of the alternator to generate that power and the greater must be the cross-sectional area of the conductor to transmit it; in other words, the greater is the cost of generation and transmission of the electrical energy. This is the reason why supply authorities do all they can to improve the power factor of their loads either by the installation of capacitors or special machines or by the use of tariffs which encourage consumers to do so.

11.23 Measurement of power in a single-phase circuit
Since the product of the voltage and current in an a.c. circuit must be multiplied by the power factor to give the power in watts, the most convenient method of measuring the power is to use a wattmeter, the simplest form of which is the dynamometer type described in section 22.8.

11.24 Current locus for a circuit having constant inductance and variable resistance
Suppose a circuit having a constant inductive reactance of X ohms and a variable resistance to be supplied at a constant voltage V having a frequency f hertz, as in fig. 11.46.

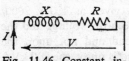

Fig. 11.46 Constant inductance and variable resistance in series.

When $R = 0$, current $= V/X$ and is represented by OA in fig. 11.47, lagging the supply voltage OV by 90°.

For the general case of R equal to some finite value,

$$I = \frac{V}{\sqrt{(R^2 + X^2)}}$$

and is represented by OI in fig. 11.47.

Draw VB perpendicular to OI and complete the rectangle OBVC; then, OB $= IR$, and BV $=$ OC $= IX$.

Also, since VB and OB are at right-angles to each other, the locus of B must be a semicircle on OV.

Join IA, as shown dotted in fig. 11.47.

For triangles IOA and VOB,

$$\angle IOA = 90° - \angle VOB = \angle OVB.$$

Also,

$$\frac{OV}{BV} = \frac{V}{XI} = \frac{V}{X} \cdot \frac{1}{I} = \frac{OA}{OI}.$$

Hence the two triangles are similar,

so that

$$\angle OIA = \angle OBV = 90°,$$

and the locus of the extremity of the current phasor must therefore be a semicircle on OA. Also, it is evident from fig. 11.47 that the power component of the current is a maximum when the phase

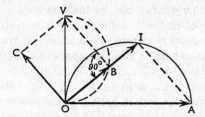

Fig. 11.47 Phasor diagram for fig. 11.46.

difference between the current and the supply voltage is 45°, i.e. when the value of the resistance is equal to that of the reactance.

11.25 Current locus for a circuit having constant resistance and variable inductance

The circuit and phasor diagrams are given in figs 11.48 and 11.49 respectively.

When $X = 0$, current $= V/R$ and is represented by OA in phase with OV.

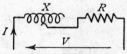

Fig. 11.48 Constant resistance and variable inductance in series.

If OI represents the current for the general case of X having some finite value, then OB and OC represent the corresponding voltages across R and X respectively. Since

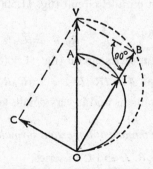

Fig. 11.49 Phasor diagram for fig. 11.48.

$\angle$OBV is a right-angle, the locus of B is a semicircle on OV; and since OI is proportional to and in phase with OB, the locus of the extremity of the current phasor must be a semicircle on OA.

11.26 Imperfect capacitors and loss angle

Losses in capacitors having a solid dielectric are due to: (*a*) leakage current through the dielectric and along surface paths between the terminals, (*b*) energy absorbed by the dielectric when subjected to an alternating electric field – analogous to hysteresis loss in magnetic materials. This effect is referred to as *dielectric hysteresis*.

An imperfect capacitor can be represented by resistance R_p in *parallel* with capacitor C_p as in fig. 11.50(*a*), or by resistance R_s in *series* with capacitor C_s, as in fig. 11.50(*c*). The values of R_p and R_s must be such that the loss in each is equal to the total loss in the imperfect capacitor. The respective phasor diagrams are given in figs. 11.50(*b*) and (*d*). The angle by which the angle of lead of the current falls short of 90° is termed the *loss angle* and is represented by the Greek letter δ.

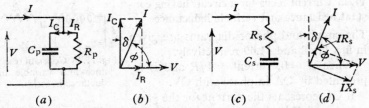

Fig. 11.50. Circuit and phasor diagrams for an imperfect capacitor.

For the equivalent parallel circuit (fig. 11.50(*a*)),

$$\tan \delta = \frac{I_\mathrm{R}}{I_\mathrm{c}} = \frac{V}{R_\mathrm{p}} \times \frac{1}{2\pi f C_\mathrm{p} V} = \frac{1}{2\pi f C_\mathrm{p} R_\mathrm{p}}$$

and for the equivalent series circuit (fig. 11.50(*c*)),

$$\tan \delta = I R_\mathrm{s}/I X_\mathrm{s} = 2\pi f C_\mathrm{s} R_\mathrm{s}.$$

If δ is expressed in radians and is very small, $\tan \delta \simeq \delta \simeq \cos \phi$.

Summary of important formulae

For a circuit with R, L and C in series,

$$\text{Impedance} = Z = \sqrt{\left\{ R^2 + \left(2\pi f L - \frac{1}{2\pi f C} \right)^2 \right\}} \qquad (11.13)$$

where $2\pi f L =$ inductive reactance in ohms
and $1/2\pi f C =$ capacitive reactance in ohms

If ϕ = phase difference between current and supply voltage,

$$\tan \phi = \frac{2\pi fL - 1/(2\pi fC)}{R} \qquad (11.14)$$

$$\cos \phi = R/Z \qquad (11.15)$$

$$\sin \phi = \frac{2\pi fL - 1/(2\pi fC)}{Z} \qquad (11.16)$$

For purely resistive circuit,

$$I = V/R \quad \text{and} \quad \phi = 0.$$

For purely inductive circuit,

$$I = V/2\pi fL$$

and $\qquad \phi = 90°$, current lagging.

For purely capacitive circuit,

$$I = 2\pi fCV$$

and $\qquad \phi = 90°$, current leading.

For resonance in a series circuit,

$$f = \frac{1}{2\pi\sqrt{(LC)}} \qquad (11.17)$$

and $\qquad Q \text{ factor} = \frac{2\pi fL}{R} = \frac{1}{R}\sqrt{\left(\frac{L}{C}\right)} \qquad (11.19)$

For resonance in a parallel circuit,

$$C = \frac{L}{R^2 + (2\pi fL)^2}$$

and $\qquad f = \frac{1}{2\pi\sqrt{(LC)}}$ when $R \ll 2\pi fL \qquad (11.23)$

$$\left.\begin{array}{l}\text{Dynamic impedance of reson-} \\ \text{ant parallel circuit}\end{array}\right\} = \frac{L}{CR} \qquad (11.24)$$

In an a.c. circuit, power in watts $= VI \times$ power factor $\qquad (11.26)$

or $\qquad$ power factor $= \dfrac{\text{power in watts}}{\text{r.m.s. volts} \times \text{r.m.s. amperes}}$

$$= \frac{\text{kilowatts}}{\text{kilovoltamperes}} \qquad (11.27)$$

For sinusoidal voltage and current having phase difference ϕ,

$$\text{power factor} = \cos \phi \qquad (11.28)$$

$$\text{and power (in watts)} = VI \cos \phi \qquad (11.25)$$

Active or power component of current $= I \cos \phi$ (11.29)

Reactive or wattless component of current $= I \sin \phi$ (11.30)

Reactive voltamperes (or vars) $= VI \sin \phi$ (11.31)

EXAMPLES 11

Note. In questions reproduced from examination papers, the term 'vector' has been replaced by 'phasor' by kind permission of the respective institutions.

1. A closed-circuit, 500-turn coil, of resistance 100 Ω and negligible inductance, is wound on a square frame of 40 cm side. The frame is pivoted at the mid-points of two opposite sides and is rotated at 250 rev/min in a uniform magnetic field of 60 milliteslas. The field direction is at right-angles to the axis of rotation.

   For the instant that the e.m.f. is maximum: (*a*) draw a diagram of the coil, and indicate the direction of rotation, and of the current flow, the magnetic flux, the e.m.f. and the force exerted by the magnetic field on the conductors; (*b*) calculate the e.m.f., the current and the torque, and hence verify that the mechanical power supplied balances the electric power produced.   (S.A.N.C., O.2)

2. An alternating p.d. of 100 V (r.m.s.), at 50 Hz, is maintained across a 20-Ω non-reactive resistor. Plot to scale the waveforms of p.d. and current over one cycle. Deduce the curve of power and state its mean value.

3. An inductor having a reactance of 10 Ω and negligible resistance is connected to a 100-V (r.m.s.) supply. Draw to scale for one half cycle, curves of voltage and current, and deduce and plot the power curve. What is the mean power over the half cycle?

4. A coil having an inductance of 0·2 H and negligible resistance is connected across a 100-V a.c. supply. Calculate the current when the frequency is: (*a*) 30 Hz, and (*b*) 500 Hz.

5. A coil of inductance 0·1 H and negligible resistance is connected in series with a 25-Ω resistor. The circuit is energized from a 250-V, 50-Hz source. Calculate: (*a*) the current in the circuit; (*b*) the p.d. across the coil; (*c*) the p.d. across the resistor; (*d*) the phase angle of the circuit. Draw to scale a phasor diagram representing the current and the component voltages.   (U.L.C.I., O.1)

6. A coil connected to a 250-V, 50-Hz sinusoidal supply takes a current of 10 A at a phase angle of 30°. Calculate the resistance and inductance of, and the power taken by the coil.

   Draw, for one half cycle, curves of voltage and current and deduce and plot the power curve. Comment on the power curve.   (S.A.N.C., O.1)

7. A 15-Ω non-reactive resistor is connected in series with a coil of inductance 0·08 H and negligible resistance. The combined circuit is connected to a 240-V, 50-Hz supply. Calculate: (*a*) the reactance of the coil; (*b*) the impedance of the circuit; (*c*) the current in the circuit;

(*d*) the power factor of the circuit; (*e*) the power absorbed by the circuit. (N.C.T.E.C., O.1)

8. The potential difference measured across a coil is 20 V when a direct current of 2 A is passed through it. With an alternating current of 2 A at 40 Hz, the p.d. across the coil is 140 V. If the coil is connected to a 230-V, 50-Hz supply, calculate: (*a*) the current; (*b*) the power; (*c*) the power factor. (S.A.N.C., O.1)

9. A non-inductive load takes a current of 15 A at 125 V. An inductor is then connected in series in order that the same current shall be supplied from 240-V, 50-Hz mains. Ignore the resistance of the inductor and calculate: (*a*) the inductance of the inductor; (*b*) the impedance of the circuit; (*c*) the phase difference between the current and the applied voltage. Assume the waveform to be sinusoidal.

(U.L.C.I., O.1)

10. A series a.c. circuit, ABCD, consists of a resistor AB, an inductor BC, of resistance R and inductance L, and a resistor CD. When a current of 6·5 A flows through the circuit, the voltage drops across various points are: $V_{AB} = 65$ V; $V_{BC} = 124$ V; $V_{AC} = 149$ V. The supply voltage is 220 V at 50 Hz.

Draw a phasor diagram to scale showing all the resistive and reactive volt-drops and, from the diagram, determine: (*a*) the volt-drop V_{BD} and the phase angle between it and the current; (*b*) the resistance and inductance of the inductor.

11. A coil of 0·5 H inductance and negligible resistance and a 200-Ω resistor are connected in series to a 50-Hz supply. Calculate the circuit impedance.

An inductor in a radio set has to have a reactance of 11 kΩ at a frequency of 1·5 MHz. Calculate the inductance in millihenrys.

(N.C.T.E.C., O.2)

12. A coil takes a current of 10·0 A and dissipates 1410 W when connected to a 200-V, 50-Hz sinusoidal supply. When another coil is connected in parallel with it, the total current taken from the supply is 20·0 A at a power factor of 0·866.

Determine the current and the overall power factor when the coils are connected in series across the same supply. (S.A.N.C., O.2)

13. When an iron-cored reactor and a non-reactive resistor are connected in series to a 150-V a.c. supply, a current of 3·75 A flows in the circuit. The potential differences across the reactor and across the resistor are then observed to be 120 V and 60 V respectively. If the d.c. resistance of the reactor is 4·5 Ω, determine the iron loss in the reactor and calculate its equivalent series resistance. (N.C.T.E.C., O.2)

14. A single-phase circuit consists of three parallel branches, the currents in the respective branches being represented by:

$$i_1 = 20 \sin 314t \text{ amperes,}$$
$$i_2 = 30 \sin (314t - \pi/4) \text{ amperes,}$$
and $$i_3 = 18 \sin (314t + \pi/2) \text{ amperes.}$$

(*a*) Using a scale of 1 cm = 5 A, draw a phasor diagram and find the total maximum value of current taken from the supply and the overall phase angle.

(b) Express the total current in a form similar to that of the branch currents.

(c) If the supply voltage is represented by 200 sin 314t volts, find the impedance, resistance and reactance of the circuit.

(W.J.E.C., O.1)

15. A non-inductive resistor is connected in series with a coil across a 230-V, 50-Hz supply. The current is 1·8 A and the potential differences across the resistor and the coil are 80 V and 170 V respectively. Calculate the inductance and the resistance of the coil, and the phase difference between the current and the supply voltage. Also draw the phasor diagram representing the current and the voltages.

(App. El., L.U.)

16. An inductive circuit, in parallel with a non-inductive resistor of 20Ω, is connected across a 50-Hz supply. The currents through the inductive circuit and the non-inductive resistor are 4·3 A and 2·7 A respectively, and the current taken from the supply is 5·8 A. Find (a) the power absorbed by the inductive circuit, (b) its inductance and (c) the power factor of the combined circuits. Sketch the phasor diagram.

(App. El., L.U.)

17. A coil having a resistance of 15 Ω and an inductance of 0·2 H is connected in series with another coil having a resistance of 25 Ω and an inductance of 0·04 H to a 230-V, 50-Hz supply.

Draw to scale the complete phasor diagram for the circuit and determine: (a) the voltage across each coil; (b) the power dissipated in each coil; (c) the power factor of the circuit as a whole.

(App. El., L.U.)

18. Two identical coils, each of 25-Ω resistance, are mounted coaxially a short distance apart. When one coil is supplied at 100 V, 50 Hz, the current taken is 2·1 A and the e.m.f. induced in the other coil on open-circuit is 54 V. Calculate the self inductance of each coil and the mutual inductance between them.

What current will be taken if a p.d. of 100 V, 50 Hz, is supplied across the two coils in series?

(App. El., L.U.)

19. Two similar coils have a coupling coefficient of 0·6. Each coil has a resistance of 8 Ω and a self inductance of 2 mH. Calculate the current and the power factor of the circuit when the coils are connected in series (a) cumulatively and (b) differentially, across a 10-V, 5-kHz supply.

20. A two-wire cable, 8 km long, has a capacitance of 0·3 μF/km. If the cable is connected to a 11-kV, 60-Hz supply, calculate the value of the charging current. The resistance and inductance of the conductors may be neglected.

21. Draw, to scale, phasors representing the following voltages, taking e_1 as the reference phasor:

$e_1 = 80 \sin \omega t$ volts; $e_2 = 60 \cos \omega t$ volts; $e_3 = 100 \sin (\omega t - \pi/3)$ volts.

By phasor addition, find the sum of these three voltages and express it in the form of $E_m \sin (\omega t \pm \phi)$.

When this resultant voltage is applied to a circuit consisting of a 10-Ω resistor and a capacitor of 17·3 Ω reactance connected in series find an expression for the instantaneous value of the current flowing, expressed in the same form. (N.C.T.E.C., O.2)

22. In order to use three 100-V, 60-W lamps on a 230-V, 50-Hz supply, they are connected in parallel and a capacitor is connected in series with the group. Find: (*a*) the capacitance required to give the correct voltage across the lamps; (*b*) the power factor of the circuit.

If one of the lamps is removed, to what value will the voltage across the remaining two rise, assuming that their resistances remain unchanged? (U.L.C.I., O.2.)

23. A 130-Ω resistor and a 30-μF capacitor are connected in parallel across a 200-V, 50-Hz supply. Calculate: (*a*) the current in each circuit; (*b*) the resultant current; (*c*) the phase difference between the resultant current and the applied voltage; (*d*) the power; and (*e*) the power factor. Sketch the phasor diagram.

24. A resistor and a capacitor are connected in series across a 150-V a.c. supply. When the frequency is 40 Hz the current is 5 A, and when the frequency is 50 Hz the current is 6 A. Find the resistance and capacitance of the resistor and capacitor respectively.

If they are now connected in parallel across the 150-V supply, find the total current and its power factor when the frequency is 50 Hz. (U.L.C.I., O.2)

25. A series circuit consists of a non-inductive resistor of 10 Ω, an inductor having a reactance of 50 Ω and a capacitor having a reactance of 30 Ω. It is connected to a 230-V a.c. supply. Calculate: (*a*) the current; (*b*) the power; (*c*) the power factor; (*d*) the voltage across each component circuit. Draw to scale a phasor diagram showing the supply voltage and current and the voltage across each component. (N.C.T.E.C., O.2)

26. A coil having a resistance of 20 Ω and an inductance of 0·15 H is connected in series with a 100-μF capacitor across a 230-V, 50-Hz supply. Calculate: (*a*) the active and reactive components of the current; (*b*) the voltage across the coil; and (*c*) the power factor of the circuit. Sketch the phasor diagram.

27. A p.d. of 100 V at 50 Hz is maintained across a series circuit having the following characteristics: $R = 10 \Omega$, $L = 100/\pi$ mH, $C = 500/\pi$ μF. Draw the phasor diagram and calculate: (*a*) the current; (*b*) the power factor of the circuit; (*c*) the active and reactive components of the current; (*d*) the reactive voltamperes.

28. A circuit consists of three branches in parallel. Branch A is a 10-Ω resistor, branch B is a coil of resistance 4Ω and inductance 0·02 H, and branch C is an 8-Ω resistor in series with a 200-μF capacitor. The combination is connected to a 100-V, 50-Hz supply.

Find the various branch currents and then, by resolving into in-phase and quadrature components, determine the total current taken from the supply and its power factor. A phasor diagram showing the relative positions of the various circuit quantities should accompany your solution. It need not be drawn to scale.   (E.M.E.U., O.2)

29. A 50-Hz iron-cored reactor has an impedance of 10 Ω and a Q factor of 3 at 50 Hz. If the reactor is connected to a 25-V, 50-Hz supply, calculate the power loss.

What value of capacitor must be connected in series with the reactor so that the current lags by 45° with respect to the supply voltage?   (N.C.T.E.C., O.2)

30. A coil, having a resistance of 20 Ω and an inductance of 0·0382 H, is connected in parallel with a circuit consisting of a 150-μF capacitor in series with a 10-Ω resistor. The arrangement is connected to a 240-V, 50-Hz supply. Determine the current in each branch and, sketching a phasor diagram, the total supply current, power factor and power.   (E.M.E.U., O.2)

31. A 31·8-μF capacitor, a 127·5-mH inductor of resistance 30 Ω and a 100-Ω resistor are all connected in parallel to a 200-V, 50-Hz supply. Calculate the current in each branch. Draw a phasor diagram to scale to show these currents. Find the total current and its phase angle by drawing or otherwise.   (N.C.T.E.C., O.2)

32. A 200-V, 50-Hz sinusoidal supply is connected to parallel circuit comprising three branches A, B and C, as follows:

A—a coil of resistance 3 Ω and inductive reactance 4 Ω,
B—a series circuit of resistance 4 Ω and capacitive reactance 3 Ω,
C—a capacitor.

Given that the power factor of the combined circuit is unity, find: (a) the value of the capacitor in microfarads; (b) the current taken from the supply; (c) the total power absorbed.   (S.A.N.C., O.2)

33. Two circuits, A and B, are connected in parallel to a 115-V, 50-Hz supply. The total current taken by the combination is 10 A at unity power factor. Circuit A consists of a 10-Ω resistor and a 200-μF capacitor connected in series; circuit B consists of a resistor and an inductive reactor in series.

Determine the following data for circuit B: (a) the current; (b) the power factor; (c) the impedance; (d) the resistance; (e) the reactance.   (N.C.T.E.C., O.2)

34. A parallel circuit consists of two branches A and B. Branch A has a resistance of 10 Ω and an inductance of 0·1 H in series. Branch B has a resistance of 20 Ω and a capacitance of 100 μF in series. The circuit is connected to a single-phase supply of 250 V at 50 Hz.

Calculate the magnitude and phase angle of the current taken from the supply. Verify your answers by measurement from a phasor diagram drawn to scale.   (U.L.C.I., O.2)

35. A series circuit comprises an inductor, of resistance 10 Ω and inductance 159 μH, and a variable capacitor connected to a 50-mV sinusoidal supply of frequency 1 MHz. What value of capacitance will result in resonant conditions and what will then be the current?

For what values of capacitance will the current at this frequency be reduced to 10 per cent of its value at resonance? (S.A.N.C., O.2)

36. A circuit consists of a 10-Ω resistor, a 30-mH inductor and a 1-μF capacitor, and is supplied from a 10-V variable-frequency source. Find the frequency for which the voltage developed across the capacitor is a maximum and calculate the magnitude of this voltage.
(E.M.E.U., O.2)

37. Calculate the voltage magnification created in a resonant circuit connected to a 240-V, a.c. supply consisting of a choke having inductance 0·1 H and resistance 2 Ω in series with a 100-μF capacitor. Explain the effects of increasing the above resistance value.
(W.J.E.C., O.2)

38. A series circuit consists of 0·5-μF capacitor, a coil of inductance 0·32 H and resistance 40 Ω and a 20-Ω non-inductive resistor. Calculate the value of the resonant frequency of the circuit.

When the circuit is connected to a 30-V a.c. supply at this resonant frequency, determine: (a) the p.d. across each of the three components; (b) the current flowing in the circuit; (c) the power absorbed by the circuit.
(N.C.T.E.C., O.2)

39. An e.m.f. whose instantaneous value at time t is given by 283 sin (314t + π/4) volts is applied to an inductive circuit and the current in the circuit is 5·66 sin (314t − π/6) amperes. Determine: (a) the frequency of the e.m.f.; (b) the resistance and inductance of the circuit; (c) the power absorbed.

If series capacitance is added so as to bring the circuit into resonance at this frequency and the above e.m.f. is applied to the resonant circuit, find the corresponding expression for the instantaneous value of the current. Sketch a phasor diagram for this condition.

Explain why it is possible to have a much higher voltage across a capacitor than the supply voltage in a series circuit. (U.L.C.I., O.2)

40. A coil, of resistance R and inductance L, is connected in series with a capacitor C across a variable-frequency source. The voltage is maintained constant at 300 mV and the frequency is varied until a maximum current of 5 mA flows through the circuit at 6 kHz.

If, under these conditions, the Q factor of the circuit is 105, calculate: (a) the voltage across the capacitor; (b) the values of R, L and C.
(U.E.I., O.2)

41. A constant voltage at a frequency of 1 MHz is maintained across a circuit consisting of an inductor in series with a variable capacitor. When the capacitor is set to 300 pF, the current has its maximum value. When the capacitance is reduced to 284 pF, the current is 0·707 of its maximum value. Find: (a) the inductance and the resistance of the inductor, and (b) the Q factor of the inductor at 1 MHz. Sketch the phasor diagram for each condition.

42. A coil of resistance 12 Ω and inductance 0·12 H is connected in parallel with a 60-μF capacitor to a 100-V, variable-frequency supply. Calculate the frequency at which the circuit will behave as a non-reactive resistor, and also the value of the dynamic impedance. Draw

for this condition the complete phasor diagram.   (W.J.E.C., O.2)

43. Calculate, from first principles, the impedance at resonance of a circuit consisting of a coil of inductance 0·5 mH and effective resistance 20 Ω in parallel with a 0·0002-μF capacitor.   (N.C.T.E.C., O.2)

44. A coil has resistance of 400 Ω and inductance of 318 μH. Find the capacitance of a capacitor which, when connected in parallel with the coil, will produce resonance with a supply frequency of 1 MHz.

    If a second capacitor of capacitance 23·5 pF is connected in parallel with the first capacitor, find the frequency at which resonance will occur.   (S.A.N.C., O.2)

45. A single-phase motor takes 8·3 A at a power factor of 0·866 lagging when connected to a 230-V, 50-Hz supply. Two similar capacitors are connected in parallel with each other to form a capacitance bank. This capacitance bank is now connected in parallel with the motor to raise the power factor to unity. Determine the capacitance of each capacitor.   (E.M.E.U., O.2)

46. (a) A single-phase load of 5 kW operates at a power factor of 0·6 lagging. It is proposed to improve this power factor to 0·95 lagging by connecting a capacitor across the load. Calculate the kVA rating of the capacitor.

    (b) Give reasons why it is to a consumer's economic advantage to improve his power factor with respect to the supply, and explain the fact that the improvement is rarely made to unity in practice.
    (W.J.E.C., O.2)

47. A 25-kVA single-phase motor has a power factor of 0·8 lag. A 10-kVA capacitor is connected for power-factor correction. Calculate the input in kVA taken from the mains and its power factor when the motor is (a) on half load, (b) on full load. Sketch a phasor diagram for each case.   (N.C.T.E.C., O.2)

48. A single-phase motor takes 50 A at a power factor of 0·6 lagging from a 250-V, 50-Hz supply. What value of capacitance must a shunting capacitor have to raise the overall power factor to 0·9 lagging?

    How does the installation of the capacitor affect the line and motor currents?   (S.A.N.C., O.2)

49. A 240-V, single-phase supply feeds the following loads: (a) incandescent lamps taking a current of 8 A at unity power factor; (b) fluorescent lamps taking a current of 5 A at 0·8 leading power factor; (c) a motor taking a current of 7A at 0·75 lagging power factor. Sketch the phasor diagram and determine the total current, power and reactive voltamperes taken from the supply and the overall power factor.   (U.L.C.I., O.2)

50. The load taken from an a.c. supply consists of: (a) a heating load of 15 kW; (b) a motor load of 40 kVA at 0·6 power factor lagging; (c) a load of 20 kW at 0·8 power factor lagging.

    Calculate the total load from the supply in kW and kVA and its power factor. What would be the kVAr rating of a capacitor to bring the power factor to unity and how would the capacitor be connected?
    (U.E.I., O.2)

51. A coil of resistance 2 Ω and reactance 5 Ω is connected in series with a non-reactive resistor which is continuously variable between 0 and 8 Ω. Draw the current locus diagram when the circuit is connected to a 200-V supply, and hence find the values of the maximum and minimum currents and the corresponding power factors and powers. Find also the maximum power. (W.J.E.C., O.2)

52. A circuit having a constant resistance of 60 Ω and a variable inductance of 0–0·4 H is connected across a 100-V, 50-Hz supply. Derive from first principles the locus of the extremity of the current phasor. Find (*a*) the power and (*b*) the inductance of the circuit when the power factor is 0·8. (App. El., L.U.)

53. An a.c. circuit consists of a variable resistor in series with a coil for which $R = 20$ Ω and $L = 0·1$ H. Show that when this circuit is supplied at constant voltage and frequency, and the resistance is varied between zero and infinity, the locus of the current phasor is a circular arc. Calculate, when the supply voltage is 100 V and the frequency 50 Hz: (*a*) the radius (in amperes) of this arc; (*b*) the value of the variable resistor in order that the power taken from the mains may be a maximum. (App. El., L.U.)

54. A variable capacitor and a resistor of 200 Ω are connected in series across a 200-V, 50-Hz supply. Draw a phasor locus diagram showing the variation of the current as the capacitance is changed from 10 μF to 100 μF. Find: (*a*) the capacitance for a current of 0·8 A; (*b*) the current when the capacitance is 50 μF.

55. The equivalent series circuit for a certain capacitor consists of a 2-Ω resistor in series with a 200-pF capacitor. Calculate the loss angle of the capacitor at a frequency of 5 MHz. (U.L.C.I., O.2)

56. A certain capacitor has a loss angle of 0·03 radian and when it is connected across a 6·6-kV, 50-Hz supply, the power loss is 25 W. Calculate the component values of the equivalent parallel circuit. (U.L.C.I., O.2)

CHAPTER 12

Complex Notation

12.1 The j operator

In Chapter 11, problems on a.c. circuits were solved with the aid of phasor diagrams. So long as the circuits are fairly simple, this method is satisfactory; but with a more involved circuit, such as that of fig. 12.17, the calculation can be simplified by using complex algebra. This system enables equations representing alternating voltages and currents and their phase relationships to be expressed in simple algebraic form. It is based upon the idea that a phasor can be resolved into two components at right-angles to each other. For instance, in fig. 12.1 (a), phasor OA can be resolved into components OB along

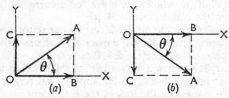

Fig. 12.1 Resolution of phasors.

the X axis and OC along the Y axis, where OB = OA cos θ and OC = OA sin θ. It would obviously be incorrect to state that OA = OB + OC, since OA is actually $\sqrt{(OB^2 + OC^2)}$; but by introducing a symbol j to denote that OC is the component along the Y axis, we can represent the phasor thus:

$$OA^* = OB + jOC = OA(\cos \theta + j \sin \theta).$$

The phasor OA may alternatively be expressed thus:

$$OA = OA \angle\theta.$$

* Symbols representing phasors are printed in bold face italics, while those representing only the magnitudes are printed in ordinary italics, thus:

    V, I, Z   phasors or complex numbers,
    V, I, Z   magnitudes or moduli.

In manuscript, a phasor may be indicated by a horizontal bar above the symbol, and the magnitude or modulus by a vertical line on each side, thus: $|V|$.

If OA is occupying the position shown in fig. 12.1 (*b*), the vertical component is negative, so that

$$OA = \text{OB} - \text{jOC} = \text{OA} \angle -\theta.$$

Fig. 12.2 represents four phasors occupying different quadrants.

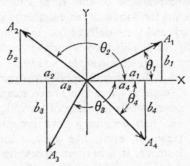

Fig. 12.2 Resolution of phasors.

These phasors may be represented thus:

$$A_1 = a_1 + jb_1 = A_1 \angle \theta_1 \tag{12.1}$$

where $A_1 = \sqrt{(a_1^2 + b_1^2)}$ and $\tan \theta_1 = b_1/a_1$.

Similarly

$$A_2 = -a_2 + jb_2 = A_2 \angle \theta_2$$
$$A_3 = -a_3 - jb_3 = A_3 \angle -\theta_3$$

and

$$A_4 = a_4 - jb_4 = A_4 \angle -\theta_4.$$

The symbol j, when applied to a phasor, alters its direction by 90° in an anticlockwise direction, without altering its length, and is consequently referred to as an *operator*. For example, if we start with a phasor A in phase with the X axis, as in fig. 12.3, then jA represents a phasor of the same length upwards along the Y axis. If we apply the operator j to jA, we turn the phasor anticlockwise through another 90°, thus giving jjA or

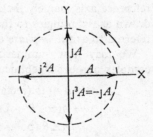

Fig. 12.3 Significance of the j operator.

j^2A in fig. 12.3. The symbol j^2 signifies that we have applied the operator j twice in succession, thereby rotating the phasor through 180°. This reversal of the phasor is equivalent to multiplying by

−1; i.e. $j^2A = -A$, so that j^2 may be regarded as being numerically equal to −1 and $j = \sqrt{-1}$.

When mathematicians were first confronted by an expression such as $A = 3 + j4$, they thought of $j4$ as being an imaginary number rather than a real number; and it was Argand, in 1806, who first suggested that an expression of this form could be represented graphically by plotting the 3 units of *real* number in the above expression along the X axis and the 4 units of *imaginary* or j number along the Y axis, as in fig. 12.4. This type of number, combining real and imaginary numbers, is termed a *complex number*.

The term *imaginary*, though still applied to numbers containing

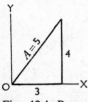

Fig. 12.4 Representation of $A = 3 + j4$.

j, such as $j4$ in the above expression, has long since lost its meaning of unreality, and the terms *real* and *imaginary* have become established as technical terms, like *positive* and *negative*. For instance, if a current represented by $I = 3 + j4$ amperes were passed through a resistor of, say, 10Ω, the power due to the 'imaginary' component of the current would be $4^2 \times 10 = 160$ W, in exactly the same way as that due to the 'real' component would be $3^2 \times 10 = 90$ W. From fig. 12.4 it is seen that the actual current would be 5 A. When this current flows through a 10-Ω resistor, the power is $5^2 \times 10 = 250$ W, namely the sum of the powers due to the real and imaginary components of the current.

Since the real component of a complex number is drawn along the reference axis, namely the X axis, and the imaginary component is drawn at right-angles to that axis, these components are sometimes referred to as the *in-phase* and *quadrature* components respectively.

We may now summarize the various ways of representing a complex number algebraically:

$$A = a + jb \text{ (rectangular or Cartesian notation),}$$
$$= A(\cos \theta + j \sin \theta) \text{ (trigonometric notation),}$$
$$= A\angle\theta \text{ (polar notation).}$$

12.2 Addition and subtraction of phasors

Suppose A_1 and A_2 in fig. 12.5 to be two phasors to be added together. From this phasor diagram, it is evident that:

$$A_1 = a_1 + jb_1 \text{ and } A_2 = a_2 + jb_2.$$

It was shown in section 10.6 that the resultant of A_1 and A_2 is given by A, the diagonal of the parallelogram drawn on A_1 and A_2. If a and b be the real and imaginary components respectively of A, then $A = a + jb$.

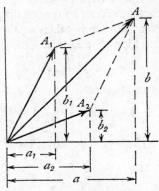

Fig. 12.5 Addition of phasors.

But it is evident from fig. 12.5 that:

$$a = a_1 + a_2 \quad \text{and} \quad b = b_1 + b_2$$

$$\therefore \quad A = a_1 + a_2 + j(b_1 + b_2)$$

$$= (a_1 + jb_1) + (a_2 + jb_2)$$

$$= A_1 + A_2.$$

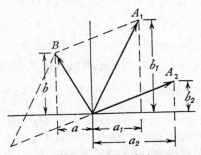

Fig. 12.6 Subtraction of phasors.

Fig. 12.6 shows the construction for subtracting phasor A_2 from phasor A_1. If B is the phasor difference of these quantities and if a_1 is assumed less than a_2, then the real component of B is negative,

$\therefore$
$$B = -a + jb = a_1 - a_2 + j(b_1 - b_2)$$
$$= (a_1 + jb_1) - (a_2 + jb_2) = A_1 - A_2.$$

12.3 Voltage, current and impedance

Let us first consider a simple circuit possessing resistance R in *series* with an inductive reactance X_L. The phasor diagram is given in fig. 12.7, where the current phasor is taken as the reference quantity and is therefore drawn along the X axis,

i.e.
$$I = I + j0 = I \angle 0°.$$

From fig. 12.7 it is evident that:
$$V = IR + jIX_L = I(R + jX_L) = IZ^* = IZ \angle \phi$$
where
$$Z = R + jX_L = Z \angle \phi \quad \text{and} \quad \tan \phi = X_L/R \qquad (12.2)$$

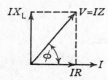

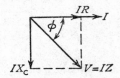

Fig. 12.7 Phasor diagram for R and L in series.

Fig. 12.8 Phasor diagram for R and C in series.

Fig. 12.8 gives the phasor diagram for a circuit having a resistance R in series with a capacitive reactance X_C. From this diagram:
$$V = IR - jIX_C = I(R - jX_C) = IZ = IZ \angle -\phi$$
where
$$Z = R - jX_C = Z \angle -\phi \quad \text{and} \quad \tan \phi = -X_C/R \qquad (12.3)$$

Hence for the general case of a circuit having R, X_L and X_C in series,
$$Z = R + j(X_L - X_C) \quad \text{and} \quad \tan \phi = (X_L - X_C)/R$$

Example 12.1 *Express in rectangular and polar notations, the impedance of each of the following circuits at a frequency of 50 Hz: (a) a resistance of 20 Ω in series with an inductance of 0·1 H; (b) a resistance*

* In the expression $V = IZ$, V and I differ from Z in that they are associated with time-varying quantities, whereas Z is a complex number independent of time and is therefore not a phasor in the sense that V and I are phasors. It has, however, become the practice to refer to all complex numbers used in a.c. circuit calculations as phasors.

of 50 Ω in series with a capacitance of 40 μF; (c) circuits (a) and (b) in series.

If the terminal voltage is 230 V, 50 Hz, calculate the value of the current in each case and the phase of each current relative to the applied voltage.

(a) For 50 Hz, $\omega = 2\pi \times 50 = 314$ rad/s,

∴ $Z = 20 + j314 \times 0.1 = 20 + j31.4 \; \Omega$

Hence $Z = \sqrt{\{(20)^2 + (31.4)^2\}} = 37.2 \; \Omega$

and $I = 230/37.2 = 6.18$ A.

If ϕ be the phase difference between the applied voltage and the current,

$$\tan \phi = 31.4/20 = 1.57,$$

∴ $\phi = 57° \; 30'$, current lagging.

The impedance can also be expressed:

$$Z = 37.2\angle 57° \; 30'* \; \Omega.$$

If the applied voltage be taken as the reference quantity, then

$$V = 230\angle 0° \text{ volts,}$$

∴ $$I = \frac{230\angle 0°}{37.2\angle 57° \; 30'} = 6.18\angle -57° \; 30' \text{ A.}$$

(b) $$Z = 50 - j\frac{10^6}{314 \times 40} = 50 - j79.6 \; \Omega$$

∴ $Z = \sqrt{\{(50)^2 + (79.6)^2\}} = 94 \; \Omega,$

and $I = 230/94 = 2.447$ A.

Tan $\phi = -79.6/50 = -1.592,$

∴ $\phi = 57° \; 52'$, current leading.

The impedance can also be expressed thus:

$$Z = 94\angle -57° \; 52' \; \Omega$$

∴ $$I = \frac{230\angle 0°}{94\angle -57° \; 52'} = 2.447\angle 57° \; 52' \text{ A.}$$

(c) $Z = 20 + j31.4 + 50 - j79.6$

$$= 70 - j48.2 \; \Omega$$

---

* This form is more convenient than that involving the j term when it is required to find the product or the quotient of two complex numbers; thus, $A\angle\alpha \times B\angle\beta = AB\angle(\alpha + \beta)$ and $A\angle\alpha/B\angle\beta = (A/B)\angle(\alpha - \beta)$.

$$\therefore \qquad Z = \sqrt{\{(70)^2 + (48 \cdot 2)^2\}} = 85 \ \Omega,$$

and $\qquad I = 230/85 = 2 \cdot 706$ A.

$$\text{Tan } \phi = -48 \cdot 2/70 = -0 \cdot 689,$$

$$\therefore \qquad \phi = 34° \ 34', \text{ current leading.}$$

The impedance can also be expressed:

$$Z = 85 \angle -34° \ 34' \ \Omega$$

so that $\qquad I = \dfrac{230 \angle 0°}{85 \angle -34° \ 34'} = 2 \cdot 706 \angle 34° \ 34'$ A.

Example 12.2 *Calculate the resistance and the inductance or capacitance in series for each of the following impedances:* (a) $10 + j15 \ \Omega$; (b) $-j80 \ \Omega$; (c) $50 \angle 30° \ \Omega$; *and* (d) $120 \angle -60° \ \Omega$. *Assume the frequency to be 50 Hz.*

(*a*) For $Z = 10 + j15 \ \Omega$, resistance $= 10 \ \Omega$,

and $\qquad$ inductive reactance $= 15 \ \Omega$,

$\therefore \qquad\qquad$ inductance $= 15/314 = 0 \cdot 0478$ H.

(*b*) For $Z = -j80 \ \Omega$, resistance $= 0$,

and $\qquad$ capacitive reactance $= 80 \ \Omega$,

$\therefore \qquad\qquad$ capacitance $= \dfrac{1}{314 \times 80}$ F $= 39 \cdot 8 \ \mu$F.

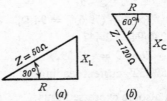

Fig. 12.9 Impedance triangles for example 12.2 (*c*) and (*d*).

(*c*) Fig. 12.9 (*a*) is an impedance triangle representing $50 \angle 30° \ \Omega$. From this diagram, it follows that the reactance is inductive and that $R = Z \cos \phi$ and $X_L = Z \sin \phi$,

$$\therefore \qquad\qquad Z = R + jX_L = Z(\cos \phi + j \sin \phi)$$
$$= 50 \ (\cos 30° + j \sin 30°) = 43 \cdot 3 + j25 \ \Omega.$$

Hence, $\qquad\qquad$ resistance $= 43 \cdot 3 \ \Omega$

and $\qquad$ inductive reactance = 25 Ω,

so that $\qquad$ inductance = 25/314 = 0·0796 H.

(d) Fig. 12.9 (b) is an impedance triangle representing $120\angle -60°$ Ω.

It will be seen that the reactance is capacitive, so that

$$\mathbf{Z} = R - jX_C = Z(\cos\phi - j\sin\phi)$$
$$= 120(\cos 60° - j\sin 60°) = 60 - j103·9 \ \Omega.$$

Hence, $\qquad$ resistance = 60 Ω

and $\qquad$ capacitive reactance = 103·9 Ω

∴ $\qquad$ capacitance $= \dfrac{10^6}{314 \times 103·9} = 30·7 \ \mu F.$

12.4 Admittance, conductance and susceptance

When resistors having resistances R_1, R_2, etc., are in parallel, the equivalent resistance R is given by:

$$\frac{1}{R} = \frac{1}{R_1} + \frac{1}{R_2} + \dots$$

In d.c. work the reciprocal of the resistance is known as *conductance* (section 1.8). It is represented by symbol G and the unit of conductance is the *siemens* (p. 10). Hence, if circuits having conductances G_1, G_2, etc., are in parallel, the total conductance G is given by:

$$G = G_1 + G_2 + \dots$$

In a.c. work the conductance is the reciprocal of the resistance *only when the circuit possesses no reactance*. This matter is dealt with more fully in section 12.5.

If circuits having impedances Z_1, Z_2, etc., are connected in parallel across a supply voltage V, then:

$$I_1 = \frac{V}{Z_1}, \ I_2 = \frac{V}{Z_2}, \text{ etc.}$$

If Z be the equivalent impedance of Z_1, Z_2, etc., in parallel and if I be the resultant current, then, using complex notation, we have:

$$\mathbf{I} = I_1 + I_2 + \dots$$

∴ $$\frac{V}{Z} = \frac{V}{Z_1} + \frac{V}{Z_2} + \dots$$

so that $$\frac{1}{Z} = \frac{1}{Z_1} + \frac{1}{Z_2} + \dots \tag{12.4}$$

The reciprocal of impedance is termed *admittance* and is represented by the symbol Y, the unit being again the *siemens* (abbreviation, S). Hence, we may write expression (12.4) thus:

$$Y = Y_1 + Y_2 + \ldots \qquad (12.5)$$

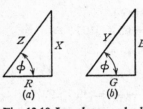

Fig. 12.10 Impedance and admittance triangles.

It has already been shown that impedance can be resolved into a real component R and an imaginary component X, as in fig. 12.10 (*a*). Similarly, an admittance may be resolved into a real component termed *conductance* and an imaginary component termed *susceptance*, represented by symbols G and B respectively as in fig. 12.10 (*b*),

i.e. $\qquad Y = G + jB \quad$ and $\quad \tan \phi = B/G.$

The significance of these terms will be more obvious when we consider their application to actual circuits.

12.5 Admittance of a circuit having resistance and inductive reactance in series

The phasor diagram for this circuit has already been given in fig. 12.7. From the latter it will be seen that the resultant voltage can be represented thus:

$$V = IR + jIX_L$$

$$\therefore \qquad Z = \frac{V}{I} = R + jX_L$$

If Y be the admittance of the circuit, then:

$$Y = \frac{1}{Z} = \frac{1}{R + jX_L} = \frac{R - jX_L}{R^2 + X_L^2}^*$$

$$= \frac{R}{R^2 + X_L^2} - \frac{jX_L}{R^2 + X_L^2} = G - jB_L \qquad (12.6)$$

where $\qquad G = \text{conductance} = \dfrac{R}{R^2 + X_L^2} = \dfrac{R}{Z^2} \qquad (12.7)$

---

* This method of transferring the j term from the denominator to the numerator is known as 'rationalizing'; thus,

$$\frac{1}{a + jb} = \frac{a - jb}{(a + jb)(a - jb)} = \frac{a - jb}{a^2 + b^2} \qquad (12.9)$$

and $\qquad B_L =$ inductive susceptance

$$= \frac{X_L}{R^2 + X_L{}^2} = \frac{X_L}{Z^2} \tag{12.8}$$

From (12.7) it is evident that if the circuit has no reactance, i.e. if $X_L = 0$, then the conductance is $1/R$, namely the reciprocal of the resistance. Similarly, from (12.8) it follows that if the circuit has no resistance, i.e. if $R = 0$, the susceptance is $1/X_L$, namely the reciprocal of the reactance. In general, we may define the *conductance* of a *series* circuit as the ratio of the resistance to the square of the impedance and the *susceptance* as the ratio of the reactance to the square of the impedance.

12.6 Admittance of a circuit having resistance and capacitive reactance in series

Fig. 12.8 gives the phasor diagram for this circuit. From this diagram it follows that:

$$V = IR - jIX_C$$

$\therefore \qquad Z = \dfrac{V}{I} = R - jX_C$

and $\qquad Y = \dfrac{1}{R - jX_C} = \dfrac{R + jX_C}{R^2 + X_C{}^2}$

$$= \frac{R}{R^2 + X_C{}^2} + \frac{jX_C}{R^2 + X_C{}^2} = G + jB_C \tag{12.10}$$

where $\qquad B_C =$ capacitive susceptance

$$= \frac{X_C}{R^2 + X_C{}^2} = \frac{X_C}{Z^2} \tag{12.11}$$

It will be seen that in the complex expression for an *inductive* circuit, the *impedance* has a *positive* sign in front of the imaginary component, whereas the imaginary component of the *admittance* is preceded by a negative sign. On the other hand, for a *capacitive* circuit, the imaginary component of the *impedance* has a *negative* sign and that of the *admittance* has a *positive* sign. Thus if the impedance of a circuit is represented by $(2 - j3)$ Ω, we know immediately that the circuit is capacitive; but if the admittance is $(2 - j3)$ siemens, the circuit must be inductive.

12.7 Admittance of a circuit having resistance and reactance in parallel

(a) *Inductive reactance*

From the circuit and phasor diagrams of figs. 12.11 and 12.12 respectively, it follows that:

$$I = I_R - jI_L = \frac{V}{R} - \frac{jV}{X_L}$$

$$\therefore \qquad Y = \frac{I}{V} = \frac{1}{R} - \frac{j}{X_L} = G - jB_L \qquad (12.12)$$

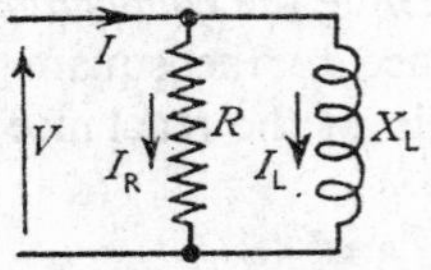

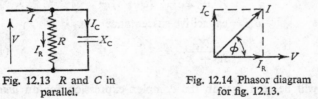

Fig. 12.11 *R* and *L* in parallel.

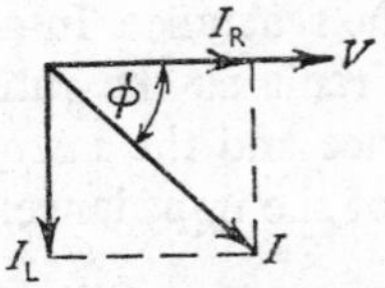

Fig. 12.12 Phasor diagram for fig. 12.11.

(b) *Capacitive reactance*

From figs. 12.13 and 12.14 it follows that:

$$I = I_R + jI_C = \frac{V}{R} + \frac{jV}{X_C}$$

$$\therefore \qquad Y = \frac{I}{V} = \frac{1}{R} + \frac{j}{X_C} = G + jB_C \qquad (12.13)$$

From expressions (12.12) and (12.13), it will be seen that if the admittance of a circuit is $(0{\cdot}2 - j0{\cdot}1)$ mho, such a circuit can be represented as a resistance of 5 Ω in *parallel* with an inductive

Fig. 12.13 *R* and *C* in parallel.

Fig. 12.14 Phasor diagram for fig. 12.13.

reactance of 10 Ω; whereas if the impedance of a circuit is $(5 + j10)$ Ω, such a circuit can be represented as a resistance of 5 Ω in *series* with an inductive reactance of 10 Ω.

Example 12.3 *Express in rectangular notation the admittance of circuits having the following impedances:*

(a) $(4 + j6)$ Ω;   (b) $20\angle-30°$ Ω.

(a) $Z = 4 + j6 \ \Omega$,

$\therefore$ $Y = \dfrac{1}{4 + j6} = \dfrac{4 - j6}{16 + 36} = 0\cdot0769 - j0\cdot1154$ siemens.

(b) $Z = 20\angle-30° = 20(\cos 30° - j \sin 30°)$
           $= 20(0\cdot866 - j0\cdot5) = 17\cdot32 - j10 \ \Omega$,

$\therefore$ $Y = \dfrac{1}{17\cdot32 - j10} = \dfrac{17\cdot32 + j10}{400}$
        $= 0\cdot0433 + j0\cdot025$ S.

Alternatively, $Y = \dfrac{1}{20\angle-30°} = 0\cdot05\angle30°$
                 $= 0\cdot0433 + j0\cdot025$ S.

Example 12.4 *The admittance of a circuit is* $(0\cdot05 - j0\cdot08)$ *S. Find the values of the resistance and the inductive reactance of the circuit if they are* (a) *in parallel,* (b) *in series.*

(a) The conductance of the circuit is $0\cdot05$ S and its inductive susceptance is $0\cdot08$ S. From (12.12) it follows that if the circuit consists of a resistance in parallel with an inductive reactance, then:

$$\text{resistance} = \frac{1}{\text{conductance}} = \frac{1}{0\cdot05} = 20 \ \Omega,$$

and inductive reactance

$$= \frac{1}{\text{inductive susceptance}} = \frac{1}{0\cdot08} = 12\cdot5 \ \Omega.$$

(b) Since $Y = 0\cdot05 - j0\cdot08$ S,

$\therefore$ $Z = \dfrac{1}{0\cdot05 - j0\cdot08} = \dfrac{0\cdot05 + j0\cdot08}{0\cdot0089} = 5\cdot62 + j8\cdot99 \ \Omega.$

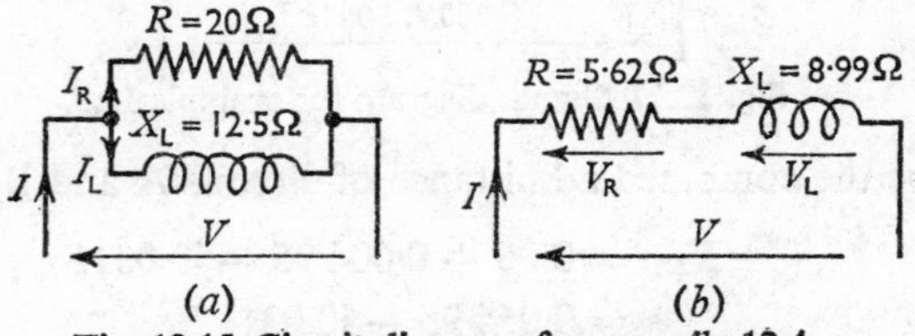

Fig. 12.15 Circuit diagrams for example 12.4.

Hence if the circuit consists of a resistance in series with an inductance, the resistance is $5\cdot62 \ \Omega$ and the inductive reactance is $8\cdot99 \ \Omega$. The two circuit diagrams are shown in figs. 12.15 (a) and (b), and their phasor diagrams are given in figs. 12.16 (a) and (b) respectively. The two circuits are equivalent in that they take the same current I for a given supply voltage V, and the phase difference ϕ between the supply voltage and current is the same in the two cases.

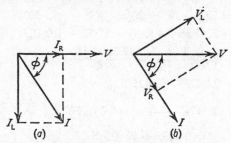

Fig. 12.16 Phasor diagrams for example 12.4.

Example 12.5 *A circuit is arranged as indicated in fig. 12.17, the values being as shown. Calculate the value of the current in each branch and its phase relative to the supply voltage. Draw the complete phasor diagram.*

$$Z_A = 20 + j0 \ \Omega,$$

$\therefore \quad Y_A = 0{\cdot}05 + j0 \ \text{S}.$

$$Z_B = 5 + j314 \times 0{\cdot}1 = 5 + j31{\cdot}4 \ \Omega,$$

$\therefore \quad Y_B = \dfrac{1}{5 + j31{\cdot}4} = \dfrac{5 - j31{\cdot}4}{1010} = 0{\cdot}004\ 95 - j0{\cdot}0311 \ \text{S}.$

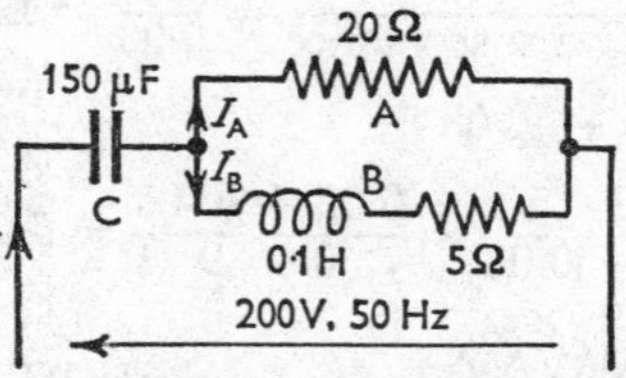

Fig. 12.17 Circuit diagram for example 12.5.

If Y_{AB} be the combined admittance of circuits A and B,

$$Y_{AB} = 0{\cdot}05 + 0{\cdot}004\ 95 - j0{\cdot}0311$$
$$= 0{\cdot}054\ 95 - j0{\cdot}0311 \ \text{S}.$$

If Z_{AB} be the equivalent impedance of circuits A and B,

$$Z_{AB} = \frac{1}{0{\cdot}054\ 95 - j0{\cdot}0311} = \frac{0{\cdot}054\ 95 + j0{\cdot}0311}{0{\cdot}003\ 987}$$
$$= 13{\cdot}78 + j7{\cdot}8 \ \Omega.$$

Hence the circuit of fig. 12.17 can be replaced by that shown in fig. 12.18.

$$Z_C = -j\frac{10^6}{314 \times 150} = -j21 \cdot 2 \ \Omega,$$

∴ total impedance = Z

$$= 13 \cdot 78 + j7 \cdot 8 - j21 \cdot 2 = 13 \cdot 78 - j13 \cdot 4$$
$$= \sqrt{\{(13 \cdot 78)^2 + (13 \cdot 4)^2\}} \angle \tan^{-1} - 13 \cdot 4/13 \cdot 78$$
$$= 19 \cdot 22 \angle - 44° \ 12' \ \Omega.$$

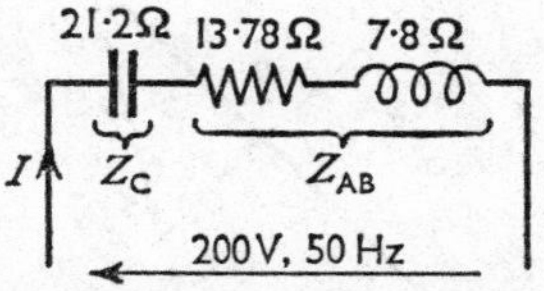

Fig. 12.18 Equivalent circuit of fig. 12.17.

If the supply voltage = $V = 200\angle0°$ volts,

∴ supply current = $I = \dfrac{200\angle0°}{19 \cdot 22 \angle -44° \ 12'} = 10 \cdot 4 \angle 44° \ 12' \ \text{A},$

i.e. the supply current is 10·4 A leading the *supply* voltage by 44° 12′.

The p.d. across circuit AB = $V_{AB} = IZ_{AB}$.

But $\quad Z_{AB} = 13 \cdot 78 + j7 \cdot 8$
$$= \sqrt{\{(13 \cdot 78)^2 + (7 \cdot 8)^2\}} \angle \tan^{-1} 7 \cdot 8/13 \cdot 78$$
$$= 15 \cdot 85 \angle 29° \ 30' \ \Omega,$$

∴ $\quad V_{AB} = 10 \cdot 4 \angle 44° \ 12' \times 15 \cdot 85 \angle 29° \ 30'$
$$= 164 \cdot 8 \angle 73° \ 42' \ \text{V}.$$

Since $\quad Z_A = 20 + j0 = 20\angle0° \ \Omega,$

∴ $\quad I_A = \dfrac{164 \cdot 8 \angle 73° \ 42'}{20\angle0°} = 8 \cdot 24 \angle 73° \ 42' \ \text{A},$

i.e. the current through branch A is 8·24 A leading the *supply* voltage by 73° 42′.

Similarly $\quad Z_B = 5 + j31 \cdot 4 = 31 \cdot 8 \angle 80° \ 58' \ \Omega,$

∴ $\quad I_B = \dfrac{164 \cdot 8 \angle 73° \ 42'}{31 \cdot 8 \angle 80° \ 58'} = 5 \cdot 18 \angle -7° \ 16' \ \text{A},$

i.e. the current through branch B is 5·18 A lagging the *supply* voltage by 7° 16'.

Impedance of $C = Z_C = -j21·2 = 21·2\angle-90°$ Ω,

∴   p.d. across $C = V_C = IZ_C = 10·4\angle44°$ 12' × 21·2∠−90°

$$= 220\angle-45°\ 48'\ V.$$

The various voltages and currents of this example are represented by the respective phasors in fig. 12.19.

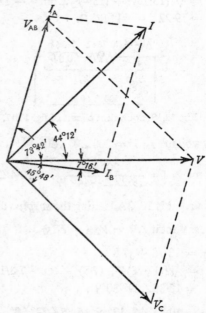

Fig. 12.19 Phasor diagram for example 12.5.

12.8 Calculation of power, using complex notation

Suppose the alternating voltage across and the current in a circuit to be represented respectively by:

$$V = V\angle\alpha = V(\cos\alpha + j\sin\alpha) = a + jb \qquad (12.14)$$

and $$I = I\angle\beta = I(\cos\beta + j\sin\beta) = c + jd \qquad (12.15)$$

Since the phase difference between the voltage and current is $(\alpha - \beta)$,

$$\begin{aligned}\text{power} &= VI\cos(\alpha - \beta) \\ &= VI(\cos\alpha\,.\,\cos\beta + \sin\alpha\,.\,\sin\beta) \\ &= ac + bd \qquad (12.16)\end{aligned}$$

i.e. the power is given by the sum of the products of the real components and of the imaginary components.

$$\text{Reactive voltamperes} = VI \sin (\alpha - \beta)$$
$$= VI (\sin \alpha \, . \cos \beta - \cos \alpha \, . \sin \beta)$$
$$= bc - ad \quad\quad\quad (12.17)$$

If we had proceeded by multiplying (12.14) by (12.15), the result would have been:

$$(ac - bd) + j(bc + ad).$$

The terms within the brackets represent neither the power nor the reactive voltamperes. The correct expressions for these quantities are derived by multiplying the voltage by the *conjugate* of the current, the conjugate of a complex number being a quantity that differs only in the sign of the imaginary component; thus the conjugate of $c + jd$ is $c - jd$. Hence:

$$(a + jb)(c - jd) = (ac + bd) + j(bc - ad)$$
$$= \text{power in watts} + j(\text{reactive voltamperes}),$$

Example 12.6 *The p.d. across and the current in a circuit are represented by* (100 + j200) *V and* (10 + j5) *A respectively. Calculate the power and the reactive voltamperes (or vars).*

From (12.16),
$$\text{power} = (100 \times 10) + (200 \times 5)$$
$$= 2000 \text{ W.}$$

From (12.17),

$$\text{reactive voltamperes (or vars)} = (200 \times 10) - (100 \times 5)$$
$$= 1500 \text{ VAr.}$$

Alternatively,
$$100 + j200 = 223 \cdot 6 \angle 63° \, 26' \text{ V}$$
and
$$10 + j5 = 11 \cdot 18 \angle 26° \, 34' \text{ A,}$$

$$\therefore \quad \left. \begin{array}{l} \text{phase difference between} \\ \text{voltage and current} \end{array} \right\} = 63° \, 26' - 26° \, 34' = 36° \, 52'.$$

Hence,
$$\text{power} = 223 \cdot 6 \times 11 \cdot 18 \cos 36° \, 52'$$
$$= 2000 \text{ W}$$

and
$$\left. \begin{array}{l} \text{reactive voltamperes} \\ \text{(or vars)} \end{array} \right\} = 223 \cdot 6 \times 11 \cdot 18 \sin 36° \, 52'$$
$$= 1500 \text{ VAr.}$$

Summary of important formulae

$$A = a + jb = A(\cos\theta + j\sin\theta) = A\angle\theta$$

where
$$A = \sqrt{(a^2 + b^2)} \quad \text{and} \quad \theta = \tan^{-1}b/a$$

$$A\angle\alpha \times B\angle\beta = AB\angle(\alpha + \beta)$$

$$\frac{A\angle\alpha}{B\angle\beta} = \frac{A}{B}\angle(\alpha - \beta)$$

$$\frac{1}{a + jb} = \frac{a - jb}{a^2 + b^2} \tag{12.9}$$

For a circuit having R and L in *series*,

$$Z = R + jX_L = Z\angle\phi \tag{12.2}$$

Admittance $= Y = \dfrac{1}{Z} = \dfrac{R}{Z^2} - \dfrac{jX_L}{Z^2} = G - jB_L = Y\angle-\phi \tag{12.6}$

For a circuit having R and C in *series*,

$$Z = R - jX_C = Z\angle-\phi \tag{12.3}$$

and
$$Y = \frac{R}{Z^2} + \frac{jX_C}{Z^2} = G + jB_C = Y\angle\phi \tag{12.10}$$

Conductance $= G = R/Z^2$ and is $1/R$ only when $X = 0$.

Susceptance $= B = X/Z^2$ and is $1/X$ only when $R = 0$.

For impedances $Z_1 = R_1 + jX_1$ and $Z_2 = R_2 + jX_2$ in *series*,

total impedance $= Z = Z_1 + Z_2 = (R_1 + R_2) + j(X_1 + X_2)$
$$= Z\angle\phi$$

where
$$Z = \sqrt{\{(R_1 + R_2)^2 + (X_1 + X_2)^2\}}$$
and
$$\phi = \tan^{-1}(X_1 + X_2)/(R_1 + R_2).$$

For a circuit having R and L in *parallel*,

$$Y = \frac{1}{R} - \frac{j}{X_L} = G - jB_L = Y\angle-\phi \tag{12.12}$$

For a circuit having R and C in *parallel*,

$$Y = \frac{1}{R} + \frac{j}{X_C} = G + jB_C = Y\angle\phi \tag{12.13}$$

For admittances $Y_1 = G_1 + jB_1$ and $Y_2 = G_2 + jB_2$ in *parallel*,

total admittance $= Y = Y_1 + Y_2 = (G_1 + G_2) + j(B_1 + B_2)$
$$= Y\angle\phi$$

where $\qquad\qquad Y = \sqrt{\{(G_1 + G_2)^2 + (B_1 + B_2)^2\}}$

and $\qquad\qquad \phi = \tan^{-1}(B_1 + B_2)/(G_1 + G_2)$.

If $\qquad\qquad\qquad\qquad V = a + jb$

and $\qquad\qquad\qquad\qquad I = c + jd$,

$$\text{power} = ac + bd \qquad\qquad (12.16)$$

and $\qquad\qquad$ reactive voltamperes $= bc - ad \qquad\qquad (12.17)$

EXAMPLES 12

1. Express in rectangular and polar notations the phasors for the following quantities: (*a*) $i = 10 \sin \omega t$; (*b*) $i = 5 \sin (\omega t - \pi/3)$; (*c*) $v = 40 \sin (\omega t + \pi/6)$.
   Draw a phasor diagram representing the above voltage and currents.
2. With the aid of a simple diagram, explain the *j*-notation method of phasor quantities.
   Four single-phase alternators whose e.m.f.s can be represented by: $e_1 = 20 \sin \omega t$; $e_2 = 40 \sin (\omega t + \pi/2)$; $e_3 = 30 \sin (\omega t - \pi/6)$; $e_4 = 10 \sin (\omega t - \pi/3)$; are connected in series so that their resultant e.m.f. is given by $e = e_1 + e_2 + e_3 + e_4$. Express each e.m.f. and the resultant in the form $a \pm jb$. Hence find the maximum value of e and its phase angle relative to e_1. (U.L.C.I., O.2)
3. Express each of the following phasors in polar notation and draw the phasor diagram: (*a*) $10 + j5$; (*b*) $3 - j8$.
4. Express each of the following phasors in rectangular notation and draw the phasor diagram: (*a*) $20\angle 60°$; (*b*) $40\angle -45°$.
5. Add the two phasors of Q. 3 and express the result in: (*a*) rectangular notation; (*b*) polar notation. Check the values by drawing a phasor diagram to scale.
6. Subtract the second phasor of Q. 3 from the first phasor and express the result in: (*a*) rectangular notation; (*b*) polar notation. Check the values by means of a phasor diagram drawn to scale.
7. Add the two phasors of Q. 4 and express the result in: (*a*) rectangular notation; (*b*) polar notation. Check the values by means of a phasor diagram drawn to scale.
8. Subtract the second phasor of Q. 4 from the first phasor and express the result in: (*a*) rectangular notation; (*b*) polar notation. Check the values by a phasor diagram drawn to scale.
9. Calculate the resistance and inductance or capacitance in *series* for each of the following impedances, assuming the frequency to be 50 Hz: (*a*) $50 + j30 \ \Omega$; (*b*) $30 - j50 \ \Omega$; (*c*) $100\angle 40° \ \Omega$; (*d*) $40\angle -60° \ \Omega$.

10. Derive expressions, in rectangular and polar notations, for the admittances of the following impedances: (*a*) $10 + j15 \ \Omega$; (*b*) $20 - j10 \ \Omega$; (*c*) $50\angle20° \ \Omega$; (*d*) $10\angle-70° \ \Omega$.

11. Derive expressions, in rectangular and polar notations, for the impedances of the following admittances: (*a*) $0·2 + j0·5$ siemens; (*b*) $0·08\angle-30°$ siemens.

12. Calculate the resistance and inductance or capacitance in *parallel* for each of the following admittances, assuming the frequency to be 50 Hz; (*a*) $0·25 + j0·6$ S; (*b*) $0·05 - j0·1$ S; (*c*) $0·8\angle30°$ S; (*d*) $0·5\angle-50°$ S.

13. A voltage, $v = 150 \sin (314t + 30°)$ volts, is maintained across a coil having a resistance of $20 \ \Omega$ and an inductance of $0·1$ H. Derive expressions for the r.m.s. values of the voltage and current phasors in: (*a*) rectangular notation; (*b*) polar notation. Draw the phasor diagram.

14. A voltage, $v = 150 \sin (314t + 30°)$ volts, is maintained across a circuit consisting of a 20-Ω non-reactive resistor in series with a loss-free 100-μF capacitor. Derive an expression for the r.m.s. value of the current phasor in: (*a*) rectangular notation; (*b*) polar notation. Draw the phasor diagram.

15. Calculate the values of resistance and reactance which, when in parallel, are equivalent to a coil having a resistance of $20 \ \Omega$ and a reactance of $10 \ \Omega$.

16. The impedances of two parallel circuits can be represented by $(24 + j18) \ \Omega$ and $(12 - j22) \ \Omega$ respectively. If the supply frequency is 50 Hz, find the resistance and inductance or capacitance of each circuit. Also, derive a symbolic expression in polar form for the admittance of the combined circuits, and thence find the phase angle between the applied voltage and the resultant current. (W.J.E.C., O.2)

17. A coil of resistance $25 \ \Omega$ and inductance $0·044$ H is connected in parallel with a circuit made up of a 50-μF capacitor in series with a 40-Ω resistor, and the whole is connected to a 200-V, 50-Hz supply. Calculate, using symbolic notation, the total current taken from the supply and its phase angle, and draw the complete phasor diagram. (W.J.E.C., O.2)

18. The current in a circuit is given by $4·5 + j12$ A when the applied voltage is $100 + j150$ V. Determine: (*a*) the complex expression for the impedance, stating whether it is inductive or capacitive; (*b*) the power; (*c*) the phase angle between voltage and current.

19. Explain how alternating quantities may be represented by complex numbers.

If the potential difference across a circuit is represented by $40 + j25$ V, and the circuit consists of a coil having a resistance of $20 \ \Omega$ and an inductance of $0·06$ H and the frequency is $79·5$ Hz, find the complex number representing the current in amperes. (App. El., L.U.)

20. The impedances of two parallel circuits can be represented by $(20 + j15) \ \Omega$ and $(10 - j60) \ \Omega$ respectively. If the supply frequency is 50 Hz, find the resistance and the inductance or capacitance of each circuit. Also, derive a complex expression for the admittance of the combined

circuits, and thence find the phase angle between the applied voltage and the resultant current. State whether this current is leading or lagging relatively to the voltage. (App. El., L.U.)

21. An alternating e.m.f. of 100 V is induced in a coil of impedance 10 + j25 Ω. To the terminals of this coil there is joined a circuit consisting of two parallel impedances, one of 30 − j20 Ω and the other of 50 + j0 Ω. Calculate the current in the coil in magnitude and phase with respect to the induced voltage. (U.L.C.I., Adv. El. Tech.)

22. A circuit consists of a 30-Ω non-reactive resistor in series with a coil having an inductance of 0·1 H and a resistance of 10 Ω. A 60-μF loss-free capacitor is connected in parallel with the *coil*. The circuit is connected across a 200-V, 50-Hz supply. Calculate the value of the current in each branch and its phase relative to the supply voltage.

23. An impedance of 2 + j6 Ω is connected in series with two impedances of 10 + j4 Ω and 12 − j8 Ω, which are in parallel. Calculate the magnitude and power factor of the main current when the combined circuit is supplied at 200 V. (U.L.C.I., Adv. El. Tech.)

24. The arms of an a.c. bridge (Maxwell) are arranged thus:

AB: a non-reactive resistor of 300 Ω.
BC: a variable resistance *R* in series with a variable inductance *L*.
CD: a coil, the resistance and reactance of which are required.
DA: a non-reactive resistor of 100 Ω.

An alternating voltage is applied across AC. Deduce the conditions for zero p.d. across BD, and calculate the resistance and the inductance of the coil if balance is obtained with *R* = 64 Ω and *L* = 0·28 H.

Note. The solution makes use of the fact that if $a + jb = c + jd$, then $a = c$ and $b = d$.

25. The arms of an a.c. bridge (de Sauty) are arranged thus:

AB: a non-reactive resistor of 1000 Ω.
BC: a variable no-loss capacitor having capacitance *C*.
CD: a capacitor X, the capacitance of which is required.
DA: a non-reactive resistor of 100 Ω.

If an alternating voltage is maintained across AC, deduce the condition for zero p.d. across BD. If the value of capacitance *C* be 0·068 μF when the bridge is balanced, calculate the capacitance of X.

26. The arms of an a.c. bridge (Owen) have the following impedances:

AB: a coil in series with a variable resistor P.
BC: a no-loss 0·5-μF capacitor in series with a variable resistor Q.
CD: a no-loss 0·3-μF capacitor.
DA: a 500-Ω non-reactive resistor.

If an alternating voltage be maintained across AC, deduce the conditions for zero p.d. across BD. If the values of P and Q to give a balance be 126 Ω and 534 Ω respectively, calculate the resistance and the inductance of the coil.

27. The p.d. across and the current in a given circuit are represented by (200 + j30) V and (5 — j2) A respectively. Calculate the power and the reactive voltamperes. State whether the reactive voltamperes are leading or lagging.
28. Had the current in Q. 27 been represented by (5 + j2) A, what would have been the power and the reactive voltamperes? Again state whether the reactive voltamperes are leading or lagging.
29. A p.d. of 200∠30° V is applied to two circuits connected in parallel. The currents in the respective branches are 20∠60° A and 40∠30° A. Find the kVA and kW in each branch and in the main circuit. Express the current in the main circuit in the form A + jB.

(W.J.E.C., O.2)

CHAPTER 13

Two- and Three-Phase Circuits

13.1 Disadvantages of the single-phase system

The earliest application of alternating current was for heating the filaments of electric lamps. For this purpose the single-phase system was perfectly satisfactory. Some years later, a.c. motors were developed, and it was found that for this application the single-phase system was not very satisfactory. For instance, the single-phase induction motor—the type most commonly employed—was not self-starting unless it was fitted with an auxiliary winding. By using two separate windings with currents differing in phase by a quarter of a cycle or three windings with currents differing in phase by a third of a cycle, it was found that the induction motor was self-starting and had better efficiency and power factor than the corresponding single-phase machine.

The system utilizing two windings is referred to as a *two-phase system* and that utilizing three windings is referred to as a *three-phase system*. We shall now consider these two systems in detail.

13.2 Generation of two-phase e.m.f.s

In fig. 13.1, AA₁ and BB₁ represent two similar loops fixed to each other at an angle of 90° and driven anticlockwise in the magnetic field due to poles NS. For the position shown in fig. 13.1, the e.m.f.

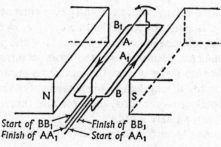

Fig. 13.1 Generation of two-phase e.m.f.s.
359

generated in AA_1 is zero and that in BB_1 is a maximum. The direction of the e.m.f. generated in BB_1 is indicated by the arrowheads. It follows that while loop AA_1 is moving through half a revolution from the position shown in fig. 13.1, the direction of the e.m.f.

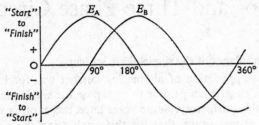

Fig. 13.2 Waveforms of two-phase e.m.f.s.

generated in AA_1 is from the 'start' terminal towards the 'finish' terminal. Let us regard this direction as positive; consequently, the e.m.f. generated in AA_1 can be represented by curve E_A in fig. 13.2.

Since the loops are rotating anticlockwise, it is evident from fig. 13.1 that the e.m.f. generated in side B has the same magnitude as that in side A but lags by 90° and is therefore represented by curve E_B.

If the two phases are kept electrically isolated from each other,

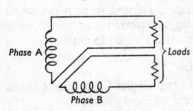

Fig. 13.3 Two-phase, four-wire system.

we have the *two-phase, four-wire system* shown in fig. 13.3. The volume of copper in the conductors between the generator and the load can be reduced by joining the phases at C and using a common return wire CC_1, as shown in fig. 13.4; thus we have the *two-phase, three-wire*

system. If the load is balanced, i.e. if the phase currents I_A and I_B are equal and differ in phase by a quarter of a cycle, as in fig. 13.5 (*a*), the current I_C in the *common* or *neutral wire* is $\sqrt{2} \times$ the phase current in each *outer conductor*. Hence, if *a* is the cross-sectional area of each conductor of the four-wire system, the total cross-sectional area of the line conductors for this system is $4a$; whereas, for the same current density, the total cross-sectional area for the three-wire system is $(a + a + 1 \cdot 414a)$, namely $3 \cdot 414a$.

The phasors $E_{CA}*$ and E_{CB} in fig. 13.5 (*b*) represent the r.m.s. values

* See note on double subscripts, section 1.19.

of the e.m.f.s in phases A and B respectively, the e.m.f. being assumed positive when its direction is from 'start' to 'finish'. It is evident from fig. 13.4 that the resultant e.m.f. acting from B via C towards A is the phasor difference of E_{CA} and E_{CB} and is represented by phasor

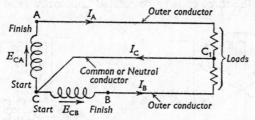

Fig. 13.4 Two-phase, three-wire system.

E_{BCA}. Hence the voltage between the outer conductors in a three-wire system is $1.414 \times$ the phase voltage.

Two-phase systems are seldom employed in practice—the four-wire system is cumbersome, and with the three-wire system there is difficulty in maintaining the voltages balanced (i.e. equal and in

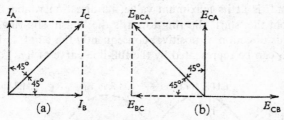

Fig. 13.5 Phasor diagrams for fig. 13.4.

quadrature) at the far end of a transmission line. On the other hand, the production of a rotating magnetic field by two-phase currents (section 16.1) has an important application in the starting of single-phase induction motors.

13.3 Generation of three-phase e.m.f.s

In fig. 13.6, RR_1, YY_1 and BB_1* represent three similar loops fixed to one another at angles of $120°$, each loop terminating in a pair of

* The letters R, Y and B are abbreviations of 'red', 'yellow' and 'blue', namely the colours used to identify the three phases. Also, 'red—yellow—blue' is the sequence that is universally adopted to denote that the e.m.f. in the yellow phase lags that in the red phase by a third of a cycle, and the e.m.f. in the blue phase lags that in the yellow phase by another third of a cycle.

slip-rings carried on the shaft as indicated in fig. 13.7. We shall refer to the slip-rings connected to sides R, Y and B as the 'finishes' of the respective phases and those connected to R_1, Y_1 and B_1 as the 'starts'.

Suppose the three coils to be rotated anticlockwise at a uniform

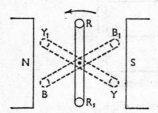

Fig. 13.6 Generation of three-phase e.m.f.s.

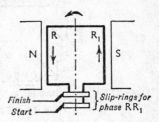

Fig. 13.7 Loop RR_1 at instant of maximum e.m.f.

speed in the magnetic field due to poles NS. The e.m.f. generated in loop RR_1 is zero for the position shown in fig. 13.6. When the loop has moved through 90° to the position shown in fig. 13.7, the generated e.m.f. is at its maximum value, its direction round the loop being from the 'start' slip-ring towards the 'finish' slip-ring. Let us regard this direction as positive; consequently the e.m.f. induced in loop RR_1 can be represented by the full-line curve of fig. 13.8.

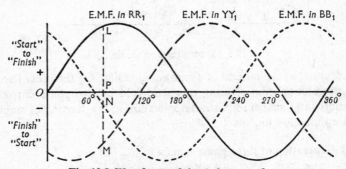

Fig. 13.8 Waveforms of three-phase e.m.f.s.

Since the loops are being rotated anticlockwise, it is evident from fig. 13.6 that the e.m.f. generated in side Y of loop YY_1 has exactly the same amplitude as that generated in side R, but lags by 120° (or 1/3 cycle). Similarly, the e.m.f. generated in side B of loop BB_1 is equal to but lags that in side Y by 120°. Hence the e.m.f.s generated in loops RR$_1$, YY$_1$ and BB$_1$ are represented by the three equally spaced curves of fig. 13.8, the e.m.f.s being assumed positive when their directions round the loops are from 'start' to 'finish' of their respective loops.

If the instantaneous value of the e.m.f. generated in phase RR$_1$ is represented by $e_R = E_m \sin \theta$,

then instantaneous e.m.f. in YY$_1 = e_Y = E_m \sin (\theta - 120°)$

and instantaneous e.m.f. in BB$_1 = e_B = E_m \sin (\theta - 240°)$.

13.4 Mesh or delta connection of three-phase windings

The three phases of fig. 13.6 can, for convenience, be represented as in fig. 13.9, where the phases are shown isolated from one another.

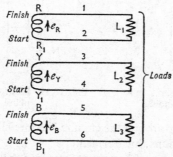

Fig. 13.9 Three-phase windings with six line conductors.

L$_1$, L$_2$ and L$_3$ represent loads connected across the respective phases. Since we have assumed the e.m.f.s to be positive when acting from 'start' to 'finish', they can be represented by the arrows e_R, e_Y and e_B in fig. 13.9. This arrangement necessitates six line conductors and is therefore cumbersome and expensive, so let us consider how it may be simplified. For instance, let us join R$_1$ and Y together as in fig. 13.10, thereby enabling conductors 2 and 3 of fig. 13.9 to be replaced by a single conductor. Similarly, let us join Y$_1$ and B together so that conductors 4 and 5 may be replaced by another single conductor. Before we can proceed to join 'start' B$_1$ to 'finish' R, we have to prove that the resultant e.m.f. between these two points is zero at every instant, so that no circulating current is set up when they are connected together.

Instantaneous value of total e.m.f. acting from B_1 to R

$$= e_R + e_Y + e_B = E_m\{\sin \theta + \sin (\theta - 120°) + \sin (\theta - 240°)\}$$
$$= E_m(\sin \theta + \sin \theta . \cos 120° - \cos \theta . \sin 120° +$$
$$\sin \theta . \cos 240° - \cos \theta . \sin 240°)$$
$$= E_m(\sin \theta - 0.5 \sin \theta - 0.866 \cos \theta - 0.5 \sin \theta + 0.866 \cos \theta)$$
$$= 0.$$

Since this condition holds for every instant, it follows that R and B_1 can be joined together, as in fig. 13.11, without any circulating

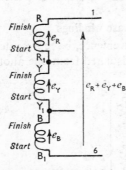

Fig. 13.10 Resultant e.m.f. in a mesh-connected winding.

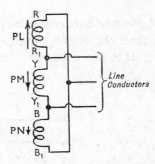

Fig. 13.11 Mesh connection of three-phase winding.

current being set up around the circuit. The three line conductors are joined to the junctions thus formed.

It may be helpful at this stage to consider the actual values and directions of the e.m.f.s at a particular instant. For instance, at

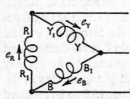

Fig. 13.12 Conventional representation of a mesh-connected winding.

instant P in fig. 13.8, the e.m.f. generated in phase R is positive and is represented by PL acting from R_1 to R in fig. 13.11. The e.m.f. in phase Y is negative and is represented by PM acting from Y to Y_1, and that in phase B is also negative and is represented by PN acting from B to B_1. But the sum of PM and PN is exactly equal numerically to PL; consequently the algebraic sum of the e.m.f.s round the closed circuit formed by the three windings is zero.

It should be noted that the directions of the arrows in fig. 13.11 represent the directions of the e.m.f. at a *particular instant*, whereas arrows placed alongside symbol *e*, as in fig. 13.10, represent the *positive* directions of the e.m.f.s.

The circuit derived in fig. 13.11 is usually drawn as in fig. 13.12, and the arrangement is referred to as *mesh* (i.e. a closed circuit) or *delta* (from the Greek capital letter Δ) connection.

It will be noticed that in fig. 13.12, R is connected to Y_1 instead of B_1 as in fig. 13.11. Actually, it is immaterial which method is used. What is of importance is that the 'start' of one phase should be connected to the 'finish' of another phase, so that the arrows representing the positive directions of the e.m.f.s point in the same direction round the mesh formed by the three windings.

13.5 Star connection of three-phase windings

Let us go back to fig. 13.9 and join together the three 'starts', R_1, Y_1 and B_1 at N, as in fig. 13.13, so that the three conductors 2, 4 and 6 of fig. 13.9 can be replaced by the single conductor NM of fig. 13.13.

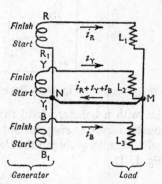

Since the generated e.m.f. has been assumed positive when acting from 'start' to 'finish', the current in each phase must also be regarded as positive when flowing in that direction, as represented by the arrows in fig. 13.13. If i_R, i_Y and i_B be the instantaneous values of the currents in the three phases, the instantaneous value of the current in the common wire MN is ($i_R + i_Y + i_B$), having its positive direction from M to N.

Fig. 13.13 Star connection of three-phase winding.

This arrangement is referred to as a *four-wire star-connected* system and is more conveniently represented as in fig. 13.14; and junction N is referred to as the *star* or *neutral point*. Three-phase motors are connected to the line conductors R, Y and B, whereas lamps, heaters, etc., are usually connected between the line and neutral conductors, as indicated by L_1, L_2 and L_3, the total load being distributed as equally as possible between the three lines. If these three loads are exactly alike, the phase currents have the same peak value, I_m, and differ in phase by 120°. Hence if the instantaneous value of the current in load L_1 be represented by:

$$i_1 = I_m \sin \theta$$

instantaneous current in $L_2 = i_2 = I_m \sin (\theta - 120°)$

and   instantaneous current in $L_3 = i_3 = I_m \sin (\theta - 240°)$.

Hence instantaneous value of the resultant current in neutral conductor MN (fig. 13.13)

$$= i_1 + i_2 + i_3$$
$$= I_m\{\sin \theta + \sin (\theta - 120°) + \sin (\theta - 240°)\}$$
$$= I_m \times 0 = 0,$$

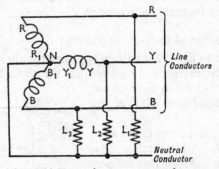

Fig. 13.14 Four-wire star-connected system.

i.e. with a balanced load the resultant current in the neutral conductor is zero at *every* instant; hence this conductor can be dispensed with, thereby giving us the *three-wire star-connected* system shown in fig. 13.15.

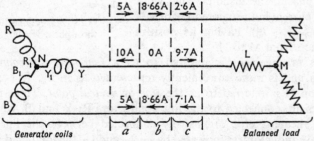

Fig. 13.15 Three-wire star-connected system with balanced load.

When we are considering the distribution of current in a three-wire three-phase system it is helpful to bear in mind: (*a*) that arrows such as those of fig. 13.13, placed alongside *symbols*, indicate the direction of the current when it is assumed to be *positive* and not the direction at a particular instant; and (*b*) that the current flowing outwards in one or two conductors is equal to that flowing back in the remaining conductor or conductors.

Let us consider the second statement in greater detail. Suppose the curves in fig. 13.16 to represent the three currents differing in phase by 120° and having a peak value of 10 A. At instant *a*, the currents in phases R and B are each 5 A, whereas the current in phase Y is −10 A. These values are indicated above *a* in fig. 13.15, i.e. 5 A are

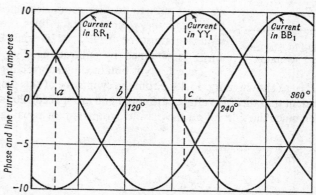

Fig. 13.16 Waveforms of current in a balanced three-phase system.

flowing outwards in phases R and B and 10 A are returning in phase Y.

At instant *b*, the current in Y is zero, that in R is 8·66 A and that in B is −8·66 A, i.e. 8·66 A are flowing outwards in phase R and returning in phase B. At instant *c*, the currents in R, Y and B are −2·6, 9·7 and −7·1 A respectively; i.e. 9·7 A flow outwards in phase Y and return via phases R (2·6 A) and B (7·1 A).

It will be seen that the distribution of currents between the three lines is continually changing; but at every instant the algebraic sum of the currents is zero.

13.6 Relationships between line and phase voltages and currents in a star-connected system

Let us again assume the e.m.f. in each phase to be positive when acting from the neutral point outwards, so that the r.m.s. values of the e.m.f.s generated in the three phases can be represented by E_{NR}, E_{NY} and E_{NB} in figs. 13.17 and 13.18.* The value of the e.m.f. acting from Y via N to R is the phasor *difference* of E_{NR} and E_{NY}. Hence E_{YN} is drawn equal and opposite to E_{NY} and added to E_{NR}, giving

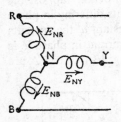

E_{YNR} as the e.m.f. acting from Y to R via N. Note that the *three* subscript letters YNR are necessary to indicate unambiguously the *positive* direction of this e.m.f.

Having decided on YNR as the positive direction of the line e.m.f. between Y and R, we must adhere to the same sequence for the e.m.f.s between the other lines, i.e. the sequence must be YNR, RNB and BNY. E_{RNB} is obtained by subtracting E_{NR} from

Fig. 13.17 Star-connected generator.

E_{NB} and E_{BNY} is obtained by subtracting E_{NB} from E_{NY} as shown in fig. 13.18. From the symmetry of this diagram it is evident that the line voltages are equal and are spaced 120° apart. Further, since the sides of all the parallelograms are of equal length,

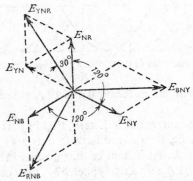

Fig. 13.18 Phasor diagram for fig. 13.17.

the diagonals bisect one another at right-angles. Also, they bisect the angles of their respective parallelograms; and since the angle between E_{NR} and E_{YN} is 60°,

* When the relationships between line and phase quantities are being derived for either the star- or the mesh-connected system, it is essential to relate the phasor diagram to a circuit diagram and to indicate on each phase the direction in which the voltage or current is assumed to be positive. A phasor diagram by itself is meaningless.

$\therefore \qquad\qquad E_{\text{YNR}} = 2E_{\text{NR}} \cos 30° = \sqrt{3}E_{\text{NR}}$

i.e. $\qquad$ line voltage $= 1.73 \times$ star (or phase) voltage.

From fig. 13.17 it is obvious that in a star-connected system the current in a line conductor is the same as that in the phase to which that line conductor is connected.

Hence, in general:

if $\quad V_{\text{L}} = $ p.d.* between any two line conductors
$\qquad\quad = $ line voltage

and $\quad V_{\text{P}} = $ p.d. between a line conductor and the neutral point
$\qquad\quad = $ star voltage (or voltage to neutral)

and if I_{L} and $I_{\text{P}} = $ line and phase currents respectively, then for a star-connected system,

$$V_{\text{L}} = 1.73\, V_{\text{P}} \qquad\qquad (13.1)$$
and $$I_{\text{L}} = I_{\text{P}} \qquad\qquad (13.2)$$

13.7 Relationships between line and phase voltages and currents in a mesh-connected system with a balanced load

Let I_1, I_2 and I_3 be the r.m.s. values of the phase currents having their positive directions as indicated by the arrows in fig. 13.19. Since the

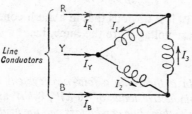

Fig. 13.19 Mesh-connected system with balanced load.

load is assumed to be balanced, these currents are equal in magnitude and differ in phase by 120°, as shown in fig. 13.20.

From fig. 13.19 it will be seen that I_1, when positive, flows away from line conductor R, whereas I_3, when positive, flows towards it.

* In practice, it is the voltage between two line conductors or between a line conductor and the neutral point that is measured. Owing to the impedance drop in the windings, this p.d. is different from the corresponding e.m.f. generated in the winding, except when the alternator is on open circuit; hence, in general, it is preferable to work with the potential difference, V, rather than with the e.m.f., E.

Consequently, I_R is obtained by subtracting I_3 from I_1, as in fig. 13.20. Similarly, I_Y is the phasor difference of I_2 and I_1, and I_B is the phasor difference of I_3 and I_2. From fig. 13.20 it is evident that the line currents are equal in magnitude and differ in phase by 120°. Also,

$$I_R = 2I_1 \cos 30° = \sqrt{3}I_1.$$

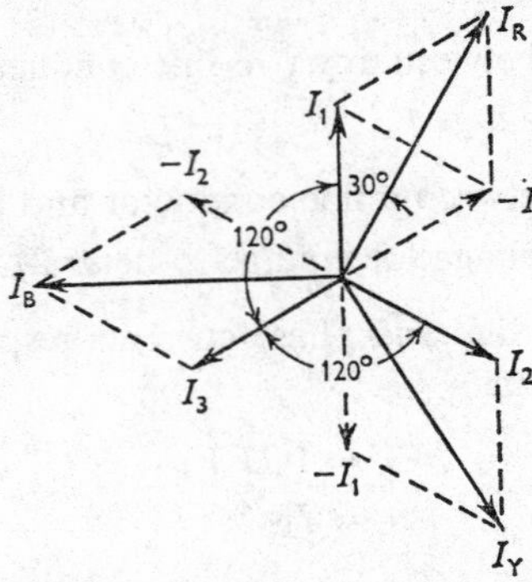

Fig. 13.20 Phasor diagram for fig. 13.19.

Hence for a mesh-connected system with a balanced load,

line current $= 1·73 \times$ phase current

i.e. $I_L = 1·73 I_P$ (13.3)

From fig. 13.19 it is obvious that in a mesh-connected system, the line and the phase voltages are the same, i.e.

$$V_L = V_P \qquad\qquad (13.4)$$

Example 13.1 *In a three-phase four-wire system the line voltage is 400 V and non-inductive loads of 10 kW, 8 kW and 5 kW are connected between the three line conductors and the neutral as in fig. 13.21. Calculate:* (a) *the current in each line, and* (b) *the current in the neutral conductor.*

(a) Voltage to neutral $= \dfrac{\text{line voltage}}{1·73} = \dfrac{400}{1·73} = 231$ V.

If I_R, I_Y and I_B be the currents taken by the 10-kW, 8-kW and 5-kW loads respectively,

$$I_R = 10 \times 1000/231 = 43·3 \text{ A},$$
$$I_Y = 8 \times 1000/231 = 34·6 \text{ A}$$

and $I_B = 5 \times 1000/231 = 21·65 \text{ A}.$

These currents are represented by the respective phasors in fig. 13.22.

(*b*) The current in the neutral is the phasor sum of the three line

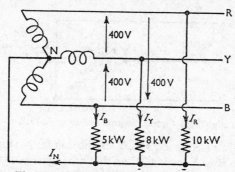

Fig. 13.21 Circuit diagram for example 13.1.

currents. In general, the most convenient method of adding such quantities is to calculate the resultant horizontal and vertical components thus:

$$\text{horizontal component} = I_H = I_Y \cos 30° - I_B \cos 30°$$
$$= 0 \cdot 866(34 \cdot 6 - 21 \cdot 65) = 11 \cdot 22 \text{ A.}$$

and $$\text{vertical component} = I_V = I_R - I_Y \cos 60° - I_B \cos 60°$$
$$= 43 \cdot 3 - 0 \cdot 5(34 \cdot 6 + 21 \cdot 65) = 15 \cdot 2 \text{ A.}$$

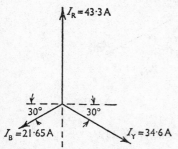

Fig. 13.22 Phasor diagram for fig. 13.21.

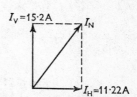

Fig. 13.23 Vertical and horizontal components of I_N.

These components are represented in fig. 13.23.

$\therefore$ current in neutral $= I_N = \sqrt{\{(11 \cdot 22)^2 + (15 \cdot 2)^2\}}$
$$= 18 \cdot 9 \text{ A.}$$

Example 13.2 *A delta-connected load is arranged as in fig. 13.24. Calculate:* (a) *the phase currents, and* (b) *the line currents. The supply voltage* * *is 400 V at 50 Hz.*

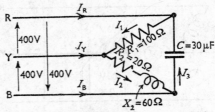

Fig. 13.24 Circuit diagram for example 13.2.

(*a*) Since the phase sequence is R, Y, B, the voltage having its positive direction from R to Y leads 120° on that having its positive direction from Y to B, i.e. V_{RY} is 120° in front of V_{YB} (see section 1.19).

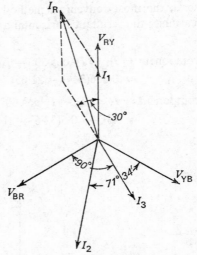

Fig. 13.25 Phasor diagram for fig. 13.24.

Similarly, V_{YB} is 120° in front of V_{BR}. Hence the phasors representing the line (and phase) voltages are as shown in fig. 13.25.

If I_1, I_2, I_3 be the phase currents in loads RY, YB and BR respectively.

* The voltage given for a three-phase system is always the line voltage unless it is stated otherwise.

$I_1 = 400/100 = 4$ A, in phase with V_{RY}.

$$I_2 = \frac{400}{\sqrt{(20^2 + 60^2)}} = \frac{400}{63 \cdot 25} = 6 \cdot 325 \text{ A}.$$

I_2 lags V_{YB} by an angle ϕ_2 such that

$$\phi_2 = \tan^{-1} 60/20 = 71° \, 34'.$$

Also $\qquad I_3 = 2 \times 3 \cdot 14 \times 50 \times 30 \times 10^{-6} \times 400$
$$= 3 \cdot 77 \text{ A, leading } V_{BR} \text{ by } 90°.$$

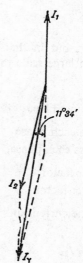

Fig. 13.26 Phasor diagram
for deriving I_Y.

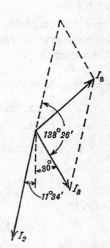

Fig. 13.27 Phasor diagram for
deriving I_B.

(*b*) If the current I_R in line conductor R be assumed to be positive when flowing towards the load, the phasor representing this current is obtained by subtracting I_3 from I_1, as in fig. 13.25.

∴ $\qquad I_R^2 = 4^2 + (3 \cdot 77)^2 + 2 \times 4 \times 3 \cdot 77 \cos 30° = 56 \cdot 3$
∴ $\qquad I_R = 7 \cdot 5$ A.

Current in line conductor Y is obtained by subtracting I_1 from I_2, as shown separately in fig. 13.26.

But angle between I_2 and I_1 reversed $= \phi_2 - 60° = 71° \, 34' - 60°$
$$= 11° \, 34'.$$

$$\therefore \quad I_Y^2 = 4^2 + (6{\cdot}325)^2 + 2 \times 4 \times 6{\cdot}325 \times \cos 11° 34' = 105{\cdot}5$$
$$\therefore \quad I_Y = 10{\cdot}27 \text{ A.}$$

Similarly, the current in line conductor B is obtained by subtracting I_2 from I_3, as shown in fig. 13.27.

Angle between I_3 and I_2 reversed $= 180° - 30° - 11° 34'$
$$= 138° 26'.$$

$$\therefore \quad I_B^2 = (3{\cdot}77)^2 + (6{\cdot}325)^2 + 2 \times 3{\cdot}77 \times 6{\cdot}325 \times \cos 138° 26'$$
$$= 18{\cdot}52.$$
$$\therefore \quad I_B = 4{\cdot}3 \text{ A.}$$

This problem could be solved graphically, but in that case it would be necessary to draw the phasors to a large scale to ensure reasonable accuracy.

13.8 Power in a three-phase system with a balanced load

If $I_P =$ r.m.s. value of the current in each phase
and $V_P =$ r.m.s. value of the p.d. across each phase,

power per phase $= I_P V_P \times$ power factor
and total power $= 3 I_P V_P \times$ power factor (13.5)

If I_L and V_L be the r.m.s. values of the line current and voltage respectively, then for a *star-connected system*,

$$V_P = V_L/1{\cdot}73 \quad \text{and} \quad I_P = I_L.$$

Substituting for I_P and V_P in (13.5), we have:

total power in watts $= 1{\cdot}73 I_L V_L \times$ power factor.

For a *mesh-connected system*,

$$V_P = V_L \quad \text{and} \quad I_P = I_L/1{\cdot}73.$$

Again, substituting for I_P and V_P in (13.5), we have:

total power in watts $= 1{\cdot}73 I_L V_L \times$ power factor.

Hence it follows that for any balanced load,

total power in watts $= 1{\cdot}73 \times$ line current $\times$ line voltage $\times$ power factor

$$= 1{\cdot}73 I_L V_L \times \text{power factor} \qquad (13.6)$$

Example 13.3 *A three-phase motor operating off a 400-V system is developing 20 kW at an efficiency of 0·87 p.u. and a power factor of 0·82. Calculate: (a) the line current and (b) the phase current if the windings are delta-connected.*

(a)   Since efficiency $= \dfrac{\text{output power in watts}}{\text{input power in watts}}$

$\qquad\qquad\qquad\quad = \dfrac{\text{output power in watts}}{1{\cdot}73 I_L V_L \times \text{p.f.}}$

∴ $\qquad\qquad 0{\cdot}87 = \dfrac{20 \times 1000}{1{\cdot}73 \times I_L \times 400 \times 0{\cdot}82}$

and $\qquad$ line current $= I_L = 40{\cdot}6$ A.

(b) For a delta-connected winding,

$\qquad$ phase current $=$ line current$/1{\cdot}73 = 40{\cdot}6/1{\cdot}73 = 23{\cdot}4$ A.

13.9 Measurement of power in a three-phase three-wire system

Case (a). *Star-connected balanced load, with neutral point accessible*
If a wattmeter W be connected with its current coil in one line and the voltage circuit between that line and the neutral point, as shown in

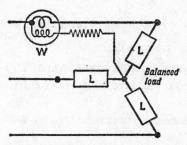

Fig. 13.28 Measurement of power in a star-connected balanced load.

fig. 13.28, the reading on the wattmeter gives the power per phase:

∴ $\qquad\qquad$ total power $= 3 \times$ wattmeter reading.

Case (b). *Balanced or unbalanced load, star- or delta-connected. The two-wattmeter method*
Suppose the three loads L_1, L_2 and L_3 to be connected in star, as in fig. 13.29. The current coils of the two wattmeters are connected in

any two lines, say the 'red' and 'blue' lines, and the voltage circuits are connected between these lines and the third line.

Suppose v_{RN}, v_{YN} and v_{BN} to be the instantaneous values of the p.d.s across the loads, these p.d.s being assumed positive when the respective line conductors are positive in relation to the neutral point. Also, suppose i_R, i_Y and i_B to be the corresponding instantaneous values of the line (and phase) currents.

∴ instantaneous power in load $L_1 = i_R v_{RN}$

instantaneous power in load $L_2 = i_Y v_{YN}$

and instantaneous power in load $L_3 = i_B v_{BN}$

∴ total instantaneous power $= i_R v_{RN} + i_Y v_{YN} + i_B v_{BN}$.

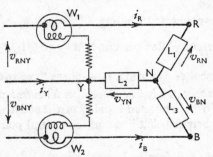

Fig. 13.29 Measurement of power by two wattmeters.

From fig. 13.29 it is seen that:

instantaneous current through current coil of $W_1 = i_R$

and instantaneous p.d. across voltage circuit of W_1 $\Big\} = v_{RN} - v_{YN}$

∴ instantaneous power measured by $W_1 = i_R(v_{RN} - v_{YN})$

Similarly,

instantaneous current through current coil of $W_2 = i_B$

and instantaneous p.d.* across voltage circuit of W_2 $\Big\} = v_{BN} - v_{YN}$

---

* It is important to note that this p.d. is not $v_{YN} - v_{BN}$. This is due to the fact that a wattmeter reads positively when the currents in the current and voltage coils are *both* flowing from the junction of these coils or *both* towards that junction; and since the positive direction of the current in the current coil of W_2 has already been taken as that of the arrowhead alongside i_B in fig. 13.29, it follows that the current in the voltage circuit of W_2 is positive when flowing from the 'blue' to the 'yellow' line.

∴ instantaneous power measured by $W_2 = i_B(v_{BN} - v_{YN})$.

Hence the sum of the instanta-⎱
neous powers of W_1 and W_2⎰ $= i_R(v_{RN} - v_{YN}) + i_B(v_{BN} - v_{YN})$

$$= i_R v_{RN} + i_B v_{BN} - (i_R + i_B)v_{YN}.$$

From Kirchhoff's First Law (section 1.14), the algebraic sum of the instantaneous currents at N is zero,

i.e. $\qquad\qquad\qquad i_R + i_Y + i_B = 0$

∴ $\qquad\qquad\qquad\qquad i_R + i_B = -i_Y$

so that sum of instantaneous powers⎱
measured by W_1 and W_2⎰ $= i_R v_{RN} + i_B v_{BN} + i_Y v_{YN}$

$$= \text{total instantaneous power.}$$

Actually, the power measured by each wattmeter varies from instant to instant, but the inertia of the moving system causes the pointer to read the average value of the power. Hence the sum of the wattmeter readings gives the average value of the total power absorbed by the three phases.

Since the above proof does not assume a balanced load or sinusoidal waveforms, it follows that the sum of the two wattmeter readings gives the total power under all conditions. The above proof was derived for a star-connected load, and it is a useful exercise to prove that the same conclusion holds for a delta-connected load.

13.10 Measurement of the power factor of a three-phase system by means of two wattmeters, assuming balanced load and sinusoidal voltages and currents

Suppose L in fig. 13.30 to represent three similar loads connected in star, and suppose V_{RN}, V_{YN} and V_{BN} to be the r.m.s. values of the phase voltages and I_R, I_Y and I_B to be the r.m.s. values of the currents. Since these voltages and currents are assumed sinusoidal, they can be represented by phasors, as in fig. 13.31, the currents being assumed to lag the corresponding phase voltages by an angle ϕ.

Current through current coil of $W_1 = I_R$.

P.d. across voltage circuit of W_1 = phasor difference of V_{RN}
and V_{YN}
$= V_{RNY}$ (see section 1.19).

Phase difference between I_R and $V_{RNY} = 30° + \phi$

∴   reading on $W_1 = P_1 = I_R V_{RNY} \cos(30° + \phi)$.

Current through current coil of $W_2 = I_B$.

P.d. across voltage circuit of W_2 = phasor difference of V_{BN}

                                     and V_{YN}

                 $= V_{BNY}$.

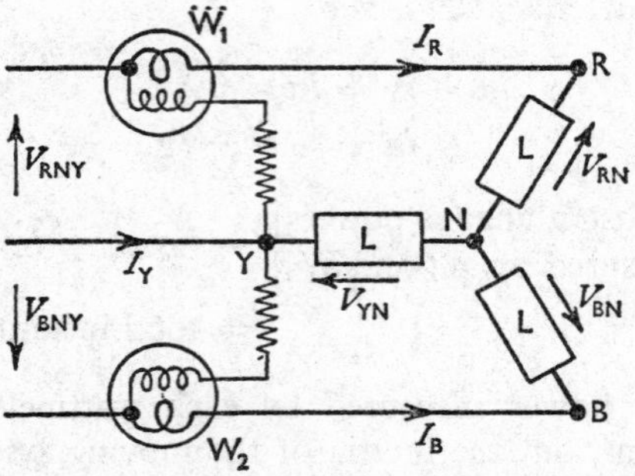

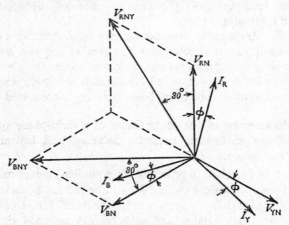

Fig. 13.30 Measurement of power and power factor by two wattmeters.

Fig. 13.31 Phasor diagram for fig. 13.30.

Phase difference between I_B and $V_{BNY} = 30° - \phi$

∴   reading on $W_2 = P_2 = I_B V_{BNY} \cos(30° - \phi)$.

Since the load is balanced,

     $I_R = I_Y = I_B = $ (say) I_L, numerically

and $V_{RNY} = V_{BNY} = $ (say) V_L, numerically.

Hence $\quad P_1 = I_L V_L \cos(30° + \phi)$ $\qquad$ (13.7)

and $\quad\;\; P_2 = I_L V_L \cos(30° - \phi)$ $\qquad$ (13.8)

$$\therefore\; P_1 + P_2 = I_L V_L\{\cos(30° + \phi) + \cos(30° - \phi)\}$$
$$= I_L V_L(\cos 30° . \cos\phi - \sin 30° . \sin\phi$$
$$+ \cos 30° . \cos\phi + \sin 30° . \sin\phi)$$
$$= 1·73 I_L V_L \cos\phi \qquad (13.9)$$

namely the expression deduced in section 13.8 for the total power in a balanced three-phase system. This is an alternative method of proving that the sum of the two wattmeter readings gives the total power, but it should be noted that this proof assumes a balanced load and sinusoidal voltages and currents.

Dividing (13.7) by (13.8), we have:

$$\frac{P_1}{P_2} = \frac{\cos(30° + \phi)}{\cos(30° - \phi)} = \text{(say) } y$$

$$\therefore \qquad y = \frac{(\sqrt{3}/2)\cos\phi - (1/2)\sin\phi}{(\sqrt{3}/2)\cos\phi + (1/2)\sin\phi}$$

so that $\quad \sqrt{3}y\cos\phi + y\sin\phi = \sqrt{3}\cos\phi - \sin\phi$

from which $\quad \sqrt{3}(1 - y)\cos\phi = (1 + y)\sin\phi$

$$\therefore \qquad 3\left(\frac{1-y}{1+y}\right)^2 \cos^2\phi = \sin^2\phi = 1 - \cos^2\phi$$

$$\left\{1 + 3\left(\frac{1-y}{1+y}\right)^2\right\}\cos^2\phi = 1$$

$$\therefore \qquad \text{power factor} = \cos\phi = \frac{1}{\sqrt{\left\{1 + 3\left(\dfrac{1-y}{1+y}\right)^2\right\}}} \qquad (13.10)$$

Since y is the ratio of the wattmeter readings, the corresponding power factor can be calculated from expression (13.10), but this procedure is very laborious. A more convenient method is to draw a graph of the power factor for various ratios of P_1/P_2; and in order that these ratios may lie between $+1$ and -1, as in fig. 13.32, it is always the practice to take P_1 as the smaller of the two readings. By adopting this practice, it is possible to derive reasonably accurate values of power factor from the graph.

When the power factor of the load is 0·5 lagging, ϕ is 60°; and from (13.7), the reading on $W_1 = I_L V_L \cos 90° = 0$. When the power factor is less than 0·5 lagging, ϕ is greater than 60° and $(30° + \phi)$ is therefore greater than 90°. Hence the reading on W_1 is negative. To measure this power it is necessary to reverse the connections to either the current or the voltage coil, but the reading thus obtained must be taken as negative when the total power and the ratio of the wattmeter readings are being calculated.

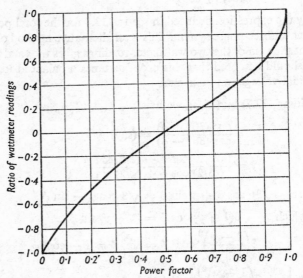

Fig. 13.32 Relationship between power factor and ratio of wattmeter readings.

An alternative method of deriving the power factor is as follows:

From (13.7), (13.8) and (13.9), $P_2 - P_1 = I_L V_L \sin \phi$

and

$$\tan \phi = \frac{\sin \phi}{\cos \phi} = 1 \cdot 73 \left(\frac{P_2 - P_1}{P_2 + P_1} \right) \qquad (13.11)$$

Hence, ϕ and $\cos \phi$ can be determined with the aid of trigonometrical tables.

Example 13.4 *The input power to a three-phase motor was measured by the two-wattmeter method. The readings were 5·2 kW and −1·7 kW, and the line voltage was 400 V. Calculate:* (a) *the total power;* (b) *the power factor; and* (c) *the line current.*

(a) Total power = 5·2 − 1·7 = 3·5 kW.

(*b*) Ratio of wattmeter readings $= -1\cdot7/5\cdot2 = -0\cdot327$.

From fig. 13.32, power factor $= 0\cdot28$.
Or alternatively, from (13.11),

$$\tan \phi = 1\cdot73\left\{\frac{5\cdot2 - (-1\cdot7)}{5\cdot2 + (-1\cdot7)}\right\} = 3\cdot41$$

$$\therefore \qquad \phi = 73°\ 39'$$

and power factor $= \cos \phi = 0\cdot281$.

From the data it is impossible to state whether the power factor is lagging or leading.

(*c*) From (13.6), $\quad 3500 = 1\cdot73 \times I_L \times 400 \times 0\cdot281$

$$\therefore \qquad\qquad I_L = 18 \text{ A.}$$

Summary of important formulae

For a star-connected system,

$$V_L = 1\cdot73\ V_P \qquad\qquad\qquad (13.1)$$

and $\qquad\qquad I_L = I_P \qquad\qquad\qquad\qquad (13.2)$

For mesh-connected system with balanced load,

$$V_L = V_P \qquad\qquad\qquad\qquad (13.4)$$

and $\qquad\qquad I_L = 1\cdot73 I_P \qquad\qquad\qquad (13.3)$

For star- or mesh-connected system with balanced load,

total power $= 1\cdot73 \times$ line current $\times$ line voltage $\times$ power factor

$$= 1\cdot73 I_L V_L \times \text{power factor} \qquad (13.6)$$

If P_1 and P_2 be the readings obtained with the two-wattmeter method on a three-wire, three-phase system,

total power $= P_1 + P_2$ under all conditions,

and for a balanced load and sinusoidal waveforms,

$$\text{power factor} = \frac{1}{\sqrt{\left\{1 + 3\left(\dfrac{P_2 - P_1}{P_2 + P_1}\right)^2\right\}}} \qquad (13.10)$$

and $\qquad\qquad \tan \phi = 1\cdot73\left(\dfrac{P_2 - P_1}{P_2 + P_1}\right) \qquad (13.11)$

EXAMPLES 13

1. In a two-phase, three-wire system, the phase voltage is 3000 V. Calculate the voltage between the outers.

   If the current in each phase is 50 A and the two currents are in quadrature, what is the value of the current in the common wire?

2. The currents in the two phases of a two-phase, three-wire system are 40 A and 70 A and are in phase with their respective voltages. Calculate the current in the neutral conductor.

3. The two-phase alternator of fig. 13.4 has a terminal voltage of 230 V per phase and the frequency is 50 Hz. A coil having a resistance of 20 Ω and an inductance of 0·1 H is connected across phase A, and a 70-μF capacitor is connected across phase B. Assuming the voltage of phase B to lag that of phase A by a quarter of a cycle, calculate the current in each outer conductor and that in the neutral conductor. What is the voltage between the outer conductors?

4. Deduce the relationship between the phase and the line voltages of a three-phase star-connected alternator.

   If the phase voltage of a three-phase star-connected alternator be 200 V, what will be the line voltages: (a) when the phases are correctly connected; (b) when the connections to one of the phases are reversed?
   (App. El., L.U.)

5. Show with the aid of a phasor diagram that for both star-connected and delta-connected balanced loads, the total power is given by $\sqrt{3}VI \cos\phi$, where V and I are the line values of voltage and current respectively and ϕ is the angle between phase values of voltage and current.

   A balanced three-phase load consists of three coils, each of resistance 4 Ω and inductance 0·02 H. Determine the total power when the coils are (a) star-connected, (b) delta-connected to a 400-V, three-phase, 50-Hz supply.
   (E.M.E.U., O.2)

6. Derive, for both star- and delta-connected systems, an expression for the total power input for a balanced three-phase load in terms of line voltage, line current and power factor.

   The star-connected secondary of a transformer supplies a delta-connected motor taking a power of 90 kW at a lagging power factor of 0·9. If the voltage between lines is 600 V, calculate the current in the transformer winding and in the motor winding. Draw circuit and phasor diagrams, properly labelled, showing all voltages and currents in the transformer secondary and the motor.
   (U.L.C.I., O.2)

7. A three-phase delta-connected load, each phase of which has an inductive reactance of 40 Ω and a resistance of 25 Ω, is fed from the secondary of a three-phase star-connected transformer, which has a phase voltage of 240 V. Draw the circuit diagram of the system and calculate: (a) the current in each phase of the load; (b) the p.d. across each phase of the load; (c) the current in the transformer secondary windings; (d) the total power taken from the supply and its power factor.
   (N.C.T.E.C., O.2)

8. Three similar coils, connected in star, take a total power of 1·5 kW, at a power factor of 0·2, from a three-phase, 400-V, 50-Hz supply. Calculate: (*a*) the resistance and inductance of each coil, and (*b*) the line currents if one of the coils is short-circuited.

9. (*a*) Three 20-μF capacitors are star-connected across a 420-V, 50-Hz, three-phase, three-wire supply, Calculate the current in each line.

   (*b*) If one of the capacitors is short-circuited, calculate the line currents.

   (*c*) If one of the capacitors is open-circuited, calculate (i) the line currents and (ii) the p.d. across each of the other two capacitors.

10. (*a*) Explain the advantages of three-phase supply for distribution purposes.

   (*b*) Assuming the relationship between the line and phase values of currents and voltages, show that the power input to a three-phase balanced load is $\sqrt{3}VI \cos \phi$, where V and I are line quantities.

   (*c*) Three similar inductors, each of resistance 10 Ω and inductance 0·019 H, are delta-connected to a three-phase, 415-V, 50-Hz sinusoidal supply. Calculate: (i) the value of the line current: (ii) the power factor; (iii) the power input to the circuit. (W.J.E.C., O.2)

11. A three-phase, 415 V, star-connected motor has an output of 50 kW, with an efficiency of 90 per cent and a power factor of 0·85. Calculate the line current. Sketch a phasor diagram showing the voltages and currents.

   If the motor windings were connected in mesh, what would be the correct voltage of a three-phase supply suitable for the motor?

12. Discuss the importance of power-factor correction in a.c. systems.

   A 415-V, 50-Hz, three-phase distribution system supplies a 20-kVA, three-phase induction motor load at a power factor of 0·8 lagging, and a star-connected set of impedances, each having a resistance of 10 Ω and an inductive reactance of 8 Ω. Calculate the capacitance of delta-connected capacitors required to improve the overall power factor to 0·95 lagging. (E.M.E.U., O.2)

13. State the advantages to be gained by raising the power factor of industrial loads.

   A 415-V, 50-Hz, three-phase motor takes a line current of 15·0 A when operating at a lagging power factor of 0·65. When a capacitor bank is connected across the motor terminals, the line current is reduced to 11·5 A. Calculate the kVAr rating and the capacitance per phase of the capacitor bank for: (*a*) star connection; (*b*) delta connection.

   Find also the new overall power factor. (S.A.N.C., O.2)

14. Three coils, each having a resistance of 20 Ω and a reactance of 15 Ω, are connected in star to a 400-V, three-phase, 50-Hz supply. Calculate: (i) the line current; (ii) the power supplied; (iii) the power factor.

   If three capacitors, each of the same capacitance, are connected in delta to the same supply so as to form a parallel circuit with the above coils, calculate: (i) the capacitance of each capacitor to obtain a resultant power factor of 0·95 lagging; (ii) the line current taken by the combined circuits. Draw to scale a phasor diagram showing: (a) the line voltage; (b) the voltage across each of the coils; (c) the line current taken by the combined circuits; (d) the current in each capacitor; (e) the current in each coil. (App. El., L.U.)

15. Derive the numerical relationship between the line and phase currents for a balanced three-phase delta-connected load.

Three coils are connected in delta to a three-phase, three-wire, 415-V, 50-Hz supply and take a line current of 5 A at 0·8 power factor lagging. Calculate the resistance and inductance of the coils.

If the coils are star-connected to the same supply, calculate the line current and the total power.

Calculate the line currents if one coil becomes open-circuited when the coils are connected in star. (E.M.E.U.)

16. The load connected to a three-phase supply comprises three similar coils connected in star. The line currents are 25 A and the kVA and kW inputs are 20 and 11 respectively. Find the line and phase voltages, the kVAr input and the resistance and reactance of each coil.

If the coils are now connected in delta to the same three-phase supply, calculate the line currents and the power taken. (U.L.C.I.)

17. Non-reactive loads of 10, 6 and 4 kW are connected between the neutral and the red, yellow and blue phases respectively of a three-phase, four-wire system. The line voltage is 400 V. Find the current in each line conductor and in the neutral. (App. El., L.U.)

18. Explain the advantage of connecting the low-voltage winding of distribution transformers in star.

A factory has the following load with power factor of 0·9 lagging in each phase. Red phase 40 A, yellow phase 50 A and blue phase 60 A. If the supply is 400-V, three-phase, four-wire, calculate the current in the neutral and the total power. Draw a phasor diagram for phase and line quantities. Assume that, relative to the current in the red phase, the current in the yellow phase lags by 120° and that in the blue phase leads by 120°.

19. A three-phase, 400-V system has the following load connected in mesh: between the red and yellow lines, a non-reactive resistor of 100 Ω; between the yellow and blue lines, a coil having a reactance of 60 Ω and negligible resistance; between the blue and red lines, a loss-free capacitor having a reactance of 130 Ω. Calculate: (a) the phase currents; (b) the line currents. Assume the phase sequence to be R–Y, Y–B and B–R. Also, draw the complete phasor diagram.

20. The phase currents in a mesh-connected three-phase load are as follows: between the red and yellow lines, 30 A at p.f. 0·707 leading; between the yellow and blue lines, 20 A at unity p.f.; between the blue and red lines, 25 A at p.f. 0·866 lagging. Calculate the line currents and draw the complete phasor diagram.

21. (a) If, in a laboratory test, you were required to measure the total power taken by a three-phase balanced load, show how to do this, using two wattmeters. Explain the principle of the method.

Draw the phasor diagram for the balanced-load case with a lagging power factor and use this to explain why the two wattmeter readings differ.

(b) The load taken by a three-phase induction motor was measured by the two-wattmeter method, and the readings were 860 W and 240 W. What is the power taken by the motor and at what power factor is it working? (W.J.E.C., O.2)

22. With the aid of a circuit diagram, show that two wattmeters can be connected to read the total power in a three-phase, three-wire system.

Two wattmeters connected to read the total power in a three-phase system supplying a balanced load read 10·5 kW and −2·5 kW respectively. Calculate the total power.

Drawing suitable phasor diagrams, explain the significance of: (a) equal wattmeter readings, and (b) a zero reading on one wattmeter. (U.L.C.I., O.2)

23. Two wattmeters are used to measure power in a three-phase, three-wire network. Show by means of connection and complexor (phasor) diagrams that the sum of the wattmeter readings will measure the total power.

Two such wattmeters read 120 W and 50 W when connected to measure the power taken by a balanced three-phase load. Find the power factor of the load.

If one wattmeter tends to read in the reverse direction, explain what changes may have occurred in the circuit. (S.A.N.C., O.2)

24. Each branch of a three-phase star-connected load consists of a coil of resistance 4·2 Ω and reactance 5·6 Ω. The load is supplied at a line voltage of 415 V, 50 Hz. The total power supplied to the load is measured by the two-wattmeter method. Draw a circuit diagram of the wattmeter connections and calculate their separate readings.

Derive any formula used in your calculations. (S.A.N.C., O.2)

25. Three non-reactive loads are connected in delta across a three-phase, three-wire, 400-V supply in the following way: (i) 10 kW across R and Y lines; (ii) 6 kW across Y and B lines; (iii) 4 kW across B and R lines.

Draw a phasor diagram showing the three line voltages and the load currents and determine: (a) the current in the B line and its phase relationship to the line voltage V_{BR}; (b) the reading of a wattmeter whose current coils are connected in the B line and whose voltage circuit is connected across the B and R lines. The phase rotation is RYB.

Where would a second wattmeter be connected for the two-wattmeter method and what would be its reading?

26. Two wattmeters connected to measure the input to a balanced three-phase circuit indicate 2500 W and 500 W respectively. Find the power factor of the circuit: (a) when both readings are positive, and (b) when the latter reading is obtained after reversing the connections to the current-coil of one instrument. Draw the phasor and connection diagrams.

27. With the aid of a phasor diagram show that the power and power factor of a balanced three-phase load can be measured by two wattmeters.

   For a certain load, one wattmeter read 20 kW and the other 5 kW after the voltage circuit of this wattmeter has been reversed. Calculate the power and the power factor of the load. (App. El., L.U.)

28. The current coil of a wattmeter is connected in the red line of a three-phase system. The voltage circuit can be connected between the red line and either the yellow line or the blue line by means of a two-way switch. Assuming the load to be balanced, show with the aid of a phasor diagram that the sum of the wattmeter readings obtained with the voltage circuit connected to the yellow and the blue lines respectively gives the total power.

29. A single wattmeter is used to measure the total power taken by a 415-V, three-phase induction motor. When the output power of the motor is 15 kW, the efficiency is 88 per cent and the power factor is 0·84 lagging. The current coil of the wattmeter is connected in the yellow line. With the aid of a phasor diagram, calculate the wattmeter reading when the voltage circuit is connected between the yellow line and (a) the red line (b) the blue line. Show that the sum of the two wattmeter readings gives the total power taken by the motor. Assume the phase sequence to be R–Y–B.

30. A wattmeter has its current coil connected in the yellow line, and its voltage circuit is connected between the red and blue lines. The line voltage is 400 V and the balanced load takes a line current of 30 A at a power factor of 0·7 lagging. Draw circuit and phasor diagrams and derive an expression for the reading on the wattmeter in terms of the line voltage and current and of the phase difference between the phase voltage and current. Calculate the value of the wattmeter reading.

CHAPTER 14

Transformers

14.1 Introductory

One of the main advantages of a.c. transmission and distribution is the ease with which an alternating voltage can be increased or reduced. For instance, the general practice in this country is to generate at voltages of about 11–22 kV, then step up by means of transformers to higher voltages, for the transmission lines. At suitable points other transformers are installed to step the voltage down to values suitable for motors, lamps, heaters, etc. A medium-size transformer has a full-load efficiency of about 97–98 per cent, so that the loss at each point of transformation is very small. Also, since there are no moving parts, the amount of supervision required is practically negligible.

14.2 Principle of action of a transformer

Fig. 14.1 shows the general arrangement of a transformer. An iron core C consists of laminated sheets, about 0·35 mm thick, insulated

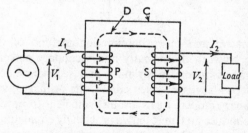

Fig. 14.1 A transformer.

from one another by thin layers of paper or varnish or by spraying the laminations with a mixture of flour, chalk and water which, when dried, adheres to the metal. The purpose of laminating the core is to reduce the loss due to eddy currents induced by the alternating magnetic flux (section 9.1 (*b*)). The vertical portions of the core are referred to as the *limbs* and the top and bottom portions are the *yokes*. Coils P and S are wound on the limbs. Coil P is connected to the supply and is therefore termed the *primary*; coil S is connected to the load and is termed the *secondary*.

An alternating voltage applied to P circulates an alternating current through P and this current produces an alternating flux in the iron core, the mean path of this flux being represented by the dotted line D. If the whole of the flux produced by P passes through S, the e.m.f. induced in each turn is the same for P and S. Hence, if N_1 and N_2 be the number of turns on P and S respectively,

$$\frac{\text{total e.m.f. induced in S}}{\text{total e.m.f. induced in P}} = \frac{N_2 \times \text{e.m.f. per turn}}{N_1 \times \text{e.m.f. per turn}} = \frac{N_2}{N_1}$$

When the secondary is on open circuit, its terminal voltage is the same as the induced e.m.f. The primary current is then very small, so that the applied voltage V_1 is practically equal and opposite to the e.m.f. induced in P. Hence:

$$\frac{V_2}{V_1} \simeq \frac{N_2}{N_1} \tag{14.1}$$

Since the full-load efficiency of a transformer is nearly 100 per cent,

$$I_1 V_1 \times \text{primary power factor} \simeq I_2 V_2 \times \text{secondary power factor}.$$

But the primary and secondary power factors at full load are nearly equal,

$$\therefore \qquad \frac{I_1}{I_2} \simeq \frac{V_2}{V_1} \tag{14.2}$$

An alternative and more illuminating method of deriving the relationship between the primary and secondary currents is based upon a comparison of the primary and secondary ampere-turns. When the secondary is on open circuit, the primary current is such that the primary ampere-turns are just sufficient to produce the flux necessary to induce an e.m.f. that is practically equal and opposite to the applied voltage. This magnetizing current is usually about 3–5 per cent of the full-load primary current.

When a load is connected across the secondary terminals, the secondary current—by Lenz's Law—produces a demagnetizing effect. Consequently the flux and the e.m.f. induced in the primary are reduced slightly. But this small change may increase the difference between the applied voltage and the e.m.f. induced in the primary from, say, 0·05 per cent to, say, 1 per cent, in which case the new primary current would be 20 times the no-load current. The demagnetizing ampere-turns of the secondary are thus *nearly* neutralized by the increase in the primary ampere-turns; and since the primary ampere-turns on no load are very small compared with the full-load ampere-turns,

∴ full-load primary ampere-turns

$\simeq$ full-load secondary ampere-turns,

i.e. $I_1 N_1 \simeq I_2 N_2$

so that
$$\frac{I_1}{I_2} \simeq \frac{N_2}{N_1} \simeq \frac{V_2}{V_1} \tag{14.3}$$

It will be seen that the magnetic flux forms the connecting link between the primary and secondary circuits and that any variation of the secondary current is accompanied by a small variation of the flux and therefore of the e.m.f. induced in the primary, thereby enabling the primary current to vary approximately proportionally to the secondary current.

This balance of primary and secondary ampere-turns is an important relationship wherever transformer action occurs.

14.3 E.m.f. equation of a transformer

Suppose the maximum value of the flux to be Φ_m webers and the frequency to be f hertz (or cycles/second). From fig. 14.2 it is seen that the flux has to change from $+\Phi_m$ to $-\Phi_m$ in half a cycle, namely in $1/2f$ second.

∴ average rate of change of flux $= 2\Phi_m \div \dfrac{1}{2f}$

$= 4f\Phi_m$ webers/second

and average e.m.f. induced/turn $= 4f\Phi_m$ volts.

But for a sinusoidal wave the r.m.s. or effective value is 1·11 times the average value,

∴ r.m.s. value of e.m.f. induced/turn $= 1·11 \times 4f\Phi_m$.

Hence, r.m.s. value of e.m.f. induced in primary

$$= E_1 = 4·44 N_1 f \Phi_m \text{ volts} \tag{14.4}$$

and r.m.s. value of e.m.f. induced in secondary

$$= E_2 = 4·44 N_2 f \Phi_m \text{ volts.} \tag{14.5}$$

An alternative method of deriving these formulae is as follows:

If Φ = instantaneous value of flux in webers

$$= \Phi_m \sin 2\pi ft$$

$\therefore$ instantaneous value of induced e.m.f. per turn

$$= -d\Phi/dt \text{ volts}$$
$$= -2\pi f \Phi_m \times \cos 2\pi ft \text{ volts}$$
$$= 2\pi f \Phi_m \times \sin (2\pi ft - \pi/2) \text{ volts} \qquad (14.6)$$

$\therefore$ maximum value of induced e.m.f./turn $= 2\pi f \Phi_m$ volts

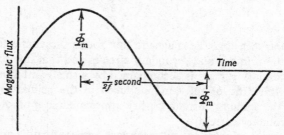

Fig. 14.2 Waveform of flux variation.

and r.m.s. value of induced e.m.f./turn

$$= 0 \cdot 707 \times 2\pi f \Phi_m = 4 \cdot 44 f \Phi_m \text{ volts.}$$

Hence r.m.s. value of primary e.m.f. $= E_1 = 4 \cdot 44 N_1 f \Phi_m$ volts

and r.m.s. value of secondary e.m.f. $= E_2 = 4 \cdot 44 N_2 f \Phi_m$ volts.

Example 14.1 *A 200-kVA, 6600-V/400-V, 50-Hz single-phase transformer has 80 turns on the secondary. Calculate:* (a) *the approximate values of the primary and secondary currents;* (b) *the approximate number of primary turns; and* (c) *the maximum value of the flux.*

(*a*) Full-load primary current $\simeq \dfrac{200 \times 1000}{6600} = 30 \cdot 3$ A,

and full-load secondary current $= \dfrac{200 \times 1000}{400} = 500$ A.

(*b*) No. of primary turns $\simeq \dfrac{80 \times 6600}{400} = 1320$.

(*c*) From expression (14.5),

$$400 = 4 \cdot 44 \times 80 \times 50 \times \Phi_m$$

$\therefore \qquad\qquad \Phi_m = 0 \cdot 0225$ Wb.

14.4 Phasor diagram for a transformer on no load

It is most convenient to commence the phasor diagram with the phasor representing the quantity that is common to the two windings, namely the flux Φ. This phasor can be made any convenient length and may be regarded merely as a reference phasor, relative to which other phasors have to be drawn.

In the preceding section, expression (14.6) shows that the e.m.f. induced by a sinusoidal flux lags the flux by a quarter of a cycle. Consequently the e.m.f.s, E_2 and E_1, induced in the secondary and primary windings respectively, are represented by phasors drawn 90° behind Φ, as in fig. 14.3. The values of E_2 and E_1 are proportional to the number of turns on the secondary and primary windings, since practically the whole of the flux set up by the primary is linked with the secondary when the latter is on open circuit. For convenience in drawing phasor diagrams for transformers, it will be assumed that N_2 and N_1 are equal, so that $E_2 = E_1$, as shown in fig. 14.3.

Since the difference between the value of the applied voltage V_1 and that of the induced e.m.f. E_1 is only about 0·05 per cent when the transformer is on no load, the phasor representing V_1 can be drawn equal and opposite to that representing E_1.

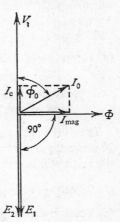

Fig. 14.3 Phasor diagram for transformer on no load.

The no-load current, I_0, taken by the primary consists of two components: (*a*) a reactive or magnetizing component*, I_{mag}, producing the flux and therefore in phase with the latter, and (*b*) an active or power component, I_c, supplying the hysteresis and eddy-current losses in the iron core and the negligible I^2R loss in the primary winding. Component I_c is in phase with the applied voltage: i.e. $I_cV_1 =$ core loss. This component is usually very small compared with I_{mag}, so that the no-load power factor is very low. From fig. 14.3, it will be seen that:

$$\text{no-load current} = I_0 = \sqrt{(I_c{}^2 + I_{mag}{}^2)} \qquad (14.7)$$

and $\qquad$ power factor on no load $= \cos \phi_0 = I_c/I_0 \qquad (14.8)$

* The waveform of this component is discussed in section 14.22.

Example 14.2 *A single-phase transformer has* 480 *turns on the primary and* 90 *turns on the secondary. The mean length of the flux path in the iron core is* 1·8 *m and the joints are equivalent to an airgap of* 0·1 *mm. If the maximum value of the flux density is to be* 1·1 *T when a p.d. of* 2200 *V at* 50 *Hz is applied to the primary, calculate:* (a) *the cross-sectional area of the core;* (b) *the secondary voltage on no load;* (c) *the primary current and power factor on no load. Assume the value of the magnetic field strength for* 1·1 *T in iron to be* 400 *A/m, the corresponding iron loss to be* 1·7 *W/kg at* 50 *Hz and the density of the iron to be* 7800 *kg/m³.*

(a) From (14·4), $2200 = 4\cdot44 \times 480 \times 50 \times \Phi_m$

$\therefore \qquad\qquad \Phi_m = 0\cdot0206$ Wb

and $\left.\begin{array}{r}\text{cross-sectional}\\ \text{area of core}\end{array}\right\} = \dfrac{0\cdot0206}{1\cdot1} = 0\cdot0187$ m².

This is the net area of the iron; the gross area of the core is about 10 per cent greater than this value to allow for the insulation between the laminations.

(b) Secondary voltage on no load $= 2200 \times \dfrac{90}{480} = 412\cdot5$ V.

(c) $\left.\begin{array}{r}\text{Total magnetomotive force}\\ \text{for the iron core}\end{array}\right\} = 400 \times 1\cdot8 = 720$ A

and $\left.\begin{array}{r}\text{magnetomotive force for}\\ \text{the equivalent airgap}\end{array}\right\} = \dfrac{1\cdot1}{4\pi \times 10^{-7}} \times 0\cdot0001 = 87\cdot5$ A

$\therefore$ $\left.\begin{array}{r}\text{total m.m.f. to produce the}\\ \text{maximum flux density}\end{array}\right\} = 720 + 87\cdot5 = 807\cdot5$ A

$\therefore$ $\left.\begin{array}{r}\text{maximum value of magnetiz-}\\ \text{ing current}\end{array}\right\} = \dfrac{807\cdot5}{480} = 1\cdot682$ A.

Assuming the current to be sinusoidal,

$\left.\begin{array}{r}\text{r.m.s. value of magnetizing cur-}\\ \text{rent}\end{array}\right\} = I_{\text{mag}} = 0\cdot707 \times 1\cdot682$

$\qquad\qquad\qquad\qquad = 1\cdot19$ A.

$\qquad\qquad$ Volume of iron $= 1\cdot8 \times 0\cdot0187$

$\qquad\qquad\qquad\qquad = 0\cdot0337$ m³

$\therefore \qquad\qquad$ mass of iron $= 0\cdot0337 \times 7800$

$\qquad\qquad\qquad\qquad = 263$ kg

and $\qquad\qquad$ iron loss $= 263 \times 1\cdot7 = 447$ W

$\therefore$ $\quad$ core-loss component of $\Big\}$ $= I_c = \dfrac{447}{2200} = 0.203$ A.
$\quad\quad\quad\quad\quad\quad$ current

From (14.7), no-load current $= I_0 = \sqrt{\{(1.19)^2 + 0.203)^2\}}$
$\quad\quad\quad\quad\quad\quad\quad\quad\quad\quad\quad\quad = 1.21$ A,

and from (14.8), power factor on $\Big\}$ $= \dfrac{0.203}{1.21} = 0.168$ lagging.
$\quad\quad\quad\quad\quad\quad\quad$ no load

14.5 Phasor diagram for a loaded transformer, assuming the voltage drop in the windings to be negligible

With this assumption, it follows that the secondary terminal voltage V_2 is the same as the e.m.f. E_2 induced in the secondary, and the primary applied voltage V_1 is equal and opposite in phase to the e.m.f. E_1 induced in the primary winding. Also, if we again assume equal number of turns on the primary and secondary windings, then $E_1 = E_2$.

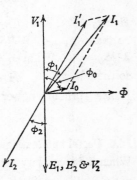

Let us consider the general case of a load having a lagging power factor $\cos \phi_2$; hence the phasor representing the secondary current I_2 lags V_2 by an angle ϕ_2, as shown in fig. 14.4. Phasor I'_1 represents the component of the primary current to neutralize the demagnetizing effect of the secondary current and is drawn equal and opposite to I_2. I_0 is the no-load current of the transformer, already discussed in section 14.4. The phasor sum of I'_1 and I_0 gives the total current I_1 taken from the supply, and the power factor on the primary side is $\cos \phi_1$, where ϕ_1 is the phase difference between V_1 and I_1.

Fig. 14.4 Phasor diagram for a loaded transformer having negligible voltage drop in windings.

In fig. 14.4, the phasor representing I_0 has, for clearness, been shown far larger relative to the other current phasors than it is in an actual transformer.

Example 14.3 *A single-phase transformer has* 1000 *turns on the primary and* 200 *turns on the secondary. The no-load current is* 3 *A at a power factor* 0.2 *lagging. Calculate the primary current and power factor when the secondary current is* 280 *A at a power factor of* 0.8 *lagging. Assume the voltage drop in the windings to be negligible.*

If I'_1 represents the component of the primary current to neutralize the demagnetizing effect of the secondary current, the ampere-turns due to I'_1 must be equal and opposite to those due to I_2, i.e.

$$I'_1 \times 1000 = 280 \times 200$$

$$\therefore \qquad\qquad I'_1 = 56 \text{ A.}$$

$$\cos \phi_2 = 0.8, \quad \therefore \sin \phi_2 = 0.6$$

and $\qquad\qquad \cos \phi_0 = 0.2, \quad \therefore \sin \phi_0 = 0.98$

From fig. 14.4 it will be seen that:

$$I_1 \cos \phi_1 = I'_1 \cos \phi_2 + I_0 \cos \phi_0$$
$$= (56 \times 0.8) + (3 \times 0.2) = 45.4 \text{ A}$$

and $\qquad I_1 \sin \phi_1 = I'_1 \sin \phi_2 + I_0 \sin \phi_0$
$$= (56 \times 0.6) + (3 \times 0.98) = 36.54 \text{ A.}$$

Hence, $\qquad\qquad I_1{}^2 = (45.4)^2 + (36.54)^2 = 3398$

so that $\qquad\qquad I_1 = 58.3 \text{ A.}$

Also, $\qquad\qquad \tan \phi_1 = 36.54/45.4 = 0.805$

so that $\qquad\qquad \phi_1 = 38° \ 50'.$

Hence   primary power factor $= \cos \phi_1 = \cos 38° \ 50'$
$$= 0.78 \text{ lagging.}$$

14.6 Useful and leakage fluxes in a transformer

When the secondary winding of a transformer is on open circuit, the current taken by the primary winding is responsible for setting up the magnetic flux and providing a very small power component to supply the iron loss in the core. To simplify matters in the present discussion, let us assume:

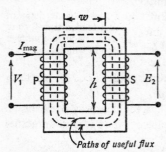

Fig. 14.5 Transformer on no load.

(*a*) the iron loss in the core and the I^2R loss in the primary winding to be negligible;

(*b*) the permeability of the iron to remain constant, so that the magnetizing current is proportional to the flux; and

(*c*) the primary and secondary windings to have the same number of turns, i.e. $N_1 = N_2$.

Fig. 14.5 shows all the flux set up by the primary winding passing through the secondary winding. There is a very small amount of flux returning through the air space around the primary winding, but

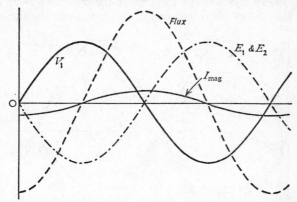

Fig. 14.6 Waveforms of voltages, current and flux of a transformer on no load.

since the relative permeability of transformer iron is of the order of 1000 or more, the reluctance of the air path may be 1000 times that of the parallel path through the iron limb carrying the secondary winding. Consequently the flux passing through the air space is negligible compared with that through the secondary. It follows that the e.m.f.s induced in the primary and secondary windings are equal and that the primary applied voltage, V_1, is equal and opposite to the e.m.f., E_1, induced in the primary, as shown in figs. 14.6 and 14.7.

Next, let us assume a load having a power factor such that the secondary current is in phase with E_2. As already explained in section 14.2, the primary current, I_1, must now have two components:

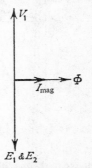

Fig. 14.7 Phasor diagram for fig. 14.6.

(a) I_{mag} to maintain the useful flux, the maximum value of which remains constant within about 2 per cent between no load and full load; and

(b) a component, I'_1, to neutralize the demagnetizing effect of the secondary current, as shown in figs. 14.8 (a) and 14.9.

At instant A in fig. 14.8 (a), the magnetizing current is *zero*, but I_2

and I'_1 are at their *maximum* values; and if the direction of the current in primary winding P is such as to produce flux upwards in the lefthand limb of fig. 14.10, the secondary current must be in such a direction as to produce flux upwards in the righthand limb, and the flux of each limb has to return through air. Since the flux of each

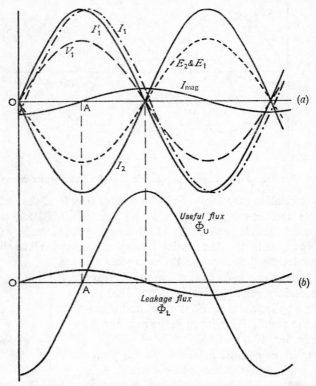

Fig. 14.8 Waveforms of induced e.m.f.s, currents and fluxes in a transformer on load.

limb is linked only with the winding by which it is produced, it is referred to as *leakage* flux and is responsible for inducing an e.m.f. of self-inductance in the winding with which it is linked. The reluctance of the paths of the leakage flux, Φ_L, is almost entirely due to the long air paths and is therefore practically constant. Consequently the value of the leakage flux is proportional to the load current, whereas the value of the useful flux remains almost independent of the load,

The reluctance of the paths of the leakage flux is very high, so that the value of this flux is relatively small even on full load when the values of I'_1 and I_2 are about 20–30 times the magnetizing current I_{mag}.

From the above discussion it follows that the actual flux in a transformer can be regarded as being due to the two components shown in fig. 14.8 (*b*), namely:

(*a*) the useful flux, Φ_U, linked with both windings and re-maining practically constant in value at all loads; and

(*b*) the leakage flux, Φ_L, half of which is linked with the primary winding and half with the secondary, and its value is proportional to the load.

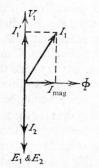

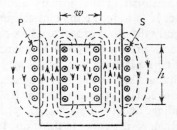

Fig. 14.9 Phasor diagram
for fig. 14.8.

Fig. 14.10 Paths of leakage
flux.

The case of the secondary current in phase with the secondary induced e.m.f. has been considered because it is easier to see that the useful and the leakage fluxes can be considered independently of each other for this condition than it is for loads of other power factor.

14.7 Leakage flux responsible for the inductive reactance of a transformer

When a transformer is on no load there is no secondary current, and the secondary winding has not the slightest effect upon the primary current. The primary winding is then behaving as a choke having a very high inductance and a very low resistance. When the transformer is supplying a load, however, this conception of the inductance of the windings is of little use and it is rather difficult at first to understand why the useful flux is not responsible for any inductive drop in the transformer. So let us first of all assume an ideal transformer, namely a transformer the windings of which have negligible resistance and in which there is no iron loss and no magnetic leakage. Also for convenience let us assume unity transformation ratio, i.e. $N_1 = N_2$.

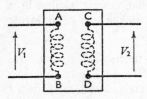

Fig. 14.11 Ideal transformer enclosed in a box.

If such a transformer were enclosed in a box and the ends of the windings brought to terminals A, B, C and D on the lid, as shown in fig. 14.11, the p.d. between C and D would be equal to that between A and B; and as far as the effect upon the output voltage is concerned, the transformer behaves as if A were connected to C and B to D. In other words, the useful flux is not responsible for any voltage drop in a transformer.

In the preceding section it was explained that the leakage flux is proportional to the primary and secondary currents and that its effect is to induce e.m.f.s of self induction in the windings. Consequently the effect of leakage flux can be considered as equivalent to inductive reactors X_1 and X_2 connected in series with a transformer having no leakage flux, as shown in fig. 14.12, these reactors being such that the flux-linkages produced by the primary current through X_1 are equal to those due to the leakage flux linked with the primary winding, and the flux-linkages produced by the secondary current through X_2 are equal to those due to the leakage flux linked with the

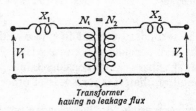

Fig. 14.12 Transformer with leakage reactances.

secondary winding of the actual transformer. The straight line drawn between the primary and secondary windings in fig. 14.12 is the symbol used to indicate that the transformer has an iron core.

14.8 Methods of reducing leakage flux

The leakage flux can be practically eliminated by winding the primary and secondary, one over the other, uniformly around a laminated iron ring of uniform cross-section. But such an arrangement is not commercially practicable except in very small sizes, owing to the cost of threading a large number of turns through the ring.

The principal methods used in practice are:

(a) *Making the transformer 'window' long and narrow*

The area of the window, $h \times w$ in fig. 14.10, is determined by the cross-sectional area of the primary and secondary windings. For a given area the reluctance of the paths of the leakage flux is increased by increasing the height h and reducing the width w of the window. But if the window is made very narrow, an excessive proportion of the width is occupied by insulation. Hence, in practice the ratio of h/w is limited to about 4.

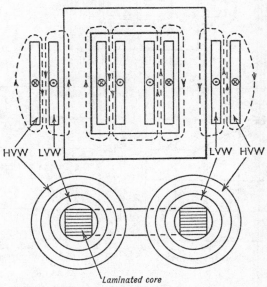

HVW LVW   LVW HVW

Laminated core

Fig. 14.13 Concentric windings.

(b) *Arranging the primary and secondary windings concentrically*

Each winding is equally divided between the two limbs, the general practice being to place the low-voltage winding LVW next to the core and the high-voltage winding HVW on the outside, as shown in fig. 14.13. At the instant when the currents are in the directions represented by the dots and crosses and their ampere-turns are exactly equal in value, the useful flux is zero and the paths where the leakage flux has its maximum density are represented by the dotted lines. With concentric windings, the leakage flux is approximately a half of that with the coils on separate limbs. Furthermore, this leakage flux is linked with only a part of the primary or secondary winding, so that the leakage reactance has been considerably reduced.

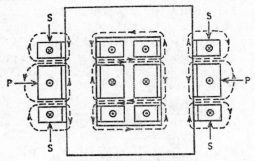

Fig. 14.14 Sandwiched windings.

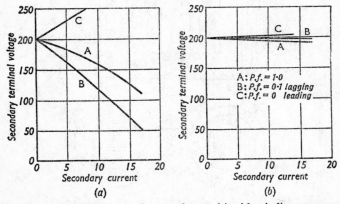

Fig. 14.15 Load characteristics of a transformer (*a*) with windings on separate limbs, (*b*) with windings sandwiched.

(c) *Sandwiching the primary and secondary windings*

Fig. 14.14 shows the primary winding P in two sections, one on each limb, each section being sandwiched between sections of the secondary windings S. From the dots and crosses representing the directions of the primary and secondary currents at the instant when their ampere-turns are equal, it follows that most of the leakage flux passes horizontally across the window. The dotted lines in fig. 14.14 represent the leakage flux in the regions where its density is highest.

The leakage flux can be reduced still further by using more sandwiched sections; but this arrangement has the disadvantage that adjacent sections must be well insulated from each other, so that with a large number of sections an excessive amount of space is taken up by insulation.

Fig. 14.15 (*a*) shows the results obtained with a certain transformer for loads of different power factors when the primary and secondary

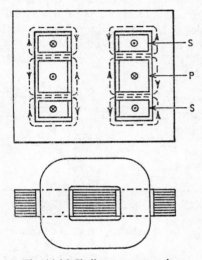

Fig. 14.16 Shell-type construction.

windings were on separate limbs, as in fig. 14.5. Fig. 14.15 (*b*) shows the results obtained with the same transformer when each limb had 10 sections of primary winding sandwiched with 10 sections of secondary winding. A comparison of these figures shows that the leakage flux can be reduced to a very small value by effective sandwiching of the primary and secondary windings.

(d) *Using shell-type construction*

In the construction shown in the above diagrams the iron core forming the limbs is surrounded by the windings. This arrangement is referred to as the *core* type. With the *shell*-type construction the windings are more completely surrounded by iron, as will be seen from fig. 14.16, where the primary and secondary windings are shown sandwiched. The width of the centre limb is about twice that of each of the outer limbs. The dotted lines are again shown in regions where the density of the leakage flux is highest.

14.9 Equivalent circuit of a transformer

The behaviour of a transformer may be conveniently considered by assuming it to be equivalent to an ideal transformer, i.e. a transformer having no losses and no magnetic leakage and an iron core of infinite permeability requiring no magnetizing current, and then allowing for the imperfections of the actual transformer by means of additional circuits or impedances inserted between the supply and the primary winding and between the secondary and the load. Thus, in fig. 14.17, P and S represent the primary and secondary

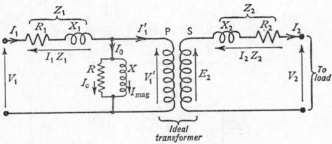

Fig. 14.17 Equivalent circuit of a transformer.

windings of the ideal transformer. R_1 and R_2 are resistances equal to the resistances of the primary and secondary windings of the actual transformer. Similarly, inductive reactances X_1 and X_2 represent the reactances of the windings due to leakage flux in the actual transformer, as already explained in section 14.7.

The inductive reactor X is such that it takes a reactive current equal to the magnetizing current I_{mag} of the actual transformer. The core losses due to hysteresis and eddy currents are allowed for by a resistor R of such value that it takes a current I_c equal to the core-loss component of the primary current, i.e. $I_c^2 R$ is equal to the core loss of the actual transformer. The resultant of I_{mag} and I_c is I_0, namely the current which the transformer takes on no load. The phasor diagram for the equivalent circuit on no load is exactly the same as that given in fig. 14.3.

14.10 Phasor diagram for a transformer on load

For convenience let us assume an equal number of turns on the primary and secondary windings, so that $E_1 = E_2$. Both E_1 and E_2 lag the flux by 90°, as shown in fig. 14.18, and V'_1 represents the voltage applied to the primary of the ideal transformer to neutralize the induced e.m.f. E_1, and is therefore equal and opposite to the

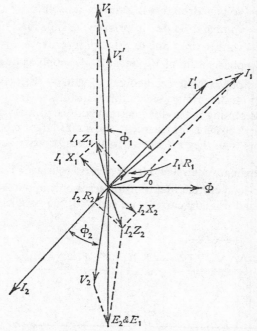

Fig. 14.18 Phasor diagram for a transformer on load.

latter. Let us also assume the general case of a load having a lagging power factor; consequently, in fig. 14.18, I_2 has been drawn lagging E_2 by about 45°. Then:

$I_2R_2 =$ voltage drop due to secondary resistance,

$I_2X_2 =$ voltage drop due to secondary leakage reactance

and $I_2Z_2 =$ voltage drop due to secondary impedance.

The secondary terminal voltage V_2 is the phasor difference of E_2 and I_2Z_2; in other words, V_2 must be such that the phasor sum of V_2

and I_2Z_2 is E_2, and the derivation of the phasor representing V_2 is evident from fig. 14.18. The power factor of the load is $\cos \phi_2$, where ϕ_2 is the phase difference between V_2 and I_2. I'_1 represents the component of the primary current to neutralize the demagnetizing effect of the secondary current and is drawn equal and opposite to I_2. I_0 is the no-load current of the transformer (section 14.4). The phasor sum of I'_1 and I_0 gives the total current I_1 taken from the supply.

I_1R_1 = voltage drop due to primary resistance,

I_1X_1 = voltage drop due to primary leakage reactance,

I_1Z_1 = voltage drop due to primary impedance

and V_1 = phasor sum of V'_1 and I_1Z_1 = supply voltage.

If ϕ_1 is the phase difference between V_1 and I_1, then $\cos \phi_1$ is the power factor on the primary side of the transformer. In fig. 14.18 the phasors representing the no-load current and the primary and secondary voltage drops are, for clearness, shown far larger relative to the other phasors than they are in an actual transformer.

14.11 Approximate equivalent circuit of a transformer

Since the no-load current of a transformer is only about 3–5 per cent of the full-load primary current, we can omit the parallel circuits R

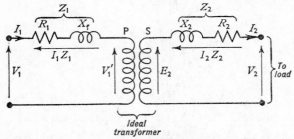

Fig. 14.19 Approximate equivalent circuit of a transformer.

and X in fig. 14.17 without introducing an appreciable error when we are considering the behaviour of the transformer on full load. Thus we have the simpler equivalent circuit of fig. 14.19

14.12 Simplification of the approximate equivalent circuit of a transformer

We can replace the resistance R_2 of the secondary of fig. 14.19 by inserting additional resistance R'_2 in the primary circuit such that the power absorbed in R'_2 when carrying the primary current is equal to that in R_2 due to the secondary current.

i.e. $$I_1^2 R'_2 = I_2^2 R_2$$
∴ $$R'_2 = R_2(I_2/I_1)^2 \simeq R_2(V_1/V_2)^2.$$

Hence if R_e be a single resistance in the primary circuit equivalent to the primary and secondary resistances of the actual transformer, then:

$$R_e = R_1 + R'_2 = R_1 + R_2(V_1/V_2)^2 \qquad (14.9)$$

Similarly, since the inductance of a coil is proportional to the square of the number of turns, the secondary leakage reactance X_2 can be replaced by an equivalent reactance X'_2 in the primary circuit, such that:

$$X'_2 = X_2(N_1/N_2)^2 \simeq X_2(V_1/V_2)^2.$$

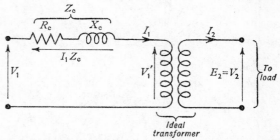

Fig. 14.20 Simplified equivalent circuit of a transformer.

If X_e be the single reactance in the primary circuit equivalent to X_1 and X_2 of the actual transformer,

$$X_e = X_1 + X'_2 = X_1 + X_2(V_1/V_2)^2 \qquad (14.10)$$

If Z_e be the equivalent impedance of the primary and secondary windings referred to the primary circuit,

$$Z_e = \surd(R_e^2 + X_e^2) \qquad (14.11)$$

If ϕ_e be the phase difference between I_1 and $I_1 Z_e$, then

$$R_e = Z_e \cos \phi_e \text{ and } X_e = Z_e \sin \phi_e.$$

The simplified equivalent circuit of the transformer is given in fig. 14.20, and fig. 14.21 (*a*) is the corresponding phasor diagram.

14.13 Voltage regulation of a transformer

The voltage regulation of a transformer is defined as the variation of the secondary voltage between no load and full load expressed as either a per-unit or a percentage of the *no-load* voltage, the primary voltage being assumed constant, i.e.

$$\left.\begin{array}{r}\text{voltage}\\ \text{regulation}\end{array}\right\} = \frac{\text{no-load voltage} - \text{full-load voltage}}{\text{no-load voltage}} \quad (14.12)$$

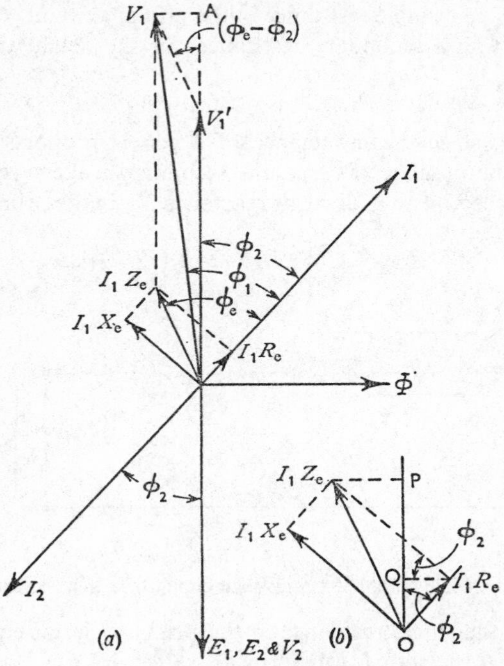

Fig. 14.21 Phasor diagram for fig. 14.20.

If V_1 = primary applied voltage,

secondary voltage on no load = $V_1 \times N_2/N_1$,

since the voltage drop in the primary winding due to the no-load current is negligible.

If V_2 = secondary terminal voltage on full load,

$$\begin{aligned}\text{voltage regulation} &= \frac{V_1(N_2/N_1) - V_2}{V_1(N_2/N_1)}\\ &= \frac{V_1 - V_2(N_1/N_2)}{V_1} \text{ per unit}\\ &= \frac{V_1 - V_2(N_1/N_2)}{V_1} \times 100 \text{ per cent}\end{aligned}$$

In the phasor diagram of fig. 14.21, N_1 and N_2 were assumed equal, so that $V'_1 = V_2$. In general, $V'_1 = V_2(N_1/N_2)$,

$$\therefore \qquad \text{per-unit voltage regulation} = \frac{V_1 - V'_1}{V_1} \qquad (14.13)$$

In fig. 14.21 (*a*), let us draw a perpendicular from V_1 to meet the extension of V'_1 at A; then:

$$V_1{}^2 = (V'_1 + V'_1 A)^2 + (V_1 A)^2$$
$$= \{V'_1 + I_1 Z_e \cos(\phi_e - \phi_2)\}^2 + \{I_1 Z_e \sin(\phi_e - \phi_2)\}^2$$

In actual practice, $I_1 Z_e \sin(\phi_e - \phi_2)$ is very small compared with V'_1, so that:

$$V_1 \simeq V'_1 + I_1 Z_e \cos(\phi_e - \phi_2).$$

Hence, per-unit voltage regulation $= \dfrac{V_1 - V'_1}{V_1}$

$$= \frac{I_1 Z_e \cos(\phi_e - \phi_2)}{V_1} \qquad (14.14)$$

Since $\quad Z_e \cos(\phi_e - \phi_2) = Z_e (\cos \phi_e . \cos \phi_2 + \sin \phi_e . \sin \phi_2)$
$$= R_e \cos \phi_2 + X_e \sin \phi_2$$

$$\therefore \quad \left.\begin{array}{c}\text{per-unit voltage}\\ \text{regulation}\end{array}\right\} = \frac{I_1(R_e \cos \phi_2 + X_e \sin \phi_2)}{V_1} \qquad (14.15)$$

This expression can also be derived by projecting $I_1 R_e$ and $I_1 Z_e$ on to OA, as shown enlarged in fig. 14.21 (*b*), from which it follows that:

$$V'_1 A \text{ in fig. 14.21 } (a) = \text{OP in fig. 14.21 (b)}$$
$$= \text{OQ} + \text{QP}$$
$$= I_1 R_e \cos \phi_2 + I_1 X_e \sin \phi_2$$

$$\therefore \quad \left.\begin{array}{c}\text{per-unit voltage}\\ \text{regulation}\end{array}\right\} = \frac{V_1 - V'_1}{V_1} \simeq \frac{V'_1 A}{V_1} = \frac{\text{OP}}{V_1}$$
$$= \frac{I_1(R_e \cos \phi_2 + X_e \sin \phi_2)}{V_1}.$$

The above expressions have been derived on the assumption that the power factor is lagging. Should the power factor be leading, the angle in expression (14.14) would be $(\phi_e + \phi_2)$ and the term in brackets in expression (14.15) would be $(R_e \cos \phi_2 - X_e \sin . \phi_2)$.

*

Example 14.4 *A* 100-*kVA transformer has* 400 *turns on the primary and* 80 *turns on the secondary. The primary and secondary resistances are* 0·3 Ω *and* 0·01 Ω *respectively, and the corresponding leakage reactances are* 1·1 Ω *and* 0·035 Ω *respectively. The supply voltage is* 2200 *V. Calculate:* (a) *the equivalent impedance referred to the primary circuit, and* (b) *the voltage regulation and the secondary terminal voltage for full load having a power factor of* (i) 0·8 *lagging and* (ii) 0·8 *leading.*

(*a*) From (14.9),

$$\left.\begin{array}{r}\text{equivalent resistance referred to}\\ \text{primary}\end{array}\right\} = R_e = 0\cdot3 + 0\cdot01(400/80)^2$$
$$= 0\cdot55 \ \Omega.$$

From (14.10),

$$\left.\begin{array}{r}\text{equivalent leakage reactance}\\ \text{referred to primary}\end{array}\right\} = X_e = 1\cdot1 + 0\cdot035(400/80)^2$$
$$= 1\cdot975 \ \Omega.$$

From (14.11),

$$\left.\begin{array}{r}\text{equivalent impedance referred to}\\ \text{primary}\end{array}\right\} = Z_e = \sqrt{\{(0\cdot55)^2 + (1\cdot975)^2\}}$$
$$= 2\cdot05 \ \Omega.$$

(*b*) (i) Since $\cos \phi_2 = 0\cdot8$, $\therefore \ \sin \phi_2 = 0\cdot6$.

$$\text{Full-load primary current} \simeq \frac{100 \times 1000}{2200} = 45\cdot45 \ \text{A}.$$

Substituting in (14.15), we have:

$$\left.\begin{array}{r}\text{voltage regulation for power}\\ \text{factor } 0\cdot8 \text{ lagging}\end{array}\right\} = \frac{45\cdot45(0\cdot55 \times 0\cdot8 + 1\cdot975 \times 0\cdot6)}{2200}$$
$$= 0\cdot0336 \ \text{per unit}$$
$$= 3\cdot36 \ \text{per cent}.$$

Secondary terminal voltage on no load = 2200 × 80/400 = 440 V

∴ decrease of secondary terminal voltage between no load and full load

$$= 440 \times 0\cdot0336 = 14\cdot8 \ \text{V}$$

∴ secondary terminal voltage on full load

$$= 440 - 14\cdot8 = 425\cdot2 \ \text{V}.$$

(ii) Voltage regulation for power factor 0·8 leading

$$= \frac{45\cdot45(0\cdot55 \times 0\cdot8 - 1\cdot975 \times 0\cdot6)}{2200} = -0\cdot0154 \text{ per unit}$$

$$= -1\cdot54 \text{ per cent.}$$

Increase of secondary terminal voltage between no load and full load

$$= 440 \times 0\cdot0154 = 6\cdot78 \text{ V}$$

∴ secondary terminal voltage on full load

$$= 440 + 6\cdot78 = 446\cdot8 \text{ V.}$$

Example 14.5 *Calculate the per-unit and the percentage resistance and leakage reactance drops of the transformer referred to in example 14.4.*

Per-unit resistance drop of a transformer

$$= \frac{\left(\begin{matrix}\text{full-load primary}\\\text{current}\end{matrix}\right) \times \left(\begin{matrix}\text{equivalent resistance re-}\\\text{ferred to primary circuit}\end{matrix}\right)}{\text{primary voltage}}$$

$$= \frac{\left(\begin{matrix}\text{full-load secondary}\\\text{current}\end{matrix}\right) \times \left(\begin{matrix}\text{equivalent resistance re-}\\\text{ferred to secondary circuit}\end{matrix}\right)}{\text{secondary voltage on no load}}$$

Thus, for example 14.4,

full-load primary current $\simeq$ 45·45 A,

and equivalent resistance referred to primary circuit$\Big\}$ = 0·55 Ω

∴ resistance drop $= \dfrac{45\cdot45 \times 0\cdot55}{2200} = 0\cdot0114$ per unit

$$= 1\cdot14 \text{ per cent.}$$

Alternatively, full-load secon-
dary current$\Big\}$ $\simeq$ 45·45 × 400/80

$$= 227\cdot2 \text{ A,}$$

and equivalent resistance re-
ferred to secondary circuit$\Big\}$ $= 0\cdot01 + 0\cdot3\left(\dfrac{80}{400}\right)^2 = 0\cdot022$ Ω.

Secondary voltage on no load = 440 V,

$$\therefore \qquad \text{resistance drop} = \frac{227 \cdot 2 \times 0 \cdot 022}{440} = 0 \cdot 0114 \text{ per unit}$$

$$= 1 \cdot 14 \text{ per cent.}$$

Similarly, leakage reactance drop of a transformer

$$= \frac{\left(\begin{array}{c}\text{full-load primary}\\ \text{current}\end{array}\right) \times \left(\begin{array}{c}\text{equivalent leakage reactance}\\ \text{referred to primary circuit}\end{array}\right)}{\text{primary voltage}}$$

$$= \frac{45 \cdot 45 \times 1 \cdot 975}{2200} = 0 \cdot 0408 \text{ per unit} = 4 \cdot 08 \text{ per cent.}$$

It is usual to refer to the per-unit or the percentage resistance and leakage reactance drops on full load as merely the per-unit or the percentage resistance and leakage reactance of the transformer; thus, the above transformer has a per-unit resistance and leakage reactance of 0·0114 and 0·0408 respectively or a percentage resistance and leakage reactance of 1·14 and 4·08 respectively.

14.14 Efficiency of a transformer

The losses which occur in a transformer on load can be divided into two groups:

(*a*) copper losses in primary and secondary windings, namely $I_1^2 R_1 + I_2^2 R_2$;

(*b*) iron losses in the core due to hysteresis and eddy currents. The factors determining these losses have already been discussed in sections 3.14 and 9.1 (*b*).

Since the maximum value of the flux in a normal transformer does not vary by more than about 2 per cent between no load and full load, it is usual to assume the iron loss constant at all loads.

Hence, if P_c = total iron loss in core,

$$\text{total losses in transformer} = P_c + I_1^2 R_1 + I_2^2 R_2$$

and efficiency $= \dfrac{\text{output power}}{\text{input power}} = \dfrac{\text{output power}}{\text{output power} + \text{losses}}$

$$= \frac{I_2 V_2 \times \text{power factor}}{I_2 V_2 \times \text{p.f.} + P_c + I_1^2 R_1 + I_2^2 R_2}. \tag{14.16}$$

Greater accuracy is possible by expressing the efficiency thus:

$$\text{efficiency} = \frac{\text{output power}}{\text{input power}} = \frac{\text{input power} - \text{losses}}{\text{input power}}$$

$$= 1 - \frac{\text{losses}}{\text{input power}} \qquad (14.17)$$

Example 14.6 *The primary and secondary windings of a 500-kVA transformer have resistances of* 0·42 Ω *and* 0·0011 Ω *respectively. The primary and secondary voltages are 6600 V and 400 V respectively and the iron loss is 2·9 kW. Calculate the efficiency on* (a) *full load and* (b) *half load, assuming the power factor of the load to be 0·8.*

(a) Full-load secondary current $= \dfrac{500 \times 1000}{400} = 1250$ A

and   full-load primary current $\simeq \dfrac{500 \times 1000}{6600} = 75 \cdot 8$ A

∴   secondary copper loss on$\left.\right\}$ $= (1250)^2 \times 0 \cdot 0011 = 1720$ W
full load

and   primary copper loss on$\left.\right\}$ $= (75 \cdot 8)^2 \times 0 \cdot 42 = 2415$ W
full load

∴ total copper loss on full load $= 4135$ W $= 4 \cdot 135$ kW
and   total loss on full load $= 4 \cdot 135 + 2 \cdot 9 = 7 \cdot 035$ kW.
     Output power on full load $= 500 \times 0 \cdot 8 = 400$ kW
∴   input power on full load $= 400 + 7 \cdot 035 = 407 \cdot 035$ kW.
From (14.17), efficiency on full$\left.\right\}$ $= \left(1 - \dfrac{7 \cdot 035}{407 \cdot 0}\right) = 0 \cdot 9827$ per unit
load
$= 98 \cdot 27$ per cent.

(b) Since the copper loss varies as the square of the current,

∴ total copper loss on half load $= 4 \cdot 135 \times (0 \cdot 5)^2 = 1 \cdot 034$ kW
and   total loss on half load $= 1 \cdot 034 + 2 \cdot 9 = 3 \cdot 934$ kW

∴   efficiency on half load $= \left(1 - \dfrac{3 \cdot 934}{203 \cdot 9}\right) = 0 \cdot 9807$ per unit
$= 98 \cdot 07$ per cent.

14.15 Condition for maximum efficiency of a transformer
If R_{e2} be the equivalent resistance of the primary and secondary windings referred to the *secondary* circuit,

$$R_{e2} = R_1 (N_2/N_1)^2 + R_2$$
$$= \text{a constant for a given transformer.}$$

Hence for any load current I_2,

$$\text{total copper loss} = I_2^2 R_{e2}$$

and

$$\text{efficiency} = \frac{I_2 V_2 \times \text{p.f.}}{I_2 V_2 \times \text{p.f.} + P_c + I_2^2 R_{e2}}$$

$$= \frac{V_2 \times \text{p.f.}}{V_2 \times \text{p.f.} + P_c/I_2 + I_2 R_{e2}} \qquad (14.18)$$

For a normal transformer, V_2 is approximately constant; hence for a load of given power factor, the efficiency is a maximum when the denominator of (14.18) is a minimum,

i.e.   when $\dfrac{d}{dI_2} (V_2 \times \text{p.f.} + P_c/I_2 + I_2 R_{e2}) = 0$

$\therefore$ $-P_c/I_2^2 + R_{e2} = 0$

or $I_2^2 R_{e2} = P_c \qquad (14.19)$

To check that this condition gives the minimum and not the maximum value of the denominator in expression (14.18), $(-P_c/I_2^2 + R_{e2})$ should be differentiated with respect to I_2, thus:

$$\frac{d}{dI_2} (-P_c/I_2^2 + R_{e2}) = 2P_c/I_2^3$$

Since this quantity is positive, expression (14.19) is the condition for the minimum value of the denominator of (14.18) and therefore the maximum value of the efficiency. Hence the efficiency is a maximum when the variable copper loss is equal to the constant iron loss.

Example 14.7 *Find the output at which the efficiency of the transformer of example 14.6 is a maximum and calculate its value, assuming the power factor of the load to be 0·8.*

With the full-load output of 500 kVA, the total copper loss is 4·135 kW.

Let $n = \begin{cases} \text{fraction of full-load kVA at which the efficiency is a} \\ \text{maximum.} \end{cases}$

Corresponding total copper loss $= n^2 \times 4·135$ kW

Hence, from (14.19), $n^2 \times 4.135 = 2·9$

$\therefore$ $n = 0·837,$

and   output at maximum efficiency $= 0·837 \times 500$

                                $= 418·5$ kVA.

It will be noted that the value of the kVA at which the efficiency is a maximum is independent of the power factor of the load.

Since the copper and iron losses are equal when the efficiency is a maximum,

$$\therefore \qquad \text{total loss} = 2 \times 2 \cdot 9 = 5 \cdot 8 \text{ kW}$$

$$\text{Output power} = 418 \cdot 5 \times 0 \cdot 8 = 334 \cdot 8 \text{ kW}$$

$$\therefore \quad \text{maximum efficiency} = \left(1 - \frac{5 \cdot 8}{334 \cdot 8 + 5 \cdot 8}\right) = 0 \cdot 983 \text{ per unit}$$

$$= 98 \cdot 3 \text{ per cent.}$$

14.16 Open-circuit and short-circuit tests on a transformer

These two tests enable the efficiency and the voltage regulation to be calculated without actually loading the transformer and with an accuracy far higher than is possible by direct measurement of input and output powers and voltages. Also, the power required to carry out these tests is very small compared with the full-load output of the transformer.

Open-circuit test

The transformer is connected as in fig. 14.22 to a supply at the rated voltage and frequency, namely the voltage and frequency given on

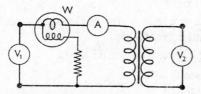

Fig. 14.22 Open-circuit test on a transformer.

the nameplate. The ratio of the voltmeter readings, V_1/V_2, gives the ratio of the number of turns. Ammeter A gives the no-load current, and its reading is a check on the magnetic quality of the iron core and joints. The primary current on no load is usually less than 5 per cent of the full-load current, so that the I^2R loss on no load is less than $1/400$ of the primary I^2R loss on full load and is therefore negligible compared with the iron loss. Hence the wattmeter reading can be taken as the iron loss of the transformer.

Short-circuit test

The secondary is short-circuited through a suitable ammeter A_2, as shown in fig. 14.23, and a *low* voltage is applied to the primary circuit. This voltage should, if possible, be adjusted to circulate full-load currents in the primary and secondary circuits. Assuming this to be the case, the copper loss in the windings is the same as that on full load. On the other hand, the iron loss is negligibly small, since the applied voltage and therefore the flux are only about one-twentieth

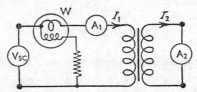

Fig. 14.23 Short-circuit test on a transformer.

to one-thirtieth of the rated voltage and flux, and the iron loss is approximately proportional to the square of the flux. Hence the power registered on wattmeter W can be taken as the copper loss in the windings.

14.17 Calculation of efficiency from the open-circuit and short-circuit tests

If P_{oc} = input power in watts on the open-circuit test,

= iron loss,

and P_{sc} = input power in watts on the short-circuit test with full-load currents,

= total copper loss on full load,

then total loss on full load $\Big\} = P_{oc} + P_{sc}$

and efficiency on full load $\Big\} = \dfrac{\text{full-load VA} \times \text{p.f.}}{(\text{full-load VA} \times \text{p.f.}) + P_{oc} + P_{sc}}$ (14.20)

Also, for any load equal to $n \times$ full load,

corresponding total loss $\Big\} = P_{oc} + n^2 P_{sc}$

and corresponding efficiency

$$= \frac{n \times \text{full-load VA} \times \text{p.f.}}{n \times \text{full-load VA} \times \text{p.f.} + P_{oc} + n^2 P_{sc}} \qquad (14.21)$$

14.18 Calculation of the voltage regulation from the short-circuit test

Since the secondary voltage is zero, the whole of the applied voltage on the short-circuit test is absorbed in sending currents through the impedances of the primary and secondary windings; and since ϕ_e in fig. 14.21 is the phase angle between the primary current and the voltage drop due to the equivalent impedance referred to the primary circuit,

$$\cos \phi_e = \text{power factor on short-circuit test,}$$

$$= \frac{P_{sc}}{I_1 V_{sc}}$$

If V_{sc} be the value of the primary applied voltage on the s.c. test. when *full-load* currents are flowing in the primary and secondary windings, then from expression (14.14),

$$\text{per-unit voltage regulation} = \frac{V_{sc} \cos (\phi_e - \phi_2)}{V_1} \qquad (14.22)$$

Example 14.8 *The following results were obtained on a 50-kVA transformer: O.C. test—primary voltage, 3300 V; secondary voltage, 400 V; primary power, 430 W. S.C. test—primary voltage, 124 V; primary current, 15·3 A; primary power, 525 W; secondary current, full-load value. Calculate:* (a) *the efficiencies at full load and at half load for 0·7 power factor;* (b) *the voltage regulations for power factor 0·7* (i) *lagging,* (ii) *leading; and* (c) *the secondary terminal voltages corresponding to* (i) *and* (ii).

(a)   Iron loss = 430 W,

copper loss on full load = 525 W

∴   total loss on full load = 955 W = 0·955 kW

and   efficiency on full load $= \dfrac{50 \times 0 \cdot 7}{(50 \times 0 \cdot 7) + 0 \cdot 955}$

$$= \left(1 - \frac{0 \cdot 955}{35 \cdot 95}\right) = 0 \cdot 9734 \text{ per unit}$$

$$= 97 \cdot 34 \text{ per cent.}$$

Copper loss on half load $= 525 \times (0 \cdot 5)^2 = 131$ W,

∴   total loss on half load = 430 + 131 = 561 W = 0·561 kW

and   efficiency on half load $= \dfrac{25 \times 0 \cdot 7}{(25 \times 0 \cdot 7) + 0 \cdot 561}$

$$= \left(1 - \frac{0 \cdot 561}{18 \cdot 06}\right) = 0 \cdot 969 \text{ per unit}$$

$$= 96 \cdot 9 \text{ per cent.}$$

(b) $$\text{Cos } \phi_e = \frac{525}{124 \times 15 \cdot 3} = 0 \cdot 2765$$

$\therefore$ $\phi_e = 73° \; 57'$

For $\cos \phi_2 = 0 \cdot 7$, $\phi_2 = 45° \; 34'$.

From expression (14.22), for power factor 0·7 lagging,

$$\text{voltage regulation} = \frac{124 \cos (73° \; 57' - 45° \; 34')}{3300}$$

$$= 0 \cdot 033 \text{ per unit} = 3 \cdot 3 \text{ per cent.}$$

For power factor 0·7 leading,

$$\text{voltage regulation} = \frac{124 \cos (73° \; 57' + 45° \; 34')}{3300}$$

$$= -0 \cdot 0185 \text{ per unit} = -1 \cdot 85 \text{ per cent.}$$

(c) Secondary voltage on open circuit = 400 V

$\therefore$ secondary voltage on full load, p.f. 0·7 lagging

$$= 400 \, (1 - 0 \cdot 033) = 386 \cdot 8 \text{ V}$$

and secondary voltage on full load, p.f. 0·7 leading

$$= 400(1 + 0 \cdot 0185) = 407 \cdot 4 \text{ V.}$$

14.19 Three-phase core-type transformers

Modern large transformers are usually of the three-phase core-type shown in fig. 14.24. Three similar limbs are connected by top and

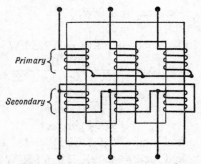

Fig. 14.24 Three-phase core-type transformer.

bottom yokes, each limb having primary and secondary windings, arranged concentrically. In fig. 14.24 the primary is shown star-connected and the secondary mesh-connected. Actually, the windings may be connected star–delta, delta–star, star–star or delta–delta, depending upon the conditions under which the transformer is to be used.

Example 14.9 *A three-phase transformer has* 420 *turns on the primary and* 36 *turns on the secondary winding. The supply voltage is* 3300 *V. Find the secondary line voltage on no load when the windings are connected* (a) *star–delta and* (b) *delta–star.*

(*a*)   Primary phase voltage $= 3300/1\cdot73 = 1908$ V,

∴   secondary phase voltage $= 1908 \times 36/420 = 163\cdot5$ V

$= $ secondary line voltage.

(*b*)   Primary phase voltage $= 3300$ V,

∴   secondary phase voltage $= 3300 \times 36/420 = 283$ V,

∴   secondary line voltage $= 283 \times 1\cdot73 = 490$ V.

14.20 Auto-transformers

An auto-transformer is a transformer having a part of its winding common to the primary and secondary circuits; thus, in fig. 14.25,

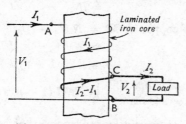

Fig. 14.25 An auto-transformer.

winding AB has a tapping at C, the load being connected across CB and the supply voltage applied across AB.

Let V_1 and $V_2 = $ primary and secondary voltages respectively,

I_1 and $I_2 = $ primary and secondary currents respectively,

$N_1 = $ no. of turns between A and B,

$N_2 = $ no. of turns between C and B

and $n = $ ratio of the *smaller* voltage to the *larger* voltage.

Neglecting the losses, the leakage reactance and the magnetizing current, we have for fig. 14.25,

$$n = \frac{V_2}{V_1} = \frac{I_1}{I_2} = \frac{N_2}{N_1}.$$

The current in section CB of the winding is the resultant of I_1 and I_2; and since these currents are practically in phase opposition and I_2 is greater than I_1, the resultant current is $(I_2 - I_1)$. Hence,

ampere-turns due to section CB $= (I_2 - I_1)N_2$

$$= (I_1/n - I_1) \times nN_1$$
$$= I_1N_1(1 - n)$$
$$= \text{ampere-turns due to section AC}$$

i.e. the ampere-turns due to sections CB and AC balance each other —a characteristic of all transformer actions (section 14.2).

Let l = mean length per turn

and J = current density throughout the winding,

$\therefore$ cross-sectional area of conductor in AC $= I_1/J$

and cross-sectional area of conductor in CB $= (I_2 - I_1)/J$

Hence, volume of copper in AC $= l(N_1 - N_2)I_1/J$

$$= lN_1(1 - n)I_1/J$$

and volume of copper in CB $= lN_2(I_2 - I_1)/J$

$$= lN_2I_2(1 - n)/J$$
$$= lN_1I_1(1 - n)/J$$

$\therefore$ total volume of copper in auto-transformer $\left.\right\} = 2lN_1I_1(1 - n)/J.$

If the two-winding transformer to perform the same duty has the same voltage per turn and therefore the same flux, we can assume the mean length per turn to remain unaltered. Hence, for the same current density J,

total volume of copper in two-winding transformer $\left.\right\} = lN_1I_1/J + lN_2I_2/J$

$$= 2lN_1I_1/J$$

Hence, $\dfrac{\text{volume of copper in auto-transformer}}{\text{volume of copper in two-winding transformer}}$

$$= \frac{2lN_1I_1(1 - n)/J}{2lN_1I_1/J} = 1 - n \qquad (14.23)$$

and saving of copper effected by using an auto-transformer $\left.\right\} = n \times \left(\begin{array}{c}\text{volume of copper in the}\\ \text{two-winding transformer}\end{array}\right)$

Thus, if $n = 0.1$, saving of copper is only 10 per cent; but if $n = 0.9$, saving of copper is 90 per cent. Hence the nearer the ratio of transformation is to unity, the greater is the economy of copper. Also, for the same current density in the windings and the same peak values of the flux and of the flux density, the I^2R loss in the auto-transformer is lower and the efficiency higher than in the two-winding transformer.

Auto-transformers are mainly used for (*a*) interconnecting systems that are operating at roughly the same voltage and (*b*) starting cage-type induction motors (section 18.11). Should an auto-transformer be used to supply a low-voltage system from a high-voltage system, it is essential to earth the common connection, for example, B in fig. 14.25, otherwise there is a risk of serious shock. In general, however, an auto-transformer should not be used for interconnecting high-voltage and low-voltage systems.

14.21 Current transformers

It is difficult to construct ammeters and the current coils of watt-meters, watthour-meters and relays to carry alternating currents greater than about 100 A. Furthermore, if the voltage of the system

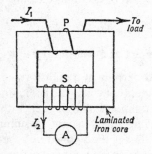

Fig. 14.26 A current transformer.

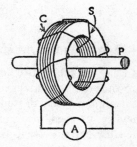

Fig. 14.27 A bar-primary
current transformer.

exceeds 500 V, it is dangerous to connect such instruments directly to the high-voltage conductors. These difficulties are overcome by using current transformers. Fig. 14.26 shows an ammeter A supplied through a current transformer. The ammeter is usually arranged to give full-scale deflection with 5 A, and the ratio of the primary to secondary turns must be such that full-scale ammeter reading is obtained with full-load current in the primary. Thus, if the primary has 4 turns and the full-load primary current is 50 A, the full-load primary ampere-turns are 200; consequently, to circulate 5 A in the secondary, the number of secondary turns must be 200/5, namely 40.

If the number of primary turns were reduced to *one* and the secondary winding had 40 turns, the primary current to give full-scale reading of 5 A on the ammeter would be 200 A. Current transformers having a single-turn primary are usually constructed as shown in fig. 14.27, where P represents the primary conductor passing through the centre of a laminated iron ring C. The secondary winding S is wound uniformly around the ring.

The secondary circuit of a current transformer must on no account be opened while the primary winding is carrying a current, since all the primary ampere-turns would then be available to produce flux. The iron loss due to the high flux density would cause excessive heating of the core and windings, and a dangerously high e.m.f. might be induced in the secondary winding. Hence if it is desired to remove the ammeter from the secondary circuit, the secondary winding must first be short circuited. This will not be accompanied by an excessive secondary current, since the latter is proportional to the primary current; and since the primary winding is in *series* with the load, the primary current is determined by the value of the load and not by that of the secondary current.

14.22 Waveform of the magnetizing current of a transformer

In fig. 14.3, the phasor for the no-load current of a transformer is shown leading the magnetic flux. A student may ask: is the flux not a maximum when this current is a maximum? The answer is that they are at their maximum values at the same instant (assuming the eddy-current loss to be negligible); but if the applied voltage is sinusoidal, then the magnetizing current of an iron-core transformer is not sinusoidal, and a non-sinusoidal quantity cannot be represented by a phasor.

Suppose the relationship between the flux and the magnetizing current for the iron core to be represented by the hysteresis loop in fig. 14.28(*a*). Also, let us assume that the waveform of the flux is sinusoidal as shown by the dotted curve in fig. 14.28(*b*). It was shown in section 14.3 that when the flux is sinusoidal, the e.m.f. induced in the primary is also sinusoidal and lags the flux by a quarter of a cycle. Hence the voltage applied to the primary must be sinusoidal

and leads the flux by a quarter of a cycle as shown in fig. 14.28(*b*).
At instant O in fig. 14.28(*b*), the flux is zero and the magnetizing
current is OA. At instant B, the flux is OB in fig. 14.28(*a*), and

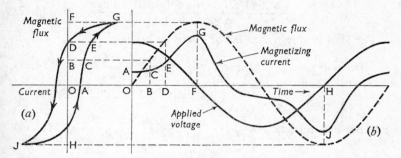

Fig. 14.28. Waveform of magnetizing current.

the current is BC. When the flux is at its maximum value OF, the
current is also at its maximum value FG. Thus, by projecting from
fig. 14.28(*a*) to the flux curve in fig. 14.28(*b*) and erecting ordinates
representing the corresponding values of the current, the waveform
of the magnetizing current can be derived. It can be shown that this
waveform contains a sinusoidal component having the same frequency as the supply voltage and referred to as the *fundamental*
component, together with other sinusoidal components having frequencies that are odd multiples of the supply frequency. The fundamental current component lags the applied voltage by an angle ϕ
that is a little less than 90°, and

hysteresis loss = r.m.s. fundamental current component × r.m.s.
voltage × cos ϕ.

Summary of important formulae

$$\frac{V_2}{V_1} \simeq \frac{N_2}{N_1} \simeq \frac{I_1}{I_2} \tag{14.3}$$

$$E_1 = 4 \cdot 44 N_1 f \Phi_\mathrm{m} \tag{14.4}$$

$$E_2 = 4 \cdot 44 N_2 f \Phi_\mathrm{m} \tag{14.5}$$

Equivalent resistance referred$\left.\right\}$ $= R_\mathrm{e} = R_1 + R_2 (V_1/V_2)^2$ (14.9)
to primary

Equivalent leakage reactance$\left.\right\}$ $= X_\mathrm{e} = X_1 + X_2 (V_1/V_2)^2$ (14.10)
referred to primary

$$\text{Equivalent impedance referred} \atop \text{to primary} \Bigg\} = Z_e = \sqrt{(R_e^2 + X_e^2)} \qquad (14.11)$$

Per-unit voltage regulation of a transformer

$$= \frac{\text{no-load voltage} - \text{full-load voltage}}{\text{no-load voltage}} \qquad (14.12)$$

For load having power factor $= \cos \phi_2$ lagging,

$$\text{per-unit voltage regulation} = \frac{I_1 Z_e \cos (\phi_e - \phi_2)}{V_1} \qquad (14.14)$$

$$= \frac{I_1(R_e \cos \phi_2 + X_e \sin \phi_2)}{V_1} \quad (14.15)$$

$$\text{and per-unit efficiency} = \frac{I_2 V_2 \times \text{power factor}}{(I_2 V_2 \times \text{p.f.}) + P_c + I_1^2 R_1 + I_2^2 R_2} \quad (14.16)$$

$$= \left(1 - \frac{\text{losses}}{\text{input power}}\right) \qquad (14.17)$$

Condition for maximum efficiency:

$$\text{copper loss} = \text{iron loss} \qquad (14.19)$$

From open-circuit and short-circuit tests:

Per-unit efficiency at $n \times$ full load

$$= \frac{n \times \text{full-load VA} \times \text{p.f.}}{n \times \text{full-load VA} \times \text{p.f.} + P_{oc} + n^2 P_{sc}} \qquad (14.21)$$

$$\text{and per-unit voltage regulation} = \frac{V_{sc} \cos (\phi_e - \phi_2)}{V_1} \qquad (14.22)$$

EXAMPLES 14

1. The design requirements of a 6600-V/400-V, 50-Hz, single-phase, core-type transformer are: approximate e.m.f./turn, 15 V; maximum flux density, 1·5 teslas. Find a suitable number of primary and secondary turns, and the net cross-sectional area of the core.

(S.A.N.C., O.2)

2. The primary winding of a single-phase transformer is connected to a 230-V, 50-Hz supply. The secondary winding has 1500 turns. If the maximum value of the core flux is 0·002 07 Wb, determine: (*a*) the number of turns on the primary winding; (*b*) the secondary induced voltage; (*c*) the net cross-sectional core area if the flux density has a maximum value of 0·465 tesla. (E.M.E.U., O.2)

3. A 6600-V/250-V, 50-Hz, single-phase, core-type transformer has a core section 25 cm $\times$ 25 cm. Allowing for a space factor of 0·9, find a suitable number of primary and secondary turns, if the flux density is not to exceed 1·2 teslas. (S.A.N.C., O.2)

4. A single-phase 50-Hz transformer has 80 turns on the primary winding and 400 turns on the secondary winding. The net cross-sectional area of the core is 200 cm². If the primary winding is connected to a 240-V, 50-Hz supply, determine: (*a*) the e.m.f. induced in the secondary winding; (*b*) the maximum value of the flux density in the core.   (N.C.T.E.C., O.2)

5. A 50-kVA single-phase transformer has a turns ratio of 300/20. The primary winding is connected to a 2200-V, 50-Hz supply. Calculate: (*a*) the secondary voltage on no load; (*d*) the approximate values of the primary and secondary currents on full load; (*c*) the maximum value of the flux.

6. A 200-kVA, 3300-V/240-V, 50-Hz, single-phase transformer has 80 turns on the secondary winding. Assuming an ideal transformer, calculate: (*a*) the primary and secondary currents on full load; (*b*) the maximum value of the flux; (*c*) the number of primary turns.

7. The following data applies to a single-phase transformer:

peak flux density in the core	$= 1.41$ T,
net core area	$= 0.01$ m²
current density in conductors	$= 2.5$ MA/m²,
conductor diameter	$= 2.0$ mm,
primary supply (assume sinusoidal)	$= 200$ V, 50 Hz.

   Calculate the kVA rating of the transformer and the number of turns on the primary winding.   (U.L.C.I., O.2)

8. A transformer for a radio receiver has a 230-V, 50-Hz primary and three secondary windings as follows: a 1000-V winding with a centre tapping, a 4-V winding with a centre tapping and a 6·3-V winding. The net cross-sectional area of the core is 14 cm². Calculate the number of turns on each winding if the maximum flux density is not to exceed 1 tesla.

   Note. Since the number of turns on a winding must be an integer, it is best to calculate the number of turns on the low-voltage winding first. In this question, the 4-V winding must have an even number of turns since it has a centre tapping.

9. The primary of a certain transformer takes 1 A at a power factor of 0·4 when connected across a 200-V, 50-Hz supply and the secondary is on open circuit. The number of turns on the primary is twice that on the secondary. A load taking 50 A at a lagging power factor of 0·8 is now connected across the secondary.

   Sketch, and explain briefly, the phasor diagram for this condition, neglecting voltage drops in the transformer. What is now the value of the primary current?   (App. El., L.U.)

10. A 4 : 1 ratio step-down transformer takes 1 A at 0·15 power factor on no load. Determine the primary current and power factor when the transformer is supplying a load of 25 A at 0·8 power factor lag. Ignore internal voltage drops.   (W.J.E.C., O.2)

11. A 3300-V/240-V, single-phase transformer, on no load, takes 2 A at power factor 0·25. Determine graphically, or otherwise, the primary

current and power factor when the transformer is supplying a load of 60 A at power factor 0·9 leading. (App. El., L.U.)

12. A three-phase transformer has its primary winding delta-connected and its secondary winding star-connected. The number of turns per phase on the primary is 4 times that on the secondary, and the secondary line voltage is 440 V. A balanced load of 20 kW, at power factor 0·8, is connected across the secondary terminals. Assuming an ideal transformer, calculate the primary voltage and the phase and line currents on the secondary and primary sides. Sketch a circuit diagram and indicate the values of the voltages and currents on the diagram.
(App. El., L.U.)

13. A 50-Hz, three-phase, core-type transformer is connected star-delta and has a line voltage ratio of 6600/440 V. The cross-section of the core is square with a circumscribing circle of 0·6 m diameter. If the maximum flux density is about 1·2 T, calculate the number of turns per phase on the low-voltage and on the high-voltage windings. Assume the insulation to occupy 10 per cent of the gross core area.

14. If three transformers, each with a turns ratio of 12 : 1, are connected star–delta and the primary line voltage is 6600 V, what is the value of the secondary no-load voltage?

If the transformers are reconnected delta–star with the same primary voltage, what is the value of the secondary line voltage?

15. A 440-V, three-phase supply is connected through a three-phase loss-free transformer of 1 : 1 ratio, which has its primary connected in mesh and secondary in star, to a load comprising three 20-Ω resistors connected in mesh. Calculate the currents in the transformer windings, in the resistors and in the lines to the supply and the load. Find also the total power supplied and the power dissipated by each resistor.

16. The no-load current of a transformer is 5·0 A at 0·3 power factor when supplied at 230 V, 50 Hz. The number of turns on the primary winding is 200. Calculate: (i) the maximum value of the flux in the core; (ii) the core loss; (iii) the magnetizing current. (N.C.T.E.C.)

17. Calculate: (a) the number of turns required for a choke to absorb 200 V on a 50-Hz circuit; (b) the length of airgap required if the coil is to take a magnetizing current of 3 A (r.m.s.); and (c) the phase difference between the current and the terminal voltage. Mean length of iron path, 500 mm; maximum flux density in core, 1 T; sectional area of core, 3000 mm^2; maximum magnetic field strength for the iron, 250 A/m; iron loss, 1·7 W/kg; density of iron, 7800 kg/m^3. Neglect the resistance of the winding and any magnetic leakage and fringing. Assume the current waveform to be sinusoidal.

18. Calculate the no-load current and power factor for the following 60-Hz transformer: mean length of iron path, 700 mm; maximum flux density, 1·1 T; maximum magnetic field strength for the iron, 300 A/m; iron loss, 2·4 W/kg. All the joints may be assumed equivalent to a single airgap of 0·2 mm. Number of primary turns, 120; primary voltage, 230 V. Neglect the resistance of the primary winding and assume the current waveform to be sinusoidal.

19. The ratio of turns of a single-phase transformer is 8, the resistances of the primary and secondary windings are 0·85 Ω and 0·012 Ω respectively, and the leakage reactances of these windings are 4·8 Ω and 0·07 Ω respectively. Determine the voltage to be applied to the primary to obtain a current of 150 A in the secondary when the secondary terminals are short-circuited. Ignore the magnetizing current. (App. El., L.U.)

20. A single-phase transformer operates from a 230-V supply. It has an equivalent resistance of 0·1 Ω and an equivalent leakage reactance of 0·5 Ω referred to the primary. The secondary is connected to a coil having a resistance of 200 Ω and a reactance of 100 Ω. Calculate the secondary terminal voltage. The secondary winding has four times as many turns as the primary.

21. A 10-kVA single-phase transformer, for 2000 V/400 V at no load, has resistances and leakage reactances as follows. *Primary winding*: resistance, 5·5 Ω; reactance, 12 Ω. *Secondary winding*: resistance, 0·2 Ω; reactance, 0·45 Ω. Determine the approximate value of the secondary voltage at full load, 0·8 power factor (lagging), when the primary supply voltage is 2000 V. (App. El., L.U.)

22. Calculate the voltage regulation at 0·8 lagging power factor for a transformer which has an equivalent resistance of 2 per cent and an equivalent leakage reactance of 4 per cent.

23. A 75-kVA transformer, rated at 6600 V/230 V on no load, requires 310 V across the primary to circulate full-load currents on short-circuit, the power absorbed being 1·6 kW. Determine (a) the percentage voltage regulation and (b) the full-load secondary terminal voltage for power factors of (i) unity, (ii) 0·8 lagging and (iii) 0·8 leading.

If the input power to the transformer on no load is 0·9 kW, calculate the per-unit efficiency at full load and at half load for power factor 0·8 and find the kVA at which the efficiency is a maximum.

24. The primary and secondary windings of a 30-kVA, 6000-V/230-V transformer have resistances of 10 Ω and 0·016 Ω respectively. The total reactance of the transformer referred to the primary is 23 Ω. Calculate the percentage regulation of the transformer when supplying full-load current at a power factor of 0·8 lagging. (W.J.E.C., O.2)

25. A 50-kVA, 6360-V/240-V transformer is tested on open and short circuit to obtain its efficiency, the results of the test being as follows:

Open circuit: primary voltage, 6360 V; primary current, 1 A; power input, 2 kW.

Short circuit: voltage across primary winding, 180 V; current in secondary winding, 175 A; power input, 2 kW.

Find the efficiency of the transformer when supplying full load at a power factor of 0·8 lagging and draw a phasor diagram (neglecting impedance drops) for this condition. (E.M.E.U., O.2)

26. A 240-V/400-V single-phase transformer absorbs 35 W when its primary winding is connected to a 240-V, 50-Hz supply, the secondary being on open circuit.

When the primary is short-circuited and a 10-V, 50-Hz supply is connected to the secondary winding, the power absorbed is 48 W when the current has the full-load value of 15 A.

Estimate the efficiency of the transformer at half load, 0·8 power factor lagging.   (U.E.I., O.2)

27. A 1-kVA transformer has an iron loss of 15 W and a full-load copper loss of 20 W. Calculate the full-load efficiency, assuming the power factor to be 0·9.

An ammeter is scaled to read 5 A, but it is to be used with a current transformer to read 15 A. Draw a diagram of connections for this, giving terminal markings and currents.   (N.C.T.E.C., O.2)

28. Discuss fully the energy losses in single-phase transformers. Such a transformer working at unity power factor has an efficiency of 90 per cent at both one-half load and at the full load of 500 W. Determine the efficiency at 75 per cent of full load.   (App. El., L.U.)

29. A single-phase transformer is rated at 10 kVA, 230 V/100 V. When the secondary terminals are open-circuited and the primary winding is supplied at normal voltage (230 V), the current input is 2·6 A at a power factor of 0·3. When the secondary terminals are short-circuited, a voltage of 18 V applied to the primary causes the full-load current (100 A) to flow in the secondary, the power input to the primary being 240 W. Calculate: (a) the efficiency of the transformer at full load, unity power factor; (b) the load at which maximum efficiency occurs; (c) the value of the maximum efficiency.   (App. El., L.U.)

30. A 400-kVA transformer has an iron loss of 2 kW and the maximum efficiency at 0·8 power factor occurs when the load is 240 kW. Calculate: (a) the maximum efficiency at unity power factor, and (b) the efficiency on full load at 0·71 power factor.   (W.J.E.C., O.2)

31. A 40-kVA transformer has a core loss of 450 W and a full-load copper loss of 850 W. If the power factor of the load is 0·8, calculate: (a) the full-load efficiency; (b) the maximum efficiency; and (c) the load at which maximum efficiency occurs.   (App. El., L.U.)

32. Each of two transformers, A and B, has an output of 40 kVA. The core losses in A and B are 500 and 250 W respectively, and the full-load copper losses are 500 and 750 W respectively. Tabulate the losses and efficiencies at quarter, half and full load for a power factor of 0·8. For each transformer, find the load at which the efficiency is a maximum.

33. If the transformers referred to in Q. 32 be used for supplying a lighting load (unity power factor) and have their primaries permanently connected to the supply system, compare the all-day efficiencies of the two transformers, assuming the output to be 4 hours at full load, 8 hours at half load and the remaining hours on no load. What would be the saving effected in 12 weeks (7 days per week) with the better transformer if the charge for electrical energy be 2p/kW·h?

Note. All-day efficiency = $\dfrac{\text{output energy in kW·h in 24 hours.}}{\text{input energy in kW·h in 24 hours}}$

34. A 100-kVA lighting transformer has a full-load loss of 3 kW, the losses being equally divided between the iron and copper. During one day the

transformer operates on full load for 3 hours, on half load for 4 hours, the output being negligible for the remainder of the day. Calculate the all-day efficiency.

35. What is meant by the term *magnetic hysteresis*? Explain why, when iron is subjected to alternating magnetization, energy losses occur due to both hysteresis and eddy currents. The iron loss in a transformer core at normal flux density was measured at frequencies of 30 and 50 Hz, the results being 30 W and 54 W respectively. Calculate: (a) the hysteresis loss; (b) the eddy-current loss at 50 Hz.   (App. El., L.U.)

Note. For a given specimen, with maximum flux density B_m, hysteresis loss $= k_h f B_m{}^x$ and eddy current loss $= k_e f^2 B_m{}^2$ (section 9.1), where k_h k_e and x are constants.

$$\text{total iron loss} = P = k_h f B_m{}^x + k_e f^2 B_m{}^2$$
For a given B_m, $$P = k'_h f + k'_e f^2$$

Hence, if the total iron loss for a given maximum flux density is known at two frequencies, k'_h and k'_e can be calculated and the hysteresis and eddy-current components of the iron loss determined at any desired frequency.

36. Explain why the ferrous magnetic circuits subject to alternating magnetism are usually laminated and give examples of typical core construction.

The total iron loss in a 460-V, 50-Hz single-phase transformer is 2400 W. When a 230-V, 25-Hz supply is applied, the total iron loss is 800 W. Calculate the eddy-current and hysteresis losses at normal voltage and frequency.   (E.M.E.U., O.2)

37. The iron loss in a certain transformer is 80 W at 25 Hz and 204 W at 60 Hz, the maximum flux density being the same. Calculate the total iron loss at 100 Hz at the same maximum flux density.

38. The following table gives the relationship between the flux and the magnetizing current in the primary winding of a transformer:

Current (A)	0	0·25	0·5	0·75	1·0	1·25	1·5	1·75
Flux (mWb)	−1·76	−1·48	−0·8	1·0	1·56	1·88	2·08	2·17
Current (A)	2·0	1·5	1·0	0·5	0			
Flux (mWb)	2·24	2·2	2·14	2·04	1·76			

Plot the above values and derive the waveform of the magnetizing current of the transformer, assuming the waveform of the flux to be sinusoidal and to have a peak value of 2·24 mWb.

39. Derive an expression for the saving in copper effected by using an auto-transformer instead of a two-winding transformer.

The primary and secondary voltages of an auto-transformer are 500 V and 400 V respectively. Show, with the aid of a diagram, the current distribution in the windings when the secondary current is 100 A, and calculate the economy of copper in this particular case.

What are the disadvantages of auto-transformers?

(N.C.T.E.C.)

40. An auto-transformer is required to step up a voltage from 220 to

250 V. The total number of turns is 2000. Determine: (*a*) the position of the tapping point; (*b*) the approximate value of the current in each part of the winding when the output is 10 kVA; and (*c*) the economy in copper over the two-winding transformer having the same peak flux and the same mean length per turn.

CHAPTER 15

Alternator Windings

15.1 General arrangement of alternators

Let us first consider why alternators are usually constructed with a stationary a.c. winding and rotating poles. Suppose we have a 20 000-kVA, 11-kV, three-phase alternator; then from expression (13.6),

$$20\,000 = 1.73 \times I_L \times 11$$

$$\therefore \qquad \text{line current} = I_L = 1050 \text{ A.}$$

Hence, if the machine was constructed with stationary poles and a rotating three-phase winding, three slip-rings would be required, each capable of dealing with 1050 A, and the insulation of each ring together with that of the brushgear would be subjected to a working voltage of 11/1.73, namely 6.35 kV. Further, it is usual to connect alternator windings in star and to join the star-point through a suitable resistor to a metal plate embedded in the ground so as to make good electrical contact with earth; consequently a fourth slip-ring would be required.

By using a stationary a.c. winding and a rotating field system, only two slip-rings are necessary and these have to deal with only the exciting current. Assuming the power required for exciting the poles of the above machine to be 150 kW and the voltage to be 400 V,

$$\text{exciting current} = 150 \times 1000/400 = 375 \text{ A.}$$

In other words, the two slip-rings and brushgear would have to deal with only 375 A and be insulated for merely 400 V. Hence, by using a stationary a.c. winding and rotating poles, the construction is considerably simplified and the slip-ring losses are reduced.

Further advantages of this arrangement are: (a) the extra space available for the a.c. winding makes it possible to use more insulation and to enable voltages up to 30 kV to be generated; (b) with the simpler and more robust mechanical construction of the rotor, a higher speed is possible, so that a greater output is obtainable from a machine of given dimensions.

The slots in the laminated stator core of an alternator are usually semi-enclosed, as shown in fig. 15.1, so as to distribute the magnetic flux as uniformly as possible in the airgap, thereby minimizing the ripple that would appear in the e.m.f. waveform if open slots were used.

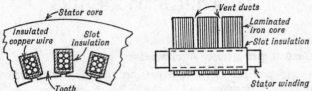

Fig. 15.1 Portion of an alternator stator.

In section 10.2 it was explained that if an alternator has p pairs of poles and the speed is n revolutions/second,

$$\text{frequency} = f = np \tag{15.1}$$

Hence for a 50-Hz supply, a two-pole alternator must be driven at 3000 rev/min, a four-pole alternator at 1500 rev/min, etc.

15.2 Types of rotor construction

Alternators can be divided into two categories: (*a*) those with salient or projecting poles; and (*b*) those with cylindrical rotors.

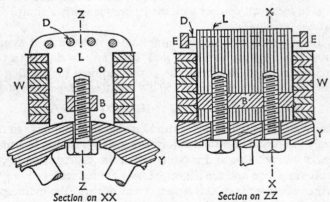

Section on XX   Section on ZZ

Fig. 15.2 Portion of a salient-pole rotor.

The salient-pole construction is used in comparatively small machines and machines driven at a relatively low speed. For instance, if a 50-Hz alternator is to be driven by an oil engine at, say,

375 rev/min, then, from expression (15.1), the machine must have 16 poles; and to accommodate all these poles, the alternator must have a comparatively large diameter. Since the output of a machine is roughly proportional to its volume, such an alternator would have a small axial length. Fig. 15.2 shows one arrangement of salient-pole construction. The poles are made of fairly thick steel laminations, L, riveted together and bolted to a steel yoke wheel Y, a bar of mild steel B being inserted to improve the mechanical strength. The exciting winding W is usually an insulated copper strip wound on edge,

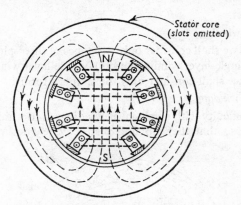

Fig. 15.3 A cylindrical rotor.

the coil being held firmly between the pole tips and the yoke wheel. The pole tips are well rounded so as to make the flux distribution around the periphery nearer a sine wave and thus improve the waveform of the generated e.m.f. Copper rods D, short-circuited at each end by copper bars E, are usually inserted in the pole shoes, their function being to improve the parallel operation of alternators.

In this country, alternators are generally driven by steam turbines, and these are essentially high-speed machines. The centrifugal force on a high-speed rotor is enormous: for instance, a mass of 1 kg on the outside of a rotor of 1 m diameter, rotating at a speed of 3000 rev/min, has a centrifugal force $(= mv^2/r)$ of about 50 kN acting upon it. To withstand such a force the rotor is usually made of a solid steel forging with longitudinal slots cut as indicated in fig. 15.3, which shows a two-pole rotor with 8 slots and 2 conductors per slot. In an actual rotor there are more slots and far more conductors per slot; and the winding is in the form of insulated copper strip held

securely in position by phosphor-bronze wedges. The regions forming the centres of the poles are usually left unslotted. The horizontal
dotted lines joining the conductors in fig. 15.3 represent the end connections. If the rotor current has the direction represented by the
dots and crosses in fig. 15.3, the flux distribution is indicated by the
light dotted lines. In addition to its mechanical robustness, this
cylindrical construction has the advantage that the flux distribution
around the periphery is nearer a sine wave than is the case with the
salient-pole machine. Consequently, a better e.m.f. waveform is
obtained.

15.3 Stator windings

In this book we shall consider only two types of three-phase winding,
namely (*a*) single-layer winding and (*b*) double-layer winding; and of
these types we shall consider only the simplest forms.

Single-layer winding

The main difficulty with single-layer windings is to arrange the end-
connections so that they do not obstruct one another. Fig. 15.4

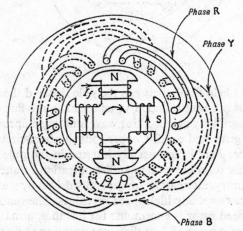

Fig. 15.4 End-connections of a three-phase single-layer winding.

shows one of the most common methods of arranging these end-
connections for a four-pole, three-phase alternator having 2 slots per
pole per phase, i.e. 6 slots per pole or a total of 24 slots. In fig. 15.4,
all the end-connections are shown bent outwards for clearness; but
in actual practice the end-connections are usually shaped as shown

in fig. 15.5 and in section in fig. 15.6. This method has the advantage that it requires only two shapes of end-connections, namely those marked C in fig. 15.6, which are brought straight out of the slots and bent so as to lie on a cylindrical plane, and those marked D. The latter, after being brought out of the slots, are bent outwards roughly

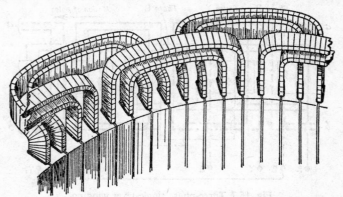

Fig. 15.5 End-connections of a three-phase single-layer winding.

at right-angles, before being again bent to form an arch alongside the core.

The connections of the various coils are more easily indicated by means of the developed diagram of fig. 15.7. The heavily lined rectangles (full and dotted lines) represent the coils, each coil consisting of a number of turns; and the thin lines—other than those representing the poles—indicate the connections between the various coils. The width of the pole face has been made two-thirds of the pole pitch, a pole pitch being the distance between the centres of adjacent poles. The poles in fig. 15.7 are assumed to be *behind* the winding and moving towards the right. From the right-hand rule (section 2.10)—bearing in mind that the thumb represents the

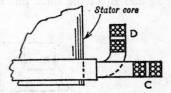

Fig. 15.6 Sectional view of end-connections.

direction of *motion of the conductor relative to the flux*, namely towards the left in fig. 15.7—the e.m.f.s in the conductors opposite the poles are as indicated by the arrowheads. The connections between the groups of coils forming any one phase must be such that all the e.m.f.s are assisting one another.

Since the alternator has 6 slots per pole and since the rotation of the poles through one pole pitch corresponds to half a cycle of the e.m.f. wave or 180 electrical degrees, it follows that the spacing between two adjacent slots corresponds to 180°/6, namely 30 electrical

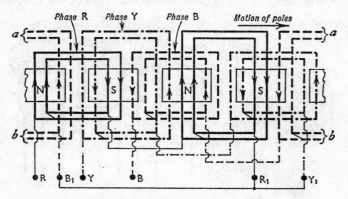

Fig. 15.7 Three-phase single-layer winding.

degrees. Hence, if the wire forming the beginning of the coil occupying the first slot is taken to the 'red' terminal R, the connection to the 'yellow' terminal Y must be a conductor from a slot 4 slot-pitches ahead, namely from the 5th slot, since this allows the e.m.f. in phase Y to lag the e.m.f. in phase R by 120°. Similarly, the connection to the 'blue' terminal B must be taken from the 9th slot in order that the e.m.f. in phase B may lag the e.m.f. in phase Y by 120°. Ends R_1, Y_1 and B_1 of the three phases can be joined together to form the

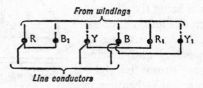

Fig. 15.8 Mesh-connection of windings.

neutral point of a star-connected system. If the windings are to be mesh-connected, end R_1 of phase R is joined to the beginning of Y, end Y_1 to the beginning of B and end B_1 to the beginning of R, as shown in fig. 15.8.

Double-layer winding

The general arrangement of a double-layer winding has been described in Chap. 6, and it is therefore only necessary to indicate the modification required to give a three-phase supply.

Let us consider a four-pole three-phase alternator with 2 slots per pole per phase and 2 conductors per slot. Fig. 15.9 shows the simplest arrangement of the end-connections of one phase, the thick lines representing the conductors (and their end-connections) forming, say, the outer layer and the thin lines representing conductors forming the inner layer of the winding. The coils are assumed full-pitch,

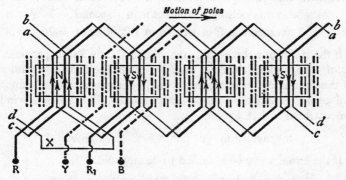

Fig. 15.9 One phase of a three-phase double-layer winding.

i.e. the spacing between the two sides of each turn is exactly a pole pitch. The main feature of the end-connections of a double-layer winding is the strap X, which enables the coils of any one phase to be connected so that all the e.m.f.s of that phase are assisting one another.

Since there are 6 slots per pole, the phase difference between the e.m.f.s of adjacent slots is 180°/6, namely 30 electrical degrees; and since there is a phase difference of 120° between the e.m.f.s of phases R and Y, there must be 4 slot pitches between the first conductor of phase R and that of phase Y. Similarly, there must be 4 slot pitches between the first conductors of phases Y and B. Hence, if the outer conductor of the third slot is connected to terminal R, the corresponding conductor in the seventh slot, in the direction of rotation of the poles, is connected to terminal Y and that in the eleventh slot to terminal B.

In general, the single-layer winding is employed where the machine has a large number of conductors per slot, whereas the double-layer winding is more convenient when the number of conductors per slot does not exceed 8.

15.4 Expression for the e.m.f. of an alternator

Let Z = no. of conductors in series per *phase*,

Φ = useful flux per pole, in webers,

p = no. of pairs of poles

and n = speed in revolutions/second.

Magnetic flux cutting a conductor in 1 revolution = $\Phi \times 2p$

Magnetic flux cutting a conductor in 1 second = $2\Phi p \times n$

∴ average e.m.f. generated in 1 conductor = $2\Phi pn$ volts.

If the stator of a three-phase machine had only 3 slots per pole, i.e. 1 slot/p/ph,* and if the coils were full-pitch, the e.m.f.s generated in all the conductors of one phase would be in phase with one another and could therefore be added arithmetically. Hence, for a winding concentrated in 1 slot/p/ph,

average e.m.f. per phase = $Z \times 2\Phi pn$ volts.

If the e.m.f. wave be assumed to be sinusoidal,

$$\left.\begin{array}{l}\text{r.m.s. value of e.m.f. per}\\ \text{phase for 1 slot/p/ph,}\end{array}\right\} = 1{\cdot}11 \times 2Z \times np \times \Phi = 2{\cdot}22Zf\Phi \tag{15.2}$$

A winding concentrated in 1 slot per pole per phase would have two disadvantages:

(a) The size of such slots and the number of conductors per slot would be so great that it would be difficult to prevent the insulation on the conductors in the centre of the slot becoming overheated, since most of the heat generated in the slots has to flow radially outwards to the iron core.

(b) The waveform of the e.m.f. would be similar to that of the flux distribution around the inner periphery of the stator; and, in general, this would not be sinusoidal.

15.5 Effect of distributing the winding: Distribution factor

By distributing the winding in 2 or more slots per pole per phase, the number of conductors per slot is reduced, thereby reducing the tem-

* '/p/ph' is an abbreviation for 'per pole per phase'.

perature rise in the centre of the slot. Also, the e.m.f. waveform is improved; for instance, consider a three-phase winding distributed in 2 slots per pole per phase as in fig. 15.4. With a salient-pole alternator having a uniform gap-length over the greater part of the pole face, the waveform of the e.m.f. generated in one conductor is

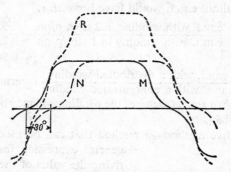

Fig. 15.10 E.m.f. waveforms of a salient-pole alternator.

represented by curve M in fig. 15.10; and that of the e.m.f. generated in the adjacent slot is represented by a similar curve N displaced 30 electrical degrees from M. The resultant waveform R is obtained by adding curves M and N; and it is evident from fig. 15.10 that curve R, though not sinusoidal, is a much closer approximation to a sine wave than the original waves M and N.

Let us consider how the magnitude of the e.m.f. is affected when the winding is distributed in 2 slots per pole per phase; and for

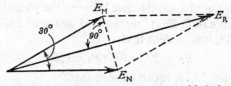

Fig. 15.11 E.m.f.s in a three-phase alternator with 2 slots/p/ph.

simplicity it will be assumed that the e.m.f. generated in each conductor is sinusoidal. The e.m.f.s, E_M and E_N, generated in coils occupying two adjacent slots M and N of any one phase are equal in magnitude but differ in phase by 30°. The resultant e.m.f. is the phasor sum of E_M and E_N and is represented by the diagonal E_R in fig. 15.11. Since the diagonals of the parallelogram bisect each other at right-angles,

$$\text{resultant e.m.f.} = E_R = 2E_M \cos 15° = 2 \times 0.966E_M$$
$$= 1.932 \, E_M.$$

Had the same number of conductors been concentrated in 1 slot/p/ph, the resultant e.m.f. would have been $2E_M$;

$$\text{hence, } \frac{\text{e.m.f. with winding in 2 slots/p/ph}}{\text{e.m.f. with winding in 1 slot/p/ph}} = \frac{1.932E_M}{2E_M}$$
$$= 0.966.$$

The ratio $\dfrac{\text{e.m.f. with a distributed winding}}{\text{e.m.f. with a concentrated winding}}$ is termed the *distribution, breadth* or *spread factor* of the winding and may be represented by the symbol k_d.

An alternative method—a method that can be used to derive a general expression for k_d—of deriving the value of the distribution factor for a three-phase winding distributed in 2 slots/p/ph is to draw the phasors representing E_M and E_N end-on as in fig. 15.12. Perpendiculars drawn at mid-points of the phasors meet at a point O, which is the centre of the circumscribing circle shown dotted. The phasors are chords of this circle, and each phasor subtends an angle of 30° at the centre, namely an angle equal to the phase difference between the phasors.

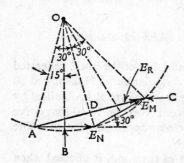

Fig. 15.12 E.m.f.s in a three-phase alternator with 2 slots/p/ph.

The resultant e.m.f., E_R, is represented by chord AC of the circle. From fig. 15.12, it is evident that:

$$E_R = 2AD = 2OA \sin 30°.$$

Also, from fig. 15.12, it can be seen that:

$$E_M = E_N = 2AB = 2OA \sin 15°$$

Hence, for a three-phase winding distributed in 2 slots/p/ph,

$$\text{distribution factor} = k_d = \frac{E_R}{2E_M} = \frac{2OA \sin 30°}{2 \times 2OA \sin 15°}$$
$$= \frac{\sin 30°}{2 \sin 15°} = 0.966.$$

In general,

if m = no. of slots per pole per phase,

       α = phase difference between e.m.f.s generated in conductors occupying adjacent slots, and

   OA = radius of the circumscribing circle for the polygon of phasors, drawn as in fig. 15.12,

angle subtended at centre of circumscribing circle by *each* phasor = α

$\therefore$ arithmetic sum of e.m.f.s = $m \times$ e.m.f. per coil
$$= m \times 2\text{OA} \sin \alpha/2.$$

Angle subtended at centre of circumscribing circle by phasor representing the resultant e.m.f. = $m\alpha$,

$\therefore$ resultant e.m.f. = $E_R = 2\text{OA} \sin m\alpha/2$,

$\therefore$ distribution factor = $k_d = \dfrac{\text{e.m.f. with distributed winding}}{\text{e.m.f. with concentrated winding}}$

$$= \frac{\text{phasor sum of e.m.f.s}}{\text{arithmetic sum of e.m.f.s}}$$

$$= \frac{2\text{OA} \sin m\alpha/2}{m \times 2\text{OA} \sin \alpha/2}$$

$$= \frac{\sin m\alpha/2}{m \sin \alpha/2} \tag{15.3}$$

Example 15.1 *Calculate the value of the distribution factor for a three-phase winding of a 4-pole alternator having 36 slots.*

$$\text{No. of slots/pole} = 36/4 = 9,$$

$\therefore$ no. of slots/p/ph = $m = 9/3 = 3$.

$$\alpha = 180°/9 = 20°,$$

$\therefore$ $m\alpha/2 = 3 \times 20°/2 = 30°.$

Substituting in expression (15.3), we have:

$$\text{distribution factor} = k_d = \frac{\sin 30°}{3 \sin 10°} = 0.960.$$

Example 15.2 *Calculate the distribution factor for a single-phase alternator having 6 slots/pole (a) when all the slots are wound and (b) when only four adjacent slots per pole are wound, the remaining slots being unwound.*

*

(*a*) When all the slots are wound,

$$m = 6; \ \alpha = 180°/6 = 30°; \text{ and } m\alpha/2 = 6 \times 30°/2 = 90°;$$

$$\therefore \qquad k_d = \frac{\sin 90°}{6 \sin 15°} = 0.644.$$

(*b*) When only 4 adjacent slots per pole are wound,

$$m = 4; \ \alpha = 30°; \text{ and } m\alpha/2 = 4 \times 30°/2 = 60°;$$

$$\therefore \qquad k_d = \frac{\sin 60°}{4 \sin 15°} = 0.837.$$

It follows from the above values of k_d that if the number of conductors per slot remained the same:

$$\frac{\text{e.m.f. with all slots wound}}{\text{e.m.f. with 4 adjacent slots/pole wound}} = \frac{6 \times 0.644}{4 \times 0.837} = 1.15.$$

This means that an increase of 50 per cent in the number of conductors gives only 15 per cent increase of e.m.f. This is the reason why it is customary to wind only about two-thirds of the slots in a single-phase alternator.

15.6 Short-pitch winding: Pitch factor

The waveform of the resultant e.m.f. generated in an alternator may be improved by making the coil pitch less than a pole pitch, as in

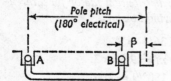

Fig. 15.13 Short-pitch coil.

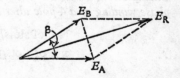

Fig. 15.14 E.m.f.s in the coil-sides of a short-pitch coil.

fig. 15.13. This practice is only possible with the two-layer type of winding shown in fig. 15.9.

With a full-pitch coil, the e.m.f.s generated in the two sides are in phase with each other as far as the resultant e.m.f. of the coil is concerned. When the coil is short-pitch by an angle β electrical degrees, as shown in fig. 15.13, the e.m.f.s generated in coil sides A and B differ in phase by an angle β, and can be represented by the phasors E_A and E_B respectively in fig. 15.14. Since the diagonals of the parallelogram drawn on E_A and E_B bisect at right-angles,

$$\text{resultant e.m.f.} = E_R = 2E_A \cos \beta/2,$$

$\therefore$ *pitch* or *coil-span factor* $= k_p = \dfrac{\text{e.m.f with short-pitch coil}}{\text{e.m.f. with full-pitch coil}}$

$$= \frac{2E_A \cos \beta/2}{2E_A} = \cos \beta/2 \qquad (15.4)$$

For a full-pitch winding, $k_p = 1\cdot0$.

Example 15.3 *An alternator has 9 slots per pole. If each coil spans 8 slot pitches, what is the value of the pitch factor?*

Since the winding pitch is one slot pitch less than the pole pitch,

$$\beta = 180°/9 = 20°.$$

Substituting in expression (15.4), we have:

$$\text{pitch factor} = k_p = \cos 10° = 0\cdot985.$$

15.7 General expression for the e.m.f. of an alternator

If k_d = distribution factor of the winding
and k_p = pitch factor of the winding,

then from expression (15.2),

$$\text{r.m.s. value of e.m.f./phase} = 2\cdot22\, k_d k_p Z f\Phi \qquad (15.5)$$

Example 15.4 *A three-phase star-connected alternator on open circuit is required to generate a line voltage of 3600 V, 50 Hz, when driven at 500 rev/min. The stator has 3 slots/p/ph and 10 conductors per slot. Calculate:* (a) *the number of poles and* (b) *the useful flux per pole. Assume all the conductors per phase to be connected in series and the coils to be full-pitch.*

(*a*) From expression (15.1), $\qquad 50 = 500 \times p/60$

$\therefore \qquad\qquad\qquad$ no. of poles $= 2p = 12$.

(*b*) $\qquad\qquad$ No. of slots per phase $= 3 \times 12 = 36$

$\therefore \qquad$ no. of conductors per phase $= 36 \times 10 = 360$

$$\text{E.m.f. per phase} = \frac{3600}{1\cdot73} = 2080 \text{ V}.$$

From example 15.1, $k_\mathrm{d} = 0.96$ for 3 slots/p/ph.
For full-pitch coils, $k_\mathrm{p} = 1.0$.

Substituting in expression (15.5), we have:

$$2080 = 2.22 \times 0.96 \times 1.0 \times 360 \times 50 \times \Phi$$
$$\therefore \qquad \Phi = 0.0543 \text{ Wb.}$$

Summary of important formulae

$$f = np \qquad (15.1)$$

R.m.s. value of e.m.f. per phase $= 2.22 k_\mathrm{d} k_\mathrm{p} Zf\Phi \qquad (15.5)$

$$\text{Distribution factor} = \frac{\text{e.m.f. with a distributed winding}}{\text{e.m.f. with a concentrated winding}}$$

$$\text{Pitch factor} = \frac{\text{e.m.f. with short-pitch coil}}{\text{e.m.f. with full-pitch coil}}.$$

EXAMPLES 15

1. Explain why the e.m.f. generated in a conductor of an alternating-current generator is seldom sinusoidal.

   A rectangular coil of 55 turns, carried by a spindle placed at right-angles to a magnetic field of uniform density, is rotated at a constant speed. The mean area per turn is 300 cm². Calculate (a) the speed in order that the frequency of the generated e.m.f. may be 60 Hz, (b) the density of the magnetic field if the r.m.s. value of the generated e.m.f. is 10 V. (App. El., L.U.)

2. The flux density in an alternator air-gap at equal intervals is as follows:

Angle (elect. degrees)	0	15	30	45	60	75	90
Flux density (teslas)	0	0.1	0.4	0.9	1.0	1.0	1.0

   Derive the waveform of the resultant e.m.f. of a single-phase alternator having 6 slots/pole when the winding is (a) concentrated in 1 slot per pole, (b) distributed in 2 adjacent slots/pole and (c) distributed in 4 adjacent slots/pole.

3. The distribution of flux density in an alternator is trapezoidal, being uniform under the pole faces and decreasing uniformly to zero at points midway between the poles. The ratio of pole arc to pole pitch is 0.6. The machine has 4 slots per pole. Derive curves representing the waveform of the generated e.m.f. when the winding is (a) con-

centrated in one slot per pole, (b) distributed in two adjacent slots per pole and (c) distributed in the four slots per pole.

4. In a certain alternator, the flux density may be assumed uniform under the poles and zero between the poles. The ratio of pole arc to pole pitch is 0·7. Calculate the form factor of the e.m.f. generated in a full-pitch coil.

5. The field-form of an alternator taken from the pole centre line, in electrical degrees, is given below, the points being joined by straight lines. Determine the form factor of the e.m.f. generated in a full-pitch coil.

Distance from pole centre, degrees:

0	20	45	60	75	90	105	120	135	
Flux density, teslas:									etc.
0·7	0·7	0·6	0·15	0	0	0	−0·15	−0·6	

6. A stator has two poles of arc equal to two-thirds of the pole pitch, producing a uniform radial flux of density 1 T. The length and diameter of the armature are both 0·2 m and the speed of rotation is 1500 rev/min. Neglecting fringing, draw to scale the waveform of e.m.f. generated in a single fully-pitched armature coil of 10 turns. If the coil is connected through slip-rings to a resistor of a value which makes the total resistance of the circuit 4 Ω, calculate the mean torque on the coil.

Explain with the aid of sketch how a torque is produced on a coil carrying a current in a magnetic field. (U.L.C.I., O.2)

7. Calculate the value of the distribution factor for a three-phase alternator having 12 slots per pole.

8. A full-pitched one-turn coil, the ends of which are connected to slip-rings, is wound on a cylindrical iron core and rotates in a two-pole field, the pole arc being 75 per cent of the pole pitch. Sketch, and account for, the waveform of the e.m.f. for one revolution of the coil, allowing for fringing of the flux at the pole tips.

The stator of a three-phase, eight-pole, 750-rev/min alternator has 72 slots, each of which contains 10 conductors. Calculate the r.m.s. value of the e.m.f. per phase if the flux per pole is 0·1 Wb, sinusoidally distributed. Assume full-pitch coils and a winding distribution factor of 0·96. (App. El., L.U.)

9. Derive an expression for the terminal voltage of an alternator in terms of the frequency, flux per pole and the number of conductors, discussing the assumptions that are made.

Define the term *winding distribution factor* and show how it may be calculated. A three-phase, four-pole alternator has a single-layer winding with 8 conductors per slot. The armature has a total of 36 slots. Calculate the distribution factor.

What is the induced voltage per phase when the alternator is driven at 1800 rev/min, with a flux of 0·041 Wb in each pole?

(App. El., L.U.)

10. A three-phase, star-connected alternator, driven at 900 rev/min is required to generate a line voltage of 460 V at 60 Hz on open circuit. The stator has 2 slots per pole per phase and 4 conductors/slot. Calculate (a) the number of poles, (b) the useful flux/pole.

11. Find the number of stator conductors per slot for a three-phase, 50-Hz alternator if the winding is star-connected and has to give a line voltage of 13 kV when the machine is on open circuit. The flux/pole is about 0·15 Wb. Assume full-pitch coils and the stator to have 3 slots per pole per phase. The speed is 300 rev/min.

12. The following figures refer to a three-phase generator: number of poles, 8; flux/pole, 0·1 Wb; number of slots, 96; conductors/slot, 6; speed, 750 rev/min. The windings are star-connected. Calculate the line voltage on open circuit.

13. A four-pole, 50-Hz, three-phase mesh-connected alternator has a single-layer stator winding distributed in 36 slots, each slot containing 16 conductors. The flux/pole is 0·04 Wb. Calculate the terminal e.m.f. on open circuit.

14. A six-pole, 50-Hz, single-phase alternator has 54 slots, only 36 of which are wound. The number of conductors per slot is 12 and the coils are full-pitch. The flux per pole is 0·04 Wb, sinusoidally distributed. Calculate (a) the distribution factor and (b) the r.m.s. value of the terminal voltage on open circuit.

15. Calculate the value of the distribution factor for a two-phase alternator having 6 slots per pole.

If this alternator has 4 poles and is driven at 1800 rev/min and if the number of conductors in series per phase is 96, calculate the flux per pole required to enable the machine to generate an e.m.f. of 550 V. Assume the e.m.f./conductor to be sinusoidal and the coils to be full-pitch.

16. A four-pole, single-phase alternator has 24 slots, 16 of which are wound. The machine has a flux of about 0·02 Wb/pole and is driven at 1500 rev/min. Calculate the number of conductors per slot in order that the generated e.m.f. may be 410 V. Assume full-pitch coils and all the conductors connected in series.

17. A certain alternator has 6 slots per pole and the coils are short-pitch by 1 slot (i.e. the coil pitch is 5 slot pitches). What is the value of the pitch factor?

18. A 50-Hz alternator has a flux of 0·1 Wb/pole, sinusoidally distributed. Calculate the r.m.s. value of the e.m.f. generated in one turn which spans $\frac{3}{4}$ of a pole pitch.

19. A 10-MVA, 11-kV, 50-Hz, three-phase star-connected alternator is driven at 300 rev/min. The winding is housed in 360 slots and has 6 conductors per slot, the coils spanning five-sixths of a pole pitch. Calculate the sinusoidally-distributed flux/pole required to give a line voltage of 11 kV on open circuit, and the full-load current per conductor.

(E.M.E.U., O.2)

20. A star-connected balanced three-phase load of 30 Ω resistance per

phase is supplied by a 415-V, three-phase alternator of efficiency 90 per cent. Calculate the power input to the alternator.

(N.C.T.E.C., O.2)

APPENDIX EXAMPLES (continued from page 696)

1. An electrodynamic milliammeter has a deflection of 50° when a direct current of 20 mA flows through it. The inductance of the instrument changes at the rate of 1·2 mH/degree over the working range. Calculate the control torque, in micronewton metres per degree, exerted by the hairsprings. (Note that $dM/d\theta$ must be expressed in henrys/*radian*).

2. The hairsprings of an electrodynamic wattmeter exert a control torque of 0·6 μN·m/degree. The voltage circuit has a resistance of 6000 Ω and negligible reactance. The mutual inductance between the fixed and moving coils is 2 μH/degree over the working range. The voltage circuit is connected across a 240-V a.c. supply and the current in the current coil is 5 A, lagging the voltage by 30°. Calculate the deflection.

3. For the generator referred to in Q.7, Examples 7, p. 224, calculate the value of M for field currents of 2 A and 5 A respectively.

4. For the machine referred to in Q.13, Examples 8, p. 247, calculate the value of M when the field current is 4 A, and therefrom find the gross torque when the machine is taking an armature current of 60 A and a field current of 4 A. If the speed is 600 rev/min, what is the value of the mechanical power developed?

5. A universal series motor exerts a gross torque of 0·4 N·m when taking a direct current of 0·6 A. Calculate the value of M. If the total resistance of the armature and field windings is 50 Ω and the speed is 3000 rev/min, what is the value of the applied voltage?

6. The motor of Q.5 is connected across a 240-V, 50-Hz supply and is exerting a torque of 0·4 N·m. The total self inductance of the field and armature windings is 0·6 H and the iron and friction losses are negligible. Determine (*a*) the current, (*b*) the rotational e.m.f., (*c*) the speed, (*d*) the output power, (*e*) the power factor and (*f*) the efficiency. Note that $V^2 = (IR + E_r)^2 + (IX)^2$, where R and X represent the total resistance and reactance respectively and E_r is the rotational e.m.f. in phase with the current.

7. A 3-phase, 50-Hz, star-connected alternator, driven at 1000 rev/min, has an open-circuit line voltage of 460 V when the field current is 16A. The stator winding has a synchronous reactance of 2 Ω/phase and negligible resistance. Calculate the value of M and therefrom determine the driving torque when the machine is supplying 50 A/phase at a p.f. of 0·8 lagging, assuming the field current and the speed to remain 16 A and 1000 rev/min respectively. Neglect the iron and friction losses. Note that $E^2 = (V \cos \phi)^2 + (V \sin \phi + IX)^2$ and $\cos \alpha = (V \cos \phi)/E$, where $\cos \phi = $ p.f. of load.

8. The alternator of Q.7 is run as a synchronous motor off a 3-phase, 400-V, 50-Hz supply, taking 40 A/phase at a p.f. of 0·9 leading. Using the value of M calculated in Q.7, determine the field current.

CHAPTER 16

Production of a Rotating Magnetic Field

16.1 Production of rotating magnetic flux by two-phase currents

Let us consider a two-pole, two-phase winding having, for simplicity, only one slot per pole per phase, as shown in fig. 16.1, where AA_1 and BB_1 represent the coils of the two phases. Let us assume the currents to be positive when they are flowing towards the paper in conductors A and B and outwards in conductors A_1 and B_1. If

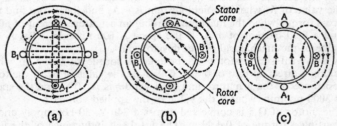

(a)   (b)   (c)

Fig. 16.1 Distribution of magnetic flux due to two-phase currents.

the currents in the two phases are represented by curves A and B in fig. 16.2, then at instant *a* the current in phase A is positive and at its maximum value, and is represented by the cross and dot on conductors A and A_1 respectively in fig. 16.1 (*a*). The current in conductors B and B_1 is zero at that instant; consequently the distribution of the magnetic flux is represented by the thin dotted lines in fig. 16.1 (*a*).

At instant *b* in fig. 16.2, the current in each phase is positive and is 0·707 of the maximum value; and the distribution of the resultant flux due to these currents is shown in fig. 16.1 (*b*). It will be seen that the axis of the resultant flux has turned clockwise through 45° from that of fig. 16.1 (*a*). At instant *c* in fig. 16.2, the current in phase A has decreased to zero; hence the distribution of the flux is as shown in fig. 16.1 (*c*) and its axis has rotated through another 45°.

It will be seen that the resultant magnetic flux rotates through a quarter of a pole pitch in an eighth of a cycle, and therefore through two pole pitches in one cycle. If the stator is wound with p pairs of poles, the flux rotates through $1/p$ revolution in one cycle, and therefore through f/p revolutions in one second. If n is the speed of the resultant magnetic flux in revolutions/second, then

$$n = f/p \quad \text{or} \quad f = np$$

which is the same as expression (10.4) derived in section 10.2 for the frequency of the e.m.f. generated in an alternator.

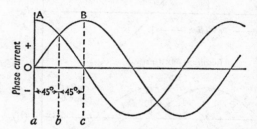

Fig. 16.2 Two-phase currents.

By re-drawing figs. 16.1 (b) and (c) with the current in, say, phase B reversed, it can be shown that the direction of rotation of the resultant magnetic flux is then anticlockwise, i.e. the direction of rotation of the resultant flux can be reversed by changing over the connections to *one* of the phases.

16.2 Mathematical derivation of the magnitude and speed of the resultant magnetic flux due to two-phase currents

In an actual machine the windings are distributed in two or more slots per pole per phase, and the distribution of the flux due to one phase depends upon the spread of the coils and the saturation of the teeth. For simplicity, we shall represent the airgap as a horizontal plane (fig. 16.3) and assume that the space distribution of the *flux density* produced by the current in each phase is sinusoidal over a pole pitch. Thus, fig. 16.3 (a) shows the distribution of the flux density over two pole pitches when the current in phase A is a maximum and that in phase B is zero, namely the values at instant a in fig. 16.2. B_m represents the maximum value of the flux density at that instant.

At instant *b* in fig. 16.2, the current in each phase is 0·707 of the maximum value, so that the distributions of the flux densities are represented by the full-line curves in fig. 16.3 (*b*). Assuming the magnetic circuit to be unsaturated, we can add together the flux densities due to the two phases and thus derive the dotted curve representing the distribution of the resultant flux density. At the point of maximum resultant density, the value of the flux density due

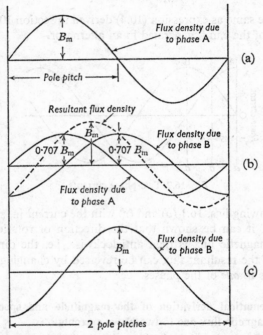

Fig. 16.3 Distribution of flux densities due to two-phase currents.

to each phase is 0·707 × 0·707 B_m, namely 0·5 B_m; hence it follows that the maximum value of the resultant density is B_m. Also, it is seen that the point of maximum resultant density has shifted a quarter of a pole pitch in an eighth of a cycle.

Fig. 16.3 (*c*) shows the flux density distribution at instant *c* in fig. 16.2. Again, it will be seen that the maximum value is B_m and that its position has shifted another quarter of a pole pitch in an eighth of a cycle. Hence, for the instants considered, the peak value of the resultant flux density remains constant at B_m, and its position moves a quarter of a pole pitch during each eighth of a cycle and therefore through two pole pitches in one cycle. It will now be shown that the peak value of the resultant flux density is constant at *every* instant and that its position moves at a *uniform* speed.

Let B_m be again the maximum value of the flux density due to one phase. Since the value of the flux density at a given point varies sinusoidally with *time*, its maximum value, t seconds after it has passed through zero from negative to positive values, is represented for phase A by $B_a = B_m \sin \omega t$, as shown in fig. 16.4. Hence the

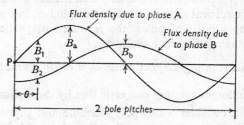

Fig. 16.4 Distribution of flux densities due to two-phase currents.

value of the flux density due to phase A at a point displaced θ electrical radians from P (namely the point of zero flux density) is represented by:

$$B_1 = B_a \sin \theta$$
$$= B_m \sin \omega t . \sin \theta.$$

Similarly, if the current in phase B lags that in phase A by a quarter of a cycle,

flux density at the centre of phase B $= B_b = B_m \sin (\omega t - \pi/2)$
$$= -B_m \cos \omega t,$$

and the value of the flux density due to phase B at a point θ radians from P is represented by:

$$B_2 = B_b \sin (\theta - \pi/2)$$
$$= B_m \cos \omega t . \cos \theta.$$

Assuming the magnetic circuit to be unsaturated,

resultant flux density at a} point θ radians from P} $= B_1 + B_2$

$$= B_m (\sin \omega t . \sin \theta + \cos \omega t . \cos \theta)$$
$$= B_m \cos (\omega t - \theta)$$

and is constant at B_m when $(\omega t - \theta) = 0$, i.e. when

$$\theta = \omega t = 2\pi f t.$$

For a value of t equal to the duration of one cycle, namely $1/f$ second,

corresponding value of $\theta = 2\pi$ radians.

Since the distribution of the resultant flux density remains sinusoidal and the peak value remains constant, it follows that the total flux over a pole pitch is constant and rotates at a uniform speed through 2π electrical radians, namely two pitches, in one cycle. If the machine has p pairs of poles, the resultant magnetic flux rotates through $1/p$ revolution in 1 cycle,

∴ speed of rotating magnetic flux $= n = f/p$ revolutions/second.

16.3 Production of rotating magnetic flux by three-phase currents

Let us again consider a two-pole machine and suppose the three-phase winding to have only one slot per pole per phase, as shown in fig. 16.5. The end-connections of the coils are not shown, but are assumed to be similar to those already shown in fig. 15.4; thus R

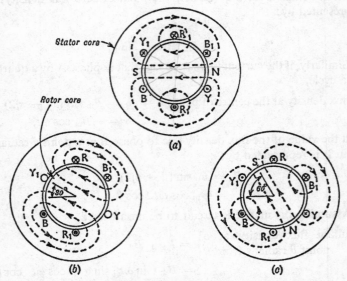

Fig. 16.5 Distribution of magnetic flux due to three-phase currents.

and R_1 represent the 'start' and the 'finish' of the 'red' phase, etc. It will be noted that R, Y and B are displaced 120 electrical degrees relative to one another. Let us also assume that the current is positive when it is flowing inwards in conductors R, Y and B, and therefore outwards in R_1, Y_1 and B_1. As far as the present discussion is concerned, the rotor core need only consist of circular iron laminations to provide a path of low reluctance for the magnetic flux.

Suppose the currents in the three phases to be represented by the curves in fig. 16.6; then at instant *a* the current in phase R is positive and at its maximum value, whereas in phases Y and B the currents are negative and each is half the maximum value. These currents,

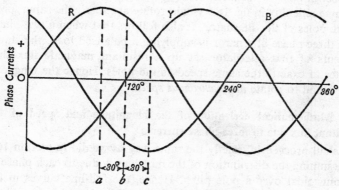

Fig. 16.6 Three-phase currents.

represented in direction by dots and crosses in fig. 16.5 (*a*), produce the magnetic flux represented by the dotted lines. At instant *b* in fig. 16.6, the currents in phases R and B are each 0·866 of the maximum; and the distribution of the magnetic flux due to these currents is shown in fig. 16.5 (*b*). It will be seen that the axis of this field is in line with coil YY_1 and therefore has turned clockwise through 30° from that of fig. 16.5 (*a*). At instant *c* in fig. 16.6, the current in phase B has attained its maximum negative value, and the currents in R and Y are both positive, each being half the maximum value. These currents produce the magnetic flux shown in fig. 16.5 (*c*), the axis of this flux being displaced clockwise by another 30° compared with that in fig. 16.5 (*b*).

These three cases are sufficient to prove that for every interval of *time* corresponding to 30° along the horizontal axis of fig. 16.6, the axis of the magnetic flux in a two-pole stator moves forward 30° in *space*. Consequently, in 1 cycle, the flux rotates through 1 revolution or 2 pole pitches. If the stator is wound for p pairs of poles, the magnetic flux rotates through $1/p$ revolution in 1 cycle and therefore through f/p revolutions in 1 second.

If n is the speed of the magnetic flux in revolutions/second,

$$n = f/p \qquad (16.1)$$

or

$$f = np$$

which is the same as expression (10.4) derived for an alternator in section 10.2. It follows that if the stator in fig. 16.5 had the same number of poles as the alternator supplying the three-phase currents, the magnetic flux in fig. 16.5 would rotate at *exactly* the same speed as the poles of the alternator. It also follows that when a two-phase or a three-phase alternator is supplying a balanced load, the stator currents of that alternator set up a resultant magnetic flux that rotates at exactly the same speed as the poles. Hence the magnetic flux is said to rotate at *synchronous speed*.

16.4 Mathematical derivation of the magnitude and speed of the resultant flux due to three-phase currents

We shall proceed in exactly the same way as we did in section 16.2 by assuming the distribution of the flux density due to each phase to be sinusoidal over a pole pitch. Hence, the full-line curves in fig. 16.7 (*a*) represent the distribution of the flux densities due to the three phases at instant *a* in fig. 16.6; and the dotted curve represents the resultant flux density on the assumption that the magnetic circuit is unsaturated. If B_m represents the maximum flux density due to the maximum current in one phase alone, it will be seen from fig. 16.7 (*a*) that the contributions made to the maximum resultant flux density by phases R, Y and B are B_m, $0.25\ B_m$ and $0.25\ B_m$ respectively, and that the peak value of the resultant flux density is therefore $1.5\ B_m$.

Fig. 16.7 (*b*) represents the distribution of the flux densities due to phases R and B at instant *b* in fig. 16.6; and again it will be seen that the peak value of the resultant flux density is $1.5\ B_m$ and that its position has shifted a sixth of a pole pitch in a twelfth of a cycle.

Similarly, if the distribution of the resultant flux density were derived for instant *c* of fig. 16.6, it would be found that the peak value of the resultant flux density would again be $1.5\ B_m$ and that it would have moved through a further sixth of a pole pitch in a twelfth of a cycle and therefore through two pole pitches in one cycle.

It will now be shown that the peak value of the resultant flux density remains constant at *every* instant and that its position rotates at a *uniform* speed.

Suppose the curves in fig. 16.8 to represent the distributions of the

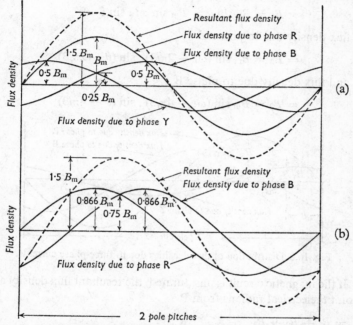

Fig. 16.7 Distribution of flux densities due to three-phase currents.

flux densities due to the three phases at an instant *t* seconds after the flux due to phase R has passed through zero from negative to positive values, and suppose B_m to be the peak density due to one phase when the current in that phase is at its maximum value. Then, at the instant under consideration,

maximum flux density due to phase R

$$= B_R = B_m \sin \omega t$$

maximum flux density due to phase Y

$$= B_Y = B_m \sin (\omega t - 2\pi/3)$$

and maximum flux density due to phase B

$$= B_{\mathrm{B}} = B_{\mathrm{m}} \sin (\omega t - 4\pi/3).$$

If P be a point of zero flux density for phase R, then for a point θ electrical radians from P,

flux density due to phase R

$$= B_1 = B_{\mathrm{R}} \sin \theta = B_{\mathrm{m}} \sin \omega t \,.\, \sin \theta$$

flux density due to phase Y

$$= B_2 = B_{\mathrm{m}} \sin (\omega t - 2\pi/3) \,.\, \sin (\theta - 2\pi/3)$$

and flux density due to phase B

$$= B_3 = B_{\mathrm{m}} \sin (\omega t - 4\pi/3) \,.\, \sin (\theta - 4\pi/3)$$

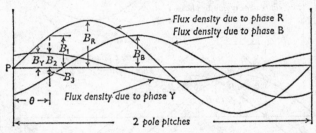

Fig. 16.8 Distribution of flux densities due to three-phase currents.

If the magnetic circuit is unsaturated, the resultant flux density at a point θ electrical radians from P

$$= B_1 + B_2 + B_3$$
$$= B_{\mathrm{m}}\{\sin \omega t \,.\, \sin \theta + \sin (\omega t - 2\pi/3) \,.\, \sin (\theta - 2\pi/3)$$
$$+ \sin (\omega t - 4\pi/3) \,.\, \sin (\theta - 4\pi/3)\}$$
$$= 1 \cdot 5\, B_{\mathrm{m}} (\sin \omega t \,.\, \sin \theta + \cos \omega t \,.\, \cos \theta)$$
$$= 1 \cdot 5\, B_{\mathrm{m}} \cos (\omega t - \theta)$$

and is constant at $1 \cdot 5\, B_{\mathrm{m}}$ when $\theta = \omega t = 2\pi f t$.

For a value of t equal to $1/f$, namely the duration of one cycle,

$$\theta = 2\pi \text{ electrical radians,}$$

i.e. the position of the peak value of the resultant flux density rotates through two pole pitches in one cycle.

Since the distribution of the resultant flux density remains sinusoidal and the peak value remains constant, it follows that the total flux over a pole pitch is constant and rotates at a uniform speed through two pole pitches in one cycle. If the machine has p pairs of poles, the resultant magnetic flux rotates through $1/p$ revolution in one cycle,

$\therefore$ speed of rotating magnetic flux $= n = f/p$ revolutions/second
or
$$f = np$$

which is the same as expression (10.4) derived in section 10.2 for the frequency of the e.m.f. generated in an alternator.

16.5 Reversal of direction of rotation of the magnetic flux produced by three-phase currents

Suppose the stator winding to be connected as in fig. 16.9 (*a*) and that this arrangement corresponds to that already shown in fig. 16.5 and discussed in section 16.3. The resultant magnetic flux was found to rotate clockwise. Let us interchange the connections between two

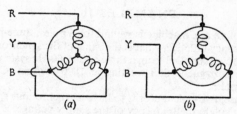

Fig. 16.9 Reversal of direction of rotation.

of the supply lines, say Y and B, and the stator windings, as shown in fig. 16.9 (*b*). The distribution of currents at instant *a* in fig. 16.6 will be exactly as shown in fig. 16.5 (*a*); but at instant *b*, the current in the winding that was originally the 'yellow' phase is now 0·866 of the maximum and the distribution of the resultant magnetic flux is as shown in fig. 16.10. From a comparison of fig. 16.5 (*a*) and 16.10 it is seen that the axis of the magnetic flux is now rotating anticlockwise. The same result may be represented thus:

| R
B ⤻ Y
Original
sequence | becomes | R
Y ⤸ B
Inter-
changing
Y and B | or | Y
B ⤸ R
Inter-
changing
R and Y | or | B
R ⤸ Y
Inter-
changing
R and B |

From the table at the bottom of page 455, it will be seen that the direction of rotation of the resultant magnetic flux can be reversed by reversing the connections to any two of the three terminals of

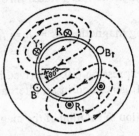

Fig. 16.10 Distribution of magnetic flux at instant *b* of fig. 16.6.

the motor. The ease with which it is possible to reverse the direction of rotation constitutes one of the advantages of three-phase motors.

EXAMPLES 16

1. The stator of an a.c. machine is wound for 6 poles, three-phase. If the supply frequency is 25 Hz, what is the value of the synchronous speed?
2. A stator winding supplied from a three-phase, 60-Hz system is required to produce a magnetic flux rotating at 1800 rev/min. Calculate the number of poles.
3. A three-phase, two-pole motor is to have a synchronous speed of 9000 rev/min. Calculate the frequency of the supply voltage.

CHAPTER 17

Characteristics of Synchronous Generators and Motors

17.1 Armature reaction in a three-phase synchronous generator (or alternator)

By 'armature reaction' is meant the influence of the stator ampere-turns upon the value and the distribution of the magnetic flux in the airgaps between the poles and the stator core. It has already been explained in Chapter 16 that balanced three-phase currents in a three-phase winding produce a resultant magnetic flux of constant magnitude rotating at synchronous speed. We shall now consider the application of this principle to an alternator.

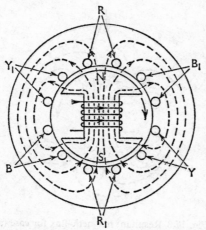

Fig. 17.1 Magnetic flux due to rotor current alone.

Case (a). *When the current and the generated e.m.f. are in phase*

Consider a two-pole, three-phase alternator with 2 slots per pole per phase. If the machine is on open circuit there is no stator current, and the magnetic flux due to the rotor current is distributed

457

symmetrically as shown in fig. 17.1. If the direction of rotation of the poles be clockwise, the e.m.f. generated in phase RR_1 is at its maximum and is towards the paper in conductors R and outwards in R_1.

Let us next consider the distribution of flux (fig. 17.2) due to the

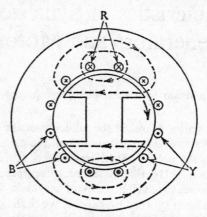

Fig. 17.2 Magnetic flux due to stator currents alone.

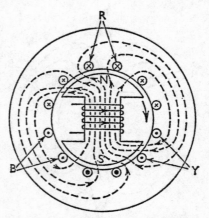

Fig. 17.3 Resultant magnetic flux for case (a).

stator currents alone at the instant when the current in phase R is at its maximum positive value (instant a in fig. 16.6) and when the rotor (unexcited) is in the position shown in fig. 17.1. This magnetic flux rotates clockwise at synchronous speed and is therefore stationary relative to the rotor.

We can now derive the resultant magnetic flux due to the rotor and stator currents by superimposing the fluxes of figs. 17.1 and 17.2 on each other. Comparison of these figures shows that over the leading half of each pole face the two fluxes are in opposition, whereas over the trailing half of each pole face they are in the same direction. Hence the effect is to distort the magnetic flux as shown in fig. 17.3. It will be noticed that the direction of most of the lines of flux in the airgaps has been skewed and thereby lengthened. But lines of flux behave like stretched elastic cords and consequently in fig. 17.3 they exert a backward pull on the rotor; and to overcome the tangential component of this pull, the engine driving the alternator has to exert a larger torque than that required on no load. Since the magnetic flux due to the stator currents rotates synchronously with the rotor, the flux distortion shown in fig. 17.3 remains the same for all positions of the rotor.

Case (b). *When the current lags the generated e.m.f. by a quarter of a cycle*

When the e.m.f. in phase R is at its maximum value, the poles are in the position shown in fig. 17.1. By the time the current in phase R

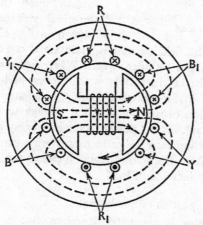

Fig. 17.4 Resultant magnetic flux for case (b).

reaches its maximum value, the poles will have moved forward through half a pole pitch to the position shown in fig. 17.4. A reference to fig. 17.2 shows that the stator ampere-turns, acting alone, would send a flux from right to left through the rotor, namely in direct opposition to the flux produced by the rotor ampere-turns. Hence it follows that the effect of armature reaction due to a current lagging the e.m.f. by 90° is to reduce the flux. The resultant distribution of the flux, however, is symmetrical over the two halves of the pole face, so that no torque is required to drive the rotor, apart from that to overcome losses.

Case (c). *When the current leads the generated e.m.f. by a quarter of a cycle*

In this case the current in phase R is a positive maximum when the N and S poles of the rotor are in the positions occupied by the S and N poles respectively in fig. 17.4. Consequently the flux due to the stator ampere-turns is now in the same direction as that due to the rotor ampere-turns, so that the effect of armature reaction due to a leading current is to increase the flux.

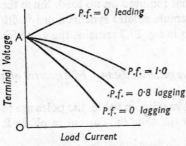

Fig. 17.5 Variation of terminal voltage with load.

The influence of armature reaction upon the variation of terminal voltage with load is shown in fig. 17.5, where it is assumed that the field current is maintained constant at a value giving an e.m.f. OA on open circuit. When the power factor of the load is unity, the fall in voltage with increase of load is comparatively small. With an inductive load, the demagnetizing effect of armature reaction causes the terminal voltage to fall much more rapidly. The graph for 0·8 power factor is roughly midway between those for unity and zero power factors. With a capacitive load, the magnetizing effect of armature reaction causes the terminal voltage to increase with increase of load.

17.2 Voltage regulation of an alternator

An a.c. generator is always designed to give a certain terminal voltage when supplying its rated current at a specified power factor—usually unity or 0·8 lagging. For instance, suppose OB in fig. 17.6 to represent the full-load current and OA the rated terminal voltage of an alternator. If the field current is adjusted to give the terminal voltage OA when the alternator is supplying current OB at unity power factor, then when the load is removed but with the field current and speed kept unaltered, the terminal voltage rises to OC. This variation of the terminal voltage between full load and no load, expressed as a per-unit value or a percentage of the full-load voltage, is termed the per-unit or the percentage *voltage regulation* of the alternator; thus:

$$\left.\begin{array}{l}\text{per-unit voltage} \\ \text{regulation}\end{array}\right\} = \frac{\begin{array}{c}\text{change of terminal voltage} \\ \text{when full load is removed}\end{array}}{\text{full-load terminal voltage}} \quad (17.1)$$

$$= \frac{\text{AC}}{\text{OA}} \text{ for unity power factor}$$

$$= \frac{\text{AD}}{\text{OA}} \text{ for p.f. of 0·8 lagging.}$$

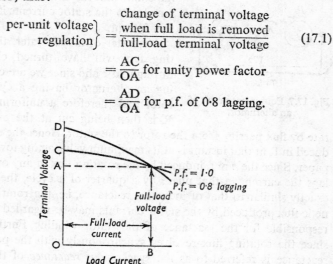

Fig. 17.6 Variation of terminal voltage with load.

The voltage regulation for a power factor of 0·8 lagging is usually far greater than that at unity power factor, and it is therefore important to include the power factor when stating the voltage regulation. (See example 17.1.)

17.3 Synchronous impedance

In figs. 17.3 and 17.4, the resultant flux was shown as the combination of the flux due to the stator ampere-turns alone and that due to the rotor ampere-turns alone. For the purpose of deriving the effect of load upon the terminal voltage, however, it is convenient to regard these two component fluxes as if they existed independently of each other and to consider the cylindrical-rotor type rather than the salient-pole type of alternator (see section 15.2). Thus the flux due to the rotor ampere-turns may be regarded as generating an e.m.f., *E*, due to the rotation of the poles, this e.m.f. being a maximum in any one phase when the conductors of that phase are opposite the centres of the poles. On the other hand, the rotating magnetic field due to the stator currents may be regarded as generating an e.m.f. lagging the current by a quarter of a cycle. For instance, in fig. 16.5 (a), the current in R is at its maximum value flowing towards the paper, but

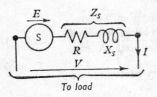

the e.m.f. induced in R by the rotating flux due to the stator currents is zero at that instant.

A quarter of a cycle later, this rotating flux will have turned clockwise through 90°; and since we are considering an alternator having a cylindrical rotor and therefore a uniform airgap, R is then being cut at the maximum

Fig. 17.7 Equivalent circuit of an alternator.

rate by flux passing from the rotor to the stator. Hence the e.m.f. induced in R at that instant is at its maximum value acting towards the paper. Since the e.m.f. induced by the rotating flux in any one phase lags the current in that phase by a quarter of a cycle, the effect is exactly similar to that of inductive reactance, i.e. the rotating magnetic flux produced by the stator currents may be regarded as being responsible for the reactance of the stator winding. Furthermore, since the rotating flux revolves synchronously with the poles, this reactance is referred to as the *synchronous reactance* of the winding.

By combining the resistance with the synchronous reactance of the winding, we obtain its *synchronous impedance*. Thus,

if X_s = synchronous reactance/phase,

R = resistance/phase

and Z_s = synchronous impedance/phase

then $$Z_s = \sqrt{(R^2 + X_s^2)}.$$

In alternator windings, R is usually very small compared with X_s, so that for many practical purposes, Z_s can be assumed to be the same as X_s.

The relationship between the terminal voltage V of the alternator and the e.m.f. E generated by the flux due to the rotor ampere-turns alone can now be derived. Thus in fig. 17.7, S represents *one* phase of the stator winding, and R and X_s represent the resistance and synchronous reactance of that phase. If the load takes a current I at a lagging power factor $\cos \phi$, the various quantities can be represented by phasors as in fig. 17.8, where:

OI = current/phase,

OV = terminal voltage/phase,

OA = IR = component of the generated e.m.f. E absorbed in sending current through R,

OB = e.m.f./phase induced by the rotating flux due to stator currents and lags OI by 90°,

OC = component of the generated e.m.f. E required to neutralize OB,

   = voltage drop due to synchronous reactance X_s

   = IX_s

OD = component of the generated e.m.f. absorbed in sending current through the synchronous impedance Z_s,

α = phase angle between OI and OD = $\tan^{-1} X_s/R$

and OE = resultant of OV and OD

   = e.m.f./phase generated by the flux due to the rotor ampere-turns.

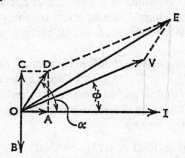

Fig. 17.8 Phasor diagram for an alternator.

From fig. 17.8,

$$OE^2 = OV^2 + OD^2 + 2 . OV . OD . \cos (\alpha - \phi)$$

i.e. $\qquad E^2 = V^2 + (IZ_s)^2 + 2V . IZ_s \cos (\alpha - \phi)$

from which the e.m.f. E generated by the flux due to the rotor ampere-turns, namely the open-circuit voltage, can be calculated. It follows that if V is the rated terminal voltage per phase of the alter-

nator and I is the full-load current per phase, the terminal voltage/
phase obtained when the load is removed, the exciting current and
speed remaining unaltered, is given by E; and from expression
(17.1),

$$\text{per-unit voltage regulation} = \frac{E - V}{V}.$$

The synchronous impedance is important when we come to deal
with the parallel operation of alternators and with synchronous
motors (section 17.8); but numerical calculations involving synchronous impedance are usually unsatisfactory, owing mainly to magnetic saturation in the poles and stator teeth and, in the case of
salient-pole machines, to the value of Z_s varying with the p.f. of the
load and with the excitation of the poles.

One method of estimating the value of the synchronous impedance
is to run the alternator on open circuit and measure the generated
e.m.f., and then short-circuit the terminals through an ammeter and
measure the short-circuit current, the exciting current and speed
being kept constant. Since the e.m.f. generated on open circuit may
be regarded as being responsible for circulating the short-circuit current through the synchronous impedance of the winding, the value
of the synchronous impedance is given by the ratio of the open-
circuit voltage/phase to the short-circuit current/phase.

Example 17.1 *A three-phase, 600-kVA alternator has a rated terminal
voltage of 3300 V (line). The stator winding is star-connected and has a
resistance of 0·37 Ω/phase and a synchronous impedance of 4·3 Ω/phase.
Calculate the voltage regulation for a load having a power factor of*
(a) *unity and* (b) 0·8 *lagging.*

(a) Since $\qquad 600 \times 1000 = 1 \cdot 73 I_L \times 3300$

∴ $\qquad\qquad$ line current $= I_L = 105$ A $=$ phase current.

$\left.\begin{array}{l}\text{Terminal voltage/phase}\\ \text{on full load}\end{array}\right\} = 3300/1 \cdot 73 = 1910$ V.

$\left.\begin{array}{l}\text{Voltage drop/phase on}\\ \text{full load due to syn-}\\ \text{chronous impedance}\end{array}\right\} = 105 \times 4 \cdot 3 = 452$ V.

From fig. 17.8 it follows that at unity power factor,

$$OV = 1910 \text{ V}; \quad OD = 452 \text{ V}; \quad \text{and} \quad \phi = 0.$$

Also, $\quad \cos \alpha = OA/OD = R/Z_s = 0.37/4.3 = 0.086,$

$\therefore \qquad OE^2 = (1910)^2 + (452)^2 + 2 \times 1910 \times 452 \times 0.086$

$\qquad\qquad\quad = 4 \times 10^6$

$\therefore \qquad OE = 2000 \text{ V},$

and voltage regulation at unity $\left.\begin{array}{r}\\ \text{power factor}\end{array}\right\} = \dfrac{2000 - 1910}{1910}$

$\qquad\qquad\qquad\qquad\qquad = 0.047 \text{ per unit}$

$\qquad\qquad\qquad\qquad\qquad = 4.7 \text{ per cent.}$

(*b*) Since the rating of the alternator is 600 kVA, the full-load current is the same whatever the power factor.

For power factor of 0·8, $\quad \phi = 36° 52'$

Also $\qquad\qquad\qquad\qquad\qquad \alpha = \cos^{-1} 0.086 = 85° 4',$

so that $(\alpha - \phi) = 48° 12'$ and $\cos 48° 12' = 0.666.$

Hence, $\quad OE^2 = (1910)^2 + (452)^2 + 2 \times 1910 \times 452 \times 0.666$

$\qquad\qquad\quad = 5.006 \times 10^6$

$\therefore \qquad OE = 2240 \text{ V},$

and voltage regulation for power $\left.\begin{array}{r}\\ \text{factor 0·8 lagging}\end{array}\right\} = \dfrac{2240 - 1910}{1910}$

$\qquad\qquad\qquad\qquad\qquad = 0.173 \text{ per unit}$

$\qquad\qquad\qquad\qquad\qquad = 17.3 \text{ per cent.}$

17.4 Synchronizing of alternators

For simplicity, let us assume single-phase alternators; thus, in fig. 17.9, A represents an alternator already connected to the busbars and B is an alternator to be connected in parallel. To enable this to be done, the following conditions must be fulfilled:

(*a*) the frequency of B must be the same as that of A,
(*b*) the e.m.f. generated in B must be equal to the busbar voltage,
(*c*) the e.m.f. of B must be in phase with the busbar voltage.

The procedure is to start up the engine driving alternator B and adjust its speed to about its rated value. The excitation of B is then adjusted so that the reading on voltmeter V_2 is the same as the busbar voltage given by voltmeter V_1. Switch S_1 is closed and voltmeter V_3 connected across switch S_2. The pointer of V_3 will then oscillate between zero and twice the busbar voltage at a frequency equal to the difference between the frequencies of A and B. This will be evident from fig. 17.10, where curves E_A and E_B represent the e.m.f.s of A

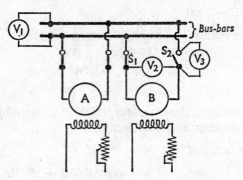

Fig. 17.9 Synchronizing of alternators.

and B respectively, the frequencies being assumed such that E_B varies through 4 cycles for every 3 cycles of E_A. The p.d. across switch S_2 is the difference between E_A and E_B and fig. 17.10 (*b*) represents two 'beats' of this voltage. If the frequency of A is 50 Hz and that of B is 50·5 Hz, the beat frequency is 0·5 Hz and the pointer of V_3 makes one complete oscillation in 2 seconds.

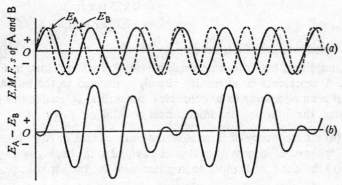

Fig. 17.10 Waveforms of E_A, E_B and ($E_A - E_B$).

The speed of B is adjusted until the pointer of voltmeter V_3 oscillates very slowly and switch S_2 is closed just as the pointer is reaching zero, i.e. when the e.m.f.s of A and B are in phase opposition relative to each other but in phase with each other relative to the busbars.

The above method of paralleling two alternators does not indicate whether the frequency of the incoming alternator B is higher or lower than that of A. In actual practice it is customary to use an instrument, called a *synchroscope*, which shows whether the incoming machine is running too fast or too slow as well as indicating the correct moment for closing switch S₂.

17.5 Parallel operation of alternators

It is not obvious why two alternators continue running in synchronism after they have been paralleled; so let us consider the case of two similar single-phase alternators, A and B, connected in parallel to the busbars, as shown in fig. 17.11, and let us, for simplicity,

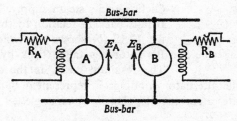

Fig. 17.11 Two alternators in parallel.

assume that there is no external load connected across the busbars. We will consider two separate cases:

(*a*) the effect of varying the torque of the driving engine, e.g. by varying the steam supply;

(*b*) the effect of varying the exciting current.

Effect of varying the driving torque

Let us first assume that each engine is exerting exactly the torque required by its own alternator and that the field resistors R_A and R_B are adjusted so that the generated e.m.f.s E_A and E_B are equal. Suppose the arrows in fig. 17.11 to represent the *positive* directions of these e.m.f.s. It will be seen that, *relative to the busbars*, these e.m.f.s are acting in the same direction, i.e. when the e.m.f. generated in each machine is positive, each e.m.f. is making the top busbar positive in relation to the bottom busbar. But *in relation to each other*, these e.m.f.s are in opposition, i.e. if we trace the closed circuit formed by the two alternators we find that the e.m.f.s oppose each other.

In the present discussion we want to find out if any current is being circulated in this closed circuit. It is therefore more convenient to consider these e.m.f.s in relation to each other rather than to the busbars. This condition is shown in fig. 17.12 (*a*) when E_A and E_B are in exact phase opposition relative to each other; and since they are equal in magnitude, their resultant is zero and consequently no current is circulated.

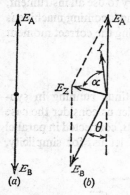

Let us next assume the driving torque of B's engine to be reduced, e.g. by a reduction of its steam supply. The rotor of B falls back in relation to that of A, and fig. 17.12 (*b*) shows the conditions when B's rotor has fallen back by an angle θ. The resultant e.m.f. in the closed circuit

Fig. 17.12 Effect of varying the driving torque.

formed by the alternator windings is represented by E_Z, and this e.m.f. circulates a current I lagging E_Z by an angle α,

where $\quad I = \dfrac{E_Z}{2Z_s} \quad$ and $\quad \alpha = \tan^{-1} X_s/R$,

$\qquad R$ = resistance of each alternator,

$\qquad X_s$ = synchronous reactance of each alternator,

and $\qquad Z_s$ = synchronous impedance of each alternator.

Since the resistance is very small compared with the synchronous reactance, α is nearly 90°, so that the current I is almost in phase with E_A and in phase opposition to E_B. This means that A is generating and B is motoring, and the power supplied from A to B compensates for the reduction of the power supplied by B's engine. If the frequency is to be maintained constant, the driving torque of A's engine has to be increased by an amount equal to the decrease in the driving torque of B's engine.

The larger the value of θ (so long as it does not exceed about 80°), the larger is the circulating current and the greater is the power supplied from A to B. Hence machine B falls back in relation to A until the power taken from the latter exactly compensates for the reduction in the driving power of B's engine. Once this balance has been attained, B and A will run at exactly the same speed.

Effect of varying the excitation

Let us again revert to fig. 17.12 (*a*) and assume that each engine is exerting the torque required by its alternator and that the e.m.f.s, E_A and E_B, are equal. The resultant e.m.f. is zero and there is therefore no circulating current.

Suppose the exciting current of alternator B to be increased so that the corresponding open-circuit e.m.f. is represented by E_B in fig. 17.13. The resultant e.m.f., E_Z ($= E_B - E_A$), circulates a current I through the synchronous impedances of the two alternators; and since the machines are assumed similar, the impedance drop per machine is $\frac{1}{2}E_Z$, so that:

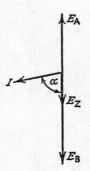

terminal voltage $= E_B - \frac{1}{2}E_Z$

or $\qquad\qquad\qquad = E_A + \frac{1}{2}E_Z$

Hence one effect has been to increase the terminal voltage. Further, since angle α is nearly 90°, the circulating current I is almost in quadrature with the generated e.m.f.s, so that very little power is circulated from one machine to the other.

Fig. 17.13 Effect of varying the excitation.

In general, we may therefore conclude:

(*a*) the distribution of load between alternators operating in parallel can be varied by varying the driving torques of the engines and only slightly by varying the exciting currents;

(*b*) the terminal voltage is controlled by varying the exciting currents.

17.6 Three-phase synchronous motor: Principle of action

In section 17.5 it was explained that when two alternators, A and B, are in parallel, with *no load* on the busbars, a reduction in the driving torque applied to B causes the latter to fall back by some angle θ (fig. 17.12) in relation to A, so that power is supplied from A to B. Machine B is then operating as a *synchronous motor*.

It will also be seen from fig. 17.12 that the current in a synchronous motor is approximately in phase opposition to the e.m.f. generated in that machine. This effect is represented in fig. 17.14, where the rotor poles are shown in the same position relative to the three-phase winding as in fig. 17.3. The latter represented a synchronous generator with the current in phase with the generated e.m.f., whereas fig. 17.14 represents a synchronous motor with the current in phase opposition to this generated e.m.f. A diagram similar to fig. 17.2 could be drawn showing the flux distribution due to the stator currents alone, but a comparison with figs. 17.1, 17.2 and 17.3 may be

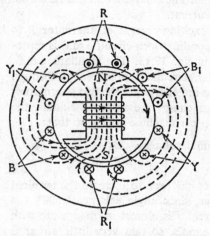

Fig. 17.14 Principle of action of a three-phase synchronous motor.

sufficient to indicate that in fig. 17.14 the effect of armature reaction is to increase the flux in the leading half of each pole and to reduce it in the trailing half. Consequently the flux is distorted in the direction of rotation and the lines of flux in the gap are skewed in such a direction as to exert a clockwise torque on the rotor. Since the resultant magnetic flux due to the stator currents rotates at synchronous speed, the rotor must also rotate at exactly the same speed for the flux distribution shown in fig. 17.14 to remain unaltered.

17.7 Effect of varying the load on a synchronous motor

Fig. 17.15 represents diagrammatically a three-phase, star-connected synchronous motor connected to the supply mains and excited by a field winding F in series with a variable resistor.

Suppose V in fig. 17.16 to represent the voltage *applied* to one phase and E the e.m.f.* *generated* in that phase. Angle θ represents

* This e.m.f. is the open-circuit voltage of the machine when driven as an alternator at the same speed with the same field current.

the displacement of E from exact phase opposition to V. Let us assume that E is exactly equal to V in magnitude. Consequently, if E had been in exact phase opposition to V, i.e. if $\theta = 0$, there would have been no stator current and therefore no power taken from the a.c. supply. The motor would have to slow down and the phase of the generated e.m.f. E would fall back in relation to the applied voltage V by some angle θ, the value of which depends upon the load.

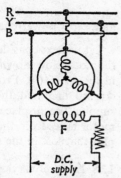

Let us first assume that the load is small so that θ is also small, as shown in fig. 17.16 (a). The resultant voltage available to send a current I through the synchronous impedance of the winding is represented by V_Z, the resultant of V and E. Alternatively, we may regard the applied voltage V as

Fig. 17.15 Connections of a three-phase synchronous motor.

having to provide two components: (*a*) V' to neutralize the generated

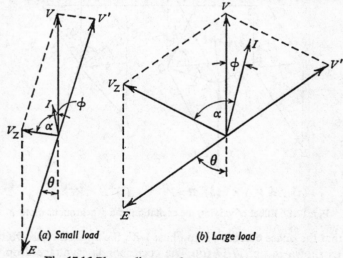

(a) Small load   (b) Large load

Fig. 17.16 Phasor diagrams for a synchronous motor.

e.m.f. E, and (*b*) V_Z to provide the voltage drop due to the synchronous impedance of the stator winding.

If the synchronous impedance per phase is Z_s, the current is V_Z/Z_s and lags V_Z by an angle α approximating 90°, since the

*

resistance is very small compared with the synchronous reactance. If the phase angle between I and the supply voltage V is ϕ,

$$\text{power factor of motor} = \cos\phi$$

and   power/phase taken by motor $= IV\cos\phi$.

Let us next consider the effect of an increase in load. The rotor of the motor slows down momentarily until the displacement θ is as shown in fig. 17.16 (b). From the latter it is seen that V_Z and I have increased and that the motor is taking more power from the a.c. supply. The displacement θ adjusts itself automatically so that the motor takes from the supply exactly the power required by the load plus that lost in the machine.

17.8 Effect of varying the excitation of a synchronous motor

Let us assume the input power to the motor to remain constant. Also, let us assume the excitation to be such that the power factor is unity,

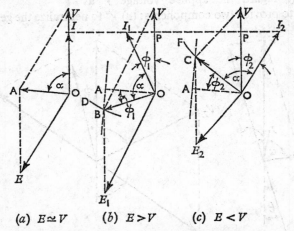

(a) $E \simeq V$ (b) $E > V$ (c) $E < V$

Fig. 17.17 Effect of varying the excitation of a synchronous motor.

so that the phase current I is in phase with the applied phase voltage V, as shown in fig. 17.17 (a). The corresponding voltage drop per phase, V_Z, is represented by OA leading the current by an angle α which is nearly 90°, and the generated or back e.m.f. is represented by E, the magnitude of which is approximately the same as that of V.

Fig. 17.17 (b) shows the result of increasing the rotor field current and thus increasing the generated e.m.f. to E_1. The effect is to swing the phasor V_Z forward through an angle ϕ_1 from OA to OB. The current, I_1, lags OB by the same angle α and therefore *leads the terminal voltage, V,* by angle ϕ_1.

The projection of OI_1 on OV, namely OP, represents the active or power component of the current, and this component is being assumed constant. Considering triangles OPI_1 and OAB, we have:

$$OA = OP \times \text{synchronous impedance/phase}$$

and $\qquad OB = OI_1 \times \text{synchronous impedance/phase,}$

$\therefore \qquad OA/OB = OP/OI_1.$

Also, $\quad \angle POI_1 = \angle AOB = \phi_1;$
hence the two triangles are similar, so that:

$$\angle OAB = \angle OPI_1 = 90°.$$

Consequently the locus of B is the dotted line drawn through A at right-angles to OA. Point B can therefore be determined by drawing

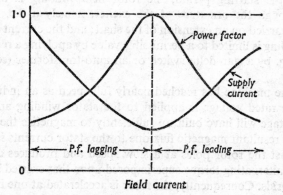

Fig. 17.18 Variation of supply current and power factor with field current for constant input power to a synchronous motor.

an arc D of radius equal to E_1, using point V as centre. Phasor E_1 is then drawn parallel and equal to the line joining V and B.

The effect of reducing the excitation is shown in fig. 17.17 (*c*). With centre V and radius equal to E_2, an arc F is drawn to cut the locus through A at point C. Phasor E_2 is then drawn parallel and equal to line VC. The current phasor OI_2 lags OC by angle α and therefore *lags the terminal voltage* by angle ϕ_2.

It follows from these phasor diagrams that the power factor of a synchronous motor can be controlled by varying the field current, as indicated in fig. 17.18, the actual range of power-factor variation being dependent upon the value of the load.

17.9 Methods of starting a three-phase synchronous motor

When the rotor of a three-phase synchronous motor is stationary, the rotating magnetic field due to the stator currents produce an alternating torque on the rotor, i.e. at one instant, the rotor is being urged clockwise, and at the next instant anticlockwise. Since the net torque is zero, a synchronous motor is not self-starting. The methods most commonly used to bring the motor up to synchronism are:

(a) *By means of damping grids in the pole shoes*

These grids consist of copper bars short-circuited at each end, as shown in fig. 15.2. The rotating magnetic flux induce currents in these grids and the machine accelerates like a cage-type induction motor (section 18.1).

During the starting period, the rotor field winding is usually closed through the armature of the exciter, namely a d.c. shunt generator carried on an extension of the shaft; and the current in the stator winding is limited to a permissible value by applying a reduced voltage, e.g. by a star–delta switch or an auto-transformer (section 18.11).

When the machine has reached nearly full speed as an induction motor, the rated voltage is applied to the stator winding and the exciter voltage will have built up sufficiently to magnetize the rotor poles. The resultant magnetic flux due to the stator currents is then moving past the rotor poles at a slow speed and produces a low-frequency alternating torque superimposed upon that exerted by the damping grids. Consequently, the rotor is accelerated at one instant and retarded at the next instant, and the fluctuation of speed may be sufficient to bring the rotor up to synchronous speed. Once this speed has been attained, the rotor continues to run in synchronism.

Owing to the starting.torque being relatively small and the starting currents being relatively large, this method is only suitable when the load—if any—is small.

(b) *By means of a wound rotor*

With this method, the rotor of the synchronous motor is similar to the wound rotor of an induction motor (section 18.8).

The machine is started by connecting the slip-rings to variable resistors—exactly as for a slip-ring induction motor. If the synchronous motor has an exciter carried on an extension of the shaft, the windings of this machine can be connected in series with one of the resistors, as shown in fig. 17.19.

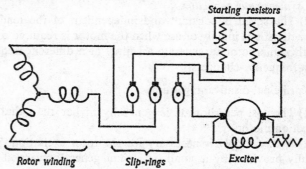

Fig. 17.19 Synchronous-induction motor.

When the speed exceeds a critical value, the exciter voltage builds up and the direct current from this exciter magnetizes the rotor of the synchronous motor with the same number of poles* as the stator. Consequently, by the time the starting resistors are short-circuited, the machine is running at full speed as an induction motor and is also being subjected to a low-frequency alternating torque superimposed upon the induction-motor torque, as already mentioned in section (a). The fluctuation of speed caused by this alternating torque enables the motor to be pulled into synchronism.

This type of machine is referred to as a *synchronous-induction motor* and is capable of exerting a large starting torque with a relatively low starting current.

17.10 Advantages and disadvantages of the synchronous motor

The principal advantages of the synchronous motor are:

(1) The ease with which the power factor can be controlled. An over-excited synchronous motor having a leading power factor can be operated in parallel with induction motors having lagging

---

* With the arrangement shown in fig. 17.19, the whole of the direct current flows through one phase of the rotor winding and half of the current flows through each of the other two phases. The distribution of the magnetic flux produced by these direct currents in a three-phase winding having 1 slot per pole per phase is exactly similar to that shown in fig. 16.5 (*a*).

power factor, thereby improving the power factor of the supply system.

Synchronous motors are sometimes run on *no load* for power-factor correction or for improving the voltage regulation of a transmission line. In such applications, the machine is referred to as a *synchronous capacitor*.

(2) The speed is constant and independent of the load. This characteristic is mainly of use when the motor is required to drive another alternator to generate a supply at a different frequency, as in frequency-changers.

The principal disadvantages are:

(1) The cost per kilowatt is generally higher than that of an induction motor.

(2) A d.c. supply is necessary for the rotor excitation. This is usually provided by a small d.c. shunt generator carried on an extension of the shaft.

(3) Some arrangement must be provided for starting and synchronizing the motor.

EXAMPLES 17

1. A single-phase alternator has a rated output of 500 kVA at a terminal voltage of 3300 V. The stator winding has a resistance of 0·6 Ω and a synchronous reactance of 4 Ω. Calculate the percentage voltage regulation at power factor of (*a*) unity, (*b*) 0·8 lagging and (*c*) 0·8 leading. Sketch the phasor diagram for each case.

2. A single-phase alternator, having a synchronous reactance of 5·5 Ω and a resistance of 0·6 Ω, delivers a current of 100 A. Calculate the e.m.f. generated in the stator winding when the terminal voltage is 2000 V and the power factor of the load is 0·8 lagging. Sketch the phasor diagram.

3. A three-phase, star-connected, 50-Hz alternator has 96 conductors per phase and a flux/pole of 0·1 Wb. The alternator winding has a synchronous reactance of 5 Ω/phase and negligible resistance. The distribution factor for the stator winding is 0·96.

   Calculate the terminal voltage when three non-inductive resistors, of 10 Ω/phase, are connected in star across the terminals. Sketch the phasor diagram for one phase.

4. A 1500-kVA, 6·6-kV, three-phase, star-connected synchronous generator has a resistance of 0·5 Ω/phase and a synchronous reactance of 5 Ω/phase. Calculate the percentage change of voltage when the rated output of 1500-kVA at power factor 0·8 lagging is switched off. Assume the speed and the exciting current to remain unaltered.

5. Two single-phase alternators are connected in parallel, and the excitation of each machine is such as to generate an open-circuit e.m.f. of 3500 V. The stator winding of each machine has a synchronous reactance of 30 Ω and negligible resistance. If there is a phase displacement of 40 electrical degrees between the e.m.f.s, calculate: (a) the current circulating between the two machines; (b) the terminal voltage; and (c) the power supplied from one machine to the other. Assume that there is no external load. Sketch the phasor diagram.

6. Two similar three-phase star-connected alternators are connected in parallel. Each machine has a synchronous reactance of 4·5 Ω/phase and negligible resistance, and is excited to generate an e.m.f. of 1910 V/phase. The machines have a phase displacement of 30 electrical degrees relative to each other. Calculate: (a) the circulating current; (b) the terminal voltage/phase; (c) the power supplied from one machine to the other. Sketch the phasor diagram for one phase.

7. If the two alternators of Q.6 are adjusted to be in exact phase opposition relative to each other and if the excitation of one alternator is adjusted to give an open-circuit voltage of 2240 V/phase and that of the other machine adjusted to give an open-circuit voltage of 1600 V/phase, calculate: (a) the circulating current; (b) the terminal voltage.

8. A single-phase synchronous motor, having a synchronous reactance of 5 Ω and negligible resistance, is connected across a 2200-V supply. If the input power remains constant at 100 kW, calculate the generated e.m.f. when the power factor is: (a) unity; (b) 0·8 lagging; and (c) 0·8 leading. Sketch the phasor diagram for each case.

9. A three-phase, star-connected synchronous motor is connected across a 400-V supply. The stator winding has a synchronous reactance of 0·8 Ω/phase and negligible resistance. If the input power remains constant at 30 kW, calculate the generated e.m.f. per phase when the power factor is: (a) unity; (b) 0·7 lagging; and (c) 0·7 leading. Sketch, for each case, the phasor diagram showing the phase current and voltages for one phase only.

10. A 200-V single-phase synchronous motor takes a current of 10 A at unity power factor from the supply. Calculate the e.m.f. and the angle of retard if the synchronous reactance is 5 Ω and the resistance is negligible.

    If the e.m.f. is decreased by 10 per cent, calculate the current and the power factor. Assume the angle of retard to remain unaltered.

    (E.M.E.U.)

11. Explain why a synchronous motor will only develop a continuous torque at synchronous speed. How does a synchronous motor reach synchronous speed?

    A three-phase synchronous motor has 12 poles and operates from a 440-V, 50-Hz supply. Calculate its speed. If it takes a line current of 100 A at 0·8 power factor lead, what torque will the machine be developing? Neglect losses. (N.C.T.E.C., O.2)

12. An alternator supplying 2800 kW at power factor 0·7 lagging is loaded to its full kVA capacity. If the power factor is raised to unity

by means of an over-excited synchronous motor, how many more kilowatts can the alternator supply and what must be the power factor of the synchronous motor, assuming that the latter absorbs all extra power obtainable from the alternator? Sketch the phasor diagram.

13. A factory takes 600 kVA at a lagging power factor of 0·6. A synchronous motor is to be installed to raise the power factor to 0·9 lagging when the motor is taking 200 kW. Calculate the corresponding kilovoltamperes taken by the synchronous motor and the power factor at which the motor will be operating.

14. A three-phase induction motor, taking 200 kW at 0·8 power factor lagging, is in parallel with a three-phase synchronous motor taking 250 kVA at 0·9 power factor leading. The supply voltage is 3300 V. Calculate the line current and the power factor of the total load. Sketch a phasor diagram.

CHAPTER 18

Three-Phase Induction Motors

18.1 Principle of action

The stator of an induction motor is similar to that of an alternator; and in the case of a machine supplied with three-phase currents, a rotating magnetic flux is produced, as already explained in section 16.3. The rotor core is laminated and the conductors often consist of

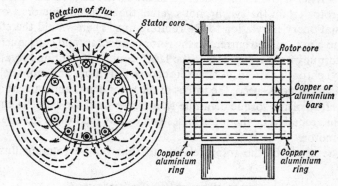

Fig. 18.1 Induction motor with a cage rotor.

uninsulated copper or aluminium bars in semi-enclosed slots, the bars being short-circuited at each end by rings or plates to which the bars are brazed or welded. In motors below about 50 kW, the aluminium rotor bars and end-rings are often cast in one operation. This type is known as the *cage* or *short-circuited* rotor. The airgap between the rotor and the stator is uniform and made as small as is mechanically possible. For simplicity, the stator slots and winding have been omitted in fig. 18.1.

If the stator is wound for two poles, the distribution of the magnetic flux due to the stator currents at a particular instant is shown in fig. 18.1. The e.m.f. generated in a rotor conductor is a maximum in the region of maximum flux density; and if the flux be assumed to rotate anticlockwise, the directions of the e.m.f.s generated in the stationary rotor conductors can be determined by the righthand rule and are indicated by the crosses and dots in fig. 18.1. The e.m.f. generated in the rotor conductor shown in fig. 18.2 circulates a current the effect of which is to strengthen the flux density on the righthand side and weaken that on the lefthand side; i.e. the flux in the

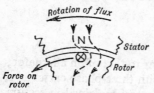

Fig. 18.2 Force on rotor.

gap is distorted as indicated by the dotted lines in fig. 18.2. Consequently, a force is exerted on the rotor tending to drag it in the direction of the rotating flux.

The higher the speed of the rotor, the lower is the speed of the rotating field relative to the rotor winding and the smaller is the e.m.f. generated in the latter. Should the speed of the rotor attain the synchronous value, the rotor conductors would be stationary in relation to the rotating flux. There would therefore be no e.m.f. and no current in the rotor conductors and consequently no torque on the rotor. Hence the latter could not continue rotating at synchronous speed. As the rotor speed falls more and more below the synchronous speed, the values of the rotor e.m.f. and current and therefore of the torque continue to increase until the latter is equal to that required by the rotor losses and by any load there may be on the motor.

The speed of the rotor relative to that of the rotating flux is

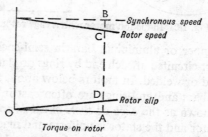

Fig. 18.3 Slip and rotor speed of an induction motor.

termed the *slip*; thus for a torque OA in fig. 18.3, the rotor speed is AC and the slip is AD, where

$$AD = AB - AC = CB.$$

For torques varying between zero and the full-load value, the slip is practically proportional to the torque. It is usual to express the slip either as a per-unit or fractional value or as a percentage of the synchronous speed; thus in fig. 18.3,

$$\text{per-unit slip} = \frac{\text{slip in rev/min}}{\text{synchronous speed in rev/min}} = \frac{AD}{AB}$$

$$= \frac{\text{synchronous speed} - \text{rotor speed}}{\text{synchronous speed}} \qquad (18.1)$$

and percentage slip = per unit or fractional slip $\times$ 100

$$= \frac{AD}{AB} \times 100.$$

The value of the slip at full load varies from about 6 per cent for small motors to about 2 per cent for large machines. The induction motor may therefore be regarded as practically a constant-speed machine; and the difficulty of varying its speed economically constitutes one of its main disadvantages (see section 18.10).

18.2 Frequency of rotor e.m.f. and current

It was shown in section 16.3 that for a three-phase winding with p pairs of poles supplied at a frequency of f hertz the speed of the rotating flux is given by n revolutions/second, where

$$f = np$$

If n_r is the rotor speed in revolutions/second, the speed at which the rotor conductors are being cut by the rotating flux is $(n - n_r)$ revolutions/second,

$$\therefore \qquad \text{frequency of rotor e.m.f.} = f_r = (n - n_r)p$$

If $\qquad s = $ per-unit or fractional slip $= (n - n_r)/n$

then $\qquad\qquad n - n_r = sn$

and $\qquad\qquad f_r = snp = sf \qquad (18.2)$

Polyphase currents in the stator winding produce resultant ampere-turns, the axes of which rotate at synchronous speed, n revolutions/second, relative to the stator. Similarly the polyphase currents in the rotor winding produce resultant ampere-turns the axes of which rotate [by expression (18.2)] at a speed sn revolutions/second relative to the rotor surface, *in the direction of rotation of the rotor*. But the rotor is revolving at a speed n_r revolutions/second relative to the stator core; hence the speed of the resultant rotor ampere-turns relative to the stator core

$$= sn + n_r$$
$$= (n - n_r) + n_r = n \text{ revolutions/second,}$$

i.e. the axes of the resultant rotor ampere-turns are travelling at the same speed as those of the resultant stator ampere-turns, so that they are stationary relative to each other. Consequently the polyphase induction motor can be regarded as being equivalent to a transformer having an airgap separating the iron portions of the magnetic circuit carrying the primary and secondary windings.

Owing to this gap, the magnetizing current and the magnetic leakage for an induction motor are large compared with the corresponding values for a transformer of the same kVA rating. Also, the friction and windage losses contribute towards making the efficiency of the induction motor less than that of the corresponding transformer. On the other hand, the stator ampere-turns have to balance the rotor ampere-turns and also provide the magnetizing and no-load loss components of the stator current, as in a transformer. Hence, an increase of slip due to increase of load is accompanied by an increase of the rotor currents and therefore by a corresponding increase of the stator currents.

18.3 Rotor e.m.f. and current

Let $\quad V_P$ = voltage per phase applied to stator winding,

$\quad\quad Z_s$ = no. of stator conductors in series/phase,

$\quad\quad k_d$ = distribution factor of winding,

$\quad\quad k_p$ = pitch factor of winding

$\quad\quad\quad$ = 1·0 for full-pitch coils,

and $\quad \Phi$ = flux/pole, i.e. total flux entering or leaving the stator over one pole pitch, namely the distance between 2 adjacent points of zero flux density.

Since the back e.m.f. generated in the stator winding is approximately equal to the applied voltage, then from expression (15.3) we have:

$$V_P \simeq 2 \cdot 22 k_d k_p Z_s f \Phi \tag{18.3}$$

When the rotor is at standstill, the rotating flux Φ cuts the rotor at the same speed as it cuts the stator winding, so that the frequency of the rotor e.m.f. is then the same as the supply frequency, namely f hertz. Hence:

if E_0 = rotor e.m.f. generated per phase* at standstill,

and Z_r = no. of rotor conductors in series/phase,

then $$E_0 = 2 \cdot 22 k_d k_p Z_r f \Phi \qquad (18.4)$$

Assuming the distribution and pitch factors to be the same for the stator and rotor windings, we have from (18.3) and (18.4):

$$E_0 \simeq V_P \times \frac{Z_r}{Z_s} \qquad (18.5)$$

If E_r is the rotor e.m.f. generated per phase when the per-unit slip is s and the rotor frequency is $f_r = sf$,

$$E_r = 2 \cdot 22 \times k_d k_p Z_r f_r \Phi$$
$$= sE_0 \qquad (18.6)$$

If R = resistance/phase of the rotor winding,

and X_0 = leakage reactance/phase of rotor winding at standstill

  $= 2\pi f \times$ leakage inductance/phase of rotor winding,

then for per-unit or fractional slip s,

corresponding reactance/phase $= X_r = sX_0$

and corresponding impedance/phase $= Z_r = \sqrt{\{R^2 + (sX_0)^2\}}$ $\qquad (18.7)$

If I_0 = rotor current/phase at standstill,

and I_r = rotor current/phase at slip s,

$$I_0 = \frac{E_0}{\sqrt{(R^2 + X_0{}^2)}}$$

and $$I_r = \frac{E_r}{\sqrt{\{R^2 + (sX_0)^2\}}} = \frac{sE_0}{\sqrt{\{R^2 + (sX_0)^2\}}} \qquad (18.8)$$

---

* In the cage rotor, the number of bars is usually a prime number, such as 47. Consequently the e.m.f.s in all the rotor bars differ from one another in phase, so that the number of phases is the same as the number of rotor bars. The rotor winding may also be of the three-phase type, star- or delta-connected, with its ends joined to 3 slip-rings. The relative advantages and disadvantages of the two types of rotor are discussed in section 18.12.

If ϕ_r be the phase difference between E_r and I_r,

$$\tan \phi_r = X_r/R = sX_0/R \qquad (18.9)$$

and

$$\cos \phi_r = \frac{R}{\sqrt{(R^2 + X_r^2)}} \qquad (18.10)$$

Example 18.1 *A three-phase induction motor is wound for 4 poles and is supplied from a 50-Hz system. Calculate: (a) the synchronous speed; (b) the speed of the rotor when the slip is 4 per cent; and (c) the rotor frequency when the speed of the rotor is 600 rev/min.*

(a) From (16.1), synchronous speed $= \dfrac{60f}{p} = \dfrac{60 \times 50}{2}$

$$= 1500 \text{ rev/min.}$$

(b) From (18.1),
$$0.04 = \frac{1500 - \text{rotor speed}}{1500}$$

$\therefore$ rotor speed $= 1440$ rev/min.

(c) Also from (18.1),   per-unit slip $= \dfrac{1500 - 600}{1500} = 0.6$

Hence, from (18.2), rotor frequency $= 0.6 \times 50 = 30$ Hz.

Example 18.2 *The stator winding of the motor of example 18.1 is delta-connected with 240 conductors per phase and the rotor winding is star-connected with 48 conductors per phase. The rotor winding has a resistance of 0.013 Ω/phase and a leakage reactance of 0.048 Ω/phase at standstill. The supply voltage is 400 V. Assuming the distribution factor to be 0.96 and the pitch factor to be 1.0 for each winding, calculate: (a) the flux per pole; (b) the rotor e.m.f. per phase at standstill with the rotor on open circuit; (c) the rotor e.m.f. and current per phase at 4 per cent slip; and (d) the phase difference between the rotor e.m.f. and current for a slip of (i) 4 per cent and (ii) 100 per cent. Assume the impedance of the stator winding to be negligible.*

(a) From (18.3), $400 = 2.22 \times 0.96 \times 240 \times 50 \times \Phi$

$\therefore$ $\Phi = 0.015\ 65$ Wb.

(b) From (18.4), $E_0 = 2.22 \times 0.96 \times 48 \times 50 \times 0.015\ 65$
$$= 80 \text{ V.}$$

Alternatively, $E_0 = 400 \times 48/240 = 80$ V.

(c) From (18.6),

rotor e.m.f. for 4 per cent slip $\Big\} = 80 \times 0.04 = 3.2 \text{ V}$

From (18.7),

impedance/phase for 4 per cent slip $\Big\} = \sqrt{\{(0.013)^2 + (0.04 \times 0.048)^2\}}$

$$= 0.013\ 14\ \Omega.$$

∴ rotor current $= 3.2/0.013\ 14 = 243.5 \text{ A}.$

(d) From (18.9), it follows that for 4 per cent slip,

$$\tan \phi_r = \frac{0.04 \times 0.048}{0.013} = 0.1477$$

$$\phi_r = 8° \ 24'.$$

For 100 per cent slip, $\tan \phi_r = 0.048/0.013 = 3.692,$

∴ $$\phi_r = 74° \ 51'.$$

This example shows that the slip has a considerable effect upon the phase difference between the rotor e.m.f. and current—a fact that is very important when we come to discuss the variation of torque with slip.

18.4 Relationship between the rotor I^2R loss and the rotor slip

The following table indicates concisely what becomes of the power supplied to the stator of the induction motor:

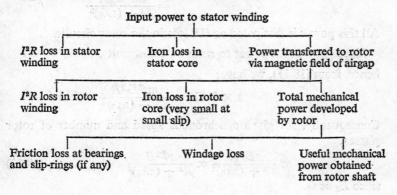

Input power to stator winding

I^2R loss in stator winding	Iron loss in stator core	Power transferred to rotor via magnetic field of airgap

I^2R loss in rotor winding	Iron loss in rotor core (very small at small slip)	Total mechanical power developed by rotor

Friction loss at bearings and slip-rings (if any)	Windage loss	Useful mechanical power obtained from rotor shaft

If T = torque, in newton-metres, exerted on the rotor by the rotating flux

and n_s = synchronous speed in revolutions/second,

power transferred from stator to rotor = $2\pi T n_s$ watts.

This input power to the rotor is often referred to as the *torque in synchronous watts*.

If n_r = rotor speed in revolutions/second,

$$\left. \begin{array}{c} \text{total mechanical power developed} \\ \text{by rotor} \end{array} \right\} = 2\pi T n_r \text{ watts.}$$

But from the table on page 485 it is seen that:

total I^2R loss in rotor

$$\simeq \left(\begin{array}{c} \text{power transferred from} \\ \text{stator to rotor} \end{array} \right) - \left(\begin{array}{c} \text{total mechanical power} \\ \text{developed by rotor} \end{array} \right)$$

$$= 2\pi T(n_s - n_r) \text{ watts}$$

$$\therefore \quad \frac{\text{total rotor } I^2R \text{ loss}}{\text{input power to rotor}} = \frac{2\pi T(n_s - n_r)}{2\pi T n_s} = s \qquad (18.11)$$

or $\qquad$ total rotor I^2R loss = s × input power to rotor
$\qquad\qquad$ (in watts) $\qquad\qquad\qquad\qquad$ (in watts)

18.5 Factors determining the torque

If $\qquad\qquad m$ = number of rotor phases,

then, using the symbols given in section 18.3, we have:

electrical power generated in rotor = $m I_r E_r \cos \phi_r$ watts

$$= \frac{m s^2 E_0^2 R}{R^2 + (sX_0)^2}.$$

All this power is dissipated as I^2R loss in the rotor circuits.

Since $\qquad$ input power to rotor = $2\pi T n_s$ watts,

hence, from (18.11), we have:

$$s \times 2\pi T n_s = \frac{m s^2 E_0^2 R}{R^2 + (sX_0)^2}.$$

Consequently, for given synchronous speed and number of rotor phases,

$$T \propto \frac{s E_0^2 R}{R^2 + (sX_0)^2} \propto \frac{s \Phi^2 R}{R^2 + (sX_0)^2} \qquad (18.12)$$

since $E_0 \propto \Phi$.

18.6 Variation of torque with slip, other factors remaining constant

If the impedance of the stator winding is assumed to be negligible, then for a given supply voltage, Φ and E_0 remain constant,

$$\therefore \qquad \text{torque} \propto \frac{sR}{R^2 + (sX_0)^2} \qquad (18.13)$$

The value of X_0 is usually far greater than the resistance of the rotor winding; so let us for simplicity assume $R = 1\ \Omega$ and $X_0 = 8\ \Omega$, and calculate the value of $sR/(R^2 + s^2X_0^2)$ for various values of

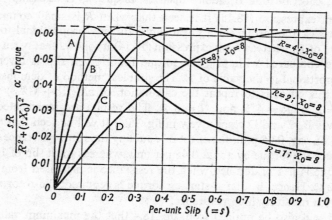

Fig. 18.4 Torque/slip curves for an induction motor.

the slip between 1 and 0. The results are represented by curve A in fig. 18.4. It will be seen that for small values of the slip, the torque is almost directly proportional to the slip; whereas for slips between about 0·2 and 1, the torque is almost inversely proportional to the slip. These relationships can be easily deduced from expression (18.13). Thus, in the case of the cage rotor, R is small compared with X_0, but for values of the slip less than about 0·1 per unit, $(sX_0)^2$ is very small compared with R^2, so that:

$$\text{torque} \propto \frac{sR}{R^2} \propto \frac{s}{R} \qquad (18.14)$$

i.e. the torque is directly proportional to the slip when the latter is very small.

For large values of the slip, R^2 is very small compared with $(sX_0)^2$

for the cage rotor and for the slip-ring rotor with no external resistance (section 18.8),

$$\therefore \qquad \text{torque} \propto \frac{sR}{(sX_0)^2} \propto \frac{R}{s} \qquad (18.15)$$

since X_0 is constant for a given motor; i.e. the torque is inversely proportional to the slip when the latter is large. The term R has been left in the above expressions as it is referred to in the next section.

18.7 Effect of rotor resistance upon the torque/slip relationship

From expression (18.15) it is seen that when R is small compared with sX_0, the torque for a given slip is directly proportional to the value of R; whereas from expression (18.14) it follows that when R is large compared with sX_0, the torque for a given slip is inversely proportional to the value of R. The simplest method of demonstrating this effect is to repeat the calculation of $sR/(R^2 + s^2X_0^2)$ with $R = 2 \; \Omega$, $R = 4 \; \Omega$ and $R = 8 \; \Omega$. The results are represented by curves B, C and D respectively in fig. 18.4. It will be seen that for a slip of, say, 0·05 p.u., the effect of doubling the rotor resistance is to reduce the torque by about 0·45 per unit, whereas for a slip of 1, the torque is nearly doubled when the resistance is increased from 1 Ω to 2 Ω. Hence, if a large starting torque is required, the rotor must have a relatively high resistance.

It will also be noticed from fig. 18.4 that the maximum value of the torque is the same for the four values of R and that the larger the resistance the greater is the slip at maximum torque. The condition for maximum torque can be derived by differentiating (18.13) with respect to s, assuming R to remain constant, or with respect to R, assuming s to remain constant. Both methods give the same result; thus with the first method, the torque is maximum when:

$$\frac{d}{ds}\left(\frac{sR}{R^2 + s^2X_0^2}\right) = \frac{(R^2 + s^2X_0^2)R - sR \times 2sX_0^2}{(R^2 + s^2X_0^2)^2} = 0$$

i.e. $\qquad\qquad R^2 - s^2X_0^2 = 0$

so that $\qquad\qquad sX_0 = R \qquad\qquad\qquad (18.16)$

Hence the torque is a maximum* when the reactance is equal to the resistance. For instance, with $R = 1 \; \Omega$ and $X_0 = 8 \; \Omega$, maximum

---

* This can be checked by differentiating expression $(R^2 - s^2X_0^2)$ with respect to s. The result, namely $-2sX_0^2$, is negative; hence expression (18.16) gives the condition for *maximum* torque.

torque occurs when $s = 0.125$ p.u.; whereas with $R = 8\,\Omega$ and $X_0 = 8\,\Omega$, maximum torque occurs when $s = 1$, namely when the rotor is at standstill.

Substituting R for sX_0 in expression (18.13), we have:

$$\text{maximum torque} \propto \frac{sR}{2R^2}$$

$$\propto \frac{1}{2X_0}.$$

But X_0 is the leakage reactance at standstill and is a constant for a given rotor; hence the maximum torque is the same whatever the value of the rotor resistance.

18.8 Starting torque

At the instant of starting, $s = 1$, and it will be seen from fig. 18.4 that with a motor having a low-resistance rotor, such as the usual type of cage rotor, the starting torque is small compared with the maximum torque available. On the other hand, if the bars of the

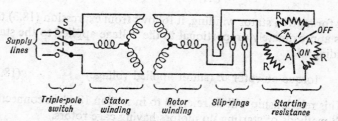

Fig. 18.5 Induction motor with slip-ring rotor.

cage rotor were made with sufficiently high resistance to give the maximum torque at standstill, the slip for full-load torque—usually about one-third to one-half of the maximum torque—would be relatively large and the I^2R loss in the rotor winding would be high, with the result that the efficiency would be low; and if this load was maintained for an hour or two, the temperature rise would be excessive. Also the variation of speed with load would be large (see section 18.10). Hence, when a motor is required to exert its maximum torque at starting, the usual practice is to insert extra resistance into the rotor circuit and to reduce this resistance as the motor accelerates. Such an arrangement involves a three-phase winding on the rotor, the three ends of the winding being connected via slip-rings on the shaft to external star-connected resistors R, as shown in fig. 18.5. The three arms, A, are mechanically and electrically connected together.

The starting procedure is to close the triple-pole switch S and then move arms A clockwise as the motor accelerates, until, at full speed, the arms are in the *ON* position shown dotted in fig. 18.5, and the starting resistors have been cut out of the rotor circuit. Large motors are often fitted with a short-circuiting and brush-lifting device which first short-circuits the three slip-rings and then lifts the brushes off the rings, thereby eliminating losses due to the brush-contact resistance and the brush friction and reducing the wear of the brushes and of the slip-rings.

18.9 Variation of torque with stator voltage, other factors remaining constant

From expression (18.12), it is seen that for given values of the slip and of the rotor resistance and reactance,

$$\text{torque on rotor} \propto \Phi^2.$$

But for a given stator winding, it follows from expression (18.3) that Φ is approximately proportional to the voltage applied to the stator winding:

$$\therefore \qquad \text{torque on rotor} \propto (\text{stator applied voltage})^2 \qquad (18.17)$$

This relationship will be referred to in section 18.11 in connection with methods of starting up motors having cage rotors.

18.10 Speed control by means of external rotor resistors

From expression (18.11), it follows that for a given input power to the rotor and therefore for a given torque exerted by the rotor, the total rotor I^2R loss is proportional to the slip. Thus, if a motor has 100 kW transferred from the stator to the rotor when the slip is 5 per cent, the total rotor I^2R loss is 5 kW and the mechanical power developed by the rotor is 95 kW. But if the slip is increased to, say, 40 per cent by the addition of external resistors in the rotor circuit and if the *torque developed by the rotor remains unaltered*, the I^2R loss in the rotor circuit increases to 40 kW and the mechanical power developed by the rotor decreases to 60 kW. Hence the efficiency of the motor has been considerably reduced.

Speed control by means of external rotor resistors has the following disadvantages:

(*a*) reduction of speed is accompanied by reduced efficiency;

(*b*) with a large resistance in the rotor circuit, the speed varies considerably with variation of torque (see fig. 18.4);

(*c*) the external rotor resistors are comparatively bulky and expensive as they may have to dissipate a good deal of power without becoming overheated.

The main advantage of this method of speed control is its simplicity.

Example 18.3 *The power supplied to a three-phase induction motor is 40 kW and the corresponding stator losses are 1·5 kW. Calculate: (a) the total mechanical power developed and the rotor I²R loss when the slip is 0·04 per unit; (b) the output power of the motor if the friction and windage losses are 0·8 kW; and (c) the efficiency of the motor. Neglect the rotor iron loss.*

(*a*) Input power to rotor = 40 − 1·5 = 38·5 kW.

From (18.11)

$$\frac{\text{rotor } I^2R \text{ loss in kW}}{38\cdot5} = 0\cdot04$$

∴   rotor I^2R loss = 1·54 kW

so that mechanical power developed by the rotor $\left.\right\}$ = 38·5 − 1·54

= 36·96 kW

(*b*)   Output power of motor = 36·96 − 0·8

= 36·16 kW

(*c*)   Efficiency of motor = 36·16/40 = 0·904 p.u.

= 90·4 per cent.

Example 18.4 *If the speed of the motor of example 18.3 is reduced to 40 per cent of its synchronous speed by means of external rotor resistors, calculate: (a) the total rotor I²R loss, and (b) the efficiency, assuming the torque and the stator losses to remain unaltered. Also, assume that the increase in the rotor iron loss is equal to the reduction in the friction and windage loss.*

(*a*)   New slip $= (100 - 40)/100 = 0.6$ p.u.
and   input power to rotor $= 38.5$ kW

From (18.11),

total rotor I^2R loss $= 0.6 \times 38.5 = 23.1$ kW.

(*b*)   Total losses in rotor $= 23.1 + 0.8 = 23.9$ kW
∴   output power of motor $= 38.5 - 23.9 = 14.6$ kW
and   efficiency of motor $= 14.6/40 = 0.365$ p.u.

$= 36.5$ per cent.

18.11 Starting of a three-phase induction motor fitted with a cage rotor

If this type of motor is started up by being switched directly across the supply, the starting current is about 4–7 times the full-load current, the actual value depending upon the size and design of the machine. Such a large current may cause a relatively large voltage drop in the cables and thereby produce an objectionable momentary dimming of the lamps in the vicinity. Consequently it is usual to start cage motors—except small machines—with a reduced voltage, using one of the following methods:

(1) *Star–delta starter*

The two ends of each phase of the stator winding are brought out to the starter which, when moved to the 'starting' position, connects the winding in star. After the motor has accelerated, the starter is quickly moved to the 'running' position, thereby changing the connections to delta. Hence the voltage per phase at starting is $1/\sqrt{3}$ of the supply voltage, and the starting torque, by expression (18.17), is one-third of that obtained if the motor were switched directly across the supply with its stator winding delta-connected. Also, the starting current per phase is $1/\sqrt{3}$ and that taken from the supply is one-third of the corresponding value with direct switching.

(2) *Auto-transformer starter*

In fig. 18.6, T represents a three-phase star-connected auto-transformer (see section 14.20) with a mid-point tapping on each phase so that the voltage applied to motor M is half the supply voltage. With such tappings, the supply current and the starting torque are only a quarter of the values when the full voltage is applied to the motor.

After the motor has accelerated, the starter is moved to the 'running' position, thereby connecting the motor directly across the supply and opening the star-connection of the auto-transformer.

Taking the general case where the output voltage per phase of the auto-transformer is n times the input voltage per phase, we have—

$$\frac{\text{starting torque with auto-transformer}}{\text{starting torque with direct switching}}$$
$$= \left(\frac{\text{output voltage}}{\text{input voltage}}\right)^2 = n^2.$$

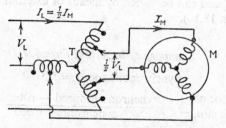

Fig. 18.6 Starting connections of an auto-transformer starter.

Also, if Z_0 be the equivalent standstill impedance/phase of a star-connected motor, referred to the stator circuit, and if V_P be the star voltage of the supply, then,

with direct switching, starting current $= V_P/Z_0$.

With auto-transformer starting,

$$\text{starting current in motor} = nV_P/Z_0$$

and   starting current from supply $= n \times$ motor current
$$= n^2V_P/Z_0$$

∴ $\dfrac{\text{starting current with auto-transformer}}{\text{starting current with direct switching}}$
$$= (n^2V_P/Z_0) \div (V_P/Z_0) = n^2.$$

The auto-transformer is usually arranged with two or three tappings per phase so that the most suitable ratio can be selected for a given motor, but an auto-transformer starter is more expensive than a star–delta starter.

18.12 Comparison of cage and slip-ring rotors

The cage rotor possesses the following advantages:

(*a*) cheaper and more robust;

(*b*) slightly higher efficiency and power factor;

(*c*) explosion proof, since the absence of slip-rings and brushes eliminates risk of sparking.

The advantages of the slip-ring rotor are:

(*a*) the starting torque is much higher and the starting current much lower;

(*b*) the speed can be varied by means of external rotor resistors (see section 18.10).

Summary of important formulae

$$\text{Synchronous speed} = f/p \tag{16.1}$$

$$\left.\begin{array}{r}\text{Fractional or per-}\\ \text{unit slip, } s\end{array}\right\} = \frac{\text{synchronous speed} - \text{rotor speed}}{\text{synchronous speed}} \tag{18.1}$$

$$\text{Rotor frequency} = f_r = sf \tag{18.2}$$

$$\text{Rotor e.m.f./phase} = E_r = sE_0 \tag{18.6}$$

$$\text{Rotor impedance/phase} = Z_r = \sqrt{\{R^2 + (sX_0)^2\}} \tag{18.7}$$

$$s = \frac{\text{total rotor } I^2R \text{ loss}}{\text{input power to rotor}} \tag{18.11}$$

$$\text{Torque on rotor} \propto \frac{\Phi^2 sR}{R^2 + (sX_0)^2} \tag{18.12}$$

For small slip,
$$\text{torque} \propto s/R \tag{18.14}$$

For large slip and low rotor resistance,
$$\text{torque} \propto R/s \tag{18.15}$$

For maximum torque,
$$R = sX_0 \tag{18.16}$$

$$\text{Starting torque} \propto (\text{stator applied voltage})^2 \tag{18.17}$$

EXAMPLES 18

1. Explain how slip-frequency currents are set up in the rotor windings of a three-phase induction motor.

   A two-pole, three-phase, 50-Hz induction motor is running on load with a slip of 4 per cent. Calculate the actual speed and the synchronous speed of the machine.

Sketch the speed–load characteristic for this type of machine and state with which kind of d.c. motor it compares.   (U.E.I., O.2)

2. Explain why an induction motor cannot develop torque when running at synchronous speed. Define the slip speed of an induction motor and deduce how the frequency of rotor currents and the magnitude of the rotor e.m.f. is related to slip.

An induction motor has four poles and is energized from a 50-Hz supply. If the machine runs on full load at 2 per cent slip, determine the running speed and the frequency of the rotor currents.
(N.C.T.E.C., O.2)

3. Give a clear explanation of the following effects in a three-phase induction motor: (*a*) the production of the rotating field; (*b*) the presence of an induced rotor current; (*c*) the development of the torque.   (S.A.N.C., O.2)

4. If a six-pole induction motor supplied from a three-phase 50-Hz supply has a rotor frequency of 2·3 Hz, calculate (*a*) the percentage slip and (*b*) the speed of the rotor in revolutions/minute.

5. Show how a rotating magnetic field can be produced by three-phase currents.

A fourteen-pole, 50-Hz induction motor runs at 415 rev/min. Deduce the frequency of the currents in the rotor winding and the slip.
(App. El., L.U.)

6. Explain the principle of action of a three-phase induction motor and the meaning of the term *slip*. How does slip vary with the load?

A centre-zero d.c. galvanometer, suitably shunted, is connected in one lead of the rotor of a three-phase, six-pole, 50-Hz slip-ring induction motor and the pointer makes 85 complete oscillations per minute. What is the rotor speed?   (U.E.I.)

7. Describe, in general terms, the principle of operation of a three-phase induction motor.

The stator winding of a three-phase, eight-pole, 50-Hz induction motor has 720 conductors, accommodated in 72 slots. Calculate: (*a*) the distribution factor of the winding; (*b*) the flux per pole of the rotating field in the airgap of the motor, needed to generate 230 V in each phase of the stator winding.   (App. El., L.U.)

8. A three-phase induction motor, at standstill, has a rotor voltage of 100 V between the slip-rings when they are open-circuited. The rotor winding is star-connected and has a leakage reactance of 1 Ω/phase at standstill and a resistance of 0·2 Ω/phase. Calculate: (*a*) the rotor current when the slip is 4 per cent and the rings are short-circuited; (*b*) the slip and the rotor current when the rotor is developing maximum torque. Assume the flux to remain constant.

9. A three-phase, 50-Hz induction motor with its rotor star-connected gives 500 V (r.m.s.) at standstill between the slip-rings on open circuit. Calculate the current and power factor at standstill when the rotor winding is joined to a star-connected external circuit, each phase of which has a resistance of 10 Ω and an inductance of 0·04 H. The resistance per phase of the rotor winding is 0·2 Ω and its inductance is

0·04 H. Also calculate the current and power factor when the slip-rings are short-circuited and the motor is running with a slip of 5 per cent. Assume the flux to remain constant. (App. El., L.U.)

10. If the star-connected rotor winding of a three-phase induction motor has a resistance of 0·01 Ω/phase and a standstill leakage reactance of 0·08 Ω/phase, what must be the value of the resistance per phase of a starter to give the maximum starting torque? What is the percentage slip when the starting resistance has been reduced to 0·02 Ω/phase, if the motor is still exerting its maximum torque?

11. A three-phase, 50-Hz, six-pole induction motor has a slip of 0·04 per unit when the output is 20 kW. The frictional loss is 250 W. Calculate: (a) the rotor speed, and (b) the rotor I^2R loss.

12. Sketch the usual form of the torque/speed curve for a polyphase induction motor and explain the factors which determine the shape of this curve.

   In a certain eight-pole, 50-Hz machine, the rotor resistance per phase is 0·04 Ω and the maximum torque occurs at a speed of 645 rev/min. Assuming that the airgap flux is constant at all loads, determine the percentage of maximum torque: (a) at starting, and (b) when the slip is 3 per cent. (App. El., L.U.)

13. Describe briefly the construction of the stator and slip-ring rotor of a three-phase induction motor. Explain the action of the motor and why the rotor is provided with slip-rings.

   A three-phase, 50-Hz induction motor has 4 poles and runs at a speed of 1440 rev/min when the total torque developed by the rotor is 70 N·m. Calculate: (a) the total input (in kilowatts) to the rotor; (b) the rotor copper loss in watts. (App. El., L.U.)

14. Explain how a rotating magnetic field may be produced by stationary coils carrying three-phase currents.

   Determine the efficiency and the output kilowatts of a three-phase, 400-V induction motor running on load with a fractional slip of 0·04 and taking a current of 50 A at a power factor of 0·86. When running light at 400 V, the motor has an input current of 15 A and the power taken is 2000 W, of which 650 W represent the friction, windage and rotor core loss. The resistance per phase of the stator winding (delta-connected) is 0·5 Ω. (App. El., L.U.)

15. Calculate the relative values of (i) the starting torque and (ii) the starting current of a three-phase cage-rotor induction motor when started by: (a) direct switching; (b) a star–delta starter; and (c) an auto-transformer having 40 per cent tappings.

CHAPTER 19

Electronic Valves*

19.1 The two-electrode vacuum valve or diode

If a metal cylinder A surrounds an incandescent filament C in an evacuated glass bulb G, as shown in fig. 19.1, and a battery B is connected in series with a milliammeter D between the cylinder and the negative end of the filament, it is found that an electric current

flows through the milliammeter when the cylinder is made positive relative to the filament; but when the connections to battery B are reversed, so as to make A negative, there is no current through D.

Let us now consider the reason for this behaviour.

An electrical conductor contains a large number of mobile or free electrons that are not attached to any particular atom of the material, but move at random from one atom

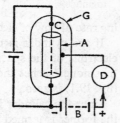

Fig. 19.1 A vacuum diode.

to another within the boundary of the conductor; and the higher the temperature of the conductor, the greater is the velocity attained by these electrons. In the case of an incandescent tungsten filament, for instance, some of these free electrons may acquire sufficient momentum to overcome the forces tending to hold them within the boundary of the filament. Consequently they escape outwards; but if there is no p.d. between the filament and the surrounding cylinder, the electrons emitted from the filament form a negatively-charged cloud or *space charge* around the wire, the latter being left positively charged. Hence the electrons near the surface experience a force urging them to re-enter the filament, and a condition of equilibrium is established in which electrons re-enter the surface at nearly the

---

* An electronic valve or tube is a device in which conduction takes place by electrons or ions between electrodes through a vacuum or gaseous medium within a gas-tight envelope. A device which conforms with this definition but is used *only* for lighting is called a lamp, and such devices are dealt with in Chapter 21.

same rate as they are being emitted, the difference being the electrons which succeed in passing through to cylinder A and represent a current of the order of microamperes. This current is reduced to zero if the potential of the cylinder is made about 1 volt negative relative to the filament.

If cylinder A is made positive in relation to the filament, electrons

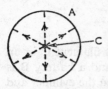

Fig. 19.2 Electron paths from cathode to anode.

are attracted outwards from the space charge, as indicated by the dotted radial lines in fig. 19.2, and fewer of the electrons emitted from the filament are repelled back into the latter. The number of electrons reaching the cylinder increases with increase in the positive potential of the cylinder, as shown in fig. 19.3, until ultimately all the electrons emitted from the filament travel to the cylinder. The corresponding rate of flow of electrons is referred to as the *saturation current*.

If the cylinder is made negative in relation to the filament, the electrons of the space charge are repelled towards the filament, so that none reaches the cylinder and the reading on milliammeter D (fig. 19.1) is zero.

Since the cylinder is normally positive in relation to the filament, the former is termed the *anode* and the latter the *cathode*; and since

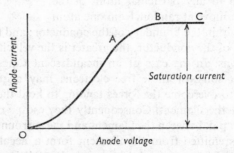

Fig. 19.3 Anode-current/anode-voltage characteristic of a vacuum diode.

current can flow in one direction only, the arrangement is termed a *thermionic valve* or merely a *valve*. The liberation of electrons from an electrode by virtue of its temperature is referred to as *thermionic emission*. When a valve contains only an anode and a cathode, it is referred to as a *diode*. The p.d. between the anode and the cathode is termed the *anode voltage* and the rate of flow of electrons from cathode to anode constitutes the *anode current*. The *conventional* direction of the anode current is from the positive of battery B via the anode to the cathode, as indicated by the arrowhead in fig. 19.1; but it is important to realize that the anode current is actually a movement of electrons in the reverse direction.

The thermionic emission from a metallic surface is given approximately by Richardson's formula:

$$I = AT^2 e^{-b/T} \text{ amperes/square metre}$$

where $\quad T$ = thermodynamic temperature, in kelvins,

$\qquad$ = 273·15 + temperature in °C,

$\qquad$ e = Napierian base = 2·718

and $\quad A$ and b = constants for a given material.

The current represented by the above expression is the saturation current, namely the anode current when all the electrons emitted from the cathode are attracted to the anode. This condition is represented by range BC of the characteristic* in fig. 19.3; and the corresponding value of the anode current is referred to as the *temperature-limited current*, since, for a given cathode, it depends only upon the temperature of the latter.

Over range OB, the same number of electrons are being emitted from the cathode, but the number which reaches the anode depends upon the combined effect of the space charge and the anode voltage. The lower the anode voltage, the more effective is the opposition of the space charge to the movement of electrons from the cathode to the anode. Over range OA, the value of this *space-charge limited current* is represented approximately by the expression:

$$i_A = kv_A^{1.5}$$

where k = a constant for a given valve.

19.2 Construction of a diode

The anode is stamped out of nickel sheet and the cathode may be either of the directly heated or of the indirectly heated type. In the former, the cathode is a filament heated by current passing through it. There are three types of directly heated filaments:

(a) *Pure tungsten filament*, which has to be operated at the relatively

---

* The characteristic shown in fig. 19.3 is for a pure tungsten cathode. The anode current for an oxide-coated cathode continues to increase, but at a reduced rate, with increase of anode voltage.

high temperature of about 2200°C in order that it may give an emission of about 5 mA per watt absorbed by the filament. It has the advantage, however, of being able to withstand positive-ion bombardment due to any residual gas (see section 19.7) far better than any other type of cathode. It is therefore used for valves operated with very high anode voltages.

(b) *Thoriated-tungsten filament*, containing a small quantity (about 1–2 per cent) of thorium oxide together with a reducing agent—usually carbon. By a special activating process the impregnated carbon reduces some of the thorium oxide to metallic thorium which diffuses to the surface, where it forms a layer, one molecule deep, capable of emitting about 60 mA/W at about 1600°C. Thoriated tungsten can withstand positive-ion bombardment better than the oxide-coated cathode and is used for valves working at moderate anode voltages of about 1000–5000 V.

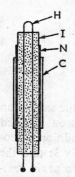

Fig. 19.4 An indirectly heated cathode.

(c) *Oxide-coated filament*, consisting of a mixture of barium and strontium oxides on a nickel wire. Such a cathode is operated at about 750°C and is capable of emitting about 100 mA/W absorbed by the filament.

In the *indirectly heated* type, the cathode C consists of a mixture of barium and strontium oxides sprayed on a hollow nickel cylinder N, as in fig. 19.4. The cathode is heated by a tungsten filament H, known as the *heater*, embedded in an insulator I to prevent the heater making electrical contact with the cathode. Oxide emitters must be activated by special heat-treatment to produce a layer of metallic molecules of barium and strontium on the surface of the oxide.

An indirectly heated cathode gives greater flexibility in the spacing of the electrodes and makes it possible to use an a.c. supply for heating the cathode. If a directly heated cathode were supplied from an a.c. source, the alternating p.d. across the filament would cause a corresponding variation in the value of the anode current. Also, in apparatus incorporating several valves, the indirectly heated cathode allows the heaters to be supplied from a common source.

The oxide-coated cathode has a higher efficiency and, in the absence of positive-ion bombardment, has a longer life than the other types of cathode and is almost universally used in small valves.

19.3 Static characteristic of a vacuum diode

The static* characteristic of a diode gives the relationship between the anode current and the anode voltage for a given filament voltage and therefore for a given filament temperature. A convenient circuit arrangement for determining the characteristic of a directly heated vacuum diode is shown in fig. 19.5. The filament voltage is given by voltmeter V_F and its value can be varied by means of R_1. The anode voltage can be varied between zero and any desired maximum by means of a resistor R_2 arranged as a voltage divider. Fig. 19.6 shows typical results obtained with different filament voltages. It will be seen that with a relatively large anode voltage, a further increase of anode

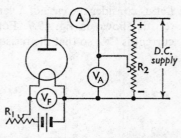

Fig. 19.5 Determination of the static characteristic of a vacuum diode.

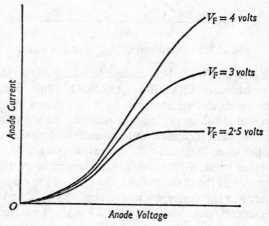

Fig. 19.6 Static characteristics of a vacuum diode.

voltage produces only a relatively small increase of anode current, i.e. the valve is reaching a state of saturation, when practically all the electrons emitted by the cathode are being attracted to the

---

* The *dynamic* characteristic of a diode gives the relationship between the anode current and the *supply* voltage when a load, such as a resistor, is connected in series with the anode (example 19.1).

anode. The smaller the filament voltage and therefore the lower the cathode temperature, the smaller is the saturation current.

19.4 Anode a.c. or slope resistance: Load line

Let us consider the case of a vacuum diode having the static characteristic shown in fig. 19.7. For an anode voltage OC, the anode current is CA, so that the resistance between the anode and cathode

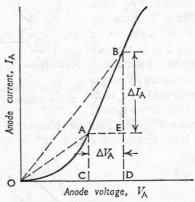

Fig. 19.7 Static characteristic of a vacuum diode.

= OC/CA = 1/tan AOC. For an anode voltage OD, the anode–cathode resistance = OD/DB = 1/tan BOD. This resistance is termed the *anode d.c. resistance*; and it is obvious that its value varies with anode voltage, being high when the anode voltage is low and decreasing with increased voltage, until when saturation is approached, the value increases with increased voltage.

On the other hand, if the diode is being operated on the straight-line portion AB of the characteristic, the ratio of the change, ΔI_A, of anode current to the corresponding change, ΔV_A, of anode voltage remains practically constant. The ratio $\Delta V_A/\Delta I_A$ is referred to as the *anode a.c. resistance** or the *anode slope resistance* of the valve and is represented by the symbol r_a.

Example 19.1 *The following table gives the corresponding values of anode voltage and current for a certain vacuum diode:*

V_A (in volts)	0	10	20	30	40	50	60
I_A (in milliamperes)	0	0·2	1	2	3	4	5

* In B.S. 204:1960, *anode a.c. resistance* is recommended as the preferred term.

A 10-kΩ *load resistor R is connected in series with the anode. Deter-*
mine (a) *the anode d.c. resistance for an anode voltage of* 50 *V and* (b)
the anode a.c. resistance of the diode. Derive the dynamic charac-
teristic (see footnote on p. 501), *namely*
the graph representing the variation of the
anode current I_A *with the supply voltage V*
in fig. 19.8.

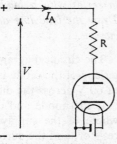

Graph S in fig. 19.9 is the static characteristic plotted from the values given in the
table.

(*a*) For $V_A = 50$ V, $I_A = 4$ mA,

∴ anode d.c. resistance = (50/4) × 1000

$$= 12\ 500\ \Omega.$$

Fig. 19.8 Circuit diagram
for example 19.1.

(*b*) For values of anode current between
1 mA and 5 mA the characteristic is linear, hence:

$$\text{anode a.c. resistance} = r_a = \frac{60 - 20}{5 - 1} \times 1000$$

$$= 10\ 000\ \Omega.$$

For an anode current of, say, 2 mA, the p.d. between anode and
cathode is 30 V and is represented by AB in fig. 19.9. The voltage
drop in R due to a current of 2 mA is 0·002 × 10 000, namely 20 V,

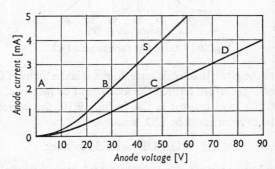

Fig. 19.9 Static and dynamic characteristics for example 19.1.

and is represented by BC. Hence the total voltage for an anode
current of 2 mA is (30 + 20), namely 50 V, and is represented by AC
in fig. 19.9. By repeating this construction for several values of the

*

anode current we can obtain sufficient number of points to enable the dynamic characteristic D to be drawn.

Example 19.2 *The diode referred to in example 19.1 is connected in series with a load resistor of 12 kΩ across a 60-V d.c. supply. Draw the load line and determine therefrom the value of the anode current.*

The circuit diagram is as shown in fig. 19.8 and the static characteristic of the diode is represented by graph S in fig. 19.10. The p.d. of 60 V across the diode and load resistor is represented by OA.

If the anode and cathode of the diode were short-circuited, the whole of the supply voltage would be applied across the 12-kΩ resistor, and the current would be 60/12 000 A, namely 5 mA. This current is represented by OB in fig. 19.10. For the given supply

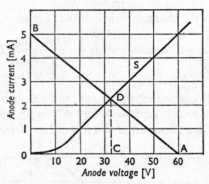

Fig. 19.10 Load line for example 19.2.

voltage and the given load resistance, line AB is the locus of the anode current and is therefore termed the *load line*. For instance, when the p.d. across the load resistor is CA in fig. 19.10, the corresponding value of the current must be CD, since

$$CA/CD = OA/OB = \text{load resistance}.$$

Since the current in the diode in fig. 19.8 is the same as that in the load resistor, the value of this current must be given by the point of intersection of the static characteristic and the load line, i.e. by point D in fig. 19.10. For this condition, OC represents the p.d. between the anode and cathode of the diode, CA represents the p.d. across the load resistor and CD represents the corresponding current. From the graph,

$$\text{anode current} = CD = 2 \cdot 28 \text{ mA.}$$

Alternatively, $\quad\quad$ anode–cathode p.d. = OC = 32·6 V,

∴ $\quad\quad\quad$ p.d. across load resistor = CA = OA − OC

$$= 60 - 32 \cdot 6 = 27 \cdot 4 \text{ V,}$$

hence $\quad\quad\quad\quad$ anode current = $(27 \cdot 4/12\ 000)$ A

$$= 2 \cdot 28 \text{ mA.}$$

The use of a load line is a very convenient method of dealing with a non-linear circuit, i.e. a circuit in which the current is not directly proportional to the applied voltage. The load line is discussed more fully in section 19.17.

19.5 The vacuum diode as a half-wave rectifier

In fig. 19.11 a transformer T is arranged with two secondary windings, one giving a low-voltage supply for the heater of diode D and

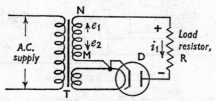

Fig. 19.11 Half-wave rectifier.

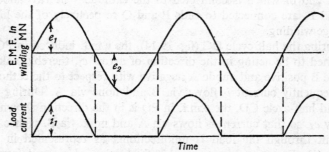

Fig. 19.12 Waveforms of input voltage and output current of a half-wave rectifier.

the other giving a suitable high-voltage supply for the anode circuit. One end of the high-voltage winding MN is connected to the cathode of the diode, and a load resistor R is connected between the anode and the other end of the high-voltage winding. When the e.m.f. induced in winding MN is in the direction represented by e_1 in figs. 19.11 and 19.12, the anode is positive in relation to the cathode, and a current i_1 flows through load resistor R. But when the e.m.f. of MN is in the reverse direction, as shown by e_2, the anode is negative in relation to the cathode and there is no current. Hence the waveform of the load current is represented by the lower curve in fig. 19.12.

19.6 Full-wave rectification by means of a double-diode

A double-diode is a valve containing within one envelope two anodes and a common cathode as shown in fig. 19.13, where the cathode is of the indirectly heated type. The heater is supplied from a low-voltage winding M.

The high-voltage secondary winding PQ in fig. 19.13 has a mid-

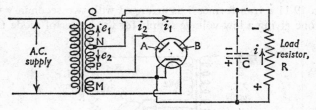

Fig. 19.13 Full-wave rectifier.

point tapping N connected to one end of the load resistor R, the other end of which is connected to the cathode. The two anodes, A and B, are connected to ends P and Q respectively of the high-voltage winding.

During the half-cycle OC (fig. 19.14), the e.m.f. induced in PQ is assumed to be acting in the direction of arrow e_1, thereby making anode B positive and anode A negative with respect to the cathode. Consequently current i_1 flows via B and none via A. During the second half-cycle CD, the e.m.f. in PQ is in the direction shown by arrow e_2, so that current i_2 flows via A and none via B. Hence the current through the load is unidirectional as represented in fig. 19.14 (*b*). For many purposes this fluctuation of current is very undesirable.

One method of reducing this fluctuation of the output current and voltage is to connect a capacitor C in parallel with the load, as shown dotted in fig. 19.13. Let us consider the effect of this capacitor *with the load disconnected.* During the first half-cycle, Q is assumed to be positive in relation to P, so that electrons are being withdrawn from the lower plate of C and flow via anode B and winding QN towards the upper plate of C. Thus the capacitor is charged to the maximum value of the e.m.f. induced in winding NQ. As the e.m.f. of NQ decreases to zero, the electrons assembled on the upper plate of C are unable to return to the lower plate (the insulation resistance of C and the valve being assumed infinite). Consequently the p.d. across

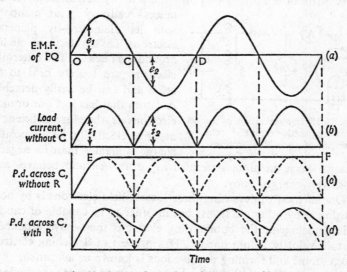

Fig. 19.14 Waveforms for a full-wave rectifier.

C remains constant at the peak value of the alternating e.m.f., as shown by the horizontal line EF in fig. 19.14 (*c*).

When a load R is connected in parallel with C, the latter discharges through R during the time when the e.m.f. induced in NQ or NP is less than the p.d. across C and is recharged when the induced e.m.f. exceeds the p.d. across C. Consequently the terminal voltage varies as shown in fig. 19.14 (*d*). The larger the capacitance of C and the higher the resistance of load R, the smaller is the fluctuation of the terminal voltage. This fluctuation can be practically eliminated by inserting an iron-core choke in series with the load and connecting a capacitor across the rectifier output and another capacitor in parallel with the load.

19.7 Gas-filled diode or rectifier

This type of rectifier usually consists of a glass bulb with an oxide-coated cathode and an anode, as already described for the vacuum diode, but the bulb contains either a small amount of an inert gas such as argon or a small globule of mercury which is vaporized by the heat of the filament.

It was stated in section 5.2 that an atom of any material normally contains equal amounts of positive and negative electricity, and that it consists of a nucleus of positive electricity surrounded—at dis-

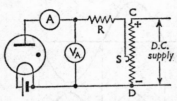

Fig. 19.15 Determination of anode-voltage/anode-current characteristic of a gas-filled rectifier.

tances relatively great compared with its diameter—by planetary electrons, i.e. by electrons circling around the nucleus. The outermost electrons are loosely held to the atom and can be easily detached. An atom that has lost one or more electrons is therefore deficient in negative electricity, i.e. its positive charge is greater than its negative charge. Such an atom is termed a *positive ion* and behaves as a positively-charged body.

One method of detaching these loosely held electrons is by bombardment by other electrons; but in order to be capable of causing this detachment, the bombarding electrons must have attained a certain velocity before impact. This process of detaching electrons from atoms and forming positive ions is known as *ionization*.

In the gas-filled rectifier, the electrons emitted from the cathode are attracted towards the anode when the latter is positive with respect to the cathode; and the greater the p.d. between the anode and the cathode, the greater is the acceleration of the electrons. When an electron has attained a certain velocity, its collision with an atom results in the liberation of one or more electrons and the formation of a positive ion.

Once ionization commences, the electrons released from atoms join those emitted from the cathode and thus increase the number of electrons travelling towards the anode. Consequently the number of collisions between these electrons and other atoms is increased, with the result that still more electrons are released and more positive ions formed. For this reason the ionization tends to increase indefinitely and has to be limited by a resistor R in series with the anode, as shown in fig. 19.15. It will be seen that the anode current in a gas-filled diode consists of electrons moving towards the anode and of relatively heavy positive ions moving towards the cathode. At the instant of impact with the cathode, these positive ions give up their kinetic energy and combine with electrons from the cathode to form neutral atoms of the gaseous element.

The large dot inserted inside the circle representing the glass bulb is the convention used to distinguish a gas-filled tube from a vacuum tube.

The relationship between the anode–cathode p.d. and the anode current is shown in fig. 19.16. When the p.d. is less than the critical value to produce ionization, the anode current is very small, the value being limited by the negative space charge formed by the electrons emitted from the cathode; but once ionization commences, the p.d. remains practically independent of the anode current. Consequently the effect of moving slider S towards C in fig. 19.15 is merely to cause such an increase of anode current that the increase of voltage drop in R absorbs the increase of voltage between S and D.

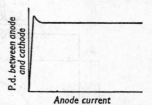

Fig. 19.16 Anode-voltage/ anode-current characteristic of a gas-filled rectifier.

The ionizing p.d. is usually about 15–20 V and the presence of ionization is indicated by a glow in the space between the anode and the cathode. The bombardment of the cathode by the positive ions may be sufficiently severe to disintegrate the electrode; and in order that the cathode may have a reasonably long life, it is necessary to limit the maximum value of the anode current, the actual value being dependent upon the area of the cathode surface.

The circuit diagrams for a gas-filled diode rectifier are similar to those already given for the vacuum diode rectifier. The advantages of the gas-filled diode over the vacuum diode are:

(*a*) the anode–cathode voltage is much smaller than that normally required for the vacuum valve, so that the efficiency of the gas-filled diode is much higher than that of the vacuum diode;

(*b*) owing to ionization, the anode current obtainable is far greater than that obtainable when there is no gas present.

The main disadvantage of the gas-filled rectifier is its shorter life due to the disintegration of the cathode by positive ion bombardment.

19.8 Full-wave mercury-arc rectifier

A glass bulb B, shaped as in fig. 19.17, has a mercury pool C and two electrodes A_1 and A_2, made of graphite in small rectifiers and iron in large rectifiers. The bulb is thoroughly evacuated before being sealed. These electrodes are connected to the two ends of the secondary winding of transformer T and a mid-point tapping M is connected through an inductor L to one side of the load, the other side of which is connected to pool C.

The essential features of one method of starting the arc between

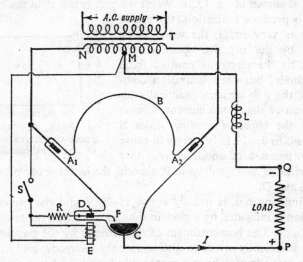

Fig. 19.17 Full-wave mercury-arc rectifier.

the anodes A_1A_2 and the cathode C is shown in fig. 19.17. A flexible metal electrode F carries an iron armature D, and an electromagnet E is situated directly under D. When switch S is closed, D is attracted downwards each time the flux in E grows from zero to a maximum. Consequently, strip F vibrates at twice the supply frequency, and the tip of F makes and breaks contact with the mercury pool C, the current being limited by resistor R to a relatively small value. Each time the circuit is broken, the arc between F and C vaporizes some of the mercury and also forms an incandescent spot on the surface of C, sufficiently to emit electrons. The latter are attracted towards the main electrode that happens to be positive with respect to cathode C and thus produce ionization. The positive ions produced by this ionization bombard the surface of the mercury sufficiently to maintain the incandescent spot and enable the supply of electrons to be continued from the cathode. Switch S is then opened. If the load current is less than a certain critical value, the ionization is insufficient to maintain the incandescent spot; consequently, the arc is extinguished.

If inductor L were not in circuit, the current through the load would vary between zero and a maximum every half-cycle, as shown in fig. 19.18. With L in circuit, the effect of the e.m.f. induced in L is to delay the decrease of the current at one anode until the potential

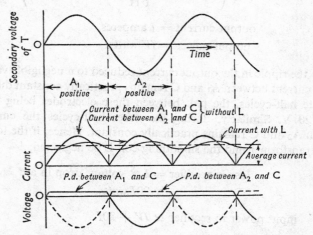

Fig. 19.18 Waveforms of voltages and currents for a full-wave mercury-arc rectifier.

of the other anode is sufficiently high to take over the arc from the first anode, thereby reducing the fluctuation of the current, as indicated in fig. 19.18.

The function of the large glass dome B is to provide the cooling surface necessary to prevent excessive temperature rise of the rectifier. In practice, the arms containing the anodes A_1 and A_2 are made with an elbow to reduce the risk of an arc directly between these electrodes.

The p.d. across the arc remains practically constant at about 20 V over a wide variation of current, as represented by the positive portions of the bottom curves in fig. 19.18 (see also section 19.7 and fig. 19.16); hence:

if V = output voltage, namely the p.d. between Q and C,

p.d. between Q and the conducting anode $\simeq V + 20$.

But this voltage is the average value of the e.m.f. induced in one-half of the secondary winding of T, the impedance drop in the latter being assumed negligible,

i.e. $V + 20 \simeq \left(\dfrac{\text{r.m.s. value of e.m.f. induced in MN}}{1{\cdot}11} \right)$

so that $V \simeq \left(\dfrac{\text{r.m.s. value of e.m.f. in MN}}{1{\cdot}11} - 20 \right)$

If   output current = I amperes,

output power = IV watts.

With the ripple in the output current reduced to a negligible value, the arc current between A_1 and C remains practically constant during alternate half-cycles, the p.d. between these electrodes being then about 20 V. Similarly, during the other half-cycles the current between A_2 and C remains practically constant. Hence, if the losses in the transformer and the smoothing coil L be neglected,

$$\text{loss in rectifier} = I \times \text{voltage drop in arc}$$
$$\simeq 20I \text{ watts,}$$

and   input power to rectifier $\simeq IV + 20I$

$\therefore$ efficiency of rectifier $\simeq \dfrac{IV}{I(V + 20)} \simeq \dfrac{V}{V + 20}$ (19.1)

Hence the efficiency is practically independent of the load.

If V = 100 volts,

$$\text{efficiency} \simeq \frac{100}{100 + 20} \simeq 0{\cdot}83 \text{ p.u.}$$

but if V = 1000 volts,

$$\text{efficiency} \simeq \frac{1000}{1000 + 20} \simeq 0{\cdot}98 \text{ p.u.}$$

It follows that the greater the output voltage, the higher is the efficiency of the mercury-arc rectifier.

19.9 Metal rectifiers*

The action of a metal rectifier depends upon the rectifying property of a *barrier layer* between a metal conductor and a semiconductor, but the precise nature of this layer is not known. There are two commercial types of metal rectifiers: (*a*) the copper-oxide rectifier, and (*b*) the selenium rectifier.

The *copper-oxide rectifier* consists of a copper disc or plate P, one side of the disc having a very thin layer C of cuprous oxide (Cu_2O) formed by special heat treatment. Contact with

Fig. 19.19 A copper-oxide rectifier unit.

the external surface of the cuprous oxide is made either by a washer Q of a soft metal such as lead, as shown in fig. 19.19, or by spraying a layer of metal on the cuprous oxide. The barrier layer L† between the copper and the copper oxide allows current to flow easily when the oxide layer is positive relative to the copper disc, but presents a high resistance to current in the reverse direction. This effect appears

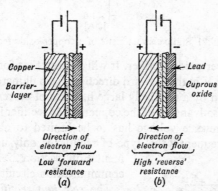

Fig. 19.20 Action of the copper-oxide rectifier.

to be due to the fact that the copper disc contains a plentiful supply of free electrons (section 5.2), many of which can be detached from the disc by the high potential gradient in the barrier layer when the copper is negative with respect to the oxide, as in fig. 19.20 (*a*). On

* Metal rectifiers are included in this chapter, since their action may be likened to that of a cold gas-filled rectifier in which the copper forms the cathode and the cuprous oxide forms the anode.

† In figs. 19.19, 19.20 and 19.22, the thicknesses of the layers are greatly exaggerated; e.g. in the actual rectifier the thickness of the oxide layer is about 0·1 mm and that of the barrier layer is estimated to be about 70×10^{-6} mm.

the other hand, the cuprous oxide, being a semiconductor, contains relatively few free electrons. Consequently, when the polarity of the applied voltage is reversed, as in fig. 19.20 (*b*), comparatively few electrons are released from the oxide layer when the latter is negative with respect to the copper disc.

The graph of fig. 19.21 shows the current/voltage characteristic of

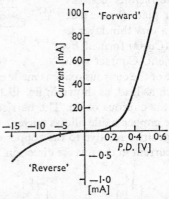

Fig. 19.21 Static characteristic of a copper-oxide rectifier.

a certain copper-oxide rectifier. It will be seen that for a p.d. greater than about 0·2 V in the forward direction, a small increase of voltage is accompanied by a relatively large increase of current.

The main disadvantage of the copper-oxide rectifier is that the peak value of the reverse voltage has to be limited to about 10 V per element. Consequently this type of rectifier is only used for instrumentation (section 22.10) and in telecommunication circuits.

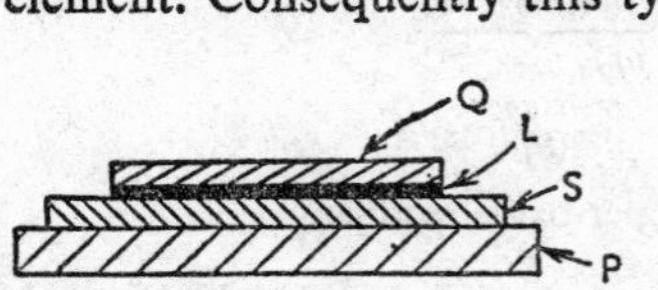

The *selenium rectifier* consists of a steel or aluminium plate or disc P (fig. 19.22) on the face of which is a thin layer of selenium S. A tin–cadmium alloy Q is sprayed on the selenium. By the aid of special heat treatment, a barrier layer L is produced between the selenium and the sprayed alloy.

Fig. 19.22 A selenium rectifier unit.

The current/voltage characteristic of the selenium rectifier is somewhat similar in shape to that of the copper-oxide rectifier, but the voltage drop in the forward direction is about 50–80 per cent higher than that of the copper-oxide rectifier. On the other hand, a selenium element can be operated with a reverse voltage up to about 30 V, so that fewer elements in series are required for a given voltage compared with the number of copper-oxide elements that would be necessary.

19.10 Full-wave rectification with metal rectifiers

The arrangement most commonly used for full-wave rectification with the aid of metal rectifiers is the bridge circuit shown in fig. 19.23. The four rectifier elements (or groups of elements in series) are

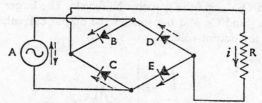

Fig. 19.23 Bridge circuit for full-wave rectification.

represented by the solid triangles butting against perpendicular lines. The apex of the triangle indicates the forward direction of the current, i.e. the direction in which the rectifier resistance is low. The elements are so arranged that an e.m.f. generated in source A in the direction represented by the full arrow sends current through B and E; but when the e.m.f. is in the reverse direction (dotted arrow), current flows through C and D. It will be evident from fig. 19.23 that the current through the load resistor R is uni-directional.

19.11 Voltage doubler

A voltage doubler is convenient when a high-voltage low-current output is required. The circuit of one type of voltage doubler is

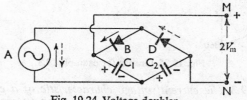

Fig. 19.24 Voltage doubler.

shown in fig. 19.24, which is similar to fig. 19.23 except that rectifiers C and E are replaced by capacitors C_1 and C_2.

When there is no load across terminals MN, capacitor C_1 is charged to the peak voltage, V_m, during the half-cycle that the e.m.f. of source A is in the direction shown by the full arrow. Capacitor C_2 is charged to the same p.d. when the voltage of the source is in the reverse direction, as indicated by the dotted arrow. From fig. 19.24, it is evident that C_1 and C_2 are in series across terminals MN and the output voltage is therefore $2V_m$.

When a load is connected across MN, the waveform of the output voltage is similar to that shown in fig. 19.14 (*d*), the charge on C_1 being replenished during the positive half-cycle of the supply voltage and that on C_2 during the negative half-cycle. The larger the capacitance of C_1 and C_2, and the smaller the output current, the more constant is the output voltage.

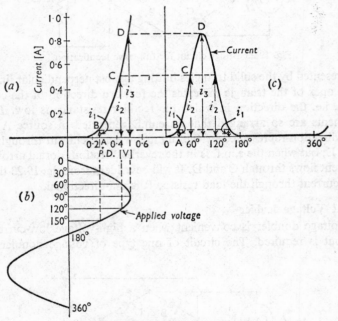

Fig. 19.25 Graphs for example 19.3.

Example 19.3 *The current/voltage characteristic of a certain metal rectifier is represented by the graph in fig. 19.25 (a). The rectifier is connected across an a.c. supply having a peak voltage of 0·5 V. The waveform of the voltage is sinusoidal. Derive a graph representing the variation of current with time over a complete cycle.*

The waveform of the applied voltage is drawn to scale on an extension of the vertical axis in fig. 19.25 (*b*), and horizontal lines are drawn at intervals corresponding to 30, 60, etc., electrical degrees. Similarly, vertical lines are drawn at the same intervals in fig. 19.25 (*c*).

At the instant corresponding to 30°, the value of the applied voltage is 0·5 sin 30°, namely 0·25 V. This p.d. is represented by OA in fig. 19.25 (*a*), and the corresponding value of the current is AB = i_1. A horizontal line is drawn from B to cut the vertical line drawn at 30° in fig. 19.25 (*c*). Hence the intercept AB in fig. 19.25 (*c*) represents the value i_1 of the current at the instant corresponding to 30°.

Similarly, at instants corresponding to 60° and 90°, the values of the current are i_2 and i_3 respectively. The same procedure applies to the values of the current at instants corresponding to 120°, 150° and 180°.

During the second half-cycle, the applied voltage is negative so that the current is zero from 180° to 360°.

The waveform of the current can now be derived by drawing a curve through points O, B, C, D, etc., in fig. 19.25 (*c*). Greater accuracy would be obtained if the values of the currents were plotted at intervals of, say, 15°.

19.12 The three-electrode vacuum valve or triode

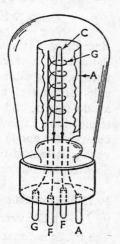

Fig. 19.26 shows the general arrangement of a triode having a directly heated cathode C. The anode cylinder A is shown cut to depict more clearly the internal construction. Grid G is usually a wire helix attached to one or two supporting rods. The pitch of this helix and the distance between the helix and the cathode are the main factors that determine the characteristics of the triode.

In a triode, the potential of the anode A is always positive with respect to the filament, so that electrons tend to be attracted towards A

Fig. 19.26 A vacuum triode.

from the space charge surrounding cathode C. The effect of making the grid G positive with respect to C is to attract more electrons from the space charge. Most of these electrons pass through the gaps between the grid wires, but some of them are caught by the grid as shown in fig. 19.26 (*a*) and return to the cathode via the grid circuit. On the other hand, the effect of making the grid negative is to neutralize, partially or wholly, the effect of the positive potential of the anode. Consequently, fewer electrons reach the anode, the paths of these electrons being as shown in fig. 19.27 (*b*). No electrons are now reaching the grid, i.e. there is no grid current when the grid is negative by more than about one volt with respect to the cathode. The paths of the electrons which are repelled back from the space charge into the cathode are not indicated in fig. 19.27.

It will be seen that the magnitude of the anode current can be controlled by varying the p.d. between the grid and the cathode; and

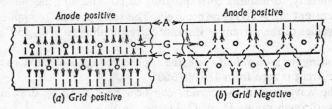

Fig. 19.27 Influence of grid potential upon electron paths.

since the grid is in close proximity to the space charge surrounding the cathode, a variation of, say, one volt in the grid potential produces a far greater change of anode current than that due to one volt variation of anode potential. The relationship between the anode current and the grid voltage for a given anode voltage is termed the *transfer** or *mutual characteristic* of the triode, and that between the anode current and the anode voltage for a given grid voltage is termed the *anode characteristic*. We shall now consider how these characteristics can be determined experimentally.

19.13 Determination of the static characteristics of a triode

(a) I_A/V_G† *Characteristics*

The triode is connected as shown in fig. 19.28. The slider on R_1 is adjusted to give a constant reading of, say, 130 V on voltmeter V_A. The readings on milliammeter A are noted for various positive and negative values of V_G, the polarity being reversed by means of switch RS. The test is repeated with V_A maintained constant at, say,

---

* This is the term recommended in B.S. 204: 1960.
† See note on subscripts, p. xvii.

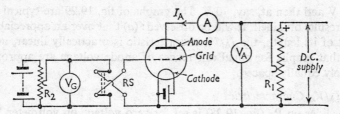

Fig. 19.28 Determination of the static characteristics of a vacuum triode.

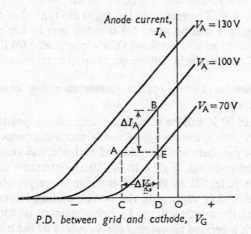

Fig. 19.29 I_A/V_G characteristics of a vacuum triode.

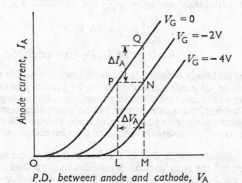

Fig. 19.30 I_A/V_A characteristics of a vacuum triode.

100 V and then at, say, 70 V. The graphs of fig. 19.29 are typical of the results obtained. It will be observed: (a) that over an appreciable part of its length, the I_A/V_G characteristic is practically linear, and (b) that graphs for equal differences of anode voltage are approximately equally spaced.

(b) I_A/V_A Characteristics

The slider on R_2 (fig. 19.28) is set for zero voltage on voltmeter V_G and the anode current is noted for various values of the anode voltage. The test is repeated with V_G equal to, say, -2 V and then with V_G equal to -4 V. The graphs in fig. 19.30 are typical of the results obtained. Again it will be seen: (a) that the graphs are practically linear over a considerable portion of their length, and (b) that graphs for equal differences of grid voltage are approximately equally spaced.

19.14 Amplification factor, mutual conductance and anode a.c. or slope* resistance

From fig. 19.29 it will be seen that for the *linear* portions of the I_A/V_G characteristics, the change of anode current is proportional to the change of p.d. between the grid and cathode, the anode voltage being assumed constant. If ΔV_G in fig. 19.29 represents the amount by which the potential of the grid is made more *positive* (or less *negative*) and if ΔI_A represents the corresponding increase in the anode current, the anode voltage V_A being maintained constant, the ratio $\Delta I_A/\Delta V_G$ is termed the *mutual conductance*† of the triode and is denoted by the symbol g_m,

i.e. mutual conductance $= g_m = \dfrac{\Delta I_A}{\Delta V_G}$ for constant V_A \hfill (19.2)

$$= \frac{\text{EB}}{\text{AE}} \text{ in fig. 19.29}$$

$$= \text{slope of } I_A/V_G \text{ characteristic.}$$

---

* *Anode slope resistance* is probably the term more commonly used, but B.S. 204 : 1960 recommends *anode a.c. resistance* as the preferred term. The latter is more descriptive of the fact that when an alternating voltage is superimposed on a direct anode voltage, the anode current has an alternating component superimposed on a steady component; and the anode a.c. resistance is the ratio of the alternating component of the anode voltage to the alternating component of the anode current.

† The term 'conductance' is due to the fact that the quantity is a ratio of current/voltage; and the term 'mutual' is intended to show that the variation of current in one circuit is due to a variation of voltage in *another* circuit, just as *mutual* inductance refers to the flux linked with one circuit due to current in another circuit (section 4.10).

Similarly, it is seen that for the linear portions of the I_A/V_A characteristics of fig. 19.30, the change of anode current is proportional to the change of p.d. between anode and cathode, the grid voltage being assumed constant. Thus, if ΔI_A be the change of anode current for a change ΔV_A of anode voltage, the grid voltage being maintained constant, the ratio $\Delta V_A/\Delta I_A$ is the *anode a.c. resistance* or the *anode slope resistance*, as already explained in section 19.4,

i.e. anode a.c. or slope resistance $= r_a = \dfrac{\Delta V_A}{\Delta I_A}$ for constant V_G

$$(19.3)$$

$$= \frac{PN}{NQ} \text{ in fig. } 19.30 = \frac{1}{NQ/PN}$$

$$= \frac{1}{\text{slope of } I_A/V_A \text{ characteristic}}$$

From expression (19.2), $\quad \Delta I_A = \Delta V_G \cdot g_m$

and from expression (19.3), $\quad \Delta I_A = \Delta V_A/r_a$.

Hence, if both V_G and V_A are varied simultaneously and if the triode is operating on the linear portions of its characteristics,

$$\Delta I_A = \Delta V_G \cdot g_m + \Delta V_A/r_a$$

If the variations of V_G and V_A be such that the anode current remains constant, i.e. if $\Delta I_A = 0$, we have:

$$0 = \Delta V_G \cdot g_m + \Delta V_A/r_a$$

$$\therefore \qquad \Delta V_A/\Delta V_G = -g_m r_a = -\mu$$

or $\qquad \mu = g_m r_a = -\Delta V_A/\Delta V_G \text{ for constant } I_A \quad (19.4)$

$$= amplification \ factor.$$

The negative sign in expression (19.4) is due to the fact that if the anode voltage is *increased*, the grid potential has to be made more *negative* in order to maintain the anode current constant. For instance, in fig. 19.29, if $V_A = 70$ V and $V_G = \text{OD}$, the anode current is DE. If the anode voltage is increased to 100 V without any change in the grid potential, the anode current increases to DB. This current can be reduced to its original value DE ($= \text{CA}$) by making the grid more *negative* by an amount represented by DC. If OD $= -1$ V and OC $= -3$ V, $\Delta V_G = -2$ V,

$$\therefore \qquad \text{amplification factor} = \mu = -\frac{100 - 70}{-2} = 15.$$

Similarly for fig. 19.30, if the anode voltage is increased from OL to OM, with $V_G = 0$, the anode current increases from LP to MQ; and in order to reduce the anode current to its initial value LP (= MN), the p.d. between grid and cathode has to be changed from 0 to −2 V. Hence, if LM = 30 V,

$$\text{amplification factor} = \mu = -30/(-2) = 15.$$

Example 19.4 *The following readings were obtained from the linear portions of the static characteristics of a vacuum triode:*

V_A (volts)	120	120	80
V_G (volts)	−1·3	−3·8	−1·3
I_A (milliamperes)	10	4	6·2

Calculate: (a) *the anode a.c. resistance,* (b) *the mutual conductance and* (c) *the amplification factor.*

(*a*) With a grid voltage of −1·3 V, a reduction of V_A from 120 to 80 V is accompanied by a reduction of I_A from 10 to 6·2 mA; i.e. $\Delta V_A = 40$ V and $\Delta I_A = 3·8$ mA.

$$\therefore \quad \text{anode a.c. resistance} = r_a = \frac{40 \times 1000}{3·8} = 10\,530 \ \Omega.$$

(*b*) With V_A constant at 120 V, I_A is increased from 4 mA to 10 mA by changing the grid voltage from −3·8 to −1·3 V, i.e. $\Delta I_A = 6$ mA and $\Delta V_G = 2·5$ V,

$$\therefore \quad \text{mutual conductance} = g_m = 6/2·5 = 2·4 \ \text{mA/V}.$$

The mutual conductance could also be expressed as 2·4 milli-siemens or 2400 microsiemens, but it is more commonly expressed as 'milliamperes per volt'.

(*c*) Amplification factor $= \mu = g_m r_a$

$$= \frac{2·4}{1000} \times 10\,530 = 25·3.$$

19.15 Voltage amplification by means of a vacuum triode

In order to obtain any useful effect from a triode, it is necessary to insert some form of load, such as a resistor or an inductor, in series with the anode. Consequently, when an alternating voltage is applied to the grid, the variation of anode current is accompanied by a variation of p.d. across the load.

Fig. 19.31 shows a non-reactive resistor R connected in series with the valve across a d.c. source having a terminal voltage V_{HT}. Between the cathode and the grid, a battery having a terminal voltage V_B is connected in series with an a.c. source S. The function of this battery is to give the grid a negative bias; and if V_B is greater than the peak value of the alternating voltage from S, the grid is never positive with respect to the cathode and therefore there is no grid current. Considerable distortion of the output-voltage waveform can

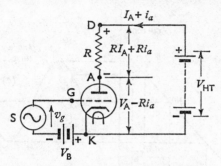

Fig. 19.31 A triode with a resistive load.

be caused by grid current if there is appreciable impedance in the grid circuit.

Fig. 19.32 (*a*) gives the *dynamic* transfer (or mutual) characteristic of the triode, namely the relationship between the anode current and the grid–cathode p.d., for a given supply voltage, when a load is connected in series with the anode.

When the alternating voltage applied to the grid is zero, the anode current I_A is represented by LM in fig. 19.32 (*a*). The corresponding p.d. across the load resistor is RI_A and that between the anode and cathode is V_A, where

$$V_A + RI_A = V_{HT}.$$

Next, let us consider the effect of applying a sinusoidal alternating voltage from source S. We shall assume that this voltage is positive when it makes the grid less negative (or more positive) with respect to the cathode, as shown by the arrow alongside S in fig. 19.31. At the instant when this alternating voltage has the value represented by v_g* in fig. 19.32 (*b*), the anode current has increased to ($I_A + i_a$), as shown in fig. 19.32 (*c*). The corresponding p.d. across the load resistor has increased to ($RI_A + Ri_a$), while the p.d. between anode and cathode has decreased to ($V_A - Ri_a$).

* See note on subscripts, p. xvii.

If the alternating voltage from source S varies as shown in fig. 19.32 (*b*), the anode current varies between a maximum NP and a minimum QS; and the waveform of the anode current can be derived by projecting points M, P and S to the right, as in fig. 19.32 (*c*). It is evident that if the portion PS of the dynamic characteristic is linear, the waveforms of the anode current and therefore of the p.d. across

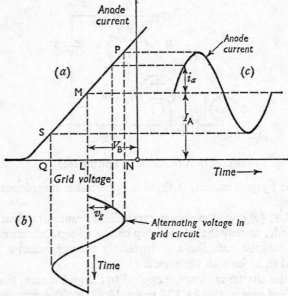

Fig. 19.32 Grid voltage and anode current for fig. 19.31.

R are exactly the same as the waveform of the alternating voltage applied to the grid.

It will be noted that when the alternating voltage applied to the grid is positive, the p.d. between anode and cathode falls below V_A; and when the alternating voltage is negative, the anode–cathode p.d. exceeds V_A. Hence the variation of the anode–cathode p.d. is in *anti-phase* to the alternating voltage applied to the grid.

The ratio of the alternating component of the anode–cathode p.d. to the alternating voltage applied to the grid is termed the *voltage amplification* and is represented by the symbol A. Hence, for a non-reactive load resistor,

$$A = \frac{-R i_a}{v_g}$$

Since the load is a non-reactive resistor and the amplification is assumed to be distortionless, we can replace the instantaneous values i_a and v_g by the r.m.s. values I_a and V_g respectively, so that—

$$\text{voltage amplification} = A = -RI_a/V_g.$$

A method of calculating the voltage amplification from the valve parameters (or constants) is given in section 19.16.

The condition of voltage amplification described above is referred to as *Class A valve amplification*, since the valve is used in such a way that the anode current does not fall below the linear region of the I_A/V_G dynamic characteristic and the grid never goes positive with respect to the cathode so that there is no grid current.

19.16 Equivalent circuit of a vacuum triode

The purpose of the direct voltages V_{HT} and V_B of fig. 19.31 is to ensure that the triode is being operated on the linear portions of its

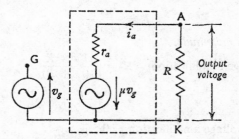

Fig. 19.33 Equivalent circuit of a valve amplifier.

characteristics. These d.c. sources have negligible impedance to alternating current so that no alternating voltages appear across them. Consequently, *as far as the alternating voltages are concerned*, the terminals of the d.c. sources can be short-circuited; thus point D of fig. 19.31 can be joined directly to the cathode K, and the grid bias battery can be omitted. Hence, as far as the alternating components of the voltages and currents are concerned, the amplifier can be represented by the simple equivalent circuit shown in fig. 19.33.

In this diagram, the valve is represented as an a.c. source having an instantaneous terminal voltage μv_g and possessing an internal resistance r_a equal to the anode a.c. resistance of the valve, as shown enclosed by the dotted rectangle in fig. 19.33. The load resistor is connected across the output terminals AK.

In order that the positive direction of i_a may be the same as in fig. 19.31, the positive direction of μv_g must be that indicated by the arrow in fig. 19.33. When the alternating current is flowing in the positive direction, i.e. when v_g is positive in fig. 19.31, terminal A is negative with respect to terminal K. Hence the output voltage is 180 electrical degrees out of phase with the alternating voltage applied to the grid of the actual valve.

$$\text{Voltage amplification} = \frac{\text{instantaneous p.d. across load}}{\text{instantaneous voltage to grid}}$$

i.e.
$$A = -Ri_a/v_g$$

As already mentioned on p. 525, we can replace the instantaneous values i_a and v_g by the r.m.s. values I_a and V_g respectively, so that

$$\text{voltage amplification} = A = -RI_a/V_g.$$

From fig. 19.33 it is evident that:

$$I_a = \frac{\mu V_g}{r_a + R}$$

Hence, in *magnitude*,

$$\text{voltage amplification} = A = \frac{\mu V_g \times R}{V_g(r_a + R)}$$

$$= \frac{\mu}{1 + r_a/R} \qquad (19.5)$$

If $R = r_a$, voltage amplification $= 0.5\ \mu$,

and if $R = 5r_a$, voltage amplification $= 0.83\ \mu$.

Hence, the higher the resistance of the load, the greater is the voltage amplification, but the latter cannot exceed the amplification factor. The disadvantage of a high load resistance is that it necessitates a correspondingly high supply voltage for the anode circuit. The main advantage of the resistive method of loading is that the voltage amplification is independent of the frequency.

Example 19.5 *A certain triode has an amplification factor of 27 and an anode a.c. resistance of 18 kΩ. Calculate the magnitude of the voltage amplification when the load is* (a) *a non-reactive resistor of 25 kΩ and* (b) *a coil having an inductance of 200 μH and a resistance of 30 Ω in parallel with a capacitor which, when adjusted to give resonance, has a capacitance of 150 pF.*

(*a*) From expression (19.5),

$$\text{voltage amplification} = \frac{27}{1 + \dfrac{18\,000}{25\,000}} = 15 \cdot 7.$$

(*b*) Dynamic impedance of the $\left.\right\}$ resonant parallel circuit $= \dfrac{L}{CR} = \dfrac{200 \times 10^{-6}}{150 \times 10^{-12} \times 30}$

$$= 44\,400 \ \Omega.$$

Since this dynamic impedance is equivalent to a resistive load of 44 400 Ω,

$$\therefore \qquad \text{voltage amplification} = \frac{27}{1 + \dfrac{18\,000}{44\,400}} = 19 \cdot 2.$$

Method (*b*) has the advantages: (i) the supply voltage for the anode circuit is lower than that required with method (*a*), since the resistance of the coil is only 30 Ω compared with 25 kΩ for the resistor, and (ii) the amplification is selective so that a voltage having a particular frequency can be separated from voltages of other frequencies.

19.17 Graphical determination of the voltage amplification: Load line

An alternative method of determining the voltage amplification of a triode having a resistive load is the graphical construction shown in fig. 19.34, where the I_A/V_A static characteristics are given for grid voltages of 0, -2 volts and -4 volts. Suppose OA to represent the terminal voltage, V_{HT}, of the d.c. source supplying the anode circuit (see fig. 19.31).

If $\quad v_A$ = p.d. between anode and cathode,

$\qquad i_A$ = corresponding anode current

and $\quad R$ = load resistance,

then $\qquad\qquad\qquad V_{HT} = v_A + Ri_A$

$\therefore \qquad\qquad\qquad i_A = (V_{HT} - v_A)/R \qquad\qquad (19.5A)$

When $v_A = 0$, i.e. when the anode is connected to the cathode, $i_A = V_{HT}/R$ and is represented by OB in fig. 19.34.

When $i_A = 0$, i.e. when the grid bias is sufficient to reduce the anode current to zero, $v_A = V_{HT}$ and is represented by OA in fig. 19.34.

Expression (19.5A) shows that the relationship between v_A and i_A is linear and is represented by the straight line AB in fig. 19.34. Hence AB is termed the *load line* and its slope is given by:

$$\frac{di_A}{dv_A} = \frac{d}{dv_A}\left(\frac{V_{HT} - v_A}{R}\right) = -\frac{1}{R}.$$

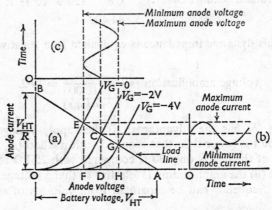

Fig. 19.34 Graphical determination of voltage amplification of a triode with a resistive load.

If DC represents the anode current for a grid voltage of -2 volts,

$$\frac{DA}{DC} = \frac{OA}{OB} = \frac{V_{HT}}{V_{HT}/R} = R$$

$$\therefore \qquad DA = DC \times R = \text{p.d. across the load resistor}$$

and $\qquad OD = OA - DA = \text{p.d. between anode and cathode.}$

If the grid voltage is reduced to zero, the anode current increases to FE and the anode–cathode p.d. falls to OF. On the other hand, if the grid voltage is increased to -4 V, the anode current falls to HG and the anode–cathode p.d. increases to OH. Hence, with a grid bias of -2 V and a sinusoidal alternating grid voltage having a peak value of 2 V, the locus of the anode current is EG, i.e. the anode current varies between FE and HG, as shown in fig. 19.34 (b), and the anode–cathode p.d. varies between OF and OH, as in fig. 19.34 (c). Consequently FH represents the total variation of anode voltage for a total variation of 4 V in the voltage applied to the grid. Also it is obvious from fig. 19.34 that the anode–cathode p.d. is a *minimum* when the alternating voltage applied to the grid is at its maximum *positive* value. Hence,

$$\begin{array}{l} \text{voltage amplification} \\ \text{(in magnitude)} \end{array} = \frac{\text{total variation of anode--cathode p.d.}}{\text{total variation of grid--cathode p.d.}}$$

$$= FH/4 \text{ in fig. 19.34.}$$

If the lengths of CE and CG in fig. 19.34 were unequal, then DF and DH would also be unequal and the output voltage would be distorted, i.e. the waveform of the output voltage would not be the same as that of the alternating voltage applied to the grid.

If the load consists of a tuned circuit as in example 19.5 (*b*), where the resistance of the inductor is negligible compared with r_a, and if the triode is to operate with an anode current DC (fig. 19.34) when the grid voltage is -2 V, then the d.c. voltage, V_{HT}, supplied to the anode circuit is represented by OD.

Point A on the horizontal axis is then such that:

$$DA = \text{current DC} \times \text{dynamic impedance of tuned circuit}$$
$$= DC \times L/(CR).$$

The load line is then the straight line joining A to C and continued to B, and the slope of the load line is $-L/(CR)$.

The case of a load supplied through a transformer is discussed in section 19.21.

19.18 Resistance–capacitance coupling of two triodes

Fig. 19.35 shows how the output voltage of triode M can be applied to the grid of a second triode N. The function of capacitor C is to prevent the d.c. component of the p.d. between the anode and cathode of M being applied to the grid of N. A bias battery B_2 is necessary to prevent the grid of N becoming positive and thus prevent any grid current flowing through resistor R_G.

The operation of the circuit may be explained qualitatively thus: with no alternating voltage applied to the grid of M, there is a steady anode current through M, and C is charged to approximately the anode–cathode p.d. of that valve. When the input alternating voltage V_1 is positive, the anode current of M increases. The *increased* p.d. across R is accompanied by the same *decrease* of p.d. between the anode and cathode of M; consequently C discharges via M and R_G, as shown by the full arrow, so that the grid end of R_G

becomes negative in relation to the cathode end, and the anode current of N is therefore reduced. When the input voltage V_i is negative, the anode current of M is reduced, the p.d. across R decreases and that between the anode and cathode of M increases.

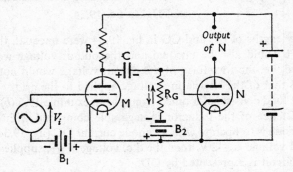

Fig. 19.35 Resistance-capacitance coupling of two triodes.

Hence C is charged by a current that flows through R_G in the direction indicated by the dotted arrow. The grid end of R_G is now positive in relation to the cathode end and the anode current of N is therefore increased. It will be noticed that the variations of the anode currents of M and N are in anti-phase, i.e. when one increases, the other decreases.

The equivalent circuit of M and its coupling to N is given in fig. 19.36. The resistance of R_G is of the order of 1 megohm and is

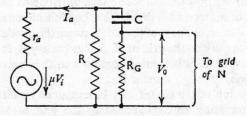

Fig. 19.36 Equivalent circuit of fig. 19.35.

normally far greater than that of R, so that the current through C and R_G is then negligible compared with that through R. Also the reactance of C is usually very small compared with the resistance of R_G; consequently, in magnitude, output voltage $= V_0 =$ p.d. across $R_G \simeq$ p.d. across R

$$\simeq I_a R \simeq \frac{\mu V_1 R}{r_a + R}$$

∴ voltage amplification due to triode M $\simeq \dfrac{\mu}{1 + r_a/R}$,

namely that given by expression (19.5).

19.19 Cathode bias

In earlier diagrams the negative bias for the grid has been shown, for simplicity, as being supplied by a battery. In practice, however, it is customary to obtain this bias by connecting a cathode resistor R_K, shunted by a capacitor C_K, as shown in fig. 19.37. If the resistance

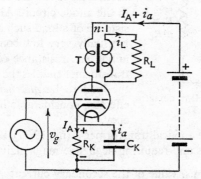

Fig. 19.37 Triode with cathode-bias resistor and transformer-coupled load.

of R_K is, say, 1 kΩ, and if the d.c. component of the anode current is, say, 6 mA, the grid bias is −6 V.

The function of capacitor C_K is to provide a low-reactance path for the alternating component i_a of the anode current. *Without* this capacitor, i_a would produce an alternating p.d., $R_K i_a$, superimposed on the grid bias, $R_K I_A$, and this alternating p.d. would be a *negative feedback* to the grid. For instance, suppose the anode current to increase from 6 to 8 mA when the v_g increases from zero to 4 V. Then, with $R_K = 1$ kΩ (as above), the p.d. across R_K would increase from 6 to 8 V, i.e. the grid end of R_K would become more *negative* by 2 V. Consequently the *actual increase* in the potential of the grid with respect to the cathode would be $(4 - 2) = 2$ V, and the effect of this feedback would be to halve the amplification.

With a capacitor of, say, 50 μF in parallel with R_K and for a frequency of 100 Hz, the reactance of C_K would be approximately 32 Ω. Hence, R_K would be practically by-passed as far as the alternating component of the anode current was concerned, and practically the whole of i_a would flow via C_K, as indicated in fig. 19.37.

19.20 Power amplifier

An amplifier may consist of one or more valves arranged for voltage amplification, but the final stage is usually a valve supplying *power* to a load, such as a loudspeaker coupled through a transformer as shown in fig. 19.37. The use of a transformer avoids the relatively

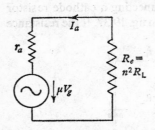

Fig. 19.38 Equivalent circuit for fig. 19.37.

large I^2R loss due to the d.c. component of the anode current that would occur if the load were connected directly in the anode circuit. Also, since the resistance of a load such as a loudspeaker is usually very low compared with the anode a.c. resistance of the valve, the transformer enables the resistance R_L of the load to be *matched* to r_a, thereby effecting considerable increase of efficiency.

Let us assume an ideal transformer T (fig. 19.37), i.e. a transformer having no losses and requiring negligible magnetizing current.

If I_L = r.m.s. value of the secondary current

and R_L = resistance of the load,

power absorbed by the load = $I_L^2 R_L$.

The resistance of the secondary circuit can be replaced by an equivalent resistance R_e in the primary circuit, the value of R_e being such that the power produced in it by the *alternating* component of the anode current is equal to that dissipated in the load due to current I_L. Thus, if I_a be the r.m.s. value of the current through R_e in the equivalent circuit of fig. 19.38, then:

$$I_a^2 R_e = I_L^2 R_L$$

$$\therefore \qquad R_e = R_L (I_L/I_a)^2 = R_L \times n^2$$

where $n = \dfrac{\text{no. of primary turns}}{\text{no. of secondary turns}} = \dfrac{I_L}{I_a}$

as already explained in section 14.2.

From fig. 19.38, it follows that:

$$I_a = \frac{\mu V_g}{r_a + R_e}$$

where V_g = r.m.s. value of the alternating voltage applied to the grid.

Hence,   power absorbed by load = $I_a^2 R_e$

$$= \left(\frac{\mu V_g}{r_a + R_e}\right)^2 \times R_e$$

$$= \left(\frac{\mu V_g}{r_a + n^2 R_L}\right)^2 \times n^2 R_L \quad (19.6)$$

It was shown in section 1.11 that the power absorbed by a load in a *resistive* circuit is a maximum when the resistance of the load is equal to the internal resistance of the source. Hence, for the circuit shown in fig. 19.38, the power absorbed by the load is a maximum when:

$$R_e = n^2 R_L = r_a$$

i.e. when $n = \sqrt{(r_a/R_L)}$ (19.7)

and maximum power absorbed by load $= \left(\frac{\mu V_g}{2r_a}\right)^2 \times r_a = \frac{(\mu V_g)^2}{4r_a}$

$$= \tfrac{1}{4}\mu g_m V_g^2 \quad (19.8)$$

When this condition is fulfilled, the load is said to be *matched* to the valve.

Example 19.6 *The parameters of a certain power triode are:* $\mu = 12$ *and* $r_a = 800\ \Omega$. *If a loudspeaker having a resistance of 10 Ω is to be supplied through a transformer from this triode, calculate:* (a) *the transformation ratio for maximum transfer of power and* (b) *the power supplied to the speaker when the r.m.s. value of the signal voltage to the grid is 5 V.*

(*a*) Substituting in expression (19.7), we have:

$$n = \sqrt{(800/10)} = 8\cdot94.$$

i.e. no. of primary turns = 8·94 × no. of secondary turns.

(*b*) Substituting in expression (19.8), we have:

$$\text{maximum power to load} = \frac{(12 \times 5)^2}{4 \times 800} = 1\cdot125\ \text{W.}$$

19.21 Graphical determination of the output power of an amplifier

The value of the power absorbed by a resistive load connected in the anode circuit can be determined from the graphical construction already described in section 19.17. Thus the peak-to-peak variation of the anode–cathode p.d. in fig. 19.34 is FH and the peak-to-peak variation of the anode current is (FE — HG). If the variations of these quantities are sinusoidal,

mean power absorbed by load, due to the alternating voltage applied to grid

$$= \begin{pmatrix} \text{r.m.s. value of alternating} \\ \text{component of anode–} \\ \text{cathode p.d.} \end{pmatrix} \times \begin{pmatrix} \text{r.m.s. value of alternating} \\ \text{component of anode current} \end{pmatrix}$$

$$= \left(0{\cdot}707 \times \frac{\text{FH}}{2} \right) \times \left(0{\cdot}707 \times \frac{\text{FE} - \text{HG}}{2} \right)$$

$$= \text{FH} \times (\text{FE} - \text{HG})/8 \tag{19.9}$$

This power is usually referred to as the a.c. power absorbed by the load to distinguish it from the power dissipated in the load due to the d.c. component of the anode current when the load resistor is connected in the anode circuit as in fig. 19.31.

If the load is supplied through a transformer T, as in fig. 19.37, and if the resistance of the primary winding be assumed negligible, the d.c. component of the anode–cathode p.d. is the same as the supply voltage.

For a supply voltage represented by OD in fig. 19.34 and for a grid bias of -2 volts, the anode current is represented by DC. The load line can be drawn by making $\text{DA} = \text{DC} \times R_\text{e}$,

where R_e = equivalent resistance of the load referred to the primary circuit

$$= n^2 R_\text{L}.$$

Points A and C are then joined and the line continued to cut the vertical axis at B. For a sinusoidal grid voltage having a peak value of 2 V, as in fig. 19.34,

$$\text{power absorbed by load} = \text{FH} \times (\text{FE} - \text{HG})/8.$$

With transformer coupling, the average power from the d.c. source supplying the anode circuit is $\text{DC} \times \text{OD}$, whereas, with the load

connected directly in the anode circuit, the voltage of the d.c. source
is OA and the average power from that source is DC × OA. In both
cases the a.c. power absorbed by the load is the same; hence the
efficiency of the anode circuit is increased considerably by the use of
transformer coupling.

19.22 Input capacitance of a triode amplifier: Miller effect

A triode has three interelectrode capacitances, namely:

(a) *grid-cathode capacitance*, C_{gk}, which is directly across the
input circuit,

(b) *grid-anode capacitance*, C_{ga}, and

(c) *anode-cathode capacitance*, C_{ak}.

These interelectrode capacitances are, for convenience, shown as
capacitances external to the triode in fig. 19.39. The values of C_{gk}

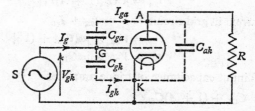

Fig. 19.39 Interelectrode capacitances.

and C_{ga} are each of the order of 5 pF, but the value of C_{ak} is much
smaller, since the electric flux existing *directly* between the anode and
the cathode is small compared with that between the anode and grid
and that between the grid and cathode for the normal potential
differences between these electrodes.

Capacitor C_{ak} is in parallel with the load resistor R, and its
reactance is so high compared with the resistance of R that its
shunting effect can be neglected except at very high frequencies.
Hence, as a first approximation, we shall omit C_{ak} in the following
analysis. Also, for simplicity, the polarizing batteries for the anode
and grid circuits have been omitted from fig. 19.39.

If V_{gk} = alternating p.d. between grid and cathode, assumed
positive when the grid is positive with respect to the cathode,

then I_{gk} = current through $C_{gk} = j\omega C_{gk} V_{gk}$.

Also, if V_{ga} = alternating p.d. between grid and anode, assumed positive when the grid is positive with respect to the anode,

$$I_{ga} = \text{current through } C_{ga} = j\omega C_{ga}V_{ga}.$$

It was shown in section 19.16 that, for a resistive load, the magnitude of the p.d. between anode and cathode is AV_{gk}, where $A = \mu/(1 + r_a/R)$ and that this p.d. is in exact phase opposition to V_{gk}; hence $V_{ak} = -AV_{gk}$.

From fig. 19.39, it is evident that—

$$V_{gk} = V_{ga} + V_{ak}$$

$$= \frac{I_{ga}}{j\omega C_{ga}} - AV_{gk}$$

$$\therefore \qquad (1 + A)V_{gk} = \frac{I_{ga}}{j\omega C_{ga}}$$

so that $\qquad I_{ga} = j\omega C_{ga}(1 + A)V_{gk}$

and total current in grid circuit $= I_g = I_{gk} + I_{ga}$

$$= V_{gk} \times j\omega\{C_{gk} + (1 + A)C_{ga}\}$$

$$= V_{gk} \times j\omega C_i$$

where C_i = input capacitance of triode amplifier

$$= C_{gk} + (1 + A)C_{ga} \qquad (19.10)$$

Hence the larger the voltage amplification A, the greater is the input capacitance and the lower the input reactance of the triode amplifier for a given frequency. This increase of the input capacitance due to voltage amplification is known as the *Miller effect*.

If $C_{gk} = C_{ga} = 5$ pF, and $A = 10$,

input capacitance $= C_i = 5 + (11 \times 5) = 60$ pF.

At 100 Hz, input reactance $= 10^{12}/(2\pi \times 100 \times 60) = 26\cdot5$ MΩ.

At 1 MHz, „ „ $= 2650$ Ω.

From these values it will be seen that at low frequencies the input reactance is very high; but at radio frequencies, the input reactance may be so low that the grid current may be sufficient to cause considerable reduction of the grid–cathode voltage if the source has a high internal impedance.

Another disadvantage of grid–anode capacitance is that it provides a coupling between the anode and grid; and when the load is inductive, this feedback may produce instability, i.e. the triode may act as an oscillator.

It was to reduce the effects of the grid–anode capacitance at radio frequencies that the tetrode and subsequently the pentode were developed.

19.23 Tetrode and pentode

The tetrode is a four-electrode valve containing a cathode, anode, control grid and an additional grid, referred to as a *screen* grid, situated between the control grid and the anode as shown in sectional elevation and plan in fig. 19.40. The conventional method of representing a tetrode having an indirectly heated cathode is given in fig. 19.41.

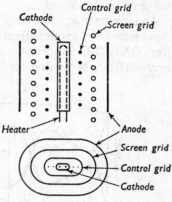

Fig. 19.40 A tetrode.

The main function of the screen grid is to provide an electrostatic shield between the anode and the control grid and thereby minimize the feedback effects discussed in section 19.22. This electrostatic shielding is so effective that the interelectrode capacitance between the anode and the control grid is reduced to a value of the order of 0·01 pF. With this value of capacitance and a voltage amplification of, say, 100, and with the same interelectrode capacitance of 5 pF between cathode and control grid assumed for the triode in section 19.22, then from expression 19.10,

input capacitance of tetrode =

$$5 + (101 \times 0 \cdot 01) = 6 \text{ pF};$$

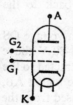

Fig. 19.41 Conventional representation of a tetrode.

hence the Miller effect in a tetrode is practically negligible.

An additional function of the screen grid is to reduce the effect of the anode voltage on the space charge at the cathode. Consequently the anode current remains nearly constant over a large variation of anode voltage.

For a given p.d. between the control grid and the cathode, the variations of anode and screen currents with variation of anode–

cathode p.d. are represented by the graphs of fig. 19.42, where OS represents the constant p.d. of about 100 V between the screen and the cathode. The chain-dotted graph represents the sum of the anode and screen currents, namely the cathode current, and indicates that this current remains approximately constant over the whole range of anode voltage.

An increase of anode voltage from zero to OA is accompanied by a rapid increase of anode current; but as the anode voltage is increased from OA to OB, the anode current decreases. This is due to *secondary emission* from the anode caused by the bombardment of the anode by

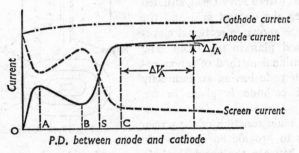

Fig. 19.42 Static characteristics of a tetrode.

the *primary* electrons emitted from the cathode. The emission of secondary electrons from the anode also takes place in a triode, but owing to the high positive potential of the anode in relation to the control grid, these secondary electrons are attracted back to the anode.

In a tetrode with an anode voltage OA and a screen voltage OS of about 100 V, the potential of the screen is about 70–80 V above that of the anode, so that the secondary electrons are attracted to the screen grid. When the anode voltage is increased in the region AB, the *increase* in the number of secondary electrons released from the anode is greater than the *increase* in the number of primary electrons reaching the anode. Consequently the net anode current is reduced. This accounts for the negative slope of the I_A/V_A characteristic over region AB.

As the anode voltage is increased above OB, the secondary electrons tend to be attracted back to the anode, and the anode current increases rapidly with increase of anode voltage in the region of OS.

When the anode voltage exceeds OC, the characteristic is practically linear and the increase of anode current is very small for a considerable increase of anode voltage. If ΔI_A be the increase of anode current for an increase ΔV_A of anode–cathode p.d.,

$$\text{anode a.c. or slope resistance} = r_a = \Delta V_A / \Delta I_A.$$

It is obvious from fig. 19.42 that for the linear or working region of the characteristic, the slope of the latter is very small and the value of r_a is therefore very high—usually of the order of 0·25 MΩ.

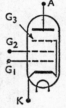

Fig. 19.43 A pentode.

The screen grid is usually fed from the H.T. supply through a resistor having the resistance necessary to give the correct screen–cathode voltage. A capacitor is connected between the screen and the cathode to bypass from the resistor any alternating component of the screen current in a manner similar to that already explained for a cathode resistor in section 19.19. In this way, the potential of the screen is maintained practically constant.

The presence of the kink in the I_A / V_A characteristic of the tetrode reduces considerably the working range available. For this reason, the tetrode, in the form described above, is now obsolete and has been replaced by the *pentode* in which a *suppressor grid*, G3, as shown in fig. 19.43, is inserted between the screen grid and the anode. This suppressor grid is usually connected to the cathode either internally, as in fig. 19.43, or externally, and its function is to prevent the secondary electrons, released from the anode, reaching the screen grid even when the anode–cathode p.d. is small. In this way, the kink of the tetrode characteristic is eliminated. The suppressor grid also provides still further electrostatic shielding of the anode from the control grid G1, so that stable operation at still higher frequencies is possible.

The full-line graphs in fig. 19.44 are typical I_A / V_A static characteristics of a pentode for various p.d.s between the control grid G1 and the cathode K. The uniformly-dotted graph shows the screen-grid current for zero p.d. between the control grid and the cathode, and the chain-dotted graph T gives the sum of the anode and screen currents, namely the cathode current, for this condition.

The screen grid, G2, of a pentode is maintained at the correct potential and decoupled by a capacitor in the way already described for the tetrode.

The value of the anode a.c. resistance for a control-grid potential of -2 volts in fig. 19.44 is given by:

$$r_a = \Delta V_A / \Delta I_A$$

and is usually of the order of 1 MΩ.

The value of the mutual conductance, g_m, for a pentode is approximately the same as that of the corresponding triode, and can be determined either from the transfer (or mutual) characteristics,

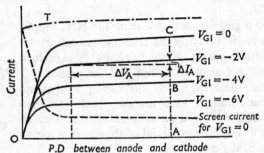

Fig. 19.44 Static characteristics of a pentode.

similar to those shown in fig. 19.29 for the triode, or it may be derived from the I_A/V_A characteristics of fig. 19.44. Thus, if a pentode is operating with a grid bias of -2 volts and has an input alternating voltage of peak value 2 volts, then for a constant anode voltage OA, the anode current varies between AB and AC;

$$\therefore \qquad \text{mutual conductance} = g_m = (AC - AB)/4$$
$$= BC/4$$

and $\qquad$ amplification factor $= \mu = r_a g_m$;

e.g. if $g_m = 1$ mA/V and $r_a = 1$ MΩ, $\mu = 1000$.

A modified form of the pentode is the *beam tetrode*, in which the suppressor grid of the pentode is replaced by two beam-forming plates, as shown in plan in fig. 19.45 (*a*). These shields are connected to the cathode. The wires forming the control and the screen grids are optically aligned so as to allow beams of electrons from the cathode to pass more easily through the grids towards the unshielded portions of the anode, as shown by the dotted lines in fig. 19.45 (*a*). With this arrangement, the negative space charge formed in the region between the screen grid and the anode prevents the secondary electrons from reaching the screen grid even when the potential of the anode is much lower than that of the screen. The result is that the characteristics of the beam tetrode have approximately the same shape as those given in fig. 19.44 for the pentode. The graphical symbol for the beam tetrode is shown in fig. 19.45 (*b*).

The mathematical and graphical methods already described for determining the voltage amplification and the power output of a triode apply equally well to the tetrode and the pentode when operated under correct conditions. Owing, however, to the anode a.c. resistance of the tetrode and the pentode being very high compared with values that are practicable for the load resistor, expression

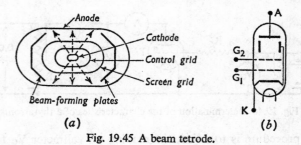

(*a*) (*b*)

Fig. 19.45 A beam tetrode.

(19.5) for the voltage amplification can usually be simplified by neglecting the resistance of the load. Hence, in magnitude,

$$\left.\begin{array}{r}\text{voltage amplification of a tetrode or a} \\ \text{pentode}\end{array}\right\} = A = \frac{\mu R}{r_a + R}$$

$$\simeq \frac{\mu R}{r_a} = g_m R \quad (19.11)$$

For instance, if $g_m = 5$ mA/V, $r_a = 1$ MΩ and $R = 20$ kΩ, the resistance of the load is very small compared with the anode a.c. resistance, so that—.

$$A \simeq 5 \times 10^{-3} \times 20 \times 10^3 = 100.$$

19.24 Thyratron

In section 19.7 it was explained that when the p.d. between the anode and cathode of a gas-filled hot-cathode tube is raised to a certain value (the actual value being dependent upon the gas pressure and the distance between the electrodes), ionization takes place. Let us now consider the effect of introducing a grid between the cathode and the anode, thereby converting the gas-filled diode into a gas-filled triode or *thyratron* (*thyra* being the Greek for a door).

The characteristics of a thyratron can be determined with the aid of the circuit shown in fig. 19.46, where T represents a thyratron having an indirectly heated cathode. There are two important features which distinguish fig. 19.46 from the arrangement shown in fig. 19.28 for determining the static characteristics of a vacuum triode, namely: (a) a resistor R_A (say, 2000 Ω) in series with the anode, and (b) a resistor R_G of the order of 0·1 MΩ in series with the grid. The function of these resistors is to limit the currents in the respective circuits to values that will prevent excessive bombardment of the cathode and grid by positive ions.

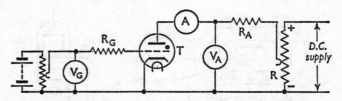

Fig. 19.46 Determination of the characteristics of a thyratron.

The procedure is to adjust the reading on voltmeter V_G to, say, zero and then move the slider on R so as to increase the reading on voltmeter V_A from *zero* up to the value that produces ionization. This value of the p.d. is termed the *striking voltage*. Immediately ionization commences, the reading on V_A falls to about 16 V, the difference between this voltage and that supplied by R being absorbed by R_A. The slider on R is then moved backwards. The anode current decreases, but the p.d. between the anode and cathode remains approximately constant, and its value at the extinction of the glow is noted.

The test is repeated with different readings on voltmeter V_G and the graphs in fig. 19.47 represent the results obtained with a certain thyratron. It is seen that the larger the negative bias applied to the grid, the greater is the striking voltage. This is due to the fact that the electrons moving outwards from the cathode must reach a certain critical velocity before they can start ionization; and the more negative the grid, the greater is the positive potential that must be applied to the anode to neutralize the retarding effect of the grid upon the movement of these electrons.

The slope of the striking-voltage characteristic is termed the *control ratio*, i.e.

$$\text{control ratio} = \frac{\text{change of anode breakdown voltage}}{\text{corresponding change of grid-bias voltage}}$$

For the thyratron having the characteristic given in fig. 19.47, the control ratio is about 34 for grid bias between −3 V and −6 V.

An important characteristic of the thyratron is that *once ionization*

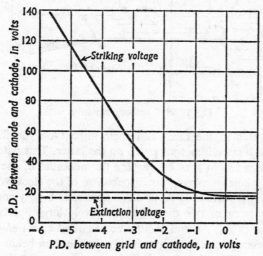

Fig. 19.47 Characteristics of a thyratron.

has commenced, the anode current cannot be controlled by varying the grid voltage. The anode current can be reduced to zero only by reducing the anode–cathode p.d. to the extinction value shown in fig. 19.47.

19.25 Cathode-ray oscilloscope

The cathode-ray oscilloscope (usually abbreviated to 'C.R.O.') is almost universally employed to display the waveforms of alternating voltages and currents and has very many applications in electrical testing—especially at high frequencies. The cathode-ray tube is an important component of both the C.R.O. and the television receiver.

Fig. 19.48 shows the principal features of the modern cathode-ray tube. C represents an indirectly heated cathode and G is a control grid with a variable negative bias by means of which the electron emission of C can be controlled, thereby varying the brilliancy of the spot on the fluorescent screen S. The anode discs A_1 and A_2 are usually connected together and maintained at a high potential relative to the cathode, so that the electrons passing through G are accelerated very rapidly. Many of these electrons shoot through the

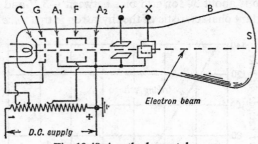

Fig. 19.48 A cathode-ray tube.

small apertures in the discs and their impact on the fluorescent screen S produces a luminous patch on the latter. This patch can be focused into a bright spot by varying the potential of the focusing electrode F, thereby varying the distribution of the electrostatic field in the space between discs A_1 and A_2. Electrode F may consist of a metal cylinder or of two discs with relatively large apertures. The combination of A_1, A_2 and F may be regarded as an *electron lens* and the system of electrodes producing the electron beam is termed an *electron gun*. The glass bulb B is thoroughly evacuated to prevent any ionization.

19.26 Deflecting systems of a cathode-ray tube

(1) *Electrostatic deflection*

The electrons after emerging through the aperture in disc A_2 pass

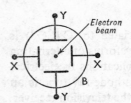

Fig. 19.49 Deflecting plates of a cathode-ray tube.

between two pairs of parallel plates, termed the X and Y plates and arranged as in fig. 19.49. One plate of each pair is usually connected to anode A_2 and to earth, as in fig. 19.48.

Suppose a d.c. supply to be applied across the Y-plates, as in fig. 19.50, then the electrons constituting the beam are attracted towards the positive plate M and the beam is deflected upwards. If an alternating voltage were applied to the Y-plates, the beam would be deflected alternately upwards and downwards and would therefore trace a *vertical* line on the screen. Similarly, an alternating voltage applied to the X-plates would cause the beam to trace a horizontal line. The method of calculating the deflection of an electron moving across an electric field is discussed in section 5.19.

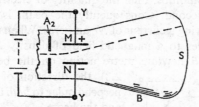

Fig. 19.50 Electrostatic deflection of an electron beam.

(2) *Electromagnetic deflection*

If two coils, P and Q, were arranged outside the tube, with their axis perpendicular to the beam, and if a direct current were passed through the coils in the direction shown in fig. 19.51, the beam would be deflected upwards. This is due to the fact that the electron beam behaves as a flexible conductor and the *conventional* direction of the current in the beam is from the screen towards the cathode. Applying either the grip or the corkscrew rule, we find that the magnetic flux underneath the beam is strengthened and that above the beam is weakened, so that the resultant flux is distorted as shown in fig. 19.51. Consequently there is a force *F* urging the beam upwards. An alternating current

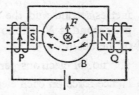

Fig. 19.51 Electromagnetic deflection of an electron beam (tube viewed from screen end).

through the coils would give a vertical line on the screen—similar to that obtained with an alternating voltage across the Y-plates.

19.27 Motion of an electron in a uniform transverse magnetic field

Suppose the area represented by the dotted rectangle ABCD in fig. 19.52 to represent the cross-section of a uniform magnetic field and suppose the direction of the magnetic flux to be outwards from the paper. Also, suppose a beam of electrons, travelling at *v* metres/ second, to enter this field as shown in fig. 19.52, the direction of the beam being perpendicular to that of the magnetic field. As explained in section 19.26, there is a force urging the electrons in the direction shown in the diagram.

To find the value of the force on each electron. If N be the number per second passing any point P and if the negative charge on each electron is **e** coulombs, then the quantity of electricity passing P is N**e** coulombs/second. Consequently the beam is equivalent to a conductor carrying a current of N**e** amperes. When an electron beam is at right-angles to a magnetic field of density B teslas, there is a force of N**e**B newtons/metre acting on the beam (see section

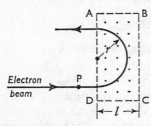

Electron beam

Fig. 19.52 Motion of an electron in a magnetic field when $l > r$.

2.8), the direction of the force being perpendicular to *both* the magnetic field and the beam.

Hence no energy is supplied to or taken from the electrons, so that the kinetic energy of each electron remains unaltered. This means that the *speed* of the electrons remains constant though the direction of the beam is continually changing.

Since the speed of the electrons is v metres/second and the number of electrons passing a given point per second is N, it follows that N electrons occupy a length v metres of the beam;

$\therefore$ no. of electrons per metre length of beam $= N/v$.

Hence force on each electron traversing the magnetic field $\Bigg\} = \dfrac{\text{force/metre length of beam}}{\text{no. of electrons/metre length of beam}}$

$$= \frac{N\mathbf{e}B}{N/v} = B\mathbf{e}v \text{ newtons} \qquad (19.12)$$

To find the locus of the electron path in the magnetic field. From Mechanics, it is known that when a constant force is acting on a body at right-angles to its direction of motion, the body moves in a circle of radius r and has an acceleration v^2/r, where v is the constant speed of the body. Hence the electron follows a circular path while it is moving through the uniform magnetic field.

Force, in newtons, on electron to maintain this circular path

$=$ mass of electron in kilograms $\times$ acceleration in metres/second^2

$= \mathbf{m}v^2/r$ newtons $\qquad (19.13)$

Equating expressions (19.12) and (19.13), we have—

$$Bev = mv^2/r$$

$$\therefore \qquad r = \frac{m}{e} \cdot \frac{v}{B} \text{ metres} \qquad (19.14)$$

If r is less than l, the electron describes a semicircle in the magnetic field, as in fig. 19.52, and emerges from the field at a distance $2r$ from the point of entry. It is then travelling in a direction directly opposite to its original direction, with the magnitude of its velocity unchanged.

If $l \ll r$, as in fig. 19.53, the path of the electron in the magnetic

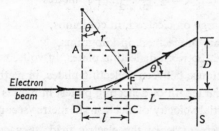

Fig. 19.53 Deflection of an electron beam in a magnetic field when $l \ll r$.

field is an arc EF of a circle of radius r and the beam is deflected through an angle θ,

where
$$\theta = \frac{\text{arc EF}}{r} \simeq \frac{l}{r} \text{ radian} \qquad (19.15)$$

Substituting for r from expression (19.14), we have:

$$\theta = \frac{lB}{v} \cdot \frac{e}{m} \text{ radian} \qquad (19.16)$$

Example 19.7 *An electron has a velocity of 10^7 m/s when it enters a magnetic field perpendicularly to the direction of the flux. If the flux density is uniform at 0·5 mT and the axial length of the magnetic field is 2 cm, calculate:* (a) *the radius of curvature of the electron path in the magnetic field, and* (b) *the angle through which the electron is deflected. Assume the ratio e/m to be 1·76 $\times$ 10^{11} C/kg.*

(a) From expression (19.14),

$$r = \frac{10^{-11}}{1·76} \times \frac{10^7}{0·5 \times 10^{-3}} = 0·1136 \text{ m}$$

$$= 11·36 \text{ cm.}$$

(b) Since $l \ll r$, we have from expression (19.15):

$$\theta \simeq l/r = 2/11.36 = 0.176 \text{ radian}$$
$$= 0.176 \times 57.3 = 10.1 \text{ degrees.}$$

19.28 Electrostatic and magnetic deflections on the screen of a cathode-ray tube

(a) *Electrostatic deflection*

It was shown in section 5.19 that the deflection of an electron passing through an electric field is given by:

$$x = \tfrac{1}{2} \cdot \frac{e}{m} \cdot \frac{V}{d} \cdot \left(\frac{l}{v}\right)^2 \text{ metres} \tag{5.19}$$

where e = charge on electron, in coulombs,

m = mass of electron, in kilograms,

V = p.d. between deflecting plates, in volts,

d = distance between deflecting plates, in metres,

l = axial length of deflecting plates, in metres

and v = initial velocity of electron, in metres/second.

When the electrons leave the electric field, they continue in a straight line to screen S, as indicated in fig. 19.54. If this straight line

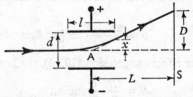

Fig. 19.54 Electrostatic deflection of electrons in a cathode-ray tube.

is produced backwards, it can be shown from the geometry of the parabolic path of the electrons in the electric field, that this line meets the original axis of the electron movement at a point A midway along the electric field.

If L = distance from A to the screen, in metres,

and D = deflection on the screen, in metres,

then $\dfrac{x}{D} = \dfrac{l/2}{L}$

$$\therefore \quad D = \frac{2L}{l} \times \frac{1}{2} \times \frac{e}{m} \times \frac{V}{d} \times \left(\frac{l}{v}\right)^2 = \frac{e}{m} \times \frac{V}{d} \times \frac{Ll}{v^2} \tag{19.17}$$

If V_A is the accelerating voltage, i.e. the p.d. between the cathode and the final anode, then, from expression (5.21),

$$\tfrac{1}{2}mv^2 = eV_A.$$

Substituting for $e/(mv^2)$ in expression (19.17), we have:

$$D = \frac{Ll}{2d} \times \frac{V}{V_A} \tag{19.18}$$

(b) *Magnetic deflection*

In a C.R.O., the length l of the magnetic field in the initial direction of the electron beam is very small compared with the radius of curvature of that beam; hence the deflection of the electron beam is given by expression (19.16),

i.e. $$\theta = \frac{lB}{v} \cdot \frac{e}{m}.$$

After emerging from the magnetic field, the electrons travel along the tangent to the arc at F, as shown in fig. 19.53. This tangent cuts the original axis of the beam at a point roughly midway along the magnetic field. Hence,

if $\quad D =$ deflection on screen, in metres

and $\quad L =$ distance of screen from mid-point of magnetic field, in metres,

$$\frac{D}{L} = \tan \theta \simeq \theta = \frac{lB}{v} \cdot \frac{e}{m}$$

$\therefore \qquad\qquad D = \frac{LlB}{v} \cdot \frac{e}{m} \tag{19.19}$

From expression (5.21), $v = \sqrt{(2V_A e/m)}$, where V_A is the p.d. between the cathode and the final anode.

Substituting for v in expression (19.19), we have:

$$D = BLl\left(\sqrt{\frac{e}{m}}\right) \times \frac{1}{\sqrt{(2V_A)}} \tag{19.20}$$

Comparison of expressions (19.18) and (19.20) shows that the electrostatic deflection is inversely proportional to the accelerating voltage V_A, whereas the magnetic deflection is inversely proportional to the square root of this voltage. Also, the magnetic deflection is a function of e/m, whereas the electrostatic deflection is independent of this ratio.

19.29 Time base for a C.R.O.

To enable the waveforms of alternating voltages and currents to be displayed on the screen of a C.R.O., it is necessary to provide a *time base*, i.e. the p.d. across the X-plates must be made to increase uniformly from M to N and then decrease quickly from N to P, as shown in fig. 19.55. In consequence, the spot on the screen travels at

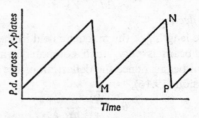

Fig. 19.55 P.d. across X-plates to give a time base.

a uniform speed from left to right across the screen (stroke) and then returns quickly (flyback). The effect is to give a horizontal line AB (fig. 19.56). When an alternating voltage is applied to the Y-plates, then during time MN (fig. 19.55), the spot traces a wave such as that represented by the full line DE; but during the short interval NP, a faint trace may be seen, as represented by the dotted line ED in

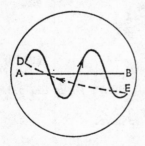

Fig. 19.56 A waveform on a C.R.O. screen.

fig. 19.56. To obtain a stationary oscillogram on the screen, it is necessary to introduce some synchronizing arrangement to ensure that the frequency of the sweep voltage applied to the X-plates is an exact sub-multiple of the frequency of the voltage applied to the Y-plates so that successive traces may be superimposed on one another.

The circuit arrangement for a synchronized time base is too complicated for inclusion in this volume, but it may be mentioned that all such circuits involve the slow charging and quick discharging of a capacitor connected in parallel with the X-plates. For instance, in fig. 19.57, C represents such a capacitor connected in series with a resistor R having very high resistance across a d.c. supply. A thyratron T is also in parallel with C and the striking voltage of T is set to any desired value by varying its grid bias, as already explained in section 19.24. As the p.d. across C and the X-plates increases, the spot on the screen moves to the right; but when this p.d. reaches the striking voltage of T (point N in fig. 19.55), ionization occurs in T so that the capacitor is quickly discharged and the p.d. across the X-plates falls to about 20 V (point P in fig. 19.55), thereby causing the spot to return rapidly to the left. After the capacitor has been discharged, the ionization in T ceases and the above cycle is repeated.

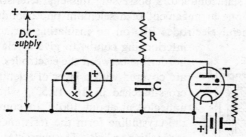

Fig. 19.57 A simple time-base circuit.

The sweep frequency depends upon the time constant (see section 5.23) of circuit RC and can be increased by reducing the value of either C or R.

The simple time-base arrangement shown in fig. 19.57 has the following disadvantages:

(*a*) the charging current does not remain constant, so that the horizontal speed of the spot is not uniform;

(*b*) the time required for deionization in T renders it unsuitable for frequencies above about 50 kHz;

(*c*) no arrangement is provided for synchronizing.

19.30 Photoelectric cells

A photoelectric cell (or photocell) is a device the electrical properties of which undergo a change when it is exposed to light. Photocells can be divided into three groups:

(*a*) *photoconductive cells*, in which the light causes a change of resistance,

(*b*) *photoemissive cells*, in which the light causes the emission of electrons from a metallic surface in an evacuated or gas-filled envelope, and (*c*) *photovoltaic cells*, in which the light causes an e.m.f. to be generated.

19.31 Photoconductive cells

In both the photoconductive cell and the photovoltaic cell, the sensitive element is a semiconductor (section 20.2). Light flux falling upon a semiconductor may displace electrons from some of the atoms, thereby increasing the conductivity of the material. Selenium is one of the semiconductors possessing this characteristic, and fig. 19.58 shows one arrangement of a selenium photoconductive cell. Two sheet-metal electrodes EE, on an insulating surface S, form interlocking combs to give a relatively long borderline between the electrodes. The latter are covered with a layer of selenium (shown cross-hatched in fig. 19.58), which is then annealed at about 100°C to convert it into the crystalline form the resistance of which is sensitive to light. The characteristics of the cell can be determined by connecting a battery in series with a microammeter A to the metal electrodes and noting the current for various values of illuminance and for radiation of different wavelengths.

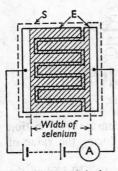

Fig. 19.58 Selenium photoconductive cell.

The resistance of such a cell, when not illuminated, is of the order of 0·1 MΩ, and this resistance is approximately halved by an illuminance of 100 lux (or lumens/metre²).

In another type of photoconductive cell, the photosensitive material is cadmium sulphide or cadmium selenide or a combination of these materials. A pure cadmium sulphide cell has its maximum sensitivity at about 0·5 μm, while a pure cadmium selenide cell is most sensitive at about 0·75 μm. Combinations of these materials have their maximum sensitivity at wavelengths between 0·5 and 0·75 μm.

Photoconductive cells have the disadvantage that their response falls off rapidly for light flux varying at frequencies above about 100 Hz.

19.32 Photoemissive cells

Some materials possess the property of emitting electrons when exposed to radiation. According to the quantum theory, electromagnetic radiation, such as light waves, behaves as if it consisted of indivisible packets or bundles, each packet—called a *photon*—containing an amount of energy that is directly proportional to the frequency of the radiation; for instance, the energy in one photon of light from a sodium lamp (wavelength 0·589 μm, section 21.1) is $33·7 \times 10^{-20}$ joule or 2·1 electronvolts (section 5.20). The energy in a photon varies between 3 and 1·8 electronvolts for the visible range of radiation.

When a photon of radiation reaches a metallic surface, the energy of the photon may be absorbed by one of the free electrons near the surface of the metal, so that the total energy of this electron may be sufficient to enable it to escape from the surface into the surrounding space. The energy required to take a free electron out of a metal against the attractive forces of the positive ions in the metal is termed the *work-function* of that metal. The work-function of copper is about 4·3 electronvolts, so that it is almost impossible for a photon to release an electron from copper unless the electron already possesses a relatively large amount of energy due to, say, high temperature.

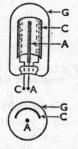

Fig. 19.59 Photo-emissive cell.

The alkali, caesium (symbol, Cs), is the metal possessing the lowest work-function, namely about 1·9 electrovolts. Hence, most photoemissive cells have caesium as one of the constituents on the cathode, the other constituents being dependent upon the purpose for which the cell is to be used and the range of frequency to which it is required to respond.

Fig. 19.59 shows one form of construction of a photoemissive cell. The anode A is a metal rod about 1 mm diameter, and cathode C is a curved plate of steel or nickel, coated on the inner surface with the photoemissive material. In one type of cell, this sensitive material consists of a layer of silver oxide covered with a very thin layer of caesium. This is referred to as the Ag–O–Cs cathode. In another type —the Sb–Cs cathode—a layer of antimony (Sb) is deposited on the plate and this is afterwards covered with a thin film of caesium. The former cathode is most sensitive in the infra-red region, while the latter is most sensitive to violet light.

Photoemissive cells are of two types: (*a*) *vacuum* type, and (*b*) *gas-filled* type. The characteristics of these cells are described in sections 19.33 and 19.34 respectively.

19.33 Characteristics of a vacuum photoemissive cell

The current/voltage characteristics of a photoemissive cell can be determined by connecting the cell in series with a microammeter A

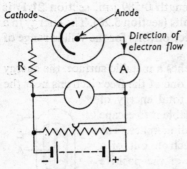

Fig. 19.60 Determination of the characteristics of a photoemissive cell.

and a resistor R, of known high resistance, across a variable-voltage supply as in fig. 19.60, and noting the current at various anode–cathode voltages for different values of light flux incident on the cathode. The p.d. between the anode and cathode of the cell is derived by subtracting the *IR* drop in R from the reading on voltmeter V. The function of resistor R is to protect the photoemissive cell in case ionization occurs.

The graphs of fig. 19.61 are typical of the characteristics obtained with a vacuum cell. For a given luminous flux, the current at first increases with increase of anode voltage, the actual variation being dependent upon the shape of the electrodes. With an anode voltage exceeding about 60 V, the current reaches its saturation value when practically all the photoelectrons (i.e. the electrons released from the

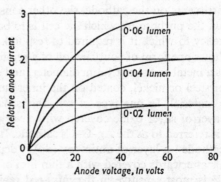

Fig. 19.61 Characteristics of a vacuum photoemissive cell.

cathode) are collected by the anode. Fig. 19.61 also shows that the saturation current is directly proportional to the light flux falling on the cathode.

Electronic Valves 555

19.34 Characteristics of a gas-filled photoemissive cell

This type of cell contains an inert gas—usually argon—at a pressure of about 0·3 mm of mercury. The graphs of fig. 19.62 are typical of the characteristics obtained with a gas-filled cell.

The increased current obtained with the gas-filled cell is due to ionization taking place when the photoelectrons emitted from the cathode have been accelerated to the required velocity (section 19.7). With a low anode voltage the velocity of the photoelectrons does not reach the critical velocity necessary to start ionization; hence the anode current is practically the same as that in the vacuum cell for

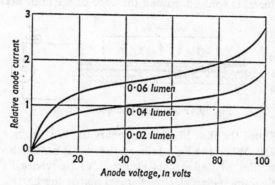

Fig. 19.62 Characteristics of a gas-filled photoemissive cell.

the same anode voltage and the same illuminance. With a higher anode voltage, however, some of the photoelectrons collide with gas atoms with sufficient velocity to displace free electrons from the atoms. The displaced electrons together with the photoelectrons travel towards the anode and cause further ionization if they acquire the necessary velocity. The positively-charged atoms (i.e. positive ions), from which the electrons have been displaced, travel towards the cathode. Consequently, a considerable increase of anode current is obtained, but if the current exceeds about ten times that in the vacuum cell under similar conditions, the ionization increases rapidly and the cathode is destroyed by excessive positive-ion bombardment.

The relationship between the anode current and the light flux is not as linear in the gas-filled cell as in the vacuum cell; hence the latter is used for precise photometric work, but the former—owing to its greater sensitivity—is used for sound reproduction from films and for energizing light-operated triggers in relay circuits.

19.35 Photovoltaic cells

It was mentioned in section 19.31 that the sensitive element in a photovoltaic cell is a semiconductor of such material that light flux falling on it displaces electrons from some of the atoms. The two semiconductors found most suitable for photovoltaic cells are selenium and cuprous oxide, and fig. 19.63 shows the arrangement of a cell having selenium as the active material. An iron plate P is coated with a thin layer S of selenium at about 200°C and annealed at about 80°C to produce the crystalline form. This selenium layer is covered with a very thin transparent film M of metal, and a collecting ring R of metal is sprayed around the edge of the film. Between the

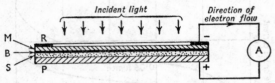

Fig. 19.63 Selenium photovoltaic cell.

selenium S and the film M, there appears to be a 'barrier layer' B (section 19.9). When the light falls on the cell, it passes through the transparent film M and causes electrons to be released from the metallic selenium. These electrons travel across the barrier layer to the metal film M, from which they are collected by ring R. A micro-ammeter A is connected between R and P. It is found that with a suitable resistance of the external circuit between R and P, the current through A is practically proportional to the illuminance and the microammeter can be calibrated to read the illuminance directly in lux (section 21.2).

Summary of important formulae

$$\text{Efficiency of mercury-arc rectifier} \simeq V/(V+20) \tag{19.1}$$

$$\text{Mutual conductance} = g_m = \frac{\Delta I_A}{\Delta V_G} \text{ for constant } V_A \tag{19.2}$$

$$\text{Anode a.c. or slope resistance} = r_a = \frac{\Delta V_A}{\Delta I_A} \text{ for constant } V_G \tag{19.3}$$

$$\text{Amplification factor} = \mu = -\frac{\Delta V_A}{\Delta V_G} \text{ for constant } I_A \tag{19.4}$$

$$\mu = g_m r_a$$

Voltage amplification with resistive load $\Big\}$ $= A = \dfrac{\mu}{1 + r_a/R}$ (19.5)

$\simeq g_m R$ for tetrodes and pentodes (19.11)

Transformer ratio for maximum power output $\Big\}$ $= n = \sqrt{(r_a/R_L)}$ (19.7)

Input capacitance of valve amplifier with resistive load $\Big\}$ $= C_i = C_{gk} + (1 + A)C_{ga}$ (19.10)

Force on electron moving across magnetic field $\Big\}$ $= Bev$ newtons (19.12)

Electrostatic deflection of cathode-ray tube $\Big\}$ $= D = \dfrac{Ll}{2d} \times \dfrac{V}{V_A}$ metres (19.18)

Magnetic deflection of cathode-ray tube $\Big\}$ $= D = BLl\sqrt{\left(\dfrac{e}{m} \times \dfrac{1}{2V_A}\right)}$ metres (19.20)

EXAMPLES 19

1. The anode current of a certain vacuum diode is 6 mA when the anode voltage is 80 V. Calculate: (a) the anode current for an anode voltage of 50 V, and (b) the anode voltage for an anode current of 4 mA, assuming that over this range the relationship between the anode voltage and current is given by $i_A = kv_A^{1.5}$.

2. The following table gives the static characteristic of a certain vacuum diode:

Anode voltage (V)	25	50	75	100	125
Anode current (mA)	0·15	0·8	1·9	3·7	6·2

Plot the characteristic and determine: (a) the anode d.c. resistance for an anode voltage of 100 V, and (b) the anode a.c. resistance for an anode voltage of 100 V.

3. A 20-kΩ non-reactive resistor is connected in series with the anode of the diode referred to in Q. 2. Derive the graph representing the dynamic characteristic for supply voltages up to 200 V.

4. The diode referred to in Q. 2 is connected in series with a 15-kΩ resistor across a 200-V d.c. supply. Draw the load line and determine the value of the anode current and the p.d. between the anode and cathode of the diode.

5. An a.c. supply having a peak voltage of 120 V is connected across the anode and cathode of the diode whose static characteristic is given in Q. 2. The waveform of the voltage is sinusoidal. Derive a graph representing the variation of anode current with time and determine

therefrom the average value of the anode current over a complete cycle.

6. Describe the construction of the indirectly heated cathode of a vacuum diode and explain how a space charge is set up in the diode.

When the anode is made positive with respect to the cathode, the anode current can be 'space-charge limited' or 'temperature limited'. Explain the meaning of these terms.

The voltage/current characteristic of a vacuum diode is as follows:

Anode voltage (V)	4	8	12	16	20	24	28	32
Anode current (mA)	2	6·1	11·6	18·6	26·8	35·7	45·5	56·6

When the current is 17 mA, find: (a) the anode d.c. resistance, and (b) the anode a.c. resistance.   (U.E.I., O.2)

7. Draw diagrams for half-wave and full-wave rectifying circuits supplying resistive loads.

Sketch the output and input voltage and current waveforms if the input voltage is sinusoidal.

A 20-Ω resistive load is supplied from: (a) a half-wave rectifier, and (b) a full-wave rectifier. Input to the rectifiers is 240 V, r.m.s., and sinusoidal in form. Calculate the mean load current in each case. (Ignore diode forward resistance.)   (U.L.C.I., O.1)

8. Sketch and explain the shape of the static characteristic for a thermionic vacuum diode.

A simple half-wave rectifier circuit is fed from a 100-V, r.m.s., sinusoidal supply. Determine the average value of the voltage developed across a resistive load. Derive any formula used. Assume zero forward and infinite reverse resistance for the diode.

(N.C.T.E.C., O.2)

9. The static characteristic of a certain vacuum diode is given by the following table:

Voltage (V)	0	10	20	30	40	60	80
Current (mA)	0	0·15	0·6	1·4	2·6	5·0	6·5

Plot the characteristic and determine the anode a.c. resistance for an anode voltage of 50 V.

10. If the diode of Q. 9 is connected in series with a 30-kΩ resistor across a 150-V d.c. supply, draw the load line and determine the value of the anode current.

11. Describe the action of conduction in a gas-filled cold-cathode diode.

What is meant by the 'striking potential' of such a diode?

A 1·8-kΩ resistor and a cold-cathode diode are connected in series to a 200-V d.c. supply. In parallel with the diode is a resistor R. If the striking potential is 125 V, determine the minimum value of R which will allow the diode to strike. Assume the diode takes no current before striking.   (N.C.T.E.C., O.2)

Note. A *cold-cathode diode* is a tube with two cold electrodes in a gas at reduced pressure. When the potential gradient in the gas is raised to a critical value, ionization takes place and electrons are then emitted from the cathode due to bombardment of the latter by positive ions (see section 19.7). Useful emission is thus obtained without the application of heat. A stabilizing resistor must be connected in series with the diode to limit the current, otherwise the cathode would be destroyed by excessive bombardment. Cold-cathode diodes are more economical than hot-cathode diodes when they have to be inoperative for long periods but must be available for instant operation.

12. A gas diode is connected in series with a 100-Ω resistor across an a.c. supply, the peak value of which is 280 V, the waveform being sinusoidal. When the diode is conducting, the p.d. between anode and cathode is practically constant at 20 V. Derive a graph representing the variation of current with time and determine the peak current and the average value of the current over a complete cycle.

13. A gas diode is connected in series with an 8-Ω resistor across a 230-V (r.m.s.) a.c. supply. Assuming the supply voltage to be sinusoidal and the voltage drop in the diode to remain constant at 15 V, calculate the peak, the average and the r.m.s. values of the current.

14. A battery having an e.m.f. of 50 V and negligible internal resistance is charged from a 230-V (r.m.s.) supply through an 8-Ω resistor in series with a gas diode. Assuming the supply voltage to be sinusoidal and the voltage drop in the diode to remain constant at 15 V, calculate the peak and the mean values of the charging current.

15. A full-wave mercury-arc rectifier is used to convert a single-phase supply to a 230-V d.c. supply, the circuit being as shown in fig. 19.17. Assuming a sinusoidal supply voltage and an arc drop of 20 V, calculate: (*a*) the r.m.s. value of the voltage between M and N, and (*b*) the efficiency of the rectifier.

16. A rectifier, consisting of two copper-oxide elements in series, has the following characteristic:

Voltage (V)	0·4	0·6	0·8	1·0	1·2
Current (mA)	0·4	2·0	6	14	24

The rectifier is connected across an a.c. supply having a peak value of 1·2 V. The waveform of the voltage is sinusoidal. Derive a graph showing the variation of current with time and determine the mean value of the current over a complete cycle. Assume the reverse resistance to be infinite.

17. Describe with sketches the construction of *either* (*a*) a metal rectifier, *or* (*b*) a small mercury-arc rectifier, giving a brief account of the principle of action of the rectifier described.

A half-wave metal rectifier is connected in series with a 50-Ω resistor, a moving-iron ammeter and a d.c. moving-coil ammeter to a sinusoidal supply voltage of 100 V, r.m.s. Calculate the readings on the instruments, assuming the resistance of the ammeters to be negligible and the rectifier to have zero resistance in the 'forward' direction and infinite resistance in the 'reverse' direction. Deduce the expressions employed. (App. El., L.U.)

18. The characteristic of each of the four elements of a copper-oxide bridge-rectifier is given by the following table:

Voltage (V)	0·1	0·15	0·2	0·24	0·28	0·34	0·38	0·41
Current (mA)	0·2	0·4	1	2	4	8	12	16

This bridge unit is connected in series with a non-reactive resistor of 50 Ω across a 1-V (r.m.s.) a.c. supply. Find the reading on a milliammeter connected across appropriate points of the bridge. Neglect the instrument resistance and any reverse current in the rectifiers. The supply voltage has a sinusoidal waveform.

19. Two anode characteristics for a triode are obtained from the following figures:

Grid voltage = 0

Anode voltage (V)	0	25	50	75	100	125
Anode current (mA)	0	2·5	5·4	8·8	12·8	17·4

Grid voltage = −4 V

Anode voltage (V)	45	75	100	125	150	175
Anode current (mA)	0	1·5	4·0	6·9	10·3	14·3

Draw the characteristics, and deduce from them the three valve constants. (W.J.E.C., O.1)

20. A triode has the following static characteristics:

Anode voltage	Anode current (mA)					
	$V_g = 0$	$V_g = -2V$	$V_g = -4V$	$V_g = -6V$	$V_g = -8V$	$V_g = -10V$
50 V	4·2	1·25				
100 V	10·6	5·6	1·7			
150 V	17·5	11·75	6·7	2·4	0·7	
200 V		17·8	12·8	7·7	3·2	1·25
250 V				14·0	8·75	4·1
300 V					14·8	9·5

Plot the characteristics and estimate the anode slope resistance and mutual conductance at the point $V_a = 125$ V and $V_g = -2$ V.

The valve is used in an amplifier with an anode supply voltage of 300 V and a 20-kΩ load resistor. Derive the dynamic mutual characteristic. (W.J.E.C., O.2)

21. Define the three parameters of a triode and state the relationship between them. Explain with the aid of a sketch how these parameters may be obtained from the static characteristics of the valve.

The slope of the I_a/V_g characteristic of a triode is 1·8 mA/V and that of the I_a/V_a characteristic is 0·2 mA/V. Calculate the amplification factor. (S.A.N.C., O.1)

22. The following readings were obtained with a certain triode:

Grid voltage, in volts	0	−1	−2	−4	−6	−8
Anode current, in mA, with anode voltage 100 volts	9·4	6·7	4·2	1·2	0·1	0
„ „ 150 volts	15·8	13·1	10·5	5·6	2·3	0·6

Plot the characteristics and estimate (*a*) the mutual conductance,

(b) the anode a.c. resistance and (c) the amplification factor for a grid bias of about −1 V and an anode voltage of about 100 V.

23. The following readings were obtained with a certain triode:

Anode voltage, in volts	120	100	80
Anode current, in mA, with			
grid voltage of 0	20·3	15	10
,, ,, ,, −2 V	12·5	7·5	3

Estimate (a) the anode a.c. resistance, (b) the amplification factor and (c) the mutual conductance for an anode voltage of 100 V and a grid bias of about −1 V.

24. The anode current of a triode was 4 mA with an anode voltage of 130 V and a grid voltage of −2·8 V. When the grid voltage was reduced to −0·4 V, with the anode voltage unaltered, the anode current increased to 8 mA. By reducing the anode voltage to 80 V, with the grid voltage at −0·4 V, the anode current was brought back to 4 mA. Calculate the corresponding values of the anode a.c. resistance, the mutual conductance and the amplification factor.

25. With the aid of a sketch, describe the construction and explain the action of the control grid in a vacuum triode.

The data in the table below refer to a triode. Plot the anode characteristics and find: (a) r_a, (b) g_m, (c) μ, in the region of $V_a = 90$ V, $V_g = 0$.

Grid voltage (V_g)		Anode voltage (V_a)							
		0	20 V	40 V	60 V	80 V	100 V	120 V	140 V
+2 V		0	1·2	2·8	4·8	6·8	9·0		
0	I_a	0	0·2	1·2	2·5	4·1	6·0	8·0	
−2 V	(mA)	0	0	0·1	0·75	1·8	3·25	5·1	7·3
−4 V		0	0	0	0	0·45	1·4	2·75	4·6

(S.A.N.C., O.1)

26. With the aid of a sketch explain how the grid potential controls the performance of a vacuum triode.

A triode has an amplification factor of 20 and an anode a.c. resistance of 10 kΩ. Determine the r.m.s. value of the a.c. component of the voltage across a purely resistive load of 50 kΩ, when an r.m.s. voltage of 1·8 V is applied between the grid and cathode. An equivalent circuit should be included. (E.M.E.U., O.2)

27. Define the terms *anode slope resistance*, *amplification factor* and *mutual conductance* as applied to a triode.

If a triode has a mutual conductance of 1·5 mA/V and an anode slope resistance of 12 kΩ, calculate the amplification factor. Determine the voltage amplification if this valve is used with a load having a resistance of 40 kΩ. (E.M.E.U., O.2)

28. Indicate, using a circuit diagram, the way in which the static characteristics of a vacuum triode would be determined. Label the diagram with approximate values of the circuit components.

A triode has the following anode characteristics:

When $V_g =$ −2 V, anode voltage (V)	75	100	125	150	175
anode current (mA)	0·7	1·5	2·7	4·3	6·2
When $V_g =$ −3 V, anode voltage (V)	125	150	175	200	
anode current (mA)	0·9	2·0	3·4	5·0	
When $V_g =$ −4 V, anode voltage (V)	150	175	200	225	
anode current (mA)	0·5	1·3	2·5	4·0	

If the high-tension supply is 300 V, the grid bias is −3 V and the load resistance is 50 kΩ, draw the load line and determine, for a signal voltage of peak value one volt: (a) the change in anode current; (b) the change in anode voltage; (c) the voltage amplification.

(U.E.I., O.2)

29. The anode-current/anode-voltage characteristics of a triode can be regarded as parallel lines having a slope of 0·12 mA/V, and for an anode voltage of 75 V, they pass through the following points:

Grid voltage (V)	0	−1	−2	−3	−4	−5	−6
Anode current (mA)	10	9	8	7	6	5	4

The valve is biased at −3 V and the grid has a sinusoidal input signal of 2·0 V peak. Determine the a.c. power output of a 12·5-kΩ resistive load fed from a 100-V d.c. supply. (S.A.N.C., O.2)

30. Explain, with reference to a connection diagram, the a.c. equivalent circuit of a simple resistance-loaded vacuum triode amplifier.

A single-stage valve amplifier with a resistive load has a mutual conductance of 0·5 mA/V and an anode slope resistance of 13 kΩ. Find: (a) the voltage amplification for a resistive load of 40 kΩ; (b) the power output when a sinusoidal signal of peak value 2 V is applied to the grid. Assume the anode voltage and grid bias to be such as to give distortionless amplification. (S.A.N.C., O.2)

31. The static characteristics of a triode are as follows:

Grid voltage = 0			
Anode voltage (V)	50	100	150	200
Anode current (mA)	3·0	6·5	10·1	13·7

Grid voltage = −3·0 V			
Anode voltage (V)	150	200	250	300
Anode current (mA)	1·4	5·0	8·5	12·1

This valve is used as an amplifier, with an anode resistor of 20 kΩ, and operates on the linear portions of its characteristics. Plot the anode-voltage/anode-current curves and determine: (a) the amplification factor, the mutual conductance and the anode slope resistance of the valve, and (b) the voltage amplification. (App. El., L.U.)

32. The slope of the I_A/V_G characteristic of a triode is 1·3 mA/V and that of the I_A/V_A characteristic is 0·2 mA/V. Calculate: (a) the amplification factor; (b) the voltage amplification obtained with an 8-kΩ nonreactive load resistor; and (c) the a.c. power dissipated in the load resistor when the signal voltage has an r.m.s. value of 1·5 V.

What resistance must the load resistor have in order that the voltage amplification may be 5?

33. A vacuum triode has the following characteristics:

Anode voltage	Grid bias:	Anode current (mA) 0	−2 V	−4 V
50 V		3	0	0
100 V		9	3	0
150 V		15	9	3
200 V		21	15	9

Plot the characteristics and draw the load line for a 15-kΩ load resistor and an H.T. supply of 250 V. From the graphs, determine: (*a*) the value of the anode current when the grid voltage is (i) zero, (ii) −4 V; (*b*) the corresponding values of the p.d. between the anode and cathode; and (*c*) the voltage amplification when the grid bias is −2 V.

Determine also the a.c. power absorbed by the load resistor when the alternating voltage applied to the grid is sinusoidal and has a peak value of 2 V.

34. A coil having an inductance of 400 μH and a resistance of 30 Ω is connected in the anode circuit of a triode having an amplification factor of 25 and an anode a.c. resistance of 20 kΩ. A variable capacitor is connected in parallel with the coil. Calculate the voltage amplification at resonance frequency if the capacitance is then 320 pF.

35. A triode has the following parameters: $r_a = 4$ kΩ and $g_m = 3$ mA/V. This triode is to supply a loudspeaker through an ideal transformer. If the resistance of the loudspeaker is 20 Ω, calculate: (*a*) the ratio of transformation for maximum power transfer, and (*b*) the power supplied to the speaker under this optimum condition when the signal voltage applied to the grid has an r.m.s. value of 2 V.

36. The interelectrode capacitances of a certain triode are: $C_{gk} = 6$ pF and $C_{ga} = 4$ pF. The triode has an anode a.c. resistance of 8 kΩ and a mutual conductance of 3 mA/V. When the valve is used with a non-reactive load having a resistance of 15 kΩ, calculate: (*a*) the input capacitance, and (*b*) the input reactance at a frequency of 50 kHz.

37. The following results were obtained with a pentode when the screen grid was maintained at a constant potential of 200 V:

Control grid voltage V_A (V)	25	50	75	100	150	200	300
0 I_A (mA)	36	56	62	65	66·8	68·6	72·2
−2 V I_A (mA)	26	38	41	42	43·2	44·4	46·8
−4 V I_A (mA)	15·5	18·5	19·8	20·5	21·5	22·5	24·5

Plot the characteristics and determine the values of the anode a.c. resistance, the mutual conductance and the amplification factor for a control grid voltage of −2 V and an anode voltage of 200 V.

For an anode supply voltage of 300 V and a load resistance of 2·5 kΩ, draw the load line. For a grid bias of −2 V and an alternating voltage of peak value 2 V applied to the control grid, determine graphically: (*a*) the voltage amplification, and (*b*) the a.c. power absorbed by the load resistor.

Compare the value of the voltage amplification determined graphically with that calculated from the value of the load resistor and the valve parameters derived from the characteristics.

38. An electron beam, after being accelerated by a p.d. of 1500 V, travels through a uniform magnetic field of density 1·5 mT, the direction of the magnetic field being normal to the initial direction of the beam. If the width of the magnetic field traversed by the electron beam is 15 mm, calculate: (*a*) the radius of curvature of the beam while it is travelling through the magnetic field, and (*b*) the angle through which the beam is deflected. Assume *e*/*m* to be 1·76 × 10^{11} C/kg.

39. It is found that an electron beam is deflected 8 degrees when it traverses a uniform magnetic field, 3 cm wide, having a density of 0·6 mT. Calculate: (*a*) the speed of the electrons, and (*b*) the force on each electron. The direction of the beam is normal to that of the flux.

40. The Y-deflecting plates of a cathode-ray tube have an axial length of 20 mm and are spaced 8 mm apart. Their centre (point A in fig. 19.54) is 150 mm from the screen. The final anode potential is 4 kV. Calculate: (*a*) the beam velocity as it enters the electric field of the Y-plates, and (*b*) the p.d. between the Y-plates to give a deflection of 30 mm on the screen.

41. A cathode-ray tube, with magnetic deflection, has its screen 20 cm from the centre of the magnetic field. The width of the uniform magnetic field is 3 cm and the final anode potential is 6 kV. Calculate the density of the magnetic field to produce a deflection of 4 cm on the screen. Assume *e*/*m* = 1·76 × 10^{11} C/kg.

CHAPTER 20

Semiconductor Devices

20.1 Introduction

The discovery of transistor action in 1948 gave a tremendous impetus to the study of semiconductors, namely materials having a resistivity roughly midway between that of a good conductor and that of a good insulator. Previous to about 1940, germanium was known merely as a rare metal but now there are extensive factories in all parts of the world manufacturing devices utilizing the electrical properties of germanium and silicon.

20.2 Atomic structure

It has already been stated in section 5.2 that an atom of a material consists of a nucleus carrying a positive charge surrounded by one or more electrons revolving around the nucleus. Electrons which are moving in orbits close to the nucleus are subject to relatively strong forces of attraction towards the protons of the nucleus, whereas those in the outer orbits are acted upon by progressively smaller forces, and the electrons in the outermost orbit can be easily detached from their atoms to become carriers of negative charges.

In semiconductor work, the materials with which we are principally concerned are germanium and silicon. These materials possess a crystalline structure, i.e. the atoms are arranged in an orderly manner. In both germanium and silicon, each atom has four electrons orbiting in the outermost shell and is therefore said to have a valency of four; or alternatively, the atoms are said to be *tetravalent*. In the case of the silicon atom, the nucleus consists of 14 protons and 14 neutrons; and when the atom is neutral, the nucleus is surrounded by 14 electrons, 4 of which

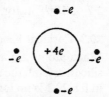

Fig. 20.1 An isolated tetravalent atom.

are *valence electrons*, one or more of which may be detached from the atom. If the four valence electrons were detached, the atom would be left with 14 units of positive charge on the protons and

565

10 units of negative charge on the 10 remaining electrons, thus giving an *ion* (i.e. an atom possessing a net positive or negative charge) carrying a net positive charge of $4e$, where e represents the magnitude of the charge on an electron, namely 1.6×10^{-19}C. The neutrons possess no resultant electric charge. A tetravalent atom, isolated from other atoms, can therefore be represented as in fig. 20.1, where the circle represents the ion carrying the net positive charge of $4e$ and the four dots represent the four valence electrons.

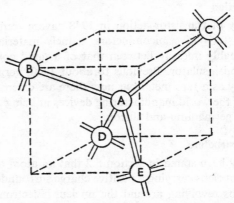

Fig. 20.2 Atomic structure of a lattice crystal.

The cubic diamond lattice arrangement of the atoms in a perfect crystal of germanium or silicon is represented by the circles in fig. 20.2, where atoms B, C, D and E are located at diagonally opposite corners of the six surfaces of an imaginary cube, shown dotted, and atom A is located at the centre of the cube. The length of each side of the dotted cube is about 2.8×10^{-10} m for germanium and about 2.7×10^{-10} m for silicon.

20.3 Covalent bonds

When atoms are as tightly packed as they are in a germanium or a silicon crystal, the simple arrangement of the valence electrons shown in fig. 20.1 is no longer applicable. The four valence electrons of each atom are now shared with the adjacent four atoms: thus in fig. 20.2, atom A shares its four valence electrons with atoms B, C, D and E. In other words, one of A's valence electrons is linked with A and B, another with A and C, etc. Similarly, one valence electron from each of atoms B, C, D and E is linked with atom A. One can imagine the arrangement to be somewhat as depicted in fig. 20.3, where the four dots, marked $-e_A$, represent the four valence electrons of atom A, and dots $-e_B$, $-e_C$, $-e_D$ and $-e_E$ represent the valence electrons of atoms B, C, D and E respectively that are linked with atom A. The dotted lines are not intended to indicate the actual paths or the relative positions of these valence electrons but merely that the electrons on the various dotted lines move around the two atoms enclosed by a given dotted line.

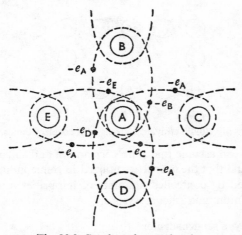

Fig. 20.3 Covalent electron bonds.

It follows that each positive ion of germanium or silicon, carrying a net charge of $4e$, has 8 electrons, i.e. 4 *electron-pairs*, surrounding it. Each electron-pair is referred to as a *covalent bond*; and in fig. 20.2, the covalent bonds are represented by the pairs of parallel lines between the respective atoms. An alternative two-dimensional method of representing the positive ions and the valence electrons forming the covalent bonds is shown in fig. 20.4; where the large circles represent the ions, each with a net positive charge of $4e$, and the bracketed dots represent the valence electrons.

These covalent bonds serve to keep the atoms together in crystal formation and are so strong that at absolute zero temperature, i.e. $-273°C$, there are no free electrons. Consequently, at that temperature, pure* germanium and silicon behave as perfect insulators. At

* A crystal can be regarded as 'pure' when impurities are less than 1 part in 10^{10}. Such a crystal is referred to as an *intrinsic* semiconductor.

*

normal atmospheric temperature, some of the covalent bonds are broken, i.e. some of the valence electrons break away from their atoms. This effect is discussed in section 20.6, but as a first approxi-

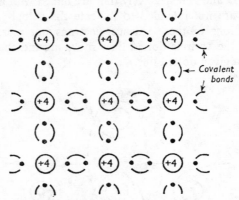

Fig. 20.4 Tetravalent atoms with covalent bonds.

mation, we can assume that pure germanium and silicon are perfect insulators and that the properties utilized in semiconductor rectifiers are produced by controlled amounts of impurities introduced into pure germanium and silicon crystals.

20.4 An n-type semiconductor

Certain elements such as phosphorus, arsenic and antimony are pentavalent, i.e. each atom has 5 valence electrons, and an isolated

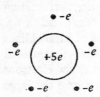

Fig. 20.5 An isolated pentavalent atom.

pentavalent atom can be represented by an ion having a net positive charge of $5e$ and five valence electrons as in fig. 20.5. When a minute* trace of the order of 1 part in 10^8 of such an element is added to pure germanium or silicon, the conductivity is considerably increased. We shall now consider the reason for this effect.

When an atom of a pentavalent element such as antimony is introduced into a crystal of pure germanium, it enters into the lattice structure by replacing one of the tetravalent germanium atoms, but *only four*

---

* It may assist us in realizing how minute this impurity is if we were to imagine a portion of a crystal magnified to such an extent that there was one atom for every cubic centimetre of a room, 8 m × 4 m × 3 m, then an impurity of 1 in 10^8 would correspond to 1 cm³ being occupied by an atom of the impurity.

of the five valence electrons of the antimony atom can join as covalent bonds. Consequently the substitution of a pentavalent atom for a germanium atom provides a free electron. This state of affairs is represented in fig. 20.6, where A is the ion of, say, an antimony atom, carrying a positive charge of 5*e*, with four of its valence electrons form-ing covalent bonds with four adjacent atoms, and B represents the unattached valence electron free to wander at ran-dom in the crystal. This random move-ment, however, is such that the density of these free or mobile electrons remains constant throughout the crystal and there-fóre there is no accumulation of free electrons in any particular region.

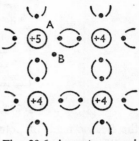

Fig. 20.6 An n-type semi-conductor.

Since the pentavalent impurity atoms are responsible for introducing or donating free electrons into the crystal, they are termed *donors*; and a crystal doped with such im-purity is referred to as an *n-type* (i.e. negative-type) semiconductor. It will be noted that each antimony ion has a *positive* charge of 5*e* and that the valence electrons of each antimony atom have a total *negative* charge of —5*e*; consequently the doped crystal is *neutral*. In other words, donors provide fixed positively-charged ions and an equal number of electrons free to move about in the crystal, as represented by the circles and dots respectively in fig. 20.7.

The greater the amount of impurity in a semiconductor, the greater is the number of free electrons per unit volume and therefore the greater is the conductivity of the semiconductor.

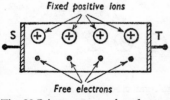

Fig. 20.7 An n-type semiconductor.

When there is no p.d. across the metal electrodes, S and T, attached to opposite ends of the semicon-ductor, the paths of the random movement of one of the free elec-trons may be represented by the full lines AB, BC, CD, etc., in fig. 20.8, i.e. the electron is accelerated in direction AB until it collides with an atom with the result that it may rebound in direction BC, etc. Different free electrons move in different directions so as to maintain the density constant; in other words, there is no resultant drift of electrons towards either S or T.

Let us next consider the effect of connecting a cell across S and T, the polarity being such that S is positive relative to T. The effect of

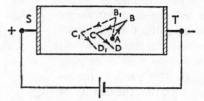

Fig. 20.8 Movement of a free electron.

the electric field (or potential gradient) in the semiconductor is to modify the random movement of the electron as shown by the dotted lines AB_1, B_1C_1 and C_1D_1 in fig. 20.8, i.e. there is superimposed on the random movement a drift of the electron towards the positive electrode S, and the number of electrons entering electrode S from the semiconductor is the same as that entering the semiconductor from electrode T.

20.5 A p-type semiconductor
Materials such as indium, gallium, boron and aluminium, are trivalent, i.e. each atom has only three valence electrons and may therefore be represented as an ion having a positive charge of $3e$

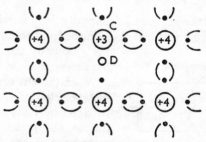

Fig. 20.9 A p-type semiconductor.

surrounded by 3 valence electrons. When a trace of, say, indium is added to pure germanium, the indium atoms replace the corresponding number of germanium atoms in the crystal structure, but each indium atom can provide only three valence electrons to join with the four valence electrons of adjacent germanium atoms, as shown in fig. 20.9, where C represents the indium atom. Consequently there is an incomplete valence bond, i.e. there is a vacancy represented by the small circle D in fig. 20.9. This vacancy is referred to as a *hole*—a term that is peculiar to semiconductors. This incomplete valent bond has the ability to attract a covalent electron from a near-by germanium atom, thereby filling the vacancy at D but creating another hole at, say, E as in fig. 20.10 (*a*). Similarly the incomplete valent bond due to hole E attracts a covalent electron from another germanium atom, thus creating a new hole at, say, F as in fig. 20.10 (*b*).

If the semiconductor is not being subjected to an external electric

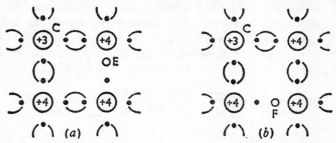

Fig. 20.10 Movement of a hole in a p-type semiconductor.

field, the position of the hole moves at random from one covalent bond to another covalent bond, the speed of this random movement being about half that of free electrons, since the latter can move about with comparative ease. Different holes move in different directions so as to maintain the density of the holes (i.e. the number of holes per unit volume) uniform throughout the crystal, otherwise there would be an accumulation of positive charge in one region with a corresponding negative charge in another region.

It will be seen that each of the *germanium* atoms associated with holes E and F in fig. 20.10 (*a*) and (*b*) respectively has a nucleus with a resultant positive charge $4e$ and three valence electrons having a total negative charge equal to $-3e$. Hence each atom associated with a hole is an ion possessing a net positive charge e, and the movement of a hole from one atom to another can be regarded as the movement of a positive charge e within the structure of the p-type semiconductor.

Germanium and silicon, doped with an impurity responsible for the formation of holes, are referred to as *p-type* (positive-type) semi-conductors; and since the trivalent impurity atoms can accept electrons from adjacent germanium or silicon atoms, they are termed *acceptors.*

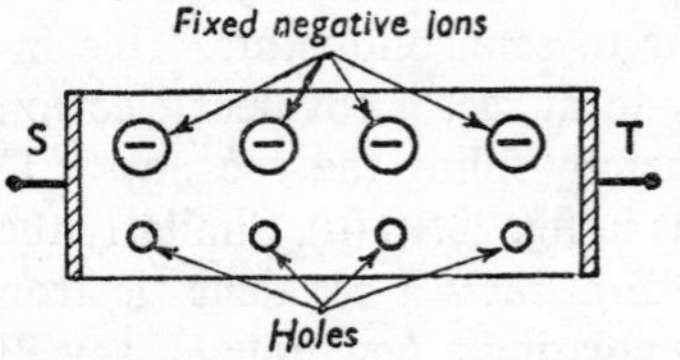

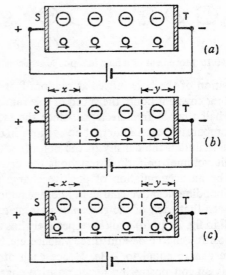

Fig. 20.11 A p-type semiconductor.

It will be seen from fig. 20.10 that the trivalent impurity atom, with the *four covalent bonds complete*, has a nucleus carrying a positive charge $3e$ and four electrons having a total negative charge $-4e$. Consequently such a trivalent atom is an ion carrying a net *negative* charge $-e$. For each such ion, however, there is a *hole* somewhere in the crystal; in other words, the function of an acceptor is to provide *fixed* negatively-charged ions and an equal number of holes as in fig. 20.11.

Let us next consider the effect of applying a p.d. across the two opposite faces of a p-type semiconductor, as in fig. 20.12 (*a*), where

Fig. 20.12 Drift of holes in a p-type semiconductor.

S and T represent metal plates, S being positive relative to T. The negative ions, being locked in the crystal structure, cannot move, but the holes drift in the direction of the electric field, namely towards T.

Consequently region x of the semiconductor acquires a net negative charge and region y acquires an equal net positive charge, as in fig. 20.12 (*b*). These charges attract electrons from T into region y and repel electrons from region x into S, as indicated in fig. 20.12 (*c*). The electrons attracted from T combine with holes in region y and electrons from covalent bonds enter S, thus creating in region x new holes which move from that region towards electrode T. The rate at which holes are being neutralized near electrode T is the same as that at which they are being created near electrode S. Hence, in a p-type semiconductor, we can regard the current as being due to the drift of holes in the conventional direction, namely from the positive electrode S to the negative electrode T.

20.6 Junction diode

Let us now consider a crystal, one half of which is doped with p-type impurity and the other half with n-type impurity. Initially, the p-type semiconductor has mobile holes and the same number of fixed negative ions carrying exactly the same total charge as the total positive charge represented by the holes. Similarly the n-type semiconductor has mobile electrons and the same number of fixed positive ions carrying the same total charge as the total negative charge on the mobile electrons. Hence each region is initially neutral.

Owing to their random movements, some of the holes will diffuse* across the boundary into the n-type semiconductor and some of the free electrons will similarly diffuse into the p-type semiconductor, as in fig. 20.13 (*a*). Consequently region A acquires an excess negative charge which repels any more electrons trying to migrate from the n-type into the p-type semiconductor. Similarly, region B acquires a surplus of positive charge which prevents any further migration of holes across the boundary. These positive and negative charges are concentrated near the junction, somewhat as indicated in fig. 20.13 (*b*), and thus form a potential barrier between the two regions.

Forward bias. Let us next consider the effect of applying a p.d. across metal electrodes S and T, S being positive relative to T, as in fig. 20.13 (*c*). The direction of the electric field in the semiconductor is such as to produce a drift of holes towards the right in the p-type semiconductor and of free electrons towards the left in the n-type

---

* It will be noted that *diffusion* takes place when there is a difference in the concentration of carriers in adjacent regions of a crystal; but *drift* of carriers takes place only when there is a difference of potential between two regions.

semiconductor. In the region of the junction, free electrons and holes combine, i.e. free electrons fill the vacancies represented by the holes. For each combination, an electron is liberated from a covalent bond in the region near positive plate S and enters that plate, thereby creating a new hole which moves through the p-type material to-

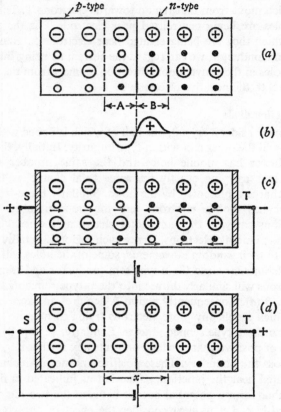

Fig. 20.13 Junction diode.

wards the junction, as described in section 20.5. Simultaneously, an electron enters the n-region from the negative plate T and moves through the n-type semiconductor towards the junction, as described in section 20.4. The current in the diode is therefore due to hole-flow in the p-region, electron-flow in the n-region and a combination of the two in the vicinity of the junction.

Reverse bias. When the polarity of the applied voltage is reversed, as shown in fig. 20.13 (*d*), the holes are attracted towards the negative electrode S and the free electrons towards the positive electrode T. This leaves a region *x*, known as a *depletion layer*, in which there are no holes or free electrons, i.e. there are no charge carriers in this region apart from the relatively few that are produced spontaneously by thermal agitation, as mentioned below. Consequently the junction behaves as an insulator.

In practice, there is a small current due to the fact that at room temperature, thermal agitation or vibration of atoms takes place in the crystal and some of the valence electrons acquire sufficient

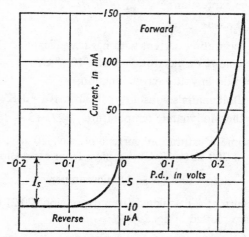

Fig. 20.14 Static characteristic for a germanium junction diode having negligible surface leakage.

velocity to break away from their atoms, thereby producing *electron-hole pairs*. An electron-hole pair has a life of about 100 micro-seconds in germanium and about 50 microseconds in silicon. The generation and recombination of electron-hole pairs is a continuous process and is a function of the temperature. The higher the temperature, the greater is the rate at which generation and recombination of electron-hole pairs take place and therefore the lower the *intrinsic resistance* of a crystal of pure germanium or silicon.

These thermally-liberated holes and free electrons are referred to as *minority carriers* because, at normal temperature, their number is very small compared with the number of *majority carriers* due to the doping of the semiconductor with donor and acceptor impurities. Hence, in a p-type semiconductor, holes form the majority carriers and electrons the minority carriers, whereas in an n-type crystal, the majority carriers are electrons and holes are the minority carriers.

When a germanium junction diode is biased in the reverse direction, the current remains nearly constant for a bias varying between about 0·1 volt and the breakdown voltage. This constant value is referred to as the *saturation current* and is represented by I_s in fig. 20.14. In practice, the reverse current increases with increase of bias, this increase being due mainly to surface leakage. In the case of a germanium junction diode in which the surface leakage is negligible, the current is given by the expression:

$$i = I_s(e^{ev/kT} - 1) \qquad\qquad (20.1)$$

where I_s = saturation current with negative bias,

e = charge on electron = $1·6 \times 10^{-19}$ C,

v = p.d., in volts, across junction,

k = Boltzmann's constant = $1·38 \times 10^{-23}$ J/K

and T = thermodynamic temperature = $273·15 + t°$C.

Let us assume a saturation current of, say, 10 μA; then for a temperature of 300 K (=27°C), we have from expression (20.1),

$$i = 10(e^{38·6v} - 1) \text{ microamperes} \qquad\qquad (20.2)$$

Values of current i, calculated from expression (20.2) for various values of v, are plotted in fig. 20.14.

20.7 Construction and static characteristics of a junction diode

Fig. 20.15 shows one arrangement of a germanium junction diode.

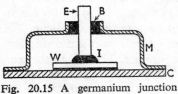

Fig. 20.15 A germanium junction diode.

A thin wafer or sheet W is cut from an n-type germanium crystal, the area of the wafer being proportional to the current rating of the diode. The lower surface of the wafer is soldered to a copper plate C and a bead of indium I is placed centrally on the upper surface. The unit is then heat-treated so that the indium forms a p-type alloy with the germanium. A copper electrode E is soldered to the bead during the heat treatment, and the whole element is hermetically sealed in a metal or other opaque container M to protect it from light and moisture. The electrode E is insulated from the container by a bush B.

Typical voltage/current characteristics of a germanium junction

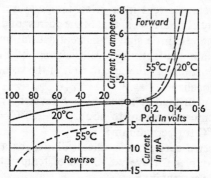

Fig. 20.16 Static characteristics of a germanium junction diode.

diode are given in fig. 20.16, the full lines being for a temperature of the surrounding air (i.e. ambient temperature) of 20°C and the dotted lines for 55°C. For a given reverse bias, the reverse current roughly doubles for every 10°C rise of temperature. This rectifier

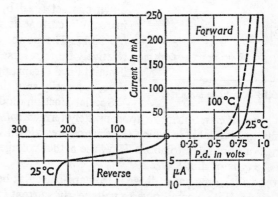

Fig. 20.17 Static characteristics of a silicon junction diode.

can withstand a peak inverse voltage of about 100 V at an ambient temperature of 20°C.

The silicon junction diode is similar in appearance to the germanium diode and typical voltage/current characteristics are given in fig. 20.17. The properties of silicon junction diodes differ from those of germanium junction diodes in the following respects:

(*a*) the forward voltage drop is roughly double that of the corresponding germanium diode;

(*b*) the reverse current at a given temperature and voltage is approximately a hundredth of that of the corresponding germanium diode, but there is little sign of current saturation as is the case with germanium—in fact, the reverse current of a silicon diode is roughly proportional to the square root of the voltage until breakdown is approached;

(*c*) it can withstand a much higher reverse voltage and can operate at temperatures up to about 150–200°C, compared with about 75–90°C for germanium;

(*d*) the reverse current of a silicon diode, for a given voltage, practically doubles for every 8°C rise of temperature, compared with 10°C for germanium.

If the reverse voltage across a p–n junction is gradually increased, a point is reached where the energy of the current carriers is sufficient to dislodge additional carriers. These carriers, in turn, dislodge more carriers and the junction goes into a form of avalanche breakdown characterized by a rapid increase in current as shown in fig. 20.17. The power due to a relatively large reverse current, if maintained for an appreciable time, can easily ruin the device.

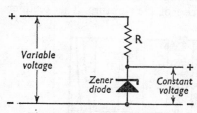

Fig. 20.18 A voltage stabilizer.

Special junction diodes, often known as *zener diodes*, but more appropriately termed *voltage regulator diodes*, are available in which the reverse breakdown voltage is in the range of about 4 V to about 75 V, the actual voltage depending upon the type of diode.

When the voltage regulator diode is forward-biased, it behaves as a normal diode. With a small reverse voltage, the current is the sum of the surface leakage current and the normal saturation current due to the thermally-generated holes and electrons (section 20.6). This current is only a few microamperes; but as the reverse voltage is increased, a value is reached at which the current suddenly increases very rapidly (Fig. 20.17). As already mentioned, this is due to the increased velocity of the carriers being sufficient to cause ionization. The carriers resulting from ionization by collision are responsible for further collisions and thus produce still more carriers. Consequently the numbers of carriers, and therefore the current, increase rapidly due to this avalanche effect.

The voltage across the regulator diode after breakdown is termed the *reference voltage* and its value for a given diode remains practically constant over a wide range of current, provided the maximum permissible junction temperature is not exceeded.

Fig. 20.18 shows how a voltage regulator diode (or zener diode) can be used as a voltage stabilizer to provide a constant voltage from a source whose voltage may vary appreciably. A resistor R is necessary to limit the reverse current through the diode to a safe value.

20.8 Junction transistor

A junction transistor* is a combination of two junction diodes and consists of either a thin layer of p-type semiconductor sandwiched between two n-type semiconductors, as in fig. 20.19 (*a*), and re-

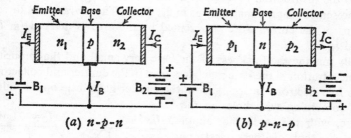

Fig. 20.19 Arrangement of a junction transistor.

ferred to as an n-p-n transistor, or a thin layer of an n-type semiconductor sandwiched between two p-type semiconductors, as in fig. 20.19 (*b*), and referred to as a p-n-p transistor. The thickness of the central layer, known as the *base*, is of the order of 25 μm (or 25×10^{-6} m).

The junction diode formed by n_1-p in fig. 20.19 (*a*) is biased in the forward direction by a battery B_1 so that free electrons are urged

* The term *transistor* is derived from the words *transfer-resistor*, namely a device for the transfer of current from a low-resistance circuit to approximately the same current in a high-resistance circuit.

from n_1 towards p. Hence n_1 is termed an *emitter*. On the other hand, the junction diode formed by n_2–p in fig. 20.19 (*a*) is biased in the reverse direction by battery B_2 so that if battery B_1 were disconnected, i.e. with zero emitter current, no current would flow between n_2 and

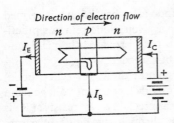

Fig. 20.20 Electron flow in an n–p–n transistor.

p apart from that due to the thermally-generated minority carriers referred to in section 20.6. However, with B_1 connected as in fig. 20.19 (*a*), the electrons from emitter n_1 enter p and diffuse through the base until they come within the influence of n_2, which is connected to the positive terminal of battery B_2. Consequently the electrons which reach n_2 are collected by the metal electrode attached to n_2; hence n_2 is termed a *collector*.

Some of the electrons, in passing through the base, combine with holes; others reach the base terminal. The electrons which do not reach collector n_2 are responsible for the current at the base terminal, and the distribution of electron flow in an n–p–n transistor can be represented diagrammatically as in fig. 20.20. The *conventional* directions of the currents are represented by the arrows marked I_E, I_B and I_C.*

By making the thickness of the base very small and the impurity concentration in the base much less than in the emitter and collector, the free electrons emerging from the emitter have little opportunity of combining with holes in the base, with the result that about 98 per cent of these electrons reach the collector.

By Kirchhoff's first law, $I_E = I_B + I_C$, so that if $I_C = 0.98I_E$, then $I_B = 0.02I_E$; thus, when $I_E = 1$ mA, $I_C = 0.98$ mA and $I_B = 0.02$ mA, as in fig. 20.21.

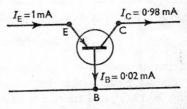

Fig. 20.21 Currents in emitter, collector and base circuits of a p–n–p transistor.

In the above explanation, we have dealt with the n–p–n transistor, but exactly the same explanation applies to the p–n–p transistor of fig. 20.19 (*b*) except that the movement of electrons is replaced by the movement of holes.

* See note on subscripts, p. xvii.

20.9 Construction of a transistor

The first step is to purify the germanium or silicon so that any impurity does not exceed about 1 part in 10^{10}. Various methods have been developed for attaining this exceptional degree of purity, and intensive research is still being carried out to develop new methods of purifying and of doping germanium and silicon.

In one form of construction of the p–n–p transistor, the purified material is grown as a single crystal, and while the material is in a molten state, an n-type impurity (e.g. antimony in the case of germanium) is added in the proportion of about 1 part in 10^8. The solidified crystal is then sawn into slices about 0·1 mm thick. Each slice is used to form the base region of a transistor; thus in the case

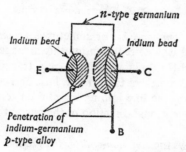

Fig. 20.22 Construction of a p–n–p germanium transistor.

of n-type germanium, a pellet of p-type impurity such as indium is placed on each side of the slice, the one which is to form the collector being about three times the size of that forming the emitter. One reason for the larger size of the collector bead is that the current carriers from the emitter spread outwards as they pass through the base, and the larger area of the collector enables the latter to collect these carriers more effectively. Another reason is that the larger area assists in dissipating the greater power loss at the collector-base junction. This greater loss is due to the p.d. between collector and base being greater than that between emitter and base.

The assembly is heated in a hydrogen atmosphere until the pellets melt and dissolve some of the germanium from the slice, as shown in fig. 20.22. Leads for the emitter and collector are soldered to the surplus material in the pellets to make non-rectifying contacts, and a nickel tab is soldered to make connection to the base. The assembly is then hermetically sealed in a metal or glass container, the glass being coated with opaque paint. The transistor is thus protected from moisture and light.

20.10 Common-base and common-emitter circuits

In fig. 20.19 the transistor is shown with the base connected directly to both the emitter and collector circuits; hence the arrangement is

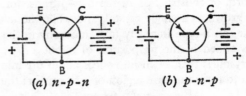

(a) *n-p-n*   (b) *p-n-p*

Fig. 20.23 Common-base circuits.

referred to as a *common-base* or sometimes *grounded-base**** circuit. The conventional way of representing this arrangement is shown in fig. 20.23 (*a*) and (*b*) for n-p-n and p-n-p transistors respectively, where E, B and C represent the emitter, base and collector terminals. The arrowhead on the line joining the emitter terminal to the base indicates the conventional direction of the current in that part of the circuit.

Fig. 20.24 shows the *common-emitter* (or *grounded-emitter*) method of connecting a transistor. In these diagrams it will be seen that the

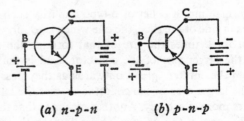

(a) *n-p-n*   (b) *p-n-p*

Fig. 20.24 Common-emitter circuits.

emitter is connected directly to the base and collector circuits. This method is more commonly used than the common-base circuit owing to its higher input resistance and the higher current and power gains (see example 20.3).

* *Common-base* is the preferable term since *grounded-base* suggests that the base is earthed. This may not be the case. Similarly, the term *common-emitter* is preferable to *grounded-emitter*.

The common-collector circuit is used only in special cases, and will therefore not be considered here.

20.11 Static characteristics for a common-base circuit

Fig. 20.25 shows an arrangement for determining the static characteristics of an n–p–n transistor used in a common-base circuit. The

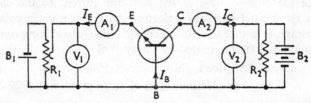

Fig. 20.25 Determination of static characteristics for a common-base n–p–n transistor circuit.

procedure is to maintain the value of the emitter current, indicated by A_1, at a constant value, say 1 mA, by means of the slider on R_1, and note the readings on A_2 for various values of the collector-base voltage given by voltmeter V_2. The test is repeated for various values of the emitter current and the results are plotted as in fig. 20.26.

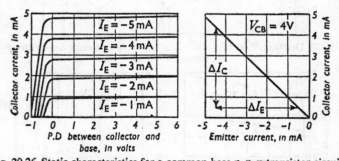

Fig. 20.26 Static characteristics for a common-base n–p–n transistor circuit.

Fig. 20.27 Relationship between collector and emitter currents for a given collector-base voltage.

In accordance with B.S. 3363, the current is assumed to be positive when its direction is from the external circuit towards the transistor terminal, and the voltage V_{CB} is positive when C is positive relative to B. Hence, for an n–p–n transistor, the collector current and collector-base voltage are positive but the emitter current is negative. For a p–n–p transistor, all the signs have to be reversed.

From fig. 20.26 it will be seen that for positive values of the collector-base voltage, the collector current remains almost constant, i.e. nearly all the electrons entering the base of an n–p–n transistor are attracted to the collector. Also, for a given collector-base voltage, the collector current is practically proportional to the emitter current. This relationship is shown in fig. 20.27 for V_{CB} equal to 4 V. The ratio of the change, ΔI_C, of the collector current to the change, ΔI_E, of the emitter current (neglecting signs), for a given collector-base voltage, is termed the *current amplification factor* for a common-base circuit* and is represented by the symbol α,

i.e. $\qquad \alpha = \Delta I_C / \Delta I_E$ for a given value of V_{CB} $\qquad$ (20.3)

$\qquad\qquad$ = slope (neglecting signs) of I_C/I_E graph in fig. 20.27.

20.12 Static characteristics for a common-emitter circuit

Fig. 20.28 shows an arrangement for determining the static characteristics of an n–p–n transistor used in a common-emitter circuit.

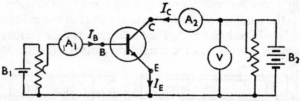

Fig. 20.28 Determination of static characteristics for a common-emitter n–p–n transistor circuit.

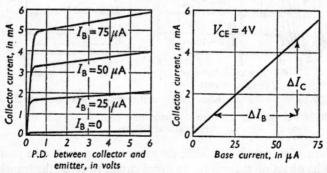

Fig. 20.29 Static characteristics for a common-emitter n–p–n transistor circuit.
Fig. 20.30 Relationship between collector and base currents for a given collector–emitter voltage.

* The term *current amplification factor* of a transistor is analogous to the *voltage amplification factor* of a thermionic valve (p. 521).

Again, the procedure is to maintain the *base* current, I_B, through a microammeter A_1 constant at, say, 25 μA, and to note the collector current, I_C, for various values of the collector–emitter voltage V_{CE}, the test being repeated for several values of the base current, and the results are plotted as shown in fig. 20.29. For a given voltage between collector and emitter, e.g. for $V_{CE} = 4$ volts, the relationship between the collector and base currents is practically linear as shown in fig. 20.30.

The ratio of the change, ΔI_C, of the collector current to the change, ΔI_B, of the base current, for a given collector–emitter voltage, is termed the *current amplification factor for a common-emitter circuit* and is represented by the symbol β.

i.e.
$$\beta = \Delta I_C/\Delta I_B \text{ for a given value of } V_{CE} \qquad (20.4)$$
$$= \text{slope of graph in fig. 20.30.}$$

20.13 Relationship between α and β

From figs. 20.25 and 20.28 it is seen that:
$$I_E = I_C + I_B$$
∴
$$\Delta I_E = \Delta I_C + \Delta I_B.$$

From expression (20.3),
$$\alpha = \Delta I_C/\Delta I_E$$
$$= \Delta I_C/(\Delta I_C + \Delta I_B)$$
∴
$$1/\alpha = 1 + \Delta I_B/\Delta I_C$$
$$= 1 + 1/\beta = (1 + \beta)/\beta$$

Hence
$$\alpha = \beta/(1 + \beta) \qquad (20.5)$$
and
$$\beta = \alpha/(1 - \alpha) \qquad (20.6)$$

Thus, if $\alpha = 0.98$, $\beta = 0.98/0.02 = 49$,
and if $\alpha = 0.99$, $\beta = 0.99/0.01 = 99$,

i.e. a small variation in α corresponds to a large variation in β. It is therefore better to determine β experimentally and calculate therefrom the corresponding value of α by means of expression (20.5).

20.14 Load line for a transistor

Let us consider an n–p–n transistor used in a common-base circuit (fig. 20.31) together with an a.c. source S having an internal resistance R_s, a load resistance R and bias batteries B_1 and B_2 giving the required transistor currents.

Let us assume that the voltage between the collector and base is to be 3 V when there is *no alternating voltage* applied to the emitter, and that with an emitter current of 3 mA, the corresponding collector

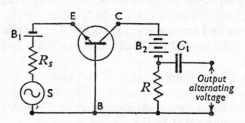

Fig. 20.31 An n–p–n transistor with load resistance R.

current is 2·9 mA. This corresponds to point D in fig. 20.32. Also suppose the resistance R of the load to be 1000 Ω. Consequently the corresponding p.d. across R is 0·0029 × 1000, namely 2·9 V, and the total bias supplied by battery B_2 must be 3 + 2·9, namely 5·9 V. This is represented by point P in fig. 20.32. The straight line PDQ drawn through P and D is the *load line*, the inverse of the slope of

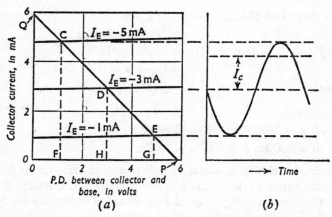

Fig. 20.32 Graphical determination of the output current of a transistor in a common-base circuit.

which is equal to the resistance of the load, e.g. HP/DH = 2·9/0·0029 = 1000 Ω. The load line is the locus of the variation of the collector current for any variation in the emitter current. Thus, if the emitter current decreases to 1 mA, the collector current decreases to GE and the p.d. across R decreases to GP. On the other hand, if the emitter current increases to 5 mA, the collector current increases to FC and the p.d. across R increases to FP. Hence, if the emitter current varies sinusoidally between 1 and 5 mA, the collector current varies as shown at (*b*) in fig. 20.32. This curve will be sinusoidal if the graphs are linear and are equally spaced over the working range.

The function of C_1 is to eliminate the d.c. component of the voltage across R from the output voltage.

If I_c* be the r.m.s. value of the *alternating* component of the collector current, then $I_c R$ gives the r.m.s. value of the alternating component of the output voltage, and $I_c^2 R$ gives the output power due to the alternating e.m.f. generated in S. Hence the larger the value of R, the greater the voltage and power gains, the maximum value of R for a given collector supply voltage being limited by the maximum permissible distortion of the output voltage.

Similar procedure can be used to determine the output voltage and power for a transistor used in a common-emitter circuit.

20.15 Equivalent T-circuit of a transistor used in a common-base circuit

So long as a transistor is being operated over the linear portions of its characteristics, the actual values of the bias voltages are of no

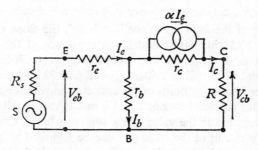

Fig. 20.33 Equivalent T-circuit of a transistor used in a common-base circuit.

consequence; hence these voltages can be omitted from an equivalent circuit used to calculate the current, voltage and power gains. Consequently the values of voltages and currents shown on an equivalent

* See note on subscripts, p. xvii.

circuit diagram refer only to the r.m.s. values of the alternating components of these quantities.

The simplest representation of a transistor is the equivalent T-circuit shown in fig. 20.33 for the common-base arrangement. The complete equivalent circuit should include capacitances between the emitter and base and between the collector and base; but at *low frequencies*, these capacitances can be neglected.

In fig. 20.33, r_e, r_b and r_c* are the resistances presented to the *alternating* components of the current by the emitter, base and collector respectively.

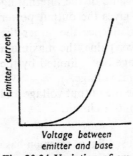

Voltage between emitter and base

Fig. 20.34 Variation of emitter current with emitter-base voltage for a given collector-base voltage.

If the emitter current is measured for different values of the voltage (V_1 in fig. 20.25) between the emitter and the base, with a *constant* voltage of, say, 4 V between collector and base, the relationship is non-linear as shown in fig. 20.34. It is therefore necessary to use a source having a relatively high resistance R_s (figs. 20.31 and 20.33) to swamp the variation of the input resistance of the transistor. The emitter current is then practically proportional to the e.m.f. generated in S and excessive distortion is avoided.

Equivalent constant-current generator in collector circuit

If the value of the load resistance is reduced, the slope of the load line of fig. 20.32 increases; and if the resistance of the load were reduced to zero, the load line would be a vertical line passing through point D. It will be evident from fig. 20.32 that the variation in the upper and lower limits of the collector current for a variation of the emitter current between 1 and 5 mA is very small for a considerable variation in the value of the load resistance. Consequently the collector circuit of the transistor can be regarded as a source of constant current αI_e, shunted by a resistance r_c as in fig. 20.33, where r_c represents the resistance of the collector to variation of collector current, and is usually of the order of a megohm.

* Symbols representing equivalent-circuit components of a transistor should be lower case letters with lower case subscripts. Symbols representing circuit components associated with but *external* to the transistor should be upper case letters.

Equivalent constant-voltage generator in collector circuit

A constant-current generator supplying a current αI_e, shunted by a resistance r_c, is equivalent* to a *constant-voltage* generator generating an e.m.f. $\alpha I_e r_c$, in series with a resistance r_c, as shown in fig. 20.36.

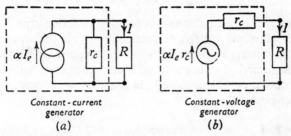

Constant - current
generator

(*a*)

Constant - voltage
generator

(*b*)

Fig. 20.35 Constant-current and constant-voltage generators.

Suppose V_{eb} and V_{cb} to be the alternating components of the voltages between E and B and between C and B respectively, the voltages being assumed positive when E and C are positive with respect to B. Also, suppose I_e, I_b and I_c to be the r.m.s. values of the

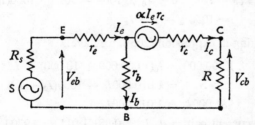

Fig. 20.36 Equivalent T-circuit of a transistor used in a common-base circuit.

alternating components of the currents at the emitter, base and collector terminals respectively, the positive direction of each current being as indicated by the corresponding arrowhead in figs. 20.33 and

* If a generator consists of a source of constant current αI_e in parallel with a resistance r_c, as in fig. 20.35 (*a*), and if the load has a resistance R, then the current I through the load is given by:

$$I = \alpha I_e \times r_c/(r_c + R) = (\alpha I_e r_c)/(r_c + R)$$

If the constant-voltage generator of fig. 20.35 (*b*) generates an e.m.f. $\alpha I_e r_c$ and has an internal resistance r_c, then:

$$I = (\alpha I_e r_c)/(r_c + R)$$

Hence the two circuits are equivalent as far as the load current is concerned.

20.36. Applying Kirchhoff's Laws to the circuits of fig. 20.36, we have:

$$I_e = I_c + I_b \tag{20.7}$$

$$V_{eb} = I_e r_e + I_b r_b \tag{20.8}$$

and

$$\alpha I_e r_c = I_c(r_c + R) - I_b r_b \tag{20.9}$$

From these three equations it is possible to derive expressions for the ratios of output voltage, current and power to the input voltage, current and power respectively. These expressions are cumbersome, and if the numerical values of r_e, r_b, r_c and α are known, it is simpler to substitute these values directly in the above equations, as in the following example.

Example 20.1 *A junction transistor has* $r_e = 50\ \Omega$, $r_b = 1\ k\Omega$, $r_c = 1\ M\Omega$ *and* $\alpha = 0.98$. *It is used in a common-base circuit with a load resistance of* 10 $k\Omega$. *Calculate the current, voltage and power gains and the input resistance.*

Substituting in expression (20.8), we have:

$$V_{eb} = I_e \times 50 + I_b \times 1000 = 50I_e + 1000(I_e - I_c)$$
$$= 1050I_e - 1000I_c \tag{20.10}$$

Substituting in expression (20.9), we have:

$$0.98I_e \times 1\,000\,000 = I_c(1\,000\,000 + 10\,000) - I_b \times 1000$$
$$= 1\,010\,000I_c - 1000(I_e - I_c)$$

so that

$$981\,000I_e = 1\,011\,000I_c$$

∴ current gain $= I_c/I_e = 0.981/1.011 = 0.9703$

i.e. the load current is 97 per cent of the emitter current; in other words, there is a current *loss* of 3 per cent. It should be remembered that I_c is equal to $0.98I_e$ only when the load resistance R is zero.

Substituting for I_c in expression (20.10), we have:

$$V_{eb} = 1050I_e - 1000 \times 0.97I_e = 80I_e$$

∴ input resistance of transistor $= V_{eb}/I_e = 80\ \Omega$.

Output voltage $= V_{cb} = I_c R$
$$= 0.97I_e \times 10\,000 = 9700 \times V_{eb}/80$$
$$= 121\ V_{eb}$$

$\therefore$ voltage gain $= V_{cb}/V_{eb} = 121$.

Output power $= I_c^2 R = 10\ 000 I_c^2$.

Input power $= I_e \times V_{eb} = I_e^2 \times$ input resistance

$= 80 I_e^2$

$\therefore$ power gain $=$ output power/input power

$= (10\ 000/80) \times (I_c/I_e)^2$

$= 125 \times (0\cdot97)^2 = 117\cdot5$.

Power gain in decibels* $= 10 \log 117\cdot5 = 20\cdot7$ dB.

Example 20.2 *If the source used for the transistor of example* 20.1 *has a resistance of* 200 Ω, *calculate the output resistance of the transistor.*

The output resistance of a transistor is the resistance offered to current when an alternating voltage is applied across the output terminals with the load removed and the source replaced by its internal

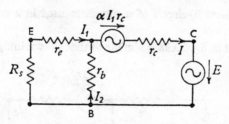

Fig. 20.37 Determination of the output resistance of a transistor in a common-base circuit.

resistance. The transistor is assumed to be correctly biased so that it is operating under normal working conditions. Thus, in the equivalent T-circuit of fig. 20.37, a resistance R_s, equal to the resistance of the source, is connected across the input terminals and an a.c. source, of negligible impedance and generating an e.m.f. E, is connected across the output terminals.

---

* If the input power to a network is P_1 and the output power is P_0, then if P_0 is greater than P_1, power gain $= \log P_0/P_1$ bels $= 10 \log P_0/P_1$ decibels. If P_0 is less than P_1, power *loss* $= 10 \log P_1/P_0$ decibels. The abbreviation for decibel is dB, but db is frequently used in telecommunications. The term *bel* commemorates Alexander Graham Bell (1847–1922), who patented the first telephone in 1876.

Applying Kirchhoff's Laws, we have:

$$E + \alpha I_1 r_c = I r_c + I_2 r_b \tag{20.11}$$

$$0 = I_1(R_s + r_e) - I_2 r_b \tag{20.12}$$

and

$$I = I_1 + I_2$$

Substituting the values of the resistances in expression (20.12), we have:

$$0 = I_1(200 + 50) - I_2 \times 1000$$
$$= 250(I - I_2) - 1000 I_2$$

$$\therefore \qquad I_2 = 0 \cdot 2I.$$

Substituting for I_1 and I_2 in expression (20.11), we have:

$$E = I \times 1\,000\,000 + I_2 \times 1000 - 0 \cdot 98(I - I_2) \times 1\,000\,000$$
$$= 20\,000 I + 981\,000 I_2 = I(20\,000 + 0 \cdot 2 \times 981\,000)$$
$$= 216\,200 I.$$

$$\therefore \qquad \text{output resistance} = E/I = 216\,200 \ \Omega.$$

20.16 Equivalent T-circuit of a transistor used in a common-emitter circuit

The T-circuit in fig. 20.38 is similar to that shown in fig. 20.36 except

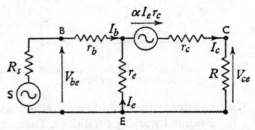

Fig. 20.38 Equivalent T-circuit of a transistor used in a common-emitter circuit.

that r_e and r_b have been interchanged. Again, assume the currents to be positive when their directions are those shown by the arrowheads. Applying Kirchhoff's Laws to fig. 20.38, we have:

$$I_c = I_e + I_b \tag{20.13}$$

$$V_{be} = I_b r_b - I_e r_e \tag{20.14}$$

and

$$\alpha I_e r_c = I_c(r_c + R) + I_e r_e \tag{20.15}$$

Example 20.3 *Using the transistor whose parameters are given in example* 20.1 *and assuming the same load resistance of* 10 $k\Omega$, *calculate the current, voltage and power gains and the input resistance when the transistor is used in a common-emitter circuit.*

Substituting in expression (20.14), we have:

$$V_{be} = I_b \times 1000 - I_e \times 50 \qquad (20.16)$$

Substituting in expression (20.15), we have:

$$0.98I_e \times 1\,000\,000 = I_c(1\,000\,000 + 10\,000) + I_e \times 50$$

$$\therefore \qquad 1.01\,I_c \simeq 0.98\,I_e = 0.98(I_c - I_b)$$

so that $\qquad 0.98I_b = -0.03I_c$

hence $\qquad$ current gain $= I_c/I_b = -32.7$.

The negative sign merely means that the phase of the collector current I_c is the reverse of that shown by the arrowhead in fig. 20.38. From expression (20.16) we have:

$$V_{be} = 1000I_b - 50(I_c - I_b)$$
$$= 1050I_b - 50 \times (-32.7)I_b = 2685I_b$$

$$\therefore \qquad \text{input resistance} = V_{be}/I_b = 2685\ \Omega.$$

Output voltage $= V_{ce} = I_cR$
$$= -32.7I_b \times 10\,000$$
$$= -327\,000 \times V_{be}/2685 = -121.8V_{be}$$

$$\therefore \qquad \text{voltage gain} = V_{ce}/V_{be} = -121.8.$$

This negative sign signifies that the phase of the output voltage is the reverse of that assumed in fig. 20.38. (With the common-base circuit, there was no reversal of phase.)

Output power $= I_c^2R = 10\,000I_c^2$
Input power $= I_bV_{be} = I_b^2 \times$ input resistance
$$= 2685I_b^2$$

$$\therefore \qquad \text{power gain} = (10\,000/2685) \times (-32.7)^2$$
$$= 3980.$$

Power gain in decibels $= 10 \log 3980 = 36$ dB.

From examples 20.1 and 20.3 it will be seen that the current and power gains for the common-emitter circuit are much higher than for the common-base circuit. Also the input resistance is higher. The phase of the voltage is reversed with the common-emitter circuit whereas the output voltage is in phase with the input voltage with the common-base circuit.

Example 20.4 *If the source used for the transistor of example 20.3 has a resistance of 200 Ω, calculate the output resistance of the transistor.*

The output resistance is given by the ratio E/I in fig. 20.39, where E is the alternating e.m.f. of a generator of negligible impedance

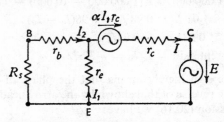

Fig. 20.39 Determination of the output resistance of a transistor in a common-emitter circuit.

connected across the output terminals. Applying Kirchhoff's Laws, we have:

$$E + \alpha I_1 r_c = I_1 r_e + I r_c \qquad (20.17)$$
$$0 = I_1 r_e - I_2 (R_s + r_b) \qquad (20.18)$$

and $$I = I_1 + I_2.$$

Substituting the values of the resistances in expression (20.18), we have:

$$0 = I_1 \times 50 - (I - I_1)(200 + 1000)$$
$$= 1250 I_1 - 1200 I$$

∴ $$I_1 = 0.96 I.$$

Substituting in expression (20.17), we have:

$$E = I(0.96 \times 50 + 1\,000\,000 - 0.98 \times 0.96 \times 1\,000\,000)$$
$$\simeq 59\,250 I$$

∴ output resistance $= E/I = 59\,250$ Ω.

Comparison of this result with that obtained for the common-base circuit in example 20.2 shows that the output resistance of the common-emitter circuit is much less than that of the common-base circuit.

20.17 Thyristor

The thyristor, formerly known as *silicon controlled rectifier* (SCR), bears the same relationship to the silicon diode as the thyratron (section 19.24) does to the gas-filled rectifier. Compared with the thyratron, the thyristor possesses the following advantages: (*a*) it has no filament; (*b*) the voltage drop in the forward direction is only about 1 to 2 volts, compared with 10 to 15 volts for the thyratron; (*c*) the triggering and recovery periods are much shorter, so that it is more suitable for high-frequency switching operations; and (*d*) being a solid-state device, it is more robust, is smaller in size and has a longer working life.

The basic part of the thyristor is its four layers of alternate p-type and n-type silicon semiconductors, forming three p–n junctions, A, B and C, as shown in fig. 20.40 (*a*). The terminals connected to the

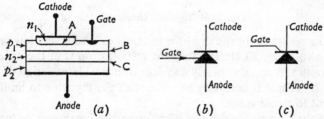

Fig. 20.40 Thyristor arrangement and symbols.

n_1 and p_2 layers are the cathode and anode respectively. A contact welded to the p_1 layer is termed the *gate*. The British Standards graphical symbol for the thyristor is given in fig. 20.40 (*b*). The direction of the arrowhead on the gate lead indicates that the gate contact is welded to a p-region and shows the direction of the gate current required to operate the device. If the gate contact is welded to an n-region, the arrowhead should point outwards from the rectifier. The graphical symbol shown in fig. 20.40 (*c*) is, however, frequently used to represent a thyristor.

When the anode is positive with respect to the cathode, junctions A and C are forward-biased and therefore have a very low resistance, whereas junction B is reverse-biased and consequently presents a very high resistance, of the order of megohms, to the passage of a current. On the other hand, if the anode terminal be made *negative* with respect to the cathode terminal, junction B is forward-biased while A and C act as two reverse-biased junctions in series.

Let us now consider the effect of increasing the voltage applied across the thyristor, *with the anode positive relative to the cathode*. At first, the forward leakage current reaches saturation value due to the action of junction B. Ultimately, a breakover value is reached

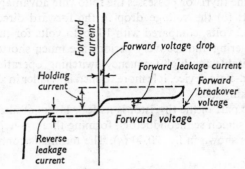

Fig. 20.41 Thyristor characteristic.

and the resistance of the thyristor instantly falls to a very low value, as shown in fig. 20.41. The forward voltage drop is of the order of 1–2 volts and remains nearly constant over a wide variation of current. A resistor is necessary in series with the thyristor to limit the current to a safe value.

We shall now consider the effect upon the breakover voltage of applying a positive potential to the gate as in fig. 20.42. When switch S is closed, a bias current, I_B, flows via the gate contact and

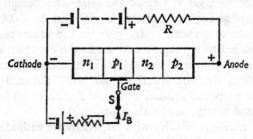

Fig. 20.42 Gate control of thyristor breakover voltage.

layers p_1 and n_1, and the value of the breakover voltage of the thyristor depends upon the magnitude of the bias current in the way shown in fig. 20.43. Thus, with $I_B = 0$, the breakover voltage is represented by OA, and remains practically constant at this value until the bias current is increased to OB. For values of bias current between OB and OD, the breakover voltage falls rapidly to nearly zero. An alternative method of representing this effect is shown in fig. 20.44.

In the triggered condition, the thyristor approximates a single p–n diode, the anode current being limited only by the resistance R of the external circuit. Once the gate has triggered the device, *it loses control over the anode current*; and the only method of restoring the device to its high-resistance condition is to reduce the anode current below the *holding* value indicated in fig. 20.41. The value of the holding current is usually very small compared with the rated forward current; for instance, the holding current may be about 10 mA for a thyristor having a forward-current rating of 40 A.

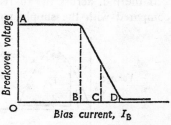

Fig. 20.43 Variation of breakover voltage with bias current.

If the thyristor is connected in series with a non-reactive load, of resistance R, across a supply voltage having a sinusoidal waveform and if it is triggered at an instant corresponding to an angle ϕ after the voltage has passed through zero from a negative to a positive value, as in fig. 20.45 (*a*), the value of the applied voltage at that instant is given by:

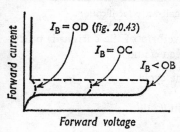

Fig. 20.44 Variation of breakover voltage with bias current, I_B.

$$v = V_m \sin \phi.$$

Up to that instant, the voltage across the thyristor has been growing from zero to v. When triggering occurs, the voltage across the thyristor instantly falls to about 1–2 volts and remains approximately constant while current flows, as shown by the slightly curved line in fig. 20.45 (*a*). Also, at the instant of triggering, the current increases immediately from zero to i,

where $$i = \frac{v - \text{p.d. across thyristor}}{R}$$

$$= v/R \text{ when the p.d. across thyristor} \ll v,$$

If ϕ is less than $\pi/2$, the current increases to a maximum I_m and then decreases to the holding value when it falls instantly to zero, as shown in fig. 20.45 (b). The average value of the current over one cycle is the shaded area enclosed by the current wave divided by 2π.

If the p.d. across the thyristor, when conducting, is very small compared with the supply voltage, and if the holding current is

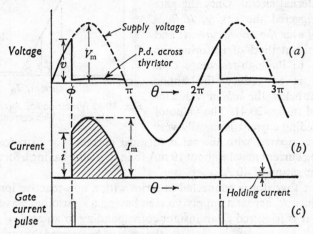

Fig. 20.45 Phase-controlled half-wave rectification.

negligible compared with I_m, the waveform of the current is practically sinusoidal for values of θ between ϕ and π,

∴ average value of current over one cycle

$$= I_{av} = \frac{1}{2\pi} \int_{\phi}^{\pi} I_m \sin \theta \,.\, d\theta$$

$$= \frac{I_m}{2\pi} \Big[-\cos \theta \Big]_{\phi}^{\pi} = \frac{I_m}{2\pi} (1 + \cos \phi).$$

When $\phi = 0$, $I_{av} = I_m/\pi = 0.3185\, I_m$.
When $\phi = \pi/2$, $I_{av} = 0.159\, I_m$.
When $\phi = \pi$, $I_{av} = 0$.

Hence, by varying the instant at which the thyristor is triggered, it is possible to control the output current over a wide range. Triggering signals for this purpose are usually in the form of a positive pulse of short duration compared with the time of one cycle of the alternating voltage. A gate current of about 50 mA applied for about 5 microseconds is all that is required to trigger a device having current ratings between a few milliamperes and hundreds of amperes.

EXAMPLES 20

1. Calculate the values of β for transistors having the following values of α: (a) 0·975, and (b) 0·96.
2. Calculate the values of α for transistors having the following values of β: (a) 30, (b) 70.
3. A junction transistor has the following equivalent T-circuit parameters: $r_e = 20\ \Omega$, $r_b = 400\ \Omega$, $r_c = 1\cdot2\ M\Omega$ and $\alpha = 0\cdot97$. The transistor is used in a common-base circuit with a load resistance of 8 kΩ. Calculate the ratios of the output current, voltage and power to the corresponding input quantities and determine the power gain in decibels. Calculate also the input resistance.
4. If the source supplying the transistor circuit of Q.3 has an internal resistance of 100 Ω, what is the value of the output resistance?
5. If the transistor whose parameters are given in Q.3 is used in a common-emitter circuit with the same load resistance of 8 kΩ, what are the current, voltage and power gains and the input resistance?
6. Calculate the output resistance of the transistor circuit of Q.5, assuming the internal resistance of the source to be 100 Ω.
7. A source generating an alternating e.m.f. of 0·02 V (r.m.s.) has an internal resistance of 300 Ω and is connected to a junction transistor having the following equivalent T-circuit parameters: $r_e = 25\ \Omega$, $r_b = 200\ \Omega$, $r_c = 0\cdot8\ M\Omega$ and $\beta = 60$. The load has a resistance of 5 kΩ. Calculate the output voltage and power when the transistor is used in: (a) a common-base circuit, and (b) a common-emitter circuit. Also calculate the power gain in decibels, due to the transistor, for each arrangement.
8. Calculate the output resistance of the transistor circuit of Q.7 for each of the two configurations stated in the question.
9. A junction transistor has the following equivalent T-network parameters: $r_e = 50\ \Omega$, $r_b = 800\ \Omega$, $r_c = 1\cdot4\ M\Omega$ and $\beta = 40$. The load has a resistance of 12 kΩ. Calculate the power gain in decibels and the input resistance when the transistor is connected: (a) in a common-base circuit, and (b) in a common-emitter circuit.
10. A generator having an open-circuit voltage of 50 mV (r.m.s.) and an internal resistance of 100 Ω is connected across the input terminals of the transistor whose parameters are given in Q.9. The load resistance remains at 12 kΩ. Calculate the power dissipated in the load for: (a) the common-base, and (b) the common-emitter circuits.
11. Draw simple circuit diagrams for single-stage transistor amplifiers using: (a) the common-base connection, and (b) the common-emitter connection.

    Derive the relationship, for small input signals, between the current gains of the transistor in the common-base and common-emitter circuits.

The characteristics of a junction transistor are given in the following table:

Collector voltage (V)	Collector current (mA)		
	$I_b = 0$	$I_b = 30$ μA	$I_b = 60$ μA
1·0	0·20	1·60	2·80
4·0	0·30	1·73	3·06
7·0	0·40	1·86	3·32

The transistor is used in a common-emitter stage with a collector load resistance of 1000 Ω and a collector supply voltage of 4·5 V. If the bias current in the base is 30 μA, draw the load line and determine: (a) the quiescent collector voltage, and (b) the total voltage swing at the collector for an alternating input signal current of 30 μA peak.

(W.J.E.C., O.2)

12. Distinguish between the common-base circuit and the common-emitter circuit of the transistor.

The common-emitter output characteristics for a transistor are as follows:

Collector voltage (V)	Collector current (mA)		
	$I_b = 20$ μA	$I_b = 40$ μA	$I_b = 60$ μA
2	1·0	2·0	3·0
6	1·25	2·4	3·6
10	1·5	2·8	4·2

Construct load lines to show the operation from a 6-V battery, with load resistors of 500 Ω, 1000 Ω and 2000 Ω respectively. If a suitable value of base bias current is 40 μA for an input signal of ±20 μA, determine the current amplification for each load.

(S.A.N.C., O.2)

13. Draw a circuit diagram showing the connections of a single-stage, resistance-loaded transistor amplifier in a common-emitter configuration. Explain how a load line is used in conjunction with the transistor characteristics to determine the gain of the amplifier.

Using the transistor characteristics given below, construct a load line corresponding to a 1000-Ω resistive load with a 10·0-V d.c. supply. If the transistor operates with an 80-μA bias current and a sinusoidal input signal of 60 μA peak, determine the r.m.s. values of the load alternating current and voltage.

Collector voltage (V)	Collector current (mA)		
	$I_b = 20$ μA	$I_b = 80$ μA	$I_b = 140$ μA
1	1·1	5·225	9·35
2	1·2	5·4	9·6
6	1·6	6·1	10·6
10	2·0	6·8	11·6

(S.A.N.C., O.2)

Note re Q. 12 and 13. In the actual examination papers, the data were in the form of graphs on work sheets to be used by the candidates. For economy of space, the data have been reproduced in tabular form.

14. Assuming the data given in Q.13 for a common-emitter circuit, calculate the a.c. power absorbed by the load resistor due to the sinusoidal input signal of 60 μA peak.

15. An n–p–n transistor has the following static characteristics when used in a common-base circuit:

Collector voltage (V)	Collector current (mA)		
	$I_E = -2\,\text{mA}$	$I_E = -4\,\text{mA}$	$I_E = -6\,\text{mA}$
1	1·8	3·6	5·4
4	1·9	3·75	5·6
7	2·0	3·9	5·8

Construct a load line when the transistor is used with a 1000-Ω non-reactive load across an 8-V supply.

If the transistor operates with an emitter bias current of −4 mA and the sinusoidal input signal has a peak value of 2 mA, determine: (a) the peak-to-peak variation of the p.d. across the load, and (b) the a.c. power absorbed by the load.

16. A 15-kΩ resistor and a thyristor are connected in series across a 240-V a.c. supply. Calculate the average value of the current if the thyristor is triggered at an instant corresponding to an angle of 40° after the voltage has passed through zero from a negative to a positive value. Assume the waveform of the supply voltage to be sinusoidal and the voltage drop in the thyristor, when conducting, to be negligible.

CHAPTER 21

Electric Lamps and Illumination

21.1 The spectrum

Newton discovered that when white light, such as that given by the sun, is passed through a glass prism, a band of light is obtained having colours in the following sequence: red, orange, yellow, green, blue, violet. It is known that light is a form of electromagnetic radiation having a velocity of 3×10^8 m/s and that the various colours have different frequencies ranging from about 4×10^{14} Hz for the extreme visible red end of the spectrum to about $7 \cdot 5 \times 10^{14}$ Hz for the extreme visible violet end.

$$\text{Since velocity in metres/second} \quad \left\{ \begin{array}{l} \text{wavelength in metres} \times \\ \text{frequency in hertz} \end{array} \right.$$

$$\therefore \quad \text{wavelength of the extreme visible red end of the spectrum} = \frac{3 \times 10^8}{4 \times 10^{14}} = 0 \cdot 75 \times 10^{-6} \text{ m}$$

$$= 0 \cdot 75 \text{ micrometre*}.$$

where 1 micrometre = one-millionth of a metre. Similarly,

$$\text{wavelength of the extreme visible violet end of the spectrum} = \frac{3 \times 10^8}{7 \cdot 5 \times 10^{14}} = 0 \cdot 4 \times 10^{-6} \text{ m}$$

$$= 0 \cdot 4 \text{ } \mu\text{m}.$$

Electromagnetic radiations that are immediately below the red and above the violet ends of the spectrum are termed *infra-red* and *ultra-violet* respectively. In the case of light emitted by incandescent solid bodies, the spectrum is continuous, but the relative intensity of the different colours depends upon the temperature of the body—the higher the temperature, the more pronounced is the violet end of the spectrum compared with the red end. When light is obtained from a gaseous discharge, the spectrum is discontinuous, i.e. it consists of

* The use of the term micron for micrometre is deprecated, and the unit angström ($= 10^{-10}$ m) is non-SI.

one or more coloured lines. Thus in the case of the sodium lamp, the spectrum consists mainly of two yellow lines very close together with wavelengths of 0·5890 and 0·5896 μm. These two wavelengths are so close to each other that the light from a sodium lamp is said to be 'monochromatic', namely a light having only one wavelength.

Fig. 21.1 shows how the sensitivity or the brightness sensation of the average human eye varies at different wavelengths for ordinary levels of illumination, the power radiated as light being assumed constant for the different wavelengths. It will be seen that the eye is most sensitive to light having a wavelength of about 0·555 μm in the

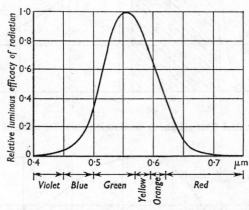

Fig. 21.1 Relative luminous efficacy of the human eye.

green portion of the spectrum. The ratio of the visual effect of light of a given wavelength to that at a wavelength of 0·555 μm is termed the *relative luminous efficacy of radiation*; thus the relative luminous efficacy of light having a wavelength of either 0·51 and 0·61 μm is about 0·5.

The most common and useful source of light is the sun, and the graph in fig. 21.2 shows the relative values of the power radiated at different wavelengths for an average summer midday sunlight. It will be seen that the maximum power is being radiated at about 0·5 μm, which is approximately the wavelength at which the human eye is most sensitive. Fig. 21.3 gives the relative values of the power radiated from an incandescent filament at different temperatures. These curves show that the lower the temperature, the lower is the amount of energy radiated in the visible range and emphasize the

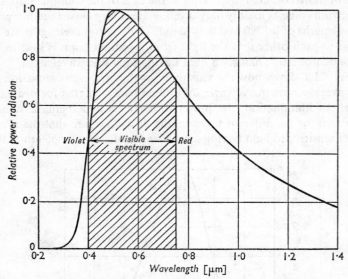

Fig. 21.2 Spectral power distribution curve for sunlight.

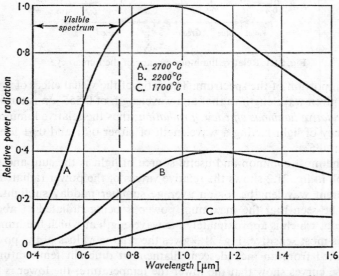

Fig. 21.3 Spectral power distribution curves for an incandescent filament.

necessity of operating incandescent lamps at the highest practicable temperature.

21.2 Units of luminous intensity and illuminance

(*a*) The standard originally used in photometry was a wax candle, but owing to its unreliability, it was replaced by a lamp burning vaporized pentane. The luminous intensity of this lamp was equal to about ten of the original candles. Even this standard was difficult to reproduce accurately, and in 1909 the incandescent filament lamp was adopted as the standard. The luminous intensity of a number of such lamps was measured by comparison with the pentane lamp and these lamps were then used as standards for determining the luminous intensity of other lamps.

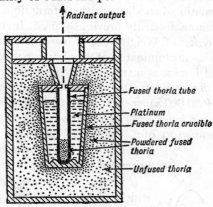

Fig. 21.4 Primary standard of light.

It is essential that a primary standard should be capable of being accurately reproduced from a specification; and in 1948 it was decided to base the unit of luminous intensity upon the luminance (or objective brightness) of a small aperture due to light emitted from a radiator maintained at the temperature of solidification of platinum, namely 1773°C. The construction of this primary standard is shown in fig. 21.4. The radiator consists of a tube of fused thoria (thorium oxide), about 45 mm long, with an internal diameter of about 2·5 mm, the bottom of the tube being packed with powdered fused thoria. The tube is supported vertically in pure platinum contained in a fused thoria crucible. The latter has a lid with a small hole in the centre, about 1·5 mm in diameter, and is almost embedded in powdered fused thoria in a larger refractory container having a funnel-shaped opening. The reason for the use of pure fused thoria is that this material is unaffected at the temperature of melting platinum and does not contaminate the latter. The presence of any impurity in the platinum would alter the melting-point temperature and therefore affect the luminance of the aperture.

The platinum is first melted by eddy currents induced in it by a high-frequency current in a coil surrounding the outer container and is then allowed to cool very slowly. While the platinum is changing from the liquid to the solid state, the temperature remains constant sufficiently long for measurements to be made of the luminous intensity of the light beam passing through the aperture of known diameter. The luminance of this new primary standard was found to be 589 000 international candles/m², the international candle being the unit of luminous intensity previously used. In 1948 the International Conference of Weights and Measures decided to adopt 600 000 units/m² as the luminance of the platinum primary standard. This new unit of luminous intensity, termed the *candela*,* is therefore defined as *the luminous intensity, in the perpendicular direction, of a surface of* 1/600 000 *square metre of a black body at the temperature of freezing platinum under standard atmospheric pressure*.

The *luminous intensity* of a lamp is defined as the light-radiating capacity of a source in a given direction, expressed in candelas (abbreviation: cd).

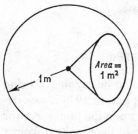

Fig. 21.5 Relationship between the candela and the lumen.

(*b*) A *point source* of light is a source which, for photometric purposes, can with sufficient accuracy be considered as concentrated at a point.

(*c*) A *uniform point source* is a point source emitting light uniformly in all directions.

(*d*) If a uniform point source of 1 candela is placed at the centre of a perfectly transparent sphere of, say, 1 m radius (fig. 21.5), then the solid angle subtended at the centre by 1 m² of area on the surface of the sphere is termed a *unit solid angle* or *steradian*; and the quantity of light emitted through a unit solid angle and therefore passing through

* Pronounced with the first syllable slightly accentuated, as in 'candle'.

1 m² of the surface area of the sphere is termed a *lumen* (abbreviation: lm).

(*e*) The *luminous flux* from a light source is the radiant power evaluated according to its ability to produce visual sensation on the basis of the relative luminous efficacy (section 21.1). The unit of luminous flux is the *lumen*, namely the light flux emitted in unit solid angle by a uniform point source having a luminous intensity of 1 candela. Thus for a uniform point source of 1 candela at the centre of the transparent sphere of 1 m radius (fig. 21.5), the luminous flux passing outwards through each square metre of the surface is 1 lumen. Since the surface area of the sphere is 4π square metres, it follows that the total luminous flux from a uniform point source of 1 candela is 4π lumens. Hence, for a point source having a luminous intensity of I candelas,

$$\left.\begin{array}{r}\text{total luminous flux emitted} \\ \text{in solid angle } d\omega\end{array}\right\} = d\Phi = I \,.\, d\omega \text{ lumens}$$

or

$$I = d\Phi/d\omega \text{ candelas,}$$

i.e. the luminous intensity of a point source can be defined as the luminous flux per unit solid angle. In the case of an actual lamp, it is usually necessary to estimate the optical centre of the lamp and to regard the light as being emitted from that point.

(*f*) The *mean spherical luminous intensity* of a luminous source is the average value of the luminous intensity in all directions. Hence if a luminous source has a mean spherical luminous intensity of I candelas, the total luminous flux emitted by that source is $4\pi I$ lumens. It is not correct to say that I candelas are equal to $4\pi I$ lumens, since the 'candela' is the unit of 'intensity', whereas the 'lumen' is the unit of 'flux'.

(*g*) The *illuminance* at a point of a surface is the luminous flux per unit area of the surface. The unit of illuminance is the *lumen per square metre* and is termed the *lux* (abbreviation: lx). With a uniform point source of 1 candela at the centre of a hollow sphere of 1 metre internal radius, the illuminance on the surface* is 1 lm/m² (or 1 lx) and the rays of light reach the surface normally, i.e. at right-angles.

If the internal radius of the sphere is increased from 1 metre to r

---

* It is assumed that no light is reflected by the internal surface. This condition can be fulfilled by using a blackened surface.

metres, the surface area is increased from 4π to $4\pi r^2$ square metres. With a uniform point source of 1 candela at the centre, number of lumens/square metre on a sphere of radius r metres

$$= \frac{4\pi}{4\pi r^2} = \frac{1}{r^2}.$$

Hence the illuminance of a surface is inversely proportional to the square of its distance from the source. It follows that if the luminous intensity of a source S in direction SA (fig. 21.6) is I candelas and if the distance between S and A is d metres, then for a surface at right-angles to the incident ray SA,

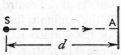

Fig. 21.6 Illuminance on a surface normal to the incident rays.

$$\text{illuminance at A} = \frac{I}{d^2} \text{ lux.}$$

(*h*) *The cosine law of illuminance.* Suppose that the surface to be illuminated is so placed that the angle between the incident ray and the normal (or perpendicular) to the surface is θ (fig. 21.7). The

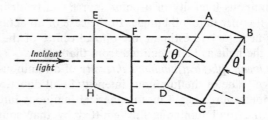

Fig. 21.7. Cosine law of illuminance.

luminous flux falling on ABCD is exactly the same as that which would fall on a surface EFGH, normal to the rays.

But

$$\frac{\text{area of surface EFGH}}{\text{area of surface ABCD}} = \cos\theta$$

$$\therefore \quad \frac{\text{illuminance on ABCD}}{\text{illuminance on EFGH}} = \frac{\text{area of EFGH}}{\text{area of ABCD}} = \cos\theta$$

so that illuminance on ABCD $= \dfrac{I}{d^2} \cdot \cos\theta$ (21.1)

where d is the distance of the surface from a source having a luminous intensity of I candelas in the direction of the surface. If d is in metres, the illuminance is in lux (or lumens per square metre).

(*i*) The *mean horizontal luminous intensity* is the average value of the luminous intensity in all directions in a plane through the centre of the source and perpendicular to its axis.

(*j*) The *reduction factor* of a luminous source is the ratio of the mean spherical luminous intensity to the mean horizontal luminous intensity.

(*k*) The *mean hemispherical luminous intensity* is the average value of the luminous intensity in all directions within the hemisphere, either above or below the horizontal plane through the centre of the source.

(*l*) The *luminance** of a source in a given direction is the luminous intensity in that direction per unit of projected area. It is usual to express the luminance in candelas/metre^2 of projected area. The candela metre^2 is often termed the *nit*. Approximate values of the luminance of different sources are given in the following table:

Source	Luminance (in cd/m^2)
Zenith sun	16×10^8
Crater of carbon arc (25 A)	2×10^8
Tungsten, gas-filled, clear (100 W)	$6 \cdot 5 \times 10^6$
Tungsten, gas-filled, pearl (100 W)	8×10^4
Mercury, high-pressure, clear (400 W)	120×10^4
Mercury, low-pressure, fluorescent (80 W)	$0 \cdot 9 \times 10^4$
Sodium, low-pressure, clear (140 W)	8×10^4
Clear blue sky	$0 \cdot 4 \times 10^4$

It is extremely important in practice to avoid considerable contrast of brightness in the line of vision of the eye, since this causes glare and an intense eyestrain. From the above table it will be seen that the luminance of a pearl lamp is only about $\frac{1}{80}$th of that of the corresponding lamp with a clear bulb; consequently there is far less risk

* The term 'brightness' by itself is ambiguous, since it is necessary to distinguish between 'objective' and 'subjective' brightness. The brightness of a source, as judged by the eye, depends upon a number of factors such as the brightness of the surrounding surface; for instance, a light source appears brighter if the surroundings are dark. Such interpretation of brightness is referred to as *subjective brightness* or *luminosity*. If the brightness of a source is measured photometrically in terms of the candelas per unit of projected area, the value is a definite quantity for a given source and is independent of such factors as the surrounding surfaces. Hence, this interpretation of brightness is referred to as *objective brightness* or *luminance*. Luminosity and luminance are not proportional to each other; for instance, a surface having twice the luminance of another surface may not *look* twice as bright.

of discomfort or glare with the former than with the latter. The very low luminance of the fluorescent lamp constitutes one of its main advantages.

21.3 Illuminance on a surface

Suppose A in fig. 21.8 to represent a luminous source suspended h metres above the ground; then:

$$\text{illuminance at B} = \frac{\text{luminous intensity in direction AB}}{h^2}$$

$$\left.\begin{array}{c}\text{and illumin-}\\ \text{ance at C}\end{array}\right\} = \frac{\text{luminous intensity in direction AC} \times \cos\theta}{AC^2}$$

$$= \frac{(\text{luminous intensity in direction AC}) \times h}{(h^2 + d^2)^{3/2}}$$

If the luminous intensity of the lamp is uniform in the lower hemisphere, then:

$$\text{illuminance at C} = \text{illuminance at B} \times \cos^3\theta$$

$$= \frac{\text{illuminance at B}}{\{1 + (d/h)^2\}^{3/2}}.$$

The above relationship is represented graphically in fig. 21.9,

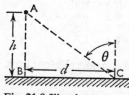

Fig. 21.8 Illuminance on a surface.

assuming the luminous intensity of the lamp to be uniform in the lower hemisphere. It will be seen that the illuminance on the ground falls off very rapidly with increase of distance from the point directly underneath the lamp. This effect can be reduced by using a glass fitting or a reflector to distribute most of the light at an angle of about 60–75° to the vertical.

By using a number of lamps, suitably spaced, it is possible to obtain fairly even illuminance over the working surface. With such an arrangement, the following method may be used to calculate the illuminance. Suppose the surface to be illuminated to have an area of A square metres and the average illuminance on that surface to be E lux. The useful luminous flux is therefore EA lumens. But the lamps must emit a larger number of lumens, since some of the light from the lamps is absorbed by the ceiling, the walls and the lamp reflectors (if any). The ratio of the useful number of lumens to the total lumens emitted by the lamps is termed the *utilization factor*.

Hence, total lumens from lamps $= \dfrac{EA}{\text{utilization factor}}$.

The value of this factor depends upon the colour of the walls and ceilings, the room proportions, the height of the lamp above the working plane and particularly upon the type of lamp fitting. Thus, with open reflectors, the utilization factor lies between 0·3 and 0·6; whereas with pendant fittings giving indirect lighting, the light has to be reflected from the ceiling and walls and the utilization factor lies between 0·1 and 0·4.

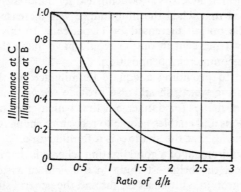

Fig. 21.9 Illuminance due to a source of uniform luminous intensity.

Finally, it is necessary to allow for the depreciation in the value of the useful luminous flux due partly to the accumulation of dust on the lamp bulb and fitting and partly to the fall in the light output of the lamp during its life. This effect is taken into account by applying a *maintenance factor* or a *depreciation factor*,* where maintenance factor

$$= \frac{\text{illuminance at any given time}}{\text{illuminance with lamps new and fittings clean}}$$

$= 0\cdot85$ to $0\cdot7$, depending upon circumstances,

and depreciation factor $= \dfrac{1}{\text{maintenance factor}}$.

Hence, total lumens from lamps when new

$$= \frac{EA}{(\text{utilization factor}) \times (\text{maintenance factor})} \qquad (21.2)$$

* This term is deprecated in B.S. 233.

The most suitable number of lamps depends upon the type of lamp fitting used and the height of the lamps above the working plane. The spacing and height of the lamps are, to a considerable extent, governed by the constructional characteristics of the room to be illuminated. In general, the spacing between the lamps should not exceed 1·5 times their height above the working plane.

The illuminance of the working interior of a factory or an office should not be less than 400 lx; and where fine work such as fine assembly is involved, the illuminance should be about 1 to 2 klx. An investigation in a factory producing leather handbags and similar articles showed that when the illuminance was increased from 300 to 1000 lx, the output increased by 11 per cent. In this case an increase of only 1 per cent in output would have made the increased illuminance a commercial proposition.

In addition to providing adequate illuminance on the working surface, a lighting system should also provide *shadow*—not the kind of shadow that makes for danger but that which gives an object its three dimensions and enables people to recognize shapes quickly and easily by ensuring that the shadow is a familiar one.

Another requirement of a good lighting scheme is that it makes the principal object the brightest thing in the field of vision, the immediate background a little less bright and the general surroundings (walls, etc.) still less bright. An illuminance ratio of 10 : 3 : 1 has been suggested as the best for task/background/surroundings. This ratio gives the conditions under which the eye works best and has the added advantage of emphasizing the importance of the workpiece and avoiding distraction from things around.

21.4 Measurement of luminous intensity

The luminous intensity of a lamp L (fig. 21.10) is determined by comparison with that of a standard lamp S, the luminous intensity of which is known. The lamps are supported on an optical bench B with a photometer-head P so mounted that its position can be varied until its two sides are equally bright. If the distances of S and L from P are d_1 and d_2 when a balance is obtained,

$$\frac{\text{luminous intensity of S}}{d_1{}^2} = \frac{\text{luminous intensity of L}}{d_2{}^2}$$

$$\therefore \quad \text{luminous intensity of L} = \left(\begin{array}{c}\text{luminous intensity}\\\text{of S}\end{array}\right) \times \left(\frac{d_2}{d_1}\right)^2.$$

For reasonably accurate comparisons, the Lummer–Brodhun photometer-head is usually employed. This is shown in its simplest form in fig. 21.11. A matt white screen W has its surfaces illuminated by the two lamps S and L. These surfaces are such as to diffuse the

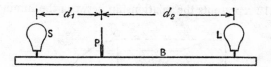

Fig. 21.10 Measurement of luminous intensity.

light, and some of this light is reflected by mirrors M_1 and M_2 on to a compound prism P_1P_2. The latter consists of two right-angled prisms, but P_1 has the outer portion of its principal surface ground away so that only the flat part at the centre makes optical contact with P_2; and of the light reflected from M_1, only the portion passing through this contact area is seen through the telescope. As to the beam of light reflected by M_2, its central core passes through the contact surface between P_1 and P_2, but the remainder of this beam is

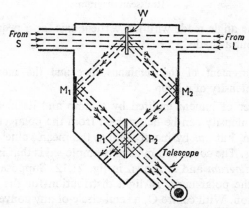

Fig. 21.11 The Lummer–Brodhun photometer-head.

totally reflected by P_2. Consequently the eye sees two zones of light as indicated in the bottom right-hand corner of fig. 21.11. The position of the photometer-head is adjusted until these zones appear equally bright.

21.5 Polar curves of light distribution

It is often of importance to know the luminous intensity of a lamp in different directions and particularly the effect of a shade or reflector upon the distribution of the luminous intensity. This effect is most conveniently represented by polar co-ordinates. Thus, curve A in fig. 21.12 represents the relationship between the luminous inten-

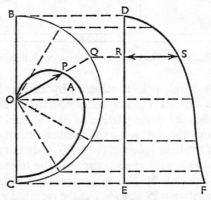

Fig. 21.12 A polar curve of light distribution and the corresponding Rousseau diagram.

sity and the angle made with the vertical axis for a pearl-type lamp used without a shade.

21.6 Measurement of the luminous flux and the mean spherical luminous intensity of a lamp

The number of lumens emitted by a lamp and its mean spherical luminous intensity can be determined from the polar curve of light distribution, but *not* by merely taking the mean value of the polar co-ordinates. The construction usually employed is that known as the *Rousseau diagram* and is shown in fig. 21.12. Suppose curve A to represent the polar curve of light distribution for the lamp under consideration. With centre O, a semi-circle of any convenient radius OB is drawn and a number of radii are inserted. From the ends of these radii, horizontal lines are drawn to cut a vertical line DE. These lines are extended to the right of DE to represent to scale the luminous intensities, in candelas, in the direction of the respective radii; thus, for radius OQ, RS is made equal to the luminous intensity OP in direction OQ. The points on the horizontal lines are then joined by a smooth curve DSF. It can be shown* that the average width between the vertical line DE and the curve DSF gives the mean spherical luminous intensity of the lamp.

Once the mean spherical luminous intensity of a lamp has been determined as described above, that lamp can then be used as a standard for measuring the mean spherical luminous intensity and the lumen output of other lamps by means of an *integrating photometer* (or *photometric integrator*). In its most accurate form the integrating photometer consists of a hollow sphere S (fig. 21.13), with

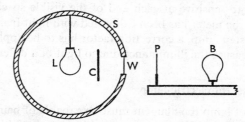

Fig. 21.13 An integrating photometer.

matt white surface on the inside, so that the light from lamp L is thoroughly diffused and the internal surface of the sphere thereby uniformly illuminated. A milk-glass window W is shielded by a screen C from the direct rays of the lamp. Consequently the illumination of W is entirely due to diffused light, and this is proportional to the total luminous flux of the lamp.

The relative luminance of the window with different lamps in the sphere can be determined by means of a photometer-head P and a comparison lamp B. Thus, if d_1 is the distance between B and P with a standard lamp of known mean spherical luminous intensity (m.s.l.i.) in the sphere, and if d_2 is the distance when the standard lamp is replaced by the lamp the m.s.l.i. of which is required, then assuming the distance between P and W to remain unaltered, we have:

$$\frac{\text{m.s.l.i. of new lamp}}{\text{m.s.l.i. of standard lamp}} = \frac{\text{luminance of W with new lamp}}{\text{luminance of W with standard lamp}} = \left(\frac{d_1}{d_2}\right)^2$$

The number of lumens emitted by the new lamp is 4π times its mean spherical luminous intensity.

* See *Electrical Measurements and Measuring Instruments*, by E. W. Golding.

21.7 Measurement of illuminance

The type of illuminance photometer most commonly employed is that which utilizes the photovoltaic cell, already described in section 19.35. With a suitable resistor in series in the external circuit between ring R and plate P in fig. 19.62, the current through the microammeter is practically proportional to the illuminance and the instrument can be calibrated to give the illuminance directly in lux.

The sensitivity of the photovoltaic cell to light of different wavelengths is not exactly the same as that of the eye, the cell being relatively more sensitive at each end of the visible spectrum. Consequently, if the meter has been calibrated with light from an incandescent tungsten lamp, a correction factor has to be applied when it is used to measure the illuminance due to light of a different colour.

21.8 Incandescent electric lamp

In this type of lamp the filament must be capable of being operated for long periods at a high temperature without appreciable deterioration; and for this duty only tungsten has proved satisfactory.

There are two types of tungsten lamp: (i) the vacuum lamp, and (ii) the gas-filled lamp. The purpose of the vacuum is to prevent loss of heat from the filament to the bulb that takes place by convection when gas is present. But the vacuum has the disadvantage that the filament vaporizes at a lower temperature than it does with the bulb filled with a gas, the effect being very similar to the variation in the boiling point of water with surface pressure.

The vaporization of the filament not only reduces the sectional area of the filament, thereby increasing its resistance and reducing the temperature and the luminous intensity of the lamp, but it also allows tungsten to condense on the internal surface of the bulb, blackening the latter and reducing the intensity still further. Consequently the highest temperature at which it is practicable to operate the filament in a vacuum lamp is limited to about 2100°C.

The introduction of an inert gas, e.g. nitrogen or argon, enables the filament temperature to be raised to about 2500°C before blackening takes place at an excessive rate. If no other change were made except to introduce a gas, it would be found that the heat lost by convection would be so great that the power required to maintain the filament at 2500°C would have increased more in proportion than the light given out by the lamp. Consequently the efficacy of the gas-filled lamp would be lower than that of the vacuum lamp. This difficulty is overcome by winding the filament as a very close helix, as shown in fig. 21.14. Langmuir discovered that a thin layer of the gas adheres to the filament and that the convection current of gas merely glides over this fixed layer. It follows that if the clearance between adjacent turns of the filament is less than the thickness of

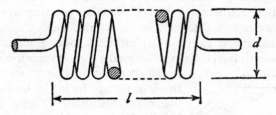

Fig. 21.14 A helical filament.

two of these layers, the latter prevent any gas passing between the turns. Hence the surface with which the gas can come into contact is practically the same as that of a rod of diameter d and length l (fig. 21.14); and since this area is far less than the surface area of the filament, the loss of heat by convection is very considerably reduced.

The latest development in this type of lamp is the tungsten halogen* lamp used for projectors, headlights, etc. The most common halogen used is iodine. At temperatures between 250°C and 750°C, iodine vapour and tungsten combine to form tungsten iodide: but at temperatures above 1250°C, tungsten iodide splits into tungsten and iodine.

The bulb wall temperature is designed to be between 250°C and 750°C. Hence the iodine vapour combines with any evaporated tungsten to form tungsten iodide vapour before the tungsten reaches the bulb wall. The iodide vapour is carried over the hot filament by gas convection where it splits into tungsten and iodine. The tungsten is deposited back on the filament. Consequently the bulb never blackens. The vapour pressure in the bulb is about 5 atmospheres (compared with about 1 atmosphere in the glass-bulb lamp), and the bulb is made of quartz to withstand the higher temperature and pressure.

* *Halogens* is the name given to a group of elements consisting of fluorine, chlorine, bromine and iodine.

21.9 Arc lamp

An arc lamp consists of two carbon rods, A and C, connected in series with a stabilizing resistor R, as in fig. 21.15. The arc is struck by bringing the electrodes into contact with each other and then drawing them apart. When the electrodes are about to be separated, the area of contact is very small and the resistance of the contact is therefore high. Consequently the heat generated is sufficient to raise the tips of the electrodes to a state of incandescence, thereby enabling the negative electrode C to emit electrons immediately the electrodes are separated. These electrons are attracted towards anode A and attain sufficiently high velocity to ionize the carbon vapour occupying

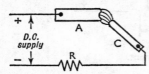

Fig. 21.15 An arc lamp.

the gap between the electrodes. The positive ions formed by his ionization travel towards the cathode C and, by their bombardment of the latter, maintain the tip sufficiently hot* to continue emitting electrons.

The temperature of the crater formed on the positive carbon is about 3500°C, with the result that about 85 per cent of the light emitted by an arc lamp is obtained from the crater. Also the positive carbon tends to be consumed much more rapidly than the negative carbon; consequently the cross-sectional area of the former is made about twice that of the latter.

The p.d. across the arc is usually about 45–60 V and the supply voltage is about 70–100 V. Arc lamps are mainly used where a concentrated source of light is required, e.g. in searchlights.

21.10 Electric discharge lamps

It was pointed out in section 19.7 that ionization in a gas tube is accompanied by a glow, and this phenomenon is utilized in the modern electric discharge lamp, of which the two main types are:

(*a*) those in which the radiation is produced directly by the discharge, as in the high-pressure mercury-vapour lamp and in the low-pressure sodium-vapour lamp;

(*b*) those in which radiation from the discharge excites a fluorescent material, as in the flourescent mercury-vapour lamp.

---

* The necessity for an incandescent cathode is emphasized because it is possible to have an arc between a cool anode and an incandescent cathode, but not between a cool cathode and an incandescent anode.

21.11 High-pressure mercury-vapour lamp

Fig. 21.16 shows one arrangement of this type of lamp. For simplicity, some of the constructional features, such as the lamp cap, have been omitted. The lamp consists of an outer bulb A and an inner bulb B, the space between the two being partially evacuated. The inner bulb contains a small quantity of mercury and argon. It has three electrodes, namely the main electrodes D and E and a starting electrode S. The latter is connected through a resistor R of about 50 000 Ω to the main electrode situated at the other end of the tube. The main electrodes consist of electron-emitting cathodes held in tungsten-wire helices.

When the lamp is switched on, the ionization for the first few

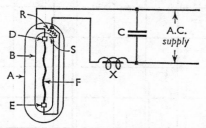

Fig. 21.16 High-pressure mercury-vapour lamp.

seconds occurs in the argon between D and S and then in the argon between D and E. As the lamp warms up, the mercury is vaporized so that the pressure inside the lamp increases and the p.d. across electrodes D and E grows from about 20 V up to about 120–140 V, the operation taking about six minutes. During this time the colour changes from the red of argon to the greenish-blue of mercury. The final vapour pressure inside the bulb is about 1 atmosphere. If the arc is examined through a dark glass, it appears as a thick flexible cord F, about 5 mm diameter, between the electrodes, as shown in fig. 21.16. The function of the choking coil X is to limit the current to a safe value; but since the presence of X produces a lagging power factor, a capacitor C is connected to neutralize this effect. The outer bulb A prevents loss of heat by convection from the inner bulb and assists in keeping the temperature of the latter as uniform as possible at about 500°C.

If a mercury lamp is operated at a low vapour pressure, a large fraction of the output is radiated at a wavelength of 0·2537 μm, which is in the ultra-violet range and is far below the visible range. When the vapour pressure is increased to about 1 atmosphere, much of the ultra-violet energy emitted by the arc is absorbed by the surrounding mercury vapour and is re-radiated at wavelengths that are in the visual range. This effect is evident from a comparison of the two diagrams of fig. 21.17, where the vertical lines indicate the wavelengths in the visible band at which radiations are emitted from low-pressure and high-pressure lamps, and the height of the lines

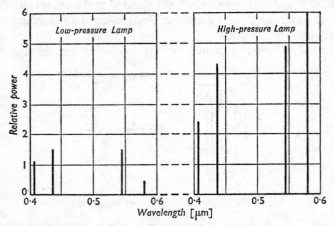

Fig. 21.17 Radiation in the visible range from a mercury-vapour lamp.

represents the relative values of the power radiated per watt supplied to the lamp. The main disadvantages of the high-pressure mercury-vapour lamp are: (i) owing to the absence of red lines the light is greenish-blue and therefore produces colour distortion; (ii) if the lamp is switched off, it cannot be restarted immediately. The vapour pressure is too high to allow the supply voltage to start ionization, and the lamp must cool sufficiently for the vapour pressure to fall to the value at which ionization between electrodes D and S can take place.

The adoption of quartz instead of glass for the inner tube has led to the use of operating pressures of about 10 atmospheres, resulting in a higher proportion of emission at longer wavelengths.

The colour quality of the light emitted from a high-pressure mercury lamp has been further improved (*a*) by coating the outer glass bulb with a fluorescent phosphor such as yttrium vanadate—

the phosphor absorbs ultra-violet radiation emitted from the mercury discharge and radiates energy in the red region of the spectrum; *or* (*b*) by the addition of a mixture of three or four metallic iodides inside the discharge tube to give a balanced multi-line spectrum.

The luminous efficacy of a 400-W fluorescent high-pressure mercury lamp is about 50 lm/W and that of a mercury-iodide lamp of the same rating is about 60 lm/W.

21.12 Low-pressure and high-pressure sodium-vapour lamps

One type of sodium-vapour lamp consists of a U-tube A (fig. 21.18) containing sodium together with neon at a pressure of about 10 mm of mercury and about 1 per cent of argon, the function of the argon being to reduce the initial ionizing voltage. The oxide-coated tungsten electrodes are connected to a step-up auto-transformer T, designed to have a relatively large leakage reactance. When this transformer is on open circuit, its output voltage is about 450 V, and this is sufficient to initiate a discharge through the gas. This discharge has a reddish colour, but as the temperature rises and the sodium is vaporized, the discharge changes to monochromatic light consisting of two yellow lines having wave-lengths of 0·5890 and 0·5896 μm.

The efficacy of the lamp decreases rapidly as the current density is increased above a certain value. Consequently, the lamp has to be operated at a low current density, and this necessitates a large surface

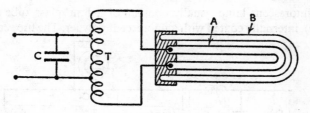

Fig. 21.18 Low-pressure sodium-vapour lamp.

area of tube compared with the power dissipated. Also, for maximum efficacy, the temperature of the tube has to be about 220°C; and in order to maintain this temperature with low power per unit length of tube, it is necessary to insulate the tube thermally. One method of doing this is to enclose the discharge tube in a double-walled vacuum jacket B—somewhat similar to that of the domestic vacuum flask—as shown in fig. 21.18.

Hot sodium vapour is very active chemically and the U-tube A (fig. 21.18) has therefore to be made of ply glass, the inner layer being a special sodium-resisting glass and the outer ordinary soda glass. Since the sodium solidifies when the tube cools, it is necessary to ensure that the sodium is deposited reasonably uniformly along the whole of the tube and not concentrated at one end. Consequently, the lamp must be used horizontally. A 140-W low-pressure sodium lamp has a luminous efficacy of about 70 lm/W.

When the pressure in a sodium lamp is increased, the spectrum of the light is broadened to include colours other than yellow. The inner discharge tube of a high-pressure sodium lamp is made of sintered aluminium oxide which is resistant to hot ionized sodium vapour up to a temperature of about 1600°C and can transmit over 90 per cent of visible radiation. The discharge tube operates at a pressure of about half an atmosphere and is enclosed in an evacuated glass envelope to maintain the tube at the correct temperature. The lamp gives a rich golden light which enables colours to be easily distinguished.

The discharge tube contains sodium and mercury, with argon or xenon at a low pressure added for starting purposes. A voltage pulse of about 2·5 kV is required to initiate the discharge. A 400-W high-pressure sodium lamp has a luminous efficacy of about 90 lm/W.

21.13 Low-pressure fluorescent mercury-vapour lamps

The fluorescent lamp usually consists of a long glass tube T (fig. 21.19), internally coated with a fluorescent powder. The tube contains

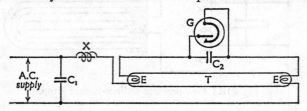

Fig. 21.19 A fluorescent mercury-vapour lamp with glow switch.

a small amount of argon together with a little mercury. At each end of the tube there is an electrode E consisting of a coiled tungsten filament coated with a mixture of barium and strontium oxides.

Each electrode has attached to it two small metal plates—one at each end of the filament. These plates act as anodes for withstanding bombardment by electrons during the half-cycles when the electrode is positive. During the other half-cycles, the adjacent hot filament acts as the cathode, emitting electrons.

Circuits for the control of fluorescent tubes can be divided into two main groups, namely *switch-start* circuits and *startless* circuits. There are two types of starter switch in general use, namely: (i) the *glow* switch—a voltage-operated device, and (ii) the *thermal* switch—a current-operated device.

Fig. 21.19 shows a tube fitted with a glow starter G. The latter consists of two electrodes enclosed in a glass bulb filled with a mixture of helium and hydrogen, one electrode being fixed and the other a U-shaped bimetallic strip. The contacts are normally open, but the application of the supply voltage starts a glow discharge between the electrodes, and the heat generated is sufficient to bend the bimetallic strip until it makes contact with the fixed electrode, thereby closing the circuit between filaments EE. A relatively large current flows through these electrodes, raising them to incandescence and the gas in their immediate vicinity is ionized. After a second or two, the bimetallic strip cools sufficiently to break contact, and the sudden reduction of current induces in choking coil X an e.m.f. of

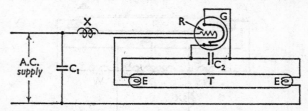

Fig. 21.20 A fluorescent mercury-vapour lamp with thermal switch.

the order of 800–1000 V. This surge is sufficient to ionize the argon in the space between electrodes EE. The heat generated in the tube vaporizes the mercury and the p.d. across the tube falls to about 100–110 V. This p.d. is not sufficient to restart the glow in the starter switch G. The final vapour pressure in the tube is about 0·01 mm of mercury for an ambient temperature of 20°C. A 0·02-μF capacitor C_2 is connected across the glow switch to suppress radio interference. Without C_2, high-frequency voltage oscillations may occur across the starter contacts.

The power factor of the lamp circuit, including choke X, is about 0·5, and a capacitor C_1 is introduced to raise the power factor to about 0·9 lagging.

Fig. 21.20 shows the circuit arrangement utilizing a thermal switch. The latter has a bimetallic strip close to a heater element R. The switch may be in free air or enclosed in a hydrogen-filled glass bulb G. The two electrodes of the switch are in contact when the lamp is not in operation. Consequently, when the lamp is switched on to the supply, the two filaments EE are connected together through the thermal switch, and the relatively large current raises them to incandescence. This current passes through the heater element R, causing the bimetallic strip to break contact and the inductive surge from choke X starts the discharge in tube T. The heat generated in R keeps the starter contacts open during the time the lamp is in operation.

Starterless circuits are referred to in various ways, e.g. instant start, rapid start, etc., and one arrangement is shown in fig. 21.21. Electrode pre-heating is provided by a small auto-transformer, the primary winding D being connected across the lamp electrodes EE. A narrow metal or metallized strip F is attached to the outer surface of the tube and connected to earth through the lamp caps.

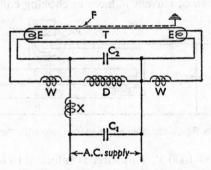

Fig. 21.21 Starterless circuit for a fluorescent lamp.

Immediately the lamp is switched across, say, a 240-V supply, almost the whole of this voltage is applied across the tube electrodes EE, and the p.d. applied from windings WW to the filaments raises the latter to incandescence. The combination of this pre-heating and of the relatively high potential gradient between each electrode and the earth strip F is sufficient to start ionization in the vicinity of the electrodes, and this ionization then spreads to the whole tube. Immediately this occurs, the p.d. across the primary winding D falls to about 100–110 V, so that the voltage across filaments EE is practically halved, the temperature of the filaments being maintained partly by the reduced current and partly by the heat of the discharge.

Hot-cathode fluorescent tubes can be operated on a d.c. supply, but with a lower efficiency owing to the power lost in the resistor which—in addition to the usual choke—must be connected in series with each tube.

It was mentioned in section 21.11 that most of the energy of a low-pressure mercury-vapour lamp is radiated at a wavelength of 0.2537 μm, which is in the ultra-violet range. The fluorescent coating absorbs this ultra-violet energy and converts it into visible radiation. Different fluorescent powders re-radiate the absorbed energy at different wavelengths; for instance, magnesium tungstate gives pale blue radiation, zinc silicate green, zinc–beryllium silicate gives yellow to orange, cadmium borate red and calcium halophosphate gives practically a white light. By using an appropriate combination of various fluorescent powders it is possible to obtain any desired colour. The average efficacy of the fluorescent lamp (warm-white type) is about 50 lm/W and its luminance is about 9000 cd/m².

The ultimate failure of a fluorescent lamp is caused by the exhaustion of the oxides with which the filaments are coated, this material being gradually disintegrated during the life of the lamp, particularly each time the lamp is switched on. The average life of the fluorescent lamp is about 5000 hours compared with about 1000 hours for the tungsten-filament lamp.

The light output of all discharge lamps fluctuates at twice the supply frequency, but with fluorescent lamps, this fluctuation is considerably reduced by the persistence of glow of the fluorescent powder. Nevertheless, this flicker can produce undesirable stroboscopic effects with rotating machinery. This trouble can be practically eliminated by: (*a*) using groups of three lamps distributed between the three phases of a three-phase supply; (*b*) using twin lamps on a single-phase supply, one being connected in series with a choke and the other in series with a choke and a capacitor of such capacitance that the current leads the supply voltage by about 60°, so that the currents in the two lamps differ in phase by about 120° and the power factor of the combined circuits is practically unity.

Mercury-vapour fluorescent tubes with cold cathodes have also been developed. The electrodes are generally plain nickel or iron cylinders, and it is usual to connect three tubes (each about 3 m long) in series across the secondary of a transformer having a relatively large leakage reactance. The secondary terminal voltage is about 3600 V on open circuit and about 1900 V under working conditions. Each tube is rated at about 70 W, with an efficacy of about 35 lm/W and a luminance of about 6000 cd/m². Since no pre-heating is required, the tubes strike immediately the supply is switched on. Their life is practically unaffected by the number of times the lamps are switched on and off, and they can be dimmed down to almost zero light output.

21.14. Phototransistor

It was mentioned in section 20.9 that the ordinary transistor is enclosed in a metal or other opaque container. If light waves are allowed to fall on a germanium or silicon crystal, the energy

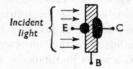

Fig. 21.22 Phototransistor.

supplied by the photons (section 19.32) is sufficient to create electron-hole pairs, exactly as created by an increase of temperature (p. 575).

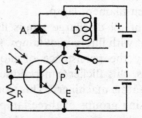

Fig. 21.23 Relay operated by a phototransistor.

The phototransistor is similar to a normal transistor except that a window in the container allows light to fall on the base, as indicated in fig. 21.22, so as to penetrate as near as possible to the collector junction, thereby creating electron-hole pairs in the base.

Fig. 21.23 shows how an n-p-n phototransistor P can be arranged to energize the coil of a relay D. The two parallel arrows outside the envelope are the symbol indicating that the device is operated by light or other radiation. The sensitivity of the phototransistor is greatly increased by connecting the base to the emitter through a resistor R of the order of 5 kΩ. The function of rectifier A is to allow the current in D to flow through A immediately after the incident light on P is cut off, thereby allowing the current in D to decay relatively slowly and thus preventing a high e.m.f. being induced in D.

The maximum sensitivity of germanium occurs at a wavelength of about 1·55 μm, which is well in the infra-red region. That of silicon occurs at about 0·8 μm (see fig. 21.1).

Phototransistors are used to operate burglar alarms, smoke detectors, counters, etc.

Summary of important relationships

Unit of luminous intensity is the candela.

Unit of luminous flux is the lumen.

Unit of illuminance is the lux ($= 1 \ \text{lm/m}^2$).

$$\left.\begin{array}{l}\text{No. of lumens emitted} \\ \text{by a lamp}\end{array}\right\} = \begin{array}{l}4\pi \times \text{mean spherical luminous in-} \\ \text{tensity of lamp.}\end{array}$$

For a surface having an angle of incidence θ, at distance d metres from a source having luminous intensity of I candelas,

$$\text{illuminance} = \frac{I}{d^2}. \cos \theta \ \text{lux} \tag{21.1}$$

To produce an average illuminance of E lux on a surface of A square metres,

total lumens from lamps when new

$$= \frac{EA}{(\text{utilization factor}) \times (\text{maintenance factor})} \tag{21.2}$$

where utilization factor

$$= \frac{\text{no. of lumens on working plane}}{\text{no. of lumens emitted by lamps}}$$

and maintenance factor

$$= \frac{\text{illuminance at any given time}}{\text{illuminance with lamps new and fittings clean}}$$

EXAMPLES 21

1. A metal-filament gas-filled lamp takes 0·42 A from a 230-V supply and emits 1120 lumens. Calculate: (a) the number of lumens per watt; (b) the mean spherical luminous intensity (or m.s.l.i.) of the lamp; and (c) the m.s.l.i. per watt.

2. Six lamps are used to illuminate a certain room. If the luminous efficacy of each lamp is 11 lm/W and the lamps have to emit a total of 10 000 lumens, calculate: (a) the m.s.l.i./lamp, and (b) the cost of the energy consumed in 4 hours if the charge for electrical energy is 1·3p per kilowatt hour.

3. The effective area of the filament of a certain pearl-type lamp, when viewed from below, was 20 cm², and the luminous intensity in the downward direction was 153 cd. Calculate the luminance of the lamp when viewed from that direction.

4. The length of the fluorescent coating of a certain fluorescent lamp is 1460 mm and the diameter is 38 mm. If the intensity in a radial direction is 340 cd, what is the luminance of the lamp?

5. A lamp having a mean spherical luminous intensity of 80 cd has 70 per cent of the light reflected uniformly on to a circular screen 3 m in diameter. Find the illuminance on the screen.

6. In a test on a photometric bench, a filament lamp was placed at a distance of 50 cm from the photometer head, the corresponding distance of a standard 60-cd lamp being 35 cm. Find the luminous intensity of the test lamp.

7. In a photometric bench test, balance is obtained when a standard lamp of 25 cd in the horizontal direction is 1 m and the lamp being tested is 1·25 m from the photometer screen. What is the luminous intensity of the test lamp?
   If the light from the test lamp is reduced by 15 per cent, what will be the respective distances of the lamps from the photometer screen? In this case, the lamps are fixed 2·5 m apart, and the photometer screen moves between them.

8. Two lamps of 16 cd and 24 cd respectively are 2 m apart. A screen is placed between them 0·8 m from the 16-cd lamp. Calculate the illuminance on each side of the screen. Where must the screen be placed in order to be equally illuminated on both sides?

9. A lamp giving 200 cd in all directions below the horizontal is suspended 2 m above the centre of a square table of 1 m side. Calculate the maximum and minimum illuminances on the surface of the table.

10. Explain what is meant by *luminous flux* and define the unit in which it is expressed. A circular area of radius 6 m is to be illuminated by a single lamp vertically above the circumference of the circle. The minimum illuminance is to be 6 lx, and the maximum illuminance 20 lx. Find the mounting height in metres and the mean spherical luminous intensity of the lamp. Assume the luminous intensity to be uniform in all directions.   (E.M.E.U.)

11. Four lamps are suspended 8 m above the ground at the corners of a

square of 4 m side. Each lamp gives 250 cd uniformly below the horizontal plane. Calculate the illuminance: (*a*) on the ground directly under each lamp; (*b*) at the centre of the square.

12. Two identical lamps, each having a uniform intensity in all directions below the horizontal, are mounted at a height of 4 m. Determine what the spacing of the supports must be in order that the illuminance of the ground midway between the supports shall be one-half of the illuminance directly beneath a lamp. **(W.J.E.C., O.2)**

Note. The simplest way of solving this problem is by trial and error, starting with the assumption that the illuminance directly under one lamp due to the other lamp is negligible.

13. A filament lamp is enclosed in a fitting that absorbs 25 per cent of the light flux from the lamp and distributes the remainder uniformly over the lower hemisphere. The lamp may be assumed to have a luminous efficacy of 12 lm/W. The fitting is suspended 2 m above the centre of a horizontal surface 3 m square. Determine the necessary rating (in watts) of the lamp if no point on the horizontal surface is to have an illuminance less than 18 lux. Reflected light from other surfaces may be ignored.
What is then the maximum illuminance on the surface?

14. Two lamps, each of the same rating and equipped with an industrial-type reflector giving a distribution curve as shown below, are suspended 4 m apart and 2 m above a horizontal working plane. Calculate the illuminance on this plane: (*a*) at the point A, vertically below one of the lamps; (*b*) at the point B, 1 m from A along the line joining A with the point vertically below the other lamps. The polar curve for a lamp and reflector is as follows:

Angle (in degrees measured from horizontal axis)	15	30	45	60	75	90
Luminous intensity (candelas)	50	125	240	190	155	140

(App. El., L.U.)

15. A 200-cd lamp emits light uniformly in all directions and is suspended 5 m above the centre of a working plane which is 7 m square. Calculate the illuminance, in lux, immediately below the lamp and also at each corner of the square. If the lamp is fitted with a reflector which distributes 60 per cent of the light emitted uniformly over a circular area 5 m in diameter, calculate the illuminance over this area.

16. A room measuring 15 m by 20 m is to be provided with an illuminance of 200 lx over the horizontal plane using fluorescent tubes, each 2 m long. Each tube gives an output of 3200 lm, 65 per cent of which is effective over the working plane. If a maintenance factor of 0·8 is to be allowed for, find the number of tubes required and sketch a plan view of a suitable arrangement for them.

17. A room, 10 m by 20 m, is to be illuminated by 8 lamps and the average illuminance is to be 300 lx. If the utilization factor is 0·48 and the maintenance factor is 0·8, calculate the mean spherical luminous intensity per lamp.

18. A room, 40 m × 15 m, is to be illuminated by 80-W fluorescent tubes mounted 3 m above the working plane on which an average illuminance of 180 lx is required. Using a maintenance factor of 0·8 and a utilization factor of 0·5, design and sketch a suitable layout. The 80-W fluorescent tube has an output of 4500 lm. (W.J.E.C., O.2)

19. Describe an experiment to determine the constant α and β in the relationship luminous intensity $= \alpha V^\beta$ for a metal-filament lamp.

If for a given lamp the value of β is 4, find the percentage change in the luminous intensity of the lamp for a change of ±5 per cent in the voltage across the lamp.

20. Two 110-V lamps, one of 60 W and the other of 75 W, are connected in series across a 220-V supply. Calculate the current taken by and the p.d. across each lamp, neglecting any variation in resistance. Assuming the luminous intensity to be proportional to the fourth power of the voltage, calculate the percentage value for each lamp compared with normal operation on a 110-V circuit. (U.E.I.)

21. A certain lamp gave the following distribution of luminous intensity, in candelas, in a vertical plane, the angle being measured from the axis through the centre of the lamp cap:

Angle, in degrees	0	15	30	45	60	90	120	150	165	180
Candelas	0	84	154	200	224	254	266	276	283	288

Find: (a) the mean spherical luminous intensity, and (b) the luminous flux of the lamp.

22. Describe briefly *three* applications of photoelectric cells and state the type of cell used for each.

The light from a certain lamp falls perpendicularly upon a photoelectric cell at a distance of 3 m from the lamp, the lamp emitting 60 cd uniformly in all directions. The photoelectric cell characteristic has a slope of 30 μA/lm and the cathode area is 10 cm². If the cell is connected in series with a resistor of 5 MΩ across a d.c. supply, calculate the difference in the p.d. across this resistor when the lamp is switched on and off. (U.L.C.I.)

CHAPTER 22

Electrical Measurements

22.1 Electrical indicating instruments

An indicating instrument is almost invariably fitted with a pointer which indicates on a scale the value of the quantity being measured. The moving system of such an instrument is usually carried by a spindle of hardened steel, having its ends tapered and highly polished* to form pivots which rest in hollow-ground bearings, usually of sapphire, set in steel screws. In some instruments, the moving system is attached to two thin ribbons of spring material such as beryllium–copper alloy, held taut by tension springs mounted on the frame of the movement. This arrangement eliminates pivot friction and the instrument is less susceptible to damage by shock or vibration.

Indicating instruments possess three essential features:

(*a*) a *deflecting device* whereby a mechanical force is produced by the electric current, voltage or power;

(*b*) a *controlling device* whereby the value of the deflection is dependent upon the magnitude of the quantity being measured; and

(*c*) a *damping device* to prevent oscillation of the moving system and enable the latter to reach its final position quickly.

The action of the deflecting device depends upon the type of instrument, and the principle of operation of each of the instruments most commonly used in practice will be described in later sections.

22.2 Controlling devices

There are two types of controlling devices, namely:

(*a*) spring control, and

(*b*) gravity control (not used in modern instruments).

---

* The necessity of handling measuring instruments with care may be realized from the fact that if a pivot has a circle of contact 0·05 mm in diameter and supports a mass of 3 grams (a normal mass for the moving system of an ammeter or voltmeter), the pressure is approximately 15 MN/m².

*

The most common arrangement of spring control utilizes two spiral hairsprings, A and B (fig. 22.1), the inner ends of which are attached to the spindle S. The outer end of B is fixed, whereas that of A is attached to one end of a lever L, pivoted at P, thereby enabling zero adjustment to be easily effected. The hairsprings are of non-magnetic alloy such as phosphor–bronze or beryllium–copper.

The two springs, A and B, are wound in opposite directions so that when the moving system is deflected, one spring winds up while the other unwinds, and the controlling torque is due to the combined torsions of the springs. Since the torsional torque of a spiral spring is

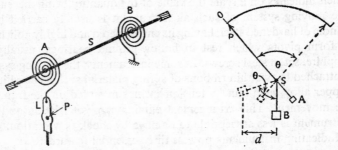

Fig. 22.1 Spring control.　　　　　Fig. 22.2 Gravity control.

proportional to the angle of twist, the controlling torque is directly proportional to the angular deflection of the pointer.

With gravity control, masses A and B are attached to the spindle S (fig. 22.2), the function of A being to balance the weight of pointer P. Mass B therefore provides the controlling torque. When the pointer is at zero, B hangs vertically downwards. When P is deflected through an angle θ, the controlling torque is equal to (weight of B × distance d) and is therefore proportional to the sine of the angular deflection. This has the disadvantage that with deflections of the order of 70° or 80°, the controlling torque increases very slowly with increase of deflection; thus:

$$\frac{\text{controlling torque for 80° deflection}}{\text{controlling torque for 70° deflection}} = \frac{\sin 80°}{\sin 70°} = 1\cdot048,$$

whereas with spring control:

$$\frac{\text{controlling torque for 80° deflection}}{\text{controlling torque for 70° deflection}} = \frac{80}{70} = 1\cdot143.$$

Hence, with gravity control, the scale at the top end is more open than with spring control and the total deflection is limited to about 80°. A further disadvantage of gravity control is that the instrument must be correctly levelled before being used. The only advantage of the gravity-controlled instrument is that it is cheaper than the corresponding spring-controlled instrument.

22.3 Damping devices

The combination of the inertia of the moving system and the controlling torque of the spiral springs or of gravity gives the moving

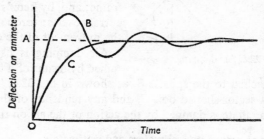

Fig. 22.3 Damping curves.

system a natural frequency of oscillation (section 11.13). Consequently, if the current through an under-damped ammeter were increased suddenly from zero to OA (fig. 22.3), the pointer would oscillate about its mean position, as shown by curve B, before coming to rest. Similarly every fluctuation of current would cause the pointer to oscillate and it might be difficult to read the instrument accurately. It is therefore desirable to provide sufficient damping to enable the pointer to reach its steady position without oscillation, as indicated by curve C. Such an instrument is said to be *dead-beat*.

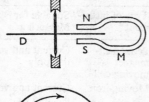

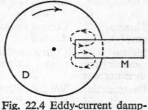

Fig. 22.4 Eddy-current damping.

The two methods of damping commonly employed are:

    (*a*) eddy-current damping, and
    (*b*) air damping.

One form of eddy-current damping is shown in fig. 22.4, where a copper or aluminium disc D, carried by a spindle, can move between the poles of a permanent magnet M. If the disc moves clockwise, the e.m.f.s induced in the disc circulate eddy currents as shown dotted. It follows from Lenz's Law that these currents exert a force opposing the motion producing them, namely the clockwise movement of the disc.

Another arrangement, used in moving-coil instruments (section 22.5), is to wind the coil on an aluminium frame. When the latter

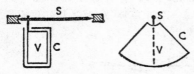

moves across the magnetic field, eddy current is induced in the frame; and, by Lenz's Law, this current exerts a torque opposing the movement of the coil.

Fig. 22.5 Air damping.

Air damping is usually obtained by means of a thin metal vane V attached to the spindle S, as shown in fig. 22.5. This vane moves in a sector-shaped box C, and any tendency of the moving system to oscillate is damped by the action of the air on the vane.

22.4 Types of ammeters, voltmeters and wattmeters

The principal types of electrical indicating instruments, together with the methods of control and damping, are summarized in the following table:

Type of instrument	Suitable for Measuring	Method of Control	Method of Damping
Permanent-magnet moving-coil	Current and voltage, d.c. only	Hairsprings	Eddy current
Moving-iron	Current and voltage, d.c. and a.c.	Hairsprings	Air
Thermo-couple*	Current and voltage, d.c. and a.c.	As for moving coil	As for moving coil
Electro-dynamic† or dynamometer	Current, voltage and power, d.c. and a.c.	Hairsprings	Air
Electrostatic	Voltage only, d.c. and a.c.	Hairsprings	Air or eddy current
Rectifier	Current and voltage, a.c. only	As for moving coil	As for moving coil

* The hot-wire instrument is obsolete and only of historic interest.

† 'Electrodynamic' is the term recommended by the British Standards Institution, but 'dynamometer' is the term more frequently used in practice.

Apart from the electrostatic type of voltmeter, all voltmeters are in effect milliammeters connected in series with a non-reactive resistor having a high resistance. For instance, if a milliammeter has a full-scale deflection with 10 mA and has a resistance of 10 Ω and if this milliammeter is connected in series with a resistor of 9990 Ω, then the p.d. required for full-scale deflection is $0\cdot01 \times 10\,000$, namely 100 V, and the scale of the milliammeter can be calibrated to give the p.d. directly in volts.

22.5 Permanent-magnet moving-coil ammeters and voltmeters

The high coercive force (section 3.11) of modern steel alloys, such as Alcomax (iron, aluminium, cobalt, nickel and copper), allows the use of relatively short magnets and has led to a variety of arrangements

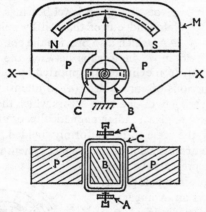

Fig. 22.6 Permanent-magnet moving-coil instrument.

of the magnetic circuit for moving-coil instruments. The front elevation and sectional plan of one arrangement are shown in fig. 22.6, where M represents a permanent magnet and PP are soft-iron pole pieces. The hardness of permanent-magnet materials makes machining difficult, whereas soft iron can be easily machined to give exact airgap dimensions. In one form of construction, the anisotropic* magnet and the pole pieces are of the sintered type, i.e. powdered

* A magnet is made *anisotropic* by being cooled at a particular rate in a powerful directional magnetic field. As it solidifies, the magnetic domains (section 3.13) remain aligned in the same direction, thereby giving the magnet a high coercive force in that direction. An *isotropic* magnet material, on the other hand, has equal magnetic properties in all directions.

magnet alloy and powdered soft iron are compressed in a die to the required shape and heat-treated so that the magnet and the pole pieces become alloyed, thereby eliminating airgaps at the junctions of the materials.

An alternative arrangement is to cast the magnet and attach the soft-iron plate, in one piece, to the two surfaces of M which have been rendered flat by grinding, the joints being made by a resin-bonding technique. This construction enables the drilling of the cylindrical hole and the machining of the gaps to be done with precision.

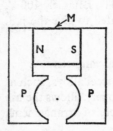

Fig. 22.7 An alternative arrangement of the magnet system.

The rectangular moving-coil C in fig. 22.6 consists of insulated copper wire wound on a light aluminium frame fitted with polished steel pivots resting on jewel bearings. Current is led into and out of the coil by spiral hairsprings AA, which also provide the controlling torque.

The coil is free* to move in airgaps between the soft-iron pole pieces PP and a soft-iron cylinder B supported by a brass plate (not shown). The functions of core B are: (*a*) to intensify the magnetic field by reducing the length of airgap across which the magnetic flux has to pass, and (*b*) to give a radial magnetic flux of uniform density, thereby enabling the scale to be uniformly divided.

An alternative arrangement of the magnet system is shown in fig. 22.7, where M represents the magnet and PP are soft-iron pole pieces.

Fig. 22.8 shows an arrangement where the central cylindrical core consists of an anisotropic magnet M, with soft-iron pole pieces PP attached to M. The return path for the magnetic flux is provided by a soft-iron ring or yoke Y, concentric with the core and separated from it by an airgap of uniform width, the distribution of the flux being as indicated by the dotted lines in fig. 22.8. The moving coil C, wound

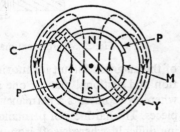

Fig. 22.8 Moving-coil instrument with centre-core magnet.

---

* Students often form the impression that the moving coil is wound on the iron cylinder. It is important to realize that this cylinder is *fixed* and that the frame, on which the coil is wound, *does not touch* the cylindrical core.

on an aluminium frame supported by jewel bearings and controlled by spiral hairsprings, is free to move in the airgap between the pole pieces and the yoke. This arrangement gives a very compact movement and is particularly suitable for small instruments.

By using a more elaborate magnet system, it is possible to construct moving-coil instruments with scales extending over about 250°, but such instruments are beyond the scope of this volume.

The manner in which a torque is produced when the coil is carrying a current may be understood more easily by considering a single turn PQ, as in fig. 22.9. Suppose P to carry current outwards from the paper; then Q is carrying current towards the paper. Current in P tends to set up a magnetic field in a counterclockwise direction around P and thus strengthens the magnetic field on the lower side and weakens it on the upper side.

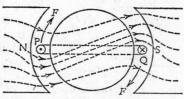

Fig. 22.9 Distribution of resultant magnetic flux.

The current in Q, on the other hand, strengthens the field on the upper side while weakening it on the lower side. Hence, the effect is to distort the magnetic flux as shown in fig. 22.9. Since the flux behaves as if it was in tension and therefore tries to take the shortest path between poles NS, it exerts forces FF on coil PQ, tending to move it out of the magnetic field.

The deflecting torque $\propto$ current through coil $\times$ flux density in gap

$$= kI \text{ for uniform flux density,}$$

where $\qquad k =$ a constant for a given instrument

and $\qquad I =$ current through coil.

The controlling torque of the spiral springs $\propto$ angular deflection

$$= c\theta$$

where $\qquad c =$ a constant for given springs

and $\qquad \theta =$ angular deflection.

For a steady deflection,

$$\text{controlling torque} = \text{deflecting torque,}$$

hence $\qquad c\theta = kI$

$\therefore \qquad \theta = \dfrac{k}{c}I,$

i.e. the deflection is proportional to the current and the scale is therefore uniformly divided.

A numerical example on the calculation of the torque on a moving coil is given in example 22.2, p. 641.

As already mentioned in section 22.3, damping is effected by eddy currents induced in the metal frame on which the coil is wound.

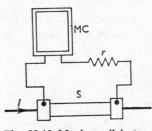

Fig. 22.10 Moving-coil instrument as an ammeter.

Owing to the delicate nature of the moving system, this type of instrument is only suitable for measuring currents up to about 50 milliamperes directly. When a larger current has to be measured, a *shunt* S (fig. 22.10), having a low resistance, is connected in parallel with the moving coil MC, and the instrument scale may be calibrated to read directly the total current *I*. Shunts are made of a material, such as manganin (copper, manganese and nickel), having negligible temperature coefficient of resistance. A 'swamping' resistor *r*, of material having negligible temperature coefficient of resistance, is connected in series with the moving coil. The latter is wound with copper wire and the function of *r* is to reduce the error due to the variation of resistance of the moving coil with variation of temperature. The resistance of *r* is usually about three times that of the coil, thereby reducing a possible error of, say, 4 per cent to about 1 per cent.

The shunt shown in fig. 22.10 is provided with four terminals, the

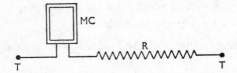

Fig. 22.11 Moving-coil instrument as a voltmeter.

milliammeter being connected across the potential terminals. If the instrument were connected across the current terminals, there might be considerable error due to the contact resistance at these terminals being appreciable compared with the resistance of the shunt.

The moving-coil instrument can be made into a voltmeter by connecting a resistor R of manganin or other similar material in series, as in fig. 22.11. The scale may be calibrated to read directly the voltage applied to the terminals TT.

The main advantages of the moving-coil instrument are:

   (i) high sensitivity;
   (ii) uniform scale;
   (iii) well shielded from any stray magnetic field.

Its main disadvantages are:

   (i) more expensive than the moving-iron instrument;
   (ii) only suitable for direct currents and voltages.

Example 22.1 *A moving coil gives full-scale deflection with 15 mA and has a resistance of 5 Ω. Calculate the resistance required:* (a) *in parallel to enable the instrument to read up to 1 A, and* (b) *in series to enable it to read up to 10 V.*

(a) $$\frac{\text{Current through coil}}{\text{(fig. 22.10)}} = \frac{\text{p.d. across coil}}{\text{resistance of coil}}$$

$$\therefore \quad \frac{15}{1000} = \frac{\text{p.d. (in volts) across coil}}{5}$$

so that p.d. across coil = 0·075 V.

For fig. 22.10,

   current through S = total current − current through coil
$$= 1 - 0·015 = 0·985 \text{ A}.$$

$$\text{Current through S} = \frac{\text{p.d. across S}}{\text{resistance of S}}$$

$$\therefore \quad 0·985 = \frac{0·075}{\text{resistance of S (in ohms)}}$$

and resistance of S $= \frac{0·075}{0·985} = 0·076\ 14\ \Omega$.

(b) For fig. 22.11,

$$\text{current through coil} = \frac{\text{p.d. across TT}}{\text{resistance between TT}}$$

$$\therefore \quad \frac{15}{1000} = \frac{10}{\text{resistance between TT}}$$

so that resistance between TT = 666·7 Ω.

Hence, resistance of resistor R required in series with coil

$$= \text{total resistance between TT} - \text{resistance of coil}$$
$$= 666{\cdot}7 - 5 = 661{\cdot}7 \ \Omega.$$

The moving-coil instrument can be arranged as a multi-range ammeter by making the shunt of different sections as shown in fig. 22.12, where A represents a milliammeter in series with a 'swamping' resistor r of material having negligible temperature coefficient of resistance.

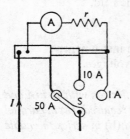

Fig. 22.12 Multi-range moving-coil ammeter.

With the selector switch S on, say, the 50-A stud, a shunt having a very low resistance is connected across the instrument, the value of its resistance being such that full-scale deflection is produced when $I = 50$ A. With S on the 10-A stud, the resistance of the two sections of the shunt is approximately five times that of the 50-A section, and full-scale deflection is obtained when $I = 10$ A. Similarly, with S on the 1-A stud, the total resistance of the three sections is such that full-scale deflection is obtained with $I = 1$ A. Such a multi-range instrument is provided with three scales so that the value of the current can be read directly.

A multi-range voltmeter is easily arranged by using a tapped resistor in series with a milliammeter A, as shown in fig. 22.13. For instance, with the data given in example 22.1, the resistance of section BC would be $661{\cdot}7 \ \Omega$ for the 10-V range. If D be the tapping for, say, 100 V, the total resistance between O and $D = 100/0{\cdot}015 = 6666{\cdot}7 \ \Omega$, so that the resistance of section CD $= 6666{\cdot}7 - 666{\cdot}7 = 6000 \ \Omega$. Similarly, if E is to be 500-V tapping, section DE must absorb 400 V at full-scale

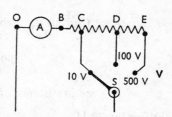

Fig. 22.13 Multi-range moving-coil voltmeter.

deflection; hence the resistance of $DE = 400/0{\cdot}015 = 26\ 667 \ \Omega$. With the aid of selector switch S, the instrument can be used on three voltage ranges, and the scales can be calibrated to enable the value of the voltage to be read directly.

Example 22.2 *The coil of a moving-coil instrument is wound with* $42\frac{1}{2}$ *turns. The mean width of the coil is 25 mm and the axial length of the magnetic field is 20 mm. If the flux density in the airgap is 0·2 T, calculate the torque, in newton metres, when the current is 15 mA.*

Since the coil has $42\frac{1}{2}$ turns, one side has 42 wires and the other side has 43 wires.

From expression (2.1), force on the side having 42 wires

$$= 0\!\cdot\!2 \text{ [T]} \times 0\!\cdot\!02 \text{ [m]} \times 0\!\cdot\!015 \text{ [A]} \times 42$$
$$= 2520 \times 10^{-6} \text{ N}$$

∴ torque on that side of coil

$$= (2520 \times 10^{-6}) \text{ [N]} \times 0\!\cdot\!0125 \text{ [m]}$$
$$= 31\!\cdot\!5 \times 10^{-6} \text{ N·m}$$

Similarly, torque on side of coil having 43 wires

$$= 31\!\cdot\!5 \times 10^{-6} \times 43/42$$
$$= 32\!\cdot\!2 \times 10^{-6} \text{ N·m}$$

∴ total torque on coil $= (31\!\cdot\!5 + 32\!\cdot\!2) \times 10^{-6}$ N·m
$$= 63\!\cdot\!7 \times 10^{-6} \text{ N·m.}$$

22.6 Moving-iron ammeters and voltmeters

Moving-iron instruments can be divided into two types:

(i) the *attraction* type, in which a sheet of soft iron is attracted towards a solenoid, and

(ii) the *repulsion* type, in which two parallel rods or strips of soft iron, magnetized inside a solenoid, are regarded as repelling each other.

These two types will now be described in greater detail.

Type (i). Fig. 22.14 shows an end elevation and a sectional front view (taken on XX) of the attracted-iron type. A soft-iron disc A is attached to a spindle S carried by jewelled centres J, and is so placed that it is attracted towards solenoid C when the latter is carrying a current.

Damping is provided by vane V attached to the spindle and moving in an air chamber, and control is provided by two spiral hairsprings SS, as shown in fig. 22.15.

Type (ii). This type is shown in fig. 22.15 where C represents the solenoid. A soft-iron rod or strip A is attached to the bobbin on which the coil is wound, and another soft-iron rod or strip B is carried by spindle D.

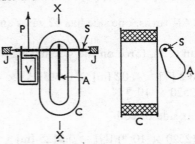

Fig. 22.14 Attraction-type moving-iron instrument.

When a current flows through coil C, A and B are magnetized in the same direction and it is usual to say that B tries to move away from A because poles of the same polarity repel each other. Such a statement, however, gives no indication as to how the deflecting force is actually produced.

In section 2.3, it was shown that when two *permanent* magnets are placed side by side, with the N poles pointing in the same direction, as in fig. 2.5, the force of repulsion between the magnets is due to lateral pressure in the magnetic field occupying the space *between* the magnets. In the case of two parallel rods, A and B, situated inside a coil C carrying a current, as in fig. 22.16,* the distribution of the magnetic flux is roughly as shown dotted. There is practically no magnetic flux in the space between A and B, and therefore there cannot be any lateral pressure in that region.

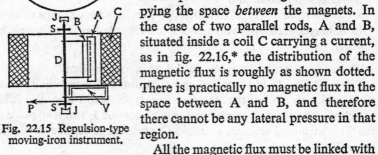

Fig. 22.15 Repulsion-type moving-iron instrument.

All the magnetic flux must be linked with the whole or part of the coil, and fig. 22.16 (*a*) shows the approximate

* The rods are shown much wider than is the case in an actual instrument. This is done to enable the dotted lines representing the flux passing through the rods to be shown clearly.

distribution of the flux passing through the rods when they are close together. It was mentioned in section 2.3 that magnetic flux behaves as if it were in tension; hence the flux passing through rod B tends to shorten its paths by pulling B towards the left, as shown in fig. 22.16 (b). (Rod A is assumed to be fixed.) In addition to this tension effect, there is also a lateral pressure in regions D and E tending to push apart the fluxes passing through A and B, thereby helping to urge rod B towards the left. These effects combine to produce a clockwise deflecting torque on the moving system.

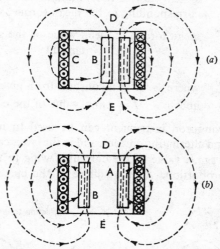

Fig. 22.16 Distribution of magnetic flux in the repulsion-type moving-iron instrument.

As a result of the shortened air paths of the flux passing through B when the latter has moved towards the left, the magnitude of the flux is increased, thereby increasing the inductance of the coil; but the implication of this effect is beyond the scope of this book.

In fig. 22.15 the controlling torque is exerted by hairsprings SS; and air damping is provided by vane V moving in an air chamber.

In commercial instruments, it is usual for the moving iron B to be in the form of a thin curved plate and for the fixed iron A to be a tapered curved sheet, as shown (without control and damping devices) in fig. 22.17. This construction can be arranged to give a longer and more uniform scale than is possible with the rods shown in fig. 22.15.

For both the attraction and the repulsion types it is found that for a given position of the moving system, the value of the deflecting torque is proportional to the square of the current, so long as the

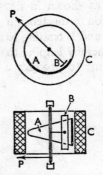

iron is working below saturation. Hence, if the current waveform is as shown in fig. 22.18, the variation of the deflecting torque is represented by the dotted wave. If the supply frequency is, say, 50 Hz, the torque varies between zero and a maximum 100 times a second, so that the moving system—owing to its inertia—takes up a position corresponding to the mean torque, where

mean torque $\propto$ mean value of the square of the current

$$= kI^2$$

where k = a constant for a given instrument

and I = r.m.s. value of the current.

Fig. 22.17 Repulsion-type moving-iron instrument.

Hence the moving-iron instrument can be used to measure both direct current and alternating current, and in the latter case the instrument gives the r.m.s. value of the current. Owing to the deflecting torque being proportional to the square of the current, the scale

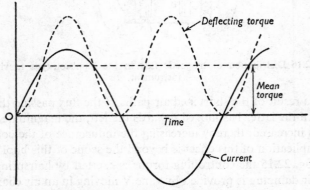

Fig. 22.18 Deflecting torque in a moving-iron instrument.

divisions are not uniform, being cramped at the beginning and open at the upper end of the scale.

Since the strength of the magnetic field, and therefore the magnitude of the deflecting torque, depend upon the number of ampere-turns on the solenoid, it is possible to arrange different instruments to have different ranges by merely winding different numbers of turns on the solenoids. For example, suppose that full-scale deflection is obtained with 400 ampere-turns, then for

full-scale reading with 100 A, no. of turns = 400/100 = 4,
and full-scale reading with 5 A, no. of turns = 400/5 = 80.

A moving-iron voltmeter is a moving-iron milliammeter connected in series with a suitable non-reactive resistor.

Example 22.2 *A moving-iron instrument requires* 250 *ampere-turns to give full-scale deflection. Calculate:* (a) *the number of turns required if the instrument is to be used as an ammeter reading up to* 50 *A, and* (b) *the number of turns and the total resistance if the instrument is to be arranged as a voltmeter reading up to* 300 *V with a current of* 20 *mA.*

(*a*) No. of turns = 250/50 = 5.
(*b*) No. of turns = 250/0·02 = 12 500.

Total resistance = 300/0·02 = 15 000 Ω.

The advantages of moving-iron instruments are:

(i) robust construction;
(ii) relatively cheap;
(iii) can be used to measure direct and alternating currents and voltages.

The disadvantages of moving-iron instruments are:

(i) Affected by stray magnetic fields. Error due to this cause is minimized by the use of a magnetic screen such as an iron casing (section 2.4).

(ii) Liable to hysteresis error when used in a d.c. circuit; i.e. for a given current, the instrument reads higher with decreasing than with increasing values of current. This error is reduced by making the iron strips of nickel–iron alloy such as Mumetal (section 3.11).

(iii) Owing to the inductance of the solenoid, the reading on moving-iron voltmeters may be appreciably affected by variation of frequency. This error is reduced by arranging for the resistance of the voltmeter to be large compared with the reactance of the solenoid.

(iv) Moving-iron voltmeters are liable to a temperature error owing to the solenoid being wound with copper wire. This error is minimized by connecting in series with the solenoid a resistor of a material, such as manganin, having a negligible temperature coefficient of resistance.

22.7 Thermocouple instruments

This type of instrument utilizes the thermoelectric effect observed by Seebeck in 1821, namely that in a closed circuit consisting of two

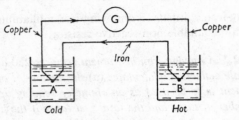

Fig. 22.19 A thermocouple.

different metals, an electric current flows when the two junctions are at different temperatures. Thus, if A and B in fig. 22.19 are junctions of copper and iron wires, each immersed in water, then if the vessel containing B is heated, it is found that an electric current flows from the iron to the copper at the cold junction and from the copper to the iron at the hot junction, as indicated by the arrowheads. A pair of metals arranged in this manner is termed a *thermocouple* and gives rise to a *thermo-e.m.f.* when the two junctions are at different temperatures.

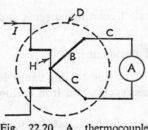

Fig. 22.20 A thermocouple ammeter.

This *thermoelectric effect* may be utilized to measure temperature. Thus, if the reading on galvanometer G be noted for different temperatures of the water in which junction B is immersed, the temperature of junction A being maintained constant, it is possible to calibrate the galvanometer in terms of the difference of temperature between A and B. The materials used in practice depend upon the temperature range to be measured; thus, copper–constantan couples are suitable for temperatures up to about 400°C and iron–constantan couples up to about 900°C, constantan being an alloy of copper and nickel. For temperatures up to about 1400°C, a couple made of platinum and platinum–iridium alloy is suitable.

A thermocouple can be used to measure the r.m.s. value of an alternating current by arranging for one of the junctions of wires of dissimilar material, B and C (fig. 22.20), to be placed near or welded to a resistor H carrying the current *I* to be measured. The current due to the thermo-e.m.f. is measured by a permanent-magnet moving-coil microammeter A. The heater and the thermocouple can be enclosed in an evacuated glass bulb D, shown dotted in fig. 22.20, to shield them from draughts. Ammeter A may be calibrated by noting its reading for various values of direct current through H, and it can then be used to measure the r.m.s. value of alternating currents of frequencies up to several megahertz.

22.8 Electrodynamic or dynamometer instruments

The action of this type of instrument depends upon the electromagnetic force exerted between fixed and moving coils carrying current. The upper diagram in fig. 22.21 shows a sectional elevation through fixed coils FF and the lower diagram represents a sectional plan on XX. The moving coil M is carried by a spindle S and the controlling torque is exerted by spiral hairsprings H, which may also serve to lead the current into and out of M.

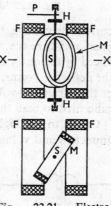

The deflecting torque is due to the interaction of the magnetic fields produced by currents in the fixed and moving coils; thus fig. 22.22 (*a*) shows the magnetic field due to current flowing through F in the direction indicated by the dots and crosses, and fig. 22.22 (*b*) shows that due to current in M. By combining these magnetic fields, it will be seen that when currents flow simultaneously

Fig. 22.21 Electrodynamic or dynamometer instrument.

through F and M, the resultant magnetic field is distorted as shown in fig. 22.22 (*c*) and the effect is to exert a clockwise torque on M.

Since M is carrying current at right-angles to the magnetic field produced by F,

deflecting force on each side of M

       ∝ (current in M)

         × (density of magnetic field due to current in F)

       ∝ current in M × current in F.

In dynamometer ammeters, the fixed and moving coils are connected in parallel, whereas in voltmeters they are in series with each other and with the usual resistor. In each case, the deflecting force is proportional to the square of the current or the voltage. Hence, when the dynamometer instrument is used to measure an alternating

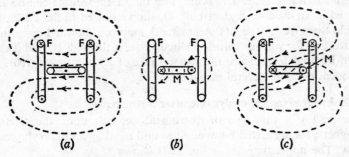

(a)   (b)   (c)

Fig. 22.22 Magnetic fields due to fixed and moving coils.

current or voltage, the moving coil—due to its inertia—takes up a position where the average deflecting torque over one cycle is balanced by the restoring torque of the spiral springs. For that position, the deflecting torque is proportional to the mean value of the square of the current or voltage, and the instrument scale can therefore be calibrated to read the r.m.s. value.

Owing to the higher cost and lower sensitivity of dynamometer ammeters and voltmeters compared with moving-iron instruments,

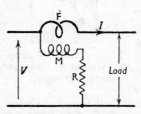

Fig. 22.23 Wattmeter connections.

the former are seldom used commercially, but *electrodynamic* or *dynamometer wattmeters* are very important because they are commonly employed for measuring the power in a.c. circuits. The fixed coils F are connected in series with the load, as shown in fig. 22.23. The moving coil M is connected in series with a non-reactive resistor R across the supply, so that the current through M is proportional to and practically in phase with the supply voltage V; hence:

instantaneous force on each side of M

$\propto$ (instantaneous current through F)

$\times$ (instantaneous current through M)

∝ (instantaneous current through load)

    × (instantaneous p.d. across load)

∝ instantaneous power taken by load

∴   average deflecting force on M

    ∝ average value of the power over a
    complete number of cycles.

When the instrument is used in an a.c. circuit, the moving coil—due to its inertia—takes up a position where the average deflecting torque over one cycle is balanced by the restoring torque of the spiral springs; hence the instrument can be calibrated to read the mean value of the power in an a.c. circuit.

22.9 Electrostatic voltmeters

In section 5.1 an experiment demonstrating the mutual attraction between positive and negative charges was described. This pheno-

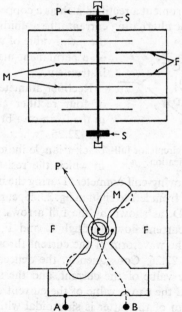

Fig. 22.24 Electrostatic voltmeter.

menon is utilized in the electrostatic voltmeter. This instrument consists of fixed metal plates F, shaped as indicated in the lower part of fig. 22.24, and very light metal vanes M attached to a spindle controlled by spiral springs S and carrying a pointer P.

The voltage to be measured is applied across terminals A and B. In section 5.27 it was shown that the force of attraction between F and M is proportional to the square of the applied voltage; hence this instrument can be used to measure either direct or alternating voltage, and when used in an a.c. circuit it reads the r.m.s. value.

The main advantages of the electrostatic voltmeter are: (*a*) it takes no current from a d.c. circuit (apart from the small initial charging current) and the current taken from an a.c. circuit is usually negligible. Hence it can be used to measure the p.d. between points in a circuit where the current taken by other types of voltmeter might considerably modify the value of that p.d. (*b*) It is particularly suitable for measuring high voltages, since the electrostatic forces are then so large that its construction can be greatly simplified.

22.10 Rectifier ammeters and voltmeters

In this type of instrument a rectifier such as a copper-oxide rectifier is used to convert the alternating current into a unidirectional current, the mean value of which is measured on a permanent-magnet moving-coil instrument.

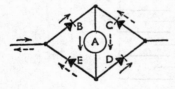

Fig. 22.25 Bridge circuit for full-wave rectification.

Rectifier ammeters usually consist of four rectifier elements arranged in the form of a bridge, as shown in fig. 22.25, where the apex of the black triangle indicates the direction in which the resistance is low, and A represents a moving-coil ammeter. During the half-cycles that the current is flowing from left to right in fig. 22.25, current flows through elements B and D, as shown by the full arrows. During the other half-cycles, the current flows through C and E, as shown by the dotted arrows. The waveform of the current through A is therefore as shown in fig. 22.26. Consequently, the deflection of A depends upon the average value of the current, and the scale of A can be calibrated to read the r.m.s. value of the current on the assumption that the waveform of the latter is sinusoidal with a form factor of 1·11.

In a rectifier voltmeter, A is a milliammeter and the bridge circuit of fig. 22.25 is connected in series with a suitable non-reactive resistor.

The main advantage of the rectifier voltmeter is that it is far more sensitive than other types of voltmeter suitable for measuring alternating voltages. Also, rectifiers can be incorporated in universal instruments, such as the Avometer, thereby enabling a moving-coil milliammeter to be used in combination with shunt and series resistors to measure various ranges of direct current and voltage, and in

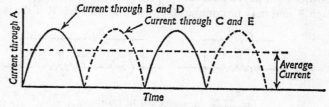

Fig. 22.26 Waveform of current through moving-coil ammeter.

combination with a bridge rectifier and suitable resistors to measure various ranges of alternating current and voltage.

If a diode or a solid-state rectifier D is connected in series with a capacitor C across an a.c. supply, as shown in fig. 22.27, the capacitor

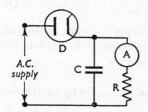

Fig. 22.27 A valve voltmeter.

is charged until its p.d. is practically equal to the peak value of the alternating voltage. The value of this voltage can be measured by connecting across the capacitor a microammeter A in series with a resistor R having a high resistance.

As explained on p. 507, capacitor C is charged during the very small fraction of each cycle that the applied voltage exceeds the p.d. across C. During the remainder of each cycle, the capacitor supplies the current flowing through A and R. The microammeter can be calibrated to read either the peak or the r.m.s. value of the voltage. In the latter case, it has to be assumed that the voltage is sinusoidal.

22.11 Measurement of resistance by the voltmeter–ammeter method

The most obvious way of measuring resistance is to measure the current through and the p.d. across the resistor and then apply Ohm's Law. This method, however, must be used with care; for instance if the instruments be connected as in fig. 22.28(*a*), and if the voltmeter V be other than the electrostatic type, the current taken by V passes through A and may be comparable with that through R if the resist-

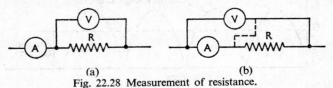

(a)   (b)

Fig. 22.28 Measurement of resistance.

ance of the latter is fairly high. If the resistance of V is known, its current can be calculated and subtracted from the reading on A to give the current through R.

A better method is to connect the voltmeter across R and A as in fig. 22.28(*b*), then:

$$\frac{\text{reading on V}}{\text{reading on A}} = \text{resistance of (R + A)}.$$

The resistance of A can be easily calculated from the p.d. across A—obtained by connecting V as shown dotted—and the corresponding current through A.

22.12 Measurement of resistance by substitution

(a) *Series method*

The unknown resistor X (fig. 22.29) is connected in series with a variable resistor R, such as a decade resistance box, which can be

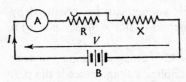

Fig. 22.29 Measurement of resistance.

varied in steps of 1 or 0·1 Ω. The total resistance of R must exceed that of X. R is adjusted to give a convenient reading on ammeter A— preferably a reading such that the pointer is exactly over a scale mark near the top end of the scale. X is then removed and R re-adjusted to give the same reading on A. The increase of R gives the resistance of X.

A modification of this arrangement is frequently used in universal instruments. A moving-coil milliammeter A is connected in series

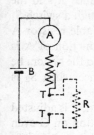

with a variable resistor r and a cell B to terminals TT, as in fig. 22.30. The procedure is to short-circuit terminals TT and adjust r to give full-scale deflection on A. The unknown resistor R is then connected across TT, and A can be calibrated to give the resistance of R directly, thereby making the instrument into an ohmmeter. Thus, suppose A to have full-scale deflection with 1 mA and B to have an e.m.f. of 1·5 V. For full-scale deflection with TT short-circuited, total resistance of r and B must be 1·5 ÷ 1/1000, namely 1500 Ω. If a 1000-Ω resistor is connected across TT, the current is 1·5 × 1000/ 2500, namely 0·6 mA; i.e. a scale deflection of 0·6 mA corresponds to a resistance of 1000 Ω, as shown in fig. 22.31.

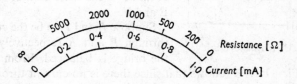

Fig. 22.31 Ammeter and ohmmeter scales.

The resistance of r in fig. 22.30 is variable to allow for variation of the e.m.f. and internal resistance of cell B. It should be mentioned, however, that if the e.m.f. of B falls appreciably below its rated value, the resistance scale will only be approximately correct.

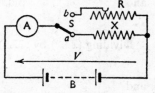

Fig. 22.32 Measurement of resistance.

(b) *Alternative circuit method*

In fig. 22.32, X is the unknown resistor, R a known variable resistor and S a two-way switch. The reading on A is noted when S is on a. The switch is then moved over to b and R adjusted to give the same reading on A. The resistance of X is obviously the same as that of R.

22.13 Measurement of resistance by the Wheatstone bridge

The four branches of the network, CDFEC, in fig. 22.33, have two known resistances P and Q, a known variable resistance R and the unknown resistance X. A battery B is connected through a switch S_1 to junctions C and F; and a galvanometer G, a variable resistor A and a switch S_2 are in series across D and E. The function of A is merely to protect G against an excessive current should the system be seriously out of balance when S_2 is closed.

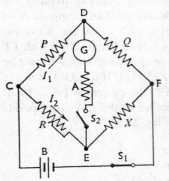

Fig. 22.33 Wheatstone bridge.

With S_1 and S_2 closed, R is adjusted until there is no deflection on G even with the resistance of A reduced to zero. Junctions D and E are then at the same potential, so that the p.d. between C and D is the same as that between C and E, and the p.d. between D and F is the same as that between E and F.

Suppose I_1 and I_2 to be the currents through P and R respectively when the bridge is balanced. From Kirchhoff's First Law it follows that since there is no current through G, the currents through Q and X are also I_1 and I_2 respectively.

But   p.d. across $P = PI_1$

and   p.d. across $R = RI_2$

$\therefore$ $$PI_1 = RI_2 \qquad (22.1)$$

Also   p.d. across $Q = QI_1$

and   p.d. across $X = XI_2$

$\therefore$ $$QI_1 = XI_2. \qquad (22.2)$$

Dividing (22.2) by (22.1), we have:

$$Q/P = X/R$$

and $$X = R \times Q/P. \qquad (22.3)$$

The resistances P and Q may take the form of the resistance of a slide-wire, in which case R may be a fixed value and balance obtained by moving a sliding contact along the wire. If the wire is homogeneous and of uniform section, the ratio of P to Q is the same as the ratio of the lengths of wire in the respective arms. A more convenient method, however, is to arrange P and Q so that each may be 10, 100 or 1000 Ω. For instance, if $P = 1000$ Ω and $Q = 10$ Ω, and if R has to be 476 Ω to give a balance, then from (20.3):

$$X = 476 \times 10/1000 = 4 \cdot 76 \ \Omega.$$

On the other hand, if P and Q had been 10 and 1000 Ω respectively, then for the same value of R:

$$X = 476 \times 1000/10 = 47 \ 600 \ \Omega.$$

Hence it is seen that with this arrangement it is possible to measure a wide range of resistance with considerable accuracy and to derive very easily and accurately the value of the resistance from that of R.

At one time, resistance boxes fitted with plugs to short-circuit the respective resistance elements were commonly employed; and in one pattern, known as the Post Office box, the ratio arms P and Q and the variable resistance R were constructed on this principle in one compact unit, complete with battery and galvanometer switches. Due partly to the trouble experienced with badly fitting plugs or plugs inadequately pressed into their sockets and partly to the labour involved in removing and replacing plugs when balancing the bridge and in the subsequent adding up of the resistances left in circuit, plugs have been superseded by rotary dial switches.

22.14 The potentiometer

One of the most useful instruments for the accurate measurement of p.d., current and resistance is the potentiometer, the principle of action being that an unknown e.m.f. or p.d. is measured by balancing it, wholly or in part, against a known difference of potential.

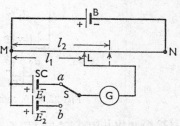

Fig. 22.34 A simple potentiometer.

In its simplest form, the potentiometer consists of a wire MN (fig. 22.34) of uniform cross-section, stretched alongside a scale and connected across a secondary cell B of ample capacity. A standard cell SC of known e.m.f. E_1, for example a cadmium cell having an e.m.f. of 1·018 59 V at 20° C (section 1.7), is connected between M and terminal a of a two-way switch S, care being taken that the corresponding terminals of B and SC are connected to M.

Slider L is then pressed momentarily against wire MN and its position adjusted until the galvanometer deflection is zero when L is making contact with MN. Let l_1 be the corresponding distance between M and L. The fall of potential over length l_1 of the wire is then the same as the e.m.f. E_1 of the standard cell.

Switch S is then moved over to b, thereby replacing the standard cell by another cell, such as a Leclanché cell, the e.m.f. E_2 of which is to be measured. Slider L is again adjusted to give zero deflection on G. If l_2 be the new distance between M and L, then:

$$E_1/E_2 = l_1/l_2$$

$$\therefore \qquad E_2 = E_1 \times l_2/l_1. \qquad (22.4)$$

22.15 A commercial form of potentiometer

The simple arrangement described in the preceding section has two disadvantages: (i) the arithmetical calculation involved in expression

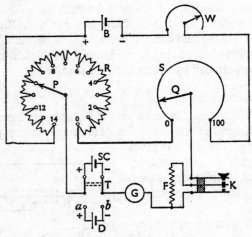

Fig. 22.35 A commercial potentiometer.

(22.4) may introduce an error, and in any case takes an appreciable time; (ii) the accuracy is limited by the length of slide-wire that is practicable and by the difficulty of ensuring exact uniformity over a considerable length.

In the commercial type of potentiometer shown in fig. 22.35 these disadvantages are practically eliminated. R consists of fourteen re-sistors in series, the resistance of each resistor being equal to that of the slide-wire S. The value of the current supplied by a lead–acid cell B is controlled by a slide-wire resistor W. A double-pole change-over switch T, closed on the upper side as in fig. 22.35, connects the stand-ard cell SC between arm P and galvanometer G. A special key K, when slightly depressed, inserts a resistor F having a high resistance in series with G; but when K is further depressed, F is short-circuited. The galvanometer is thereby protected against an excessive current should the potentiometer be appreciably out of balance when K is first depressed.

22.16 Standardization of the potentiometer

Suppose the standard cell SC to be of the cadmium type having an e.m.f. of 1·018 59 V at 20°C. Arm P is placed on stud 10 and Q on 18·59—assuming the scale alongside S to have 100 divisions. The value of W is then adjusted for zero deflection on G when K is fully depressed. The p.d. between P and Q is then exactly 1·018 59 V, so that the p.d. between two adjacent studs of R is 0·1 V and that correspond-ing to each division of S's scale is 0·001 V. Consequently, if P is moved to, say, stud 4 and Q to 78·4 on the slide-wire scale, the p.d. be-tween P and Q = $(4 \times 0·1) + (78·4 \times 0·001) = 0·4784$ V. It is there-fore a simple matter to read the p.d. directly off the potentiometer.

Since most potentiometers have fourteen steps on R, it is usually not possible to measure directly a p.d. exceeding 1·5 V.

22.17 Measurement of the e.m.f. of a cell

After the potentiometer has been standardized, switch T of fig. 22.35 is changed over to contacts *a* and *b*, across which is connected the cell D, the e.m.f. of which is required. Arms P and Q of the potentio-meter are again adjusted to give zero deflection on the galvanometer. If, at balance, P is on stud 14 and Q is on 65, then the e.m.f. of D is $(14 \times 0·1) + (65 \times 0·001)$, namely 1·465 volts.

Should the e.m.f. of the cell be greater than 1·5 V, it would be necessary to use a *volt-box* having a high resistance, as described in section 22.19.

22.18 Calibration of an ammeter by means of a potentiometer

The ammeter A to be calibrated is connected in series with a standard resistor H and a variable resistor J across a cell L of ample current capacity, as in fig. 22.36. The standard resistor H is usually provided with four terminals, namely two heavy current terminals CC and two potential terminals PP. The resistance between the potential terminals is known with a high degree of accuracy and its value must be such that with the maximum current through the ammeter, the p.d. between terminals PP does not exceed 1·5 V. For instance, suppose A to be a 10-A ammeter; then the resistance of H must not exceed 1·5/10, namely 0·15 Ω. Further, the resistance of H should

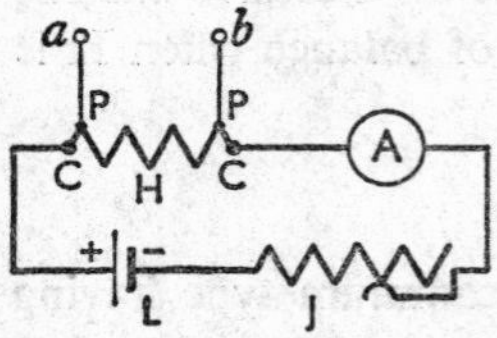

Fig. 22.36 Calibration of an ammeter.

preferably be a round figure, such as 0·1 Ω in this case, in order that the current may be quickly and accurately deduced from the potentiometer readings.

Terminals PP of the standard resistor are connected to terminals *ab* (fig. 22.35) of the potentiometer (cell D having been removed). After the potentiometer has been standardized, switch T is changed over to *ab*; and with the current adjusted to give a desired reading on the scale of ammeter A, arms P and Q are adjusted to give zero deflection on the galvanometer. For instance, suppose the current to be adjusted to give a reading of, say, 6 A on the ammeter scale, and suppose the readings on P and Q, when the potentiometer is balanced, to be 5 and 86·7 respectively, then the p.d. across terminals PP is 0·5867 V; and since the resistance between the potential terminals PP is assumed to be 0·1 Ω, the true value of the current through H is 0·5867/0·1, namely 5·867 A. Hence, the ammeter is reading high by 0·133 A.

22.19 Calibration of a voltmeter by means of a potentiometer

Suppose the voltmeter to be calibrated to have a range of 0–100 V. It is therefore necessary to use a *volt-box* to enable an accurately known fraction—not exceeding 1·5 V—to be obtained. The high-resistance volt-box *ae*, fig. 22.37, has tappings at accurately determined points. This arrangement enables voltmeters of various ranges to be calibrated; thus the 100-V voltmeter is connected across the 150-V tappings, the resistance between *ad* being 100 times that between *ab*. The 1·5-V tappings are connected to terminals *ab* of the potentiometer (fig. 22.35). Various voltages can be applied to the voltmeter by moving the slider along a resistor *f* connected across a suitable battery *g*.

Let us suppose that the voltmeter reading has been adjusted to 70 V and that the corresponding readings on P and Q to give a balance are 7 and 8·4 respectively. The p.d. across *ab* is 0·7084 V, and the true value of the p.d. across *ad* is therefore 70·84 V. Hence the voltmeter is reading low by 0·84 V.

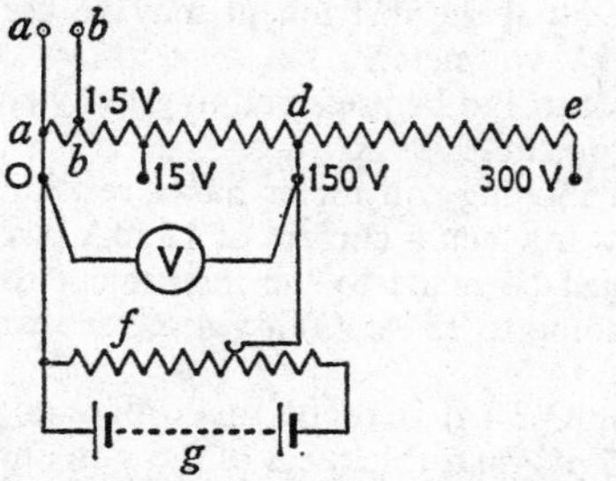

Fig. 22.37 Calibration of a voltmeter.

22.20 Measurement of resistance by means of a potentiometer

The resistor X (fig. 22.38), whose resistance is to be determined, is connected in series with a known standard resistor R, an ammeter A and a variable resistor *j* across an accumulator *h*. The purpose of A is simply to check that the value of the current is not excessive.

Connections are taken from R to, say, the upper pair of terminals of switch T in fig. 22.35 (the standard cell having been removed), and those from X are taken to the lower pair of terminals of T. With a constant current through R and X, potentiometer readings are noted, first with switch T on the upper side to measure the p.d. across R, and then with T on the lower side to measure the p.d. across X.

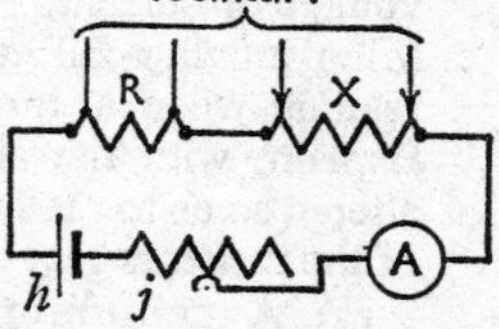

Fig. 22.38 Measurement of resistance.

If these readings were 0·648 V and 0·1242 V respectively and if the resistance of R was 0·1 Ω, then for a current *I* amperes through R and X:

$$I \times 0 \cdot 1 = 0 \cdot 648, \quad \text{and} \quad IX = 0 \cdot 1242,$$

$$\therefore \quad X/0 \cdot 1 = 0 \cdot 1242/0 \cdot 648$$

and $\qquad\qquad X = 0 \cdot 019\ 17\ \Omega.$

This method is particularly suitable for the accurate measurement of low resistances, in which case it may be necessary to use knife-edge contacts, as indicated by the arrowheads in fig. 22.38, to give the precise points between which the resistance is being determined.

EXAMPLES 22

1. Sketch and describe the construction of a moving-coil ammeter and give the principle of operation.

   A moving-coil instrument gives full-scale deflection with 15 mA and has a resistance of 5 Ω. Calculate the resistance of the necessary components in order that the instrument may be used as: (a) a 2-A ammeter; (b) a 100-V voltmeter. (App. El., L.U.)

2. Why is spring control to be preferred to gravity control in an electrical measuring instrument?

   The coil of a moving-coil meter has a resistance of 5 Ω and gives full-scale deflection when a current of 15 mA passes through it. What modification must be made to the instrument to convert it into: (a) an ammeter reading to 15 A; (b) a voltmeter reading to 15 V?
   (U.L.C.I., O.1)

3. If the shunt for Q.2 (a) is to be made of manganin strip having a resistivity of 0·5 μΩ·m, a thickness of 0·6 mm and a length of 50 mm, calculate the width of the strip.

4. Draw a diagram to show the essential parts of a modern moving-coil instrument. Label each part and state its function.

   A moving-coil milliammeter has a coil of resistance 15 Ω and full-scale deflection is given by a current of 5 mA. This instrument is to be adapted to operate: (a) as a voltmeter with a full-scale deflection of 100 V; (b) as an ammeter with a full-scale deflection of 2 A. Sketch the circuit in each case, calculate the value of any components introduced and state any precautions regarding these components. (c) Explain how the moving-coil instrument can be adapted to read alternating voltage or current. (N.C.T.E.C., O.1)

5. (a) A moving-coil galvanometer, of resistance 5 Ω, gives a full-scale reading when a current of 15 mA passes through the instrument. Explain, with the aid of circuit diagrams, how its range could be altered so as to read up to: (i) 5 A, and (ii) 150 V. Calculate the values of the resistors required.

   (b) A uniform potentiometer wire, AB, is 4 m long and has resistance 8 Ω. End A is connected to the negative terminal of a 2-V cell of negligible internal resistance, and end B is connected to the positive terminal. An ammeter of resistance 5 Ω has its negative terminal connected to A and its positive terminal to a point on the wire 3 m from A. What current will the ammeter indicate?
   (S.A.N.C., O.1)

6. A moving-coil instrument, which gives full-scale deflection with 15 mA, has a copper coil having a resistance of 1·5 Ω at 15° C, and a temperature coefficient of 1/234·5 at 0° C, in series with a swamp resistor of 3·5 Ω having a negligible temperature coefficient. Determine: (a) the resistance of shunt required for a full-scale deflection of 20 A, and (b) the resistance required for a full-scale deflection of 250 V.

   If the instrument reads correctly at 15° C, determine the percentage error in each case when the temperature is 25° C. (App. El., L.U.)

7. Describe, with the aid of sketches, the effect on a current-carrying conductor lying in and at right-angles to a magnetic field.

The coil of a moving-coil instrument is wound with $40\frac{1}{2}$ turns. The mean width of the coil is 4 cm and the axial length of the magnetic field is 5 cm. If the flux density in the gap is 0·1 T, calculate the torque in newton metres when the coil is carrying a current of 10 mA.

(U.E.I., O.1)

8. Explain, with the aid of a circuit diagram, how a d.c. voltmeter may be calibrated by means of a potentiometer method.

A moving-coil instrument, used as a voltmeter, has a coil of 150 turns with a width of 3 cm and an active length of 3 cm. The gap flux density is 0·15 T. If the full-scale reading is 150 V and the total resistance of the instrument is 100 000 Ω, find the torque exerted by the control springs at full scale. (App. El., L.U.)

9. A rectangular moving coil of a milliammeter is wound with $30\frac{1}{2}$ turns. The effective axial length of the magnetic field is 20 mm and the effective radius of the coil is 8 mm. The flux density in the gap is 0·12 T and the controlling torque of the hairsprings is $0·5 \times 10^{-6}$ newton metre per degree of deflection. Calculate the current to give a deflection of 60°.

10. A moving-iron ammeter is wound with 40 turns and gives full-scale deflection with 5 A. How many turns would be required on the same bobbin to give full-scale deflection with 20 A?

11. Describe with the aid of a diagram the construction of a repulsion-type moving-iron instrument with particular reference to the means used for: (a) deflection; (b) control; (c) damping.

A moving-iron voltmeter, in which full-scale deflection is given by 100 V, has a coil of 10 000 turns and a resistance of 2000 Ω. Calculate the number of turns required on the coil if the instrument is converted for use as an ammeter reading 20 A full-scale deflection. (U.E.I., O.1)

12. Explain the principle of operation of *one* type of moving-iron instrument, showing how it is suitable for use on d.c. and a.c. systems.

The total resistance of a moving-iron voltmeter is 1000 Ω and the coil has an inductance of 0·765 H. The instrument is calibrated with a full-scale deflection on 50 V, d.c. Calculate the percentage error when the instrument is used on (a) 25-Hz supply, (b) 250-Hz supply, the applied p.d. being 50 V in each case. (U.E.I., O.2)

13. If a rectifier-type voltmeter has been calibrated to read the r.m.s. value of a sinusoidal voltage, by what factor must the scale readings be multiplied when it is used to measure the r.m.s. value of: (a) a square-wave voltage; (b) a voltage having a form factor of 1·15?

14. A permanent-magnet moving-coil milliammeter, having a resistance of 15 Ω and giving full-scale deflection with 5 mA, is to be used with bridge-connected rectifiers (fig. 22.25) and a series resistor to measure sinusoidal alternating voltages. Assuming the 'forward' resistance of the rectifier units to be negligible and the 'reverse' resistance to be infinite, calculate the resistance of the series resistor if the instrument is to give full-scale deflection with 10 V (r.m.s.).

15. Give a summary of four different types of voltmeters commonly used in practice. State whether they can be used on a.c. or d.c. circuits. In *one* case, give a sketch showing the construction, with the method of control and damping employed.

    A d.c. voltmeter has a resistance of 28 600 Ω. When connected in series with an external resistor across a 480-V d.c. supply, the instrument reads 220 V. What is the value of the external resistance?

    (U.E.I.; O.1)

16. The resistance of a coil is measured by the ammeter–voltmeter method. With the voltmeter connected across the coil, the readings on the ammeter and voltmeter are 0·4 A and 3·2 V respectively. The resistance of the voltmeter is 500 Ω. Calculate (*a*) the true value of the resistance and (*b*) the percentage error in the value of the resistance if the voltmeter current were neglected.

17. A voltmeter is connected across a circuit consisting of a milliammeter in series with an unknown resistor R. If the readings on the instruments are 0·8 V and 12 mA respectively and if the resistance of the milliammeter is 6 Ω, calculate: (*a*) the true resistance of R, and (*b*) the percentage error had the resistance of the milliammeter been neglected.

18. A milliammeter, giving full-scale deflection with 2 mA, is used in the ohmmeter circuit of fig. 22.30. Battery B has an e.m.f. of 1·45 V. Resistance *r* is adjusted to give full-scale deflection when terminals TT are short-circuited. When an unknown resistor R is connected across TT, the milliammeter reads 0·8 mA. Calculate the resistance of R.

19. A high resistance was measured by the parallel substitution method (fig. 22.32). The resistance of R was 0·1 MΩ and that of galvanometer A was 1 kΩ. The galvanometer deflections were: (*a*) with resistor R, 65 divisions; (*b*) with the unknown resistor X, 28 divisions. Calculate the resistance of X.

20. Describe the principle of the Wheatstone bridge and derive the formula for balance conditions.

    The ratio arms of a Wheatstone bridge are 1000 and 100 Ω respectively. An unknown resistor, believed to have a resistance near 800 Ω, is to be measured, using a resistor adjustable between 60 and 100 Ω. Sketch an appropriate circuit.

    If the bridge is balanced when the adjustable resistor is set to 77·6 Ω, calculate the value of the unknown resistor.

    State, with reasons, the direction in which the current in the detector branch would flow when the bridge is slightly off balance due to the adjustable resistor being set at too *low* a value. (S.A.N.C., O.1)

21. Describe with the aid of a circuit diagram the principle of the Wheatstone bridge, and hence deduce the balance condition giving the unknown in terms of known values of resistance.

    In a Wheatstone bridge ABCD, a galvanometer is connected between B and D, and a battery of e.m.f. 10 V and internal resistance 2 Ω is connected between A and C. A resistor of unknown value is connected between A and B. When the bridge is balanced, the resistance between

B and C is 100 Ω, that between C and D is 10 Ω and that between D and A is 500 Ω. Calculate the value of the unknown resistance and the total current supplied by the battery. (U.L.C.I., O.1)

22. Describe fully, with the aid of a circuit diagram, the Murray-loop test for the location of an earth fault on one core of a two-core cable, and derive an expression for the distance of the fault from the test end of the cable. How does a high fault resistance affect the practical application of the test?

A test was carried out on a 1000-m length of twin cable on which an earth fault had occurred and the balance point on a one-metre slide wire was found to be 45 cm from the end connected to the faulty core. Determine the distance of the fault from the test end of the cable. (W.J.E.C., O.2)

23. The arms of a Wheatstone bridge have the following resistances: AB, 10 Ω; BC, 20 Ω; CD, 30 Ω;. and DA, 10 Ω. A 40-Ω galvanometer is connected between B and D and a 2-V cell, of negligible internal resistance, is connected across A and C, its positive end being connected to A. Calculate the current through the galvanometer and state its direction.

24. Describe, with the aid of a circuit diagram, how a simple potentiometer can be used to check the calibration of a d.c. ammeter..

The current through an ammeter connected in series with a standard resistor of 0·1 Ω was adjusted to 8 A. A standard cell, of e.m.f. 1·018 V, gives a balance at 78 cm, while the potential difference across the standard resistor gives a balance at 60 cm when measured with a simple potentiometer. Calculate the percentage error of the ammeter. (U.E.I., O.1)

25. Explain, with the aid of appropriate circuit diagrams, the use of a direct-reading d.c. potentiometer to calibrate a voltmeter with a full-scale reading of 250 V.

During such a calibration, the voltmeter is connected.to a volt-box with a ratio of 100/1, and its reading is adjusted to 120 V. The potentiometer is balanced with stud and slide-wire settings of 11 and 85·4 respectively. Calculate the error in the voltmeter reading and state whether the instrument reads high or low at this point. (U.L.C.I., O.2)

26. Describe with the aid of a circuit diagram how a simple potentiometer can be used to measure the e.m.f. of a cell.

A voltmeter, having a resistance of 100 Ω, registered 1·47 V when connected across the terminals of a dry cell. The simple potentiometer was used to measure the e.m.f. of the cell, and balance was obtained at 73 cm with the dry cell in circuit, and at 50 cm with a standard cell of 1·018 V. Calculate: (*a*) the e.m.f. of the dry cell, and (*b*) the internal resistance of the dry cell. (U.E.I., O.1)

27. Explain the theory of the simple slide-wire potentiometer.

A simple slide-wire potentiometer is used to check the calibration of a 5-A full-scale-deflection ammeter by measuring the voltage drop across a 0·1-Ω standard resistor. When the current is adjusted so that the ammeter reads 5 A, balance is obtained with a slide-wire setting of 24 cm. If the calibration of the potentiometer is 0·02 V/cm, what is the percentage error of the ammeter at this reading?

(E.M.E.U., O.1)

28. A simple slide-wire potentiometer and a potential divider (volt-box), of ratio 50/1, are used to check the calibration of a 50-V full-scale-deflection voltmeter. When the voltmeter reads 50 V, balance is obtained with a slide-wire setting of 49 cm. If the calibration of the potentiometer is 0·02 V/cm, what is the percentage error of the volt-meter at this reading? Draw a diagram of essential connections.

(E.M.E.U., O.1)

29. The current through an ammeter connected in series with a standard shunt of 0·01 Ω, as in fig. 22.36, is adjusted to 40 A; and the readings on the studs and the slide-wire of the potentiometer, when balanced, are 4 and 12·3 respectively. Calculate the percentage error of the ammeter and state whether the instrument is reading high or low.

30. The reading on a voltmeter connected across the 0/300-V range of a volt-box (fig. 22.37) is adjusted to 250 V, and the readings on the studs and the slide-wire of the potentiometer, when balanced, are 12 and 34·7 respectively. Calculate the percentage error of the voltmeter and state whether the instrument is reading high or low.

31. State Kirchhoff's Laws for d.c. circuits.

A 2·2-V cell is connected through a rheostat R to a slide-wire XY of resistance 50 Ω and length 100 cm. A standard cell having an e.m.f. of 1·018 V, and an 85-Ω galvanometer in series with it, are connected between points X and Z, where Z is the tapping point on the slide-wire. This potentiometer is standardized by making XZ = 50·9 cm and adjusting R until the galvanometer shows no deflection. A voltage source, of negligible internal resistance, is then substituted for the standard cell, and balance is restored by changing XZ to 72 cm. Determine the galvanometer current if XZ is now reduced to 70 cm.

(App. El., L.U.)

CHAPTER 23

Electrolysis: Primary and Secondary Cells

23.1 A simple voltaic cell

It was in 1789 that a chance observation by Luigi Galvani* (1737–98) led to the idea of generating an electric current from a source of chemical energy. Galvani, a professor of anatomy at Bologna, noticed that recently-skinned frogs' legs, hung by copper wire to an iron balcony, were convulsed whenever they touched the iron. Another Italian, Alessandro Volta* (1745–1827), a professor of physics at Pavia, subsequently showed that if a rod of two dissimilar metals, such as copper and iron, was placed so that one end was in contact with a nerve on a frog's leg and the other end in contact with a muscle on the foot, muscular contraction took place. Following his investigations of this phenomenon, Volta, in 1799, constructed a simple battery—known as Volta's pile—by assembling discs of zinc, cloth soaked with brine, and copper, piled upon one another in that order. By this means, a large number of cells were obtained in series, giving a high electromotive force, but having the disadvantage of high internal resistance.

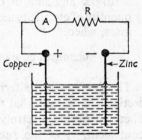

Fig. 23.1 A simple voltaic cell.

It was a comparatively short step to replace the wet cloth of Volta's pile by an electrolyte to give the simple voltaic cell shown in fig. 23.1. In this cell, plates of copper and zinc are immersed in dilute sulphuric acid contained in a glass vessel.

When a resistor R and an ammeter A are connected in series across the terminals, it is found that current flows through R from

* Galvani's name is perpetuated in such terms as 'galvanize' and 'galvano-meter', and the unit of e.m.f. and of potential difference, the 'volt', has been named after Volta.

the copper electrode to the zinc electrode and that the difference of potential between the plates is about 0·8 V, the copper plate being the positive electrode.

This type of cell is referred to as a *primary cell*, since it can only transform chemical energy into electrical energy, and the cell can be replenished only by renewal of the active materials. In a *secondary cell*, or *accumulator*, the electrode materials can be reactivated, i.e. the chemical action is reversible. This means that chemical energy is converted into electrical energy when the cell is discharging, and electrical energy is converted into chemical energy when the cell is being charged.

23.2 Electrolytic dissociation

Before we consider the various types of cells, it will be helpful to discuss what happens when an electric current flows through an electrolyte. It was in 1832 that Michael Faraday enunciated two laws relating to electrolysis, namely:

(i) the amount of chemical change produced by an electric current is proportional to the quantity of electricity, and

(ii) the amounts of different substances liberated by a given quantity of electricity are proportional to their chemical equivalent mass, where

$$\text{chemical equivalent mass} = \frac{\text{relative atomic mass}}{\text{valency}}$$

The first law can alternatively be expressed as already stated in section 1.6, namely that the mass m, in milligrams, of a substance liberated during electrolysis is the product of the electrochemical equivalent, z, of the substance, in milligrams/coulomb, and the quantity of electricity Q, in coulombs, through the electrolyte, i.e.

$$m = zQ = zIt \text{ milligrams.}$$

The above laws of Faraday give the *facts* of electrolysis; now we will consider the *mechanism* of electrolysis.

A molecule of, say, common salt consists of one atom of sodium and one atom of chlorine. When these atoms combine to form a molecule, an electron is attracted from the sodium atom to join the chlorine atom. Consequently the chlorine atom has a surplus *negative* charge and is therefore termed a *negative ion*. The sodium atom, on the other hand, is left with an excess positive charge and is therefore termed a *positive ion*. It is thought that the stability of the molecule is largely due to the electrostatic attraction between the positive sodium ion and the negative chlorine ion.

When common salt is dissolved in water, the sodium and chlorine ions of many of the molecules become separated—a phenomenon known as *electrolytic dissociation*. The separated ions are free to wander at random in the electrolyte and to recombine with other oppositely-charged free ions. In other words, dissociation goes on continuously, some molecules breaking up and oppositely-charged free ions recombining to form new molecules, thus maintaining the number of free ions constant.

The effect of dissociation in a solution of common salt in water can be represented thus:

$$NaCl \rightleftharpoons Na^+ + Cl^-.$$

The arrow pointing towards the right indicates dissociation and that pointing towards the left indicates that sodium and chlorine ions can recombine to form neutral molecules of sodium chloride.

The electrical charge on an ion is the same as the chemical valency* of the atom. For instance, the valency of both sodium and chlorine is one; hence an ion of sodium is an atom which has lost *one* electron and is represented by Na^+, whereas an ion of chlorine is an atom which has gained *one* electron and is represented by Cl^-. On the other hand, if we consider copper sulphate ($CuSO_4$), the valency of copper is two; hence an ion of copper is an atom which has lost *two* electrons and is represented by Cu^{++}.

Let us next consider a dilute solution of sulphuric acid (H_2SO_4) in water. In this case, the dissociation of one molecule of the sulphuric acid produces two positive ions of hydrogen, each having a surplus positive charge equal in *magnitude* to that of an electron, and one negative sulphate (SO_4) ion, carrying a surplus of two electrons. This dissociation can therefore be represented thus:

$$H_2SO_4 \rightleftharpoons H^+ + H^+ + SO_4^{--}.$$

Lastly, let us consider a solution of copper sulphate ($CuSO_4$) in water. As mentioned above, the copper ion has lost two electrons so that its surplus positive charge is equal in magnitude to that on two

---

* The valency of an element or a group of elements is the number of hydrogen atoms which will combine with or replace one atom of that element or group of elements.

electrons. The sulphate (SO_4) ion has, as stated in the preceding paragraph, a surplus of two electrons; hence:

$$CuSO_4 \rightleftharpoons Cu^{++} + SO_4^{--}.$$

23.3 Electrolytic cell with carbon electrodes

Fig. 23.2 shows two carbon plates, C and D, immersed in dilute sulphuric acid. Carbon is used partly because it is not affected by the products of electrolysis and partly because carbon electrodes can easily be obtained from disused Leclanché cells. The electrodes are connected to a battery B through an ammeter A, a resistor R and a switch S, and a moving-coil voltmeter is connected across the electrodes.

With switch S closed, current flows through the circuit, including

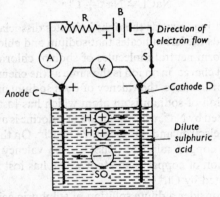

Fig. 23.2 Electrolytic cell with carbon electrodes.

the electrolyte. In the latter, the positive hydrogen ions are attracted towards the negative electrode, or *cathode*, D. When they reach this electrode, each ion absorbs an electron from the electrode to form a neutral atom of hydrogen. But neutral atoms of hydrogen cannot continue to exist independently, and pairs of such atoms combine to form molecules (H_2) of hydrogen. These molecules appear as a gas, some of which adheres to electrode D as a gaseous layer and the remainder rises to the surface in the form of tiny bubbles.

The negative sulphate (SO_4) ions are attracted towards the positive electrode, or *anode*, C. The ions which reach C give up their surplus electrons to that electrode; but SO_4 cannot exist uncharged and therefore attacks a water molecule to re-form a molecule of sulphuric acid and release an atom of oxygen. The latter combines with another oxygen atom to form an oxygen molecule (O_2). Thus we have:

$$2SO_4 + 2H_2O = 2H_2SO_4 + O_2.$$

The net result of electrolysis in this type of cell is to decompose water into its constituents, hydrogen and oxygen.

It will be seen that in the electrolyte, the positive and negative ions, by migrating towards the negative and positive electrodes respectively, act as charge carriers; but in the external circuit, the current is due to the movement of free electrons from the positive electrode, via resistor R and battery B, to the negative electrode.

When switch S is closed, it is found that the current instantly rises to its maximum value and then falls off, at first fairly rapidly. But when the switch is opened after the current has been flowing for several minutes, it is found that the voltmeter reads about 1·7 V, electrode C being positive relative to D. This p.d. falls to zero as the gases rise to the surface or are absorbed by the electrolyte.

The presence of this p.d. is due to the gaseous layers on the electrodes harbouring hydrogen and sulphate ions which have not made contact with their respective electrodes and therefore still retain their charges. These accumulated charges repel other hydrogen and sulphate ions that are approaching the cathode and anode respectively, thus giving rise to an e.m.f. that is in opposition to that of battery B. This phenomenon is known as *polarization*; and in order to send current through the cell, the p.d. applied across the electrodes has to neutralize this back e.m.f. due to polarization in addition to providing the *IR* drop due to the resistance of the electrolyte. The product of the current and the back e.m.f. is the power required to maintain the corresponding rate of electrolysis.

23.4 Electrolytic cell with copper electrodes

Let us next consider the case of two copper plates immersed in a solution of copper sulphate ($CuSO_4$) in water, the electrical circuit being as shown in fig. 23.3. As already explained in section 23.2, the effect of dissociation in a copper sulphate solution can be represented thus:

$$CuSO_4 \rightleftharpoons Cu^{++} + SO_4^{--}.$$

With switch S closed, the positive copper ions are attracted towards cathode D. On reaching this electrode, each ion absorbs two electrons from the electrode to form neutral atoms of copper, and these are deposited on the surface of the copper plate.

The negative sulphate ions are simultaneously attracted towards anode C, and each ion, on reaching the latter, gives up two electrons to that electrode, thereby compensating for the two electrons taken by each copper ion from the cathode. The neutral sulphate SO_4 cannot exist uncharged, and therefore combines with an atom of copper from anode C to form a molecule of copper sulphate which goes into

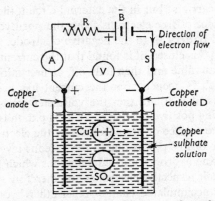

Fig. 23.3 Electrolytic cell with copper electrodes.

solution. Thus the density of the electrolyte remains constant, and the net effect is to transfer copper from the anode to the cathode.

No gas is released at either electrode. Consequently there is no polarization and the reading on voltmeter V is zero when switch S is opened. This experiment shows that one method of eliminating polarization is to substitute a harmless ion in place of the harmful hydrogen ion, i.e. to arrange for the neutralized ion to be of the same basic metal as the positive plate. Thus in the cadmium standard cell (section 1.7), the current in the region adjacent to the positive mercury electrode is carried by positive mercury ions and negative sulphate (SO_4) ions, thereby eliminating polarization. The same principle is applied in the mercury cell described in section 23.7.

23.5 Action of the simple voltaic cell

Fig. 23.4 shows a resistor R and a switch S in series across copper and zinc electrodes immersed in dilute sulphuric acid, with a voltmeter connected across the electrodes. Since the copper electrode is positive relative to the zinc electrode, current flows through R in the direction shown by the arrow. This is the conventional direction. Actually, the current in this external circuit is a flow of electrons in the opposite direction.

With switch S open, the p.d. between the two electrodes is about 0·8 V. Immediately S is closed, there is a displacement of electrons in the external circuit, with the result that some electrons are withdrawn from the zinc electrode and the same number of electrons move on to

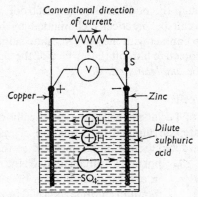

Fig. 23.4 A simple voltaic cell.

the copper electrode. This leaves the zinc plate with a surplus of positive charge and the copper plate with a surplus of negative charge. Consequently the negative sulphate ions in the electrolyte are attracted towards the zinc and the positive hydrogen ions are attracted towards the copper, as indicated in fig. 23.4.

The sulphate ions, after giving up their surplus electrons to the zinc electrode, combine with the zinc to form zinc sulphate ($ZnSO_4$), which goes into solution. The hydrogen ions, on the other hand, after absorbing electrons from the copper plate, form a gaseous layer on the surface of the copper. This gaseous layer has two disadvantages:

(*a*) it gives rise to polarization, i.e. it sets up a back e.m.f.;

(*b*) it acts as a shield reducing the active area of the electrode and thus increases the internal resistance of the cell.

The result is that the terminal voltage of the cell falls very rapidly, and this type of cell is therefore suitable only for intermittent use. Polarization can be reduced by introducing a depolarizing agent to combine with the hydrogen, e.g. manganese dioxide in the Leclanché cell (section 23.6), or it can be eliminated by arranging for the neutralized ions to be of the same basic metal as the positive electrode, e.g. the mercury cell (section 23.7).

Local action. If the negative electrode of the simple voltaic cell is made of commercial zinc, it is found to dissolve rapidly. This is due to impurities, such as lead, giving rise to local action; i.e. the zinc and the impurities form the electrodes of tiny cells, and localized currents flow through the electrolyte between these electrodes and complete their paths through the body of the zinc electrode.

Local action can be practically eliminated by bringing the zinc plate or rod into contact with mercury in dilute sulphuric acid, thereby forming a protective film on the surface of the electrode. The zinc is then said to be *amalgamated*.

23.6 Leclanché cell

The 'wet' type of Leclanché cell (now practically obsolete) consists of a carbon plate, surrounded by a mixture of manganese dioxide (MnO_2) and powdered carbon, in an unglazed earthenware pot. This

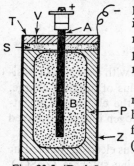

Fig. 23.5 'Dry' Leclanché cell.

pot and an amalgamated zinc rod are immersed in a saturated solution of salammoniac (ammonium chloride, NH_4Cl) in water. The carbon plate and the zinc rod form the positive and negative electrodes respectively.

The function of the manganese dioxide is to reduce polarization by combining with the hydrogen released at the carbon plate to form water and a brown oxide of manganese (Mn_2O_3), thus:

$$H_2 + 2MnO_2 = Mn_2O_3 + H_2O.$$

In the 'dry' type of Leclanché cell, the same ingredients are present, and fig. 23.5 is a sectional view of the construction most commonly used. A carbon rod A is surrounded by a black depolarizing paste B, consisting of manganese dioxide, powdered carbon, salammoniac, zinc chloride and water, the paste being usually contained in a bag of coarse linen. Around this depolarizer is a mixture P of flour, plaster of Paris, salammoniac and zinc chloride, with water added to form a white paste. The latter need only be thick enough to prevent the black paste B touching the zinc container Z. Above the depolarizer and the white paste is a layer S of sawdust or similar porous material, the cell being sealed with a layer of pitch T in which there is a vent tube V. The zinc container Z is usually covered with a cardboard case.

A Leclanché cell has an e.m.f. of about 1·5 V when new, but this e.m.f. falls fairly rapidly if the cell is in continuous use. This fall is due to polarization—the hydrogen film at the carbon electrode forms faster than can be dissipated by the depolarizer. However, if the cell is disconnected from the external circuit, depolarization continues and the e.m.f. recovers its normal value. Hence the Leclanché cell is suitable only for intermittent use, e.g. for electric torches, radio receivers, etc.

The life of a dry Leclanché cell is reduced by local action and even the shelf-life is limited to about two years owing to the local action that goes on continuously in the cell. Also, this type of cell does not lend itself to miniaturization, since the number of ampere hours obtainable from the cell falls off rapidly as the size is reduced.

23.7 Mercury cell*
This type of cell was developed to meet the requirements of miniaturization, e.g. for guided missiles, medical electronics, hearing aids, etc.,

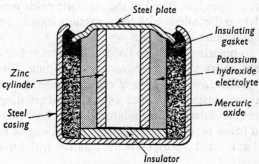

Fig. 23.6 A mercury cell.

where it is necessary to reduce the size of the cell but at the same time obtain: (*a*) a high ratio of output energy/mass; (*b*) a constant e.m.f. over a relatively long period; and (*c*) a long shelf life, i.e. absence of local action.

A cross-section of the basic type of mercury cell is shown in fig.

* The author is indebted to Mallory Batteries, Ltd, for information concerning this cell.

23.6. The negative electrode is zinc, either as a foil or as powder compressed into a hollow cylinder. This electrode is surrounded by a layer of electrolyte consisting of a concentrated aqueous solution of potassium hydroxide (KOH) and zinc oxide (ZnO). Surrounding the electrolyte is a layer of mercuric oxide (HgO). This oxide contains a small percentage of finely powdered graphite to reduce the internal resistance of the cell.

The above constituents are assembled in a nickel-plated or stainless-steel container which forms the positive electrode. The zinc cylinder and the electrolyte are supported on a disc of insulating material; and the cell is sealed by an insulating gasket between the container and a nickel-plated steel plate resting on top of the zinc cylinder.

When the cell is connected to an external circuit, the positive potassium ions (K^+) in the electrolyte move towards the mercuric oxide while the negative hydroxide ions, (OH)$^-$, move towards the negative electrode and give up their surplus electrons to the zinc, after which they combine with the zinc to form zinc oxide and water.

Simultaneously, in the mercuric oxide, the positive mercury ions (Hg^{++}) move towards the positive electrode and their charges are neutralized by the electrons absorbed from that electrode, after which the mercury is deposited on the inside of the casing. At the same time, the negative oxygen ions (O^{--}) of the mercuric oxide move towards the electrolyte and combine with the potassium ions and some water to re-form potassium hydroxide.

There is no polarization, since no gases are evolved at either electrode, except under abnormal operating conditions. Owing to the absence of polarization, this type of cell maintains its terminal voltage practically constant at about 1·2–1·3 V (depending upon the value of the load current) for a relatively long time. Also, owing to local action being practically negligible, the cell can be stored for a long period at normal atmospheric temperature without appreciable loss of capacity.

23.8 Secondary cells

Whenever a battery is required to supply a relatively large amount of power, secondary cells must be used. Such batteries may supply power in telephone exchanges, emergency lighting for hospitals, etc. They also form portable sources of power for starting, ignition and lighting of motor vehicles and as motive power for electrically propelled vehicles.

Secondary cells may be divided into two types: (*a*) the lead–acid cell, in which lead plates covered with compounds of lead are immersed in a dilute solution of sulphuric acid in water; (*b*) the nickel–cadmium and the nickel–iron alkaline cells.

These cells will now be considered in greater detail.

23.9 Lead–acid cell

The plates used in this type of cell may be grouped thus:

(*a*) *Formed* or *Planté* plates, namely those formed from lead plates by charging, discharging, charging in the reverse direction, etc., a number of times, the forming process being accelerated by the use of suitable chemicals. The main difficulty with this type of construction is to secure as large a working surface as possible for a given mass of plate. One method of increasing the surface area is to make the plates with deep corrugations, as in fig. 23.7, with reinforcing ribs at intervals.

(*b*) *Pasted* or *Faure* plates, namely those in which a paste of the active material is either pressed into recesses in a lead–antimony grid or held between two finely perforated lead sheets cast with ribs and flanges so that the two sheets, when riveted together, form in effect a number of boxes which hold the paste securely

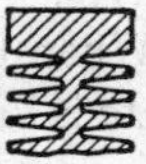

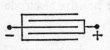

Fig. 23.7 Section of a Planté plate.

in position. The paste is usually sulphuric acid mixed with red lead (Pb_3O_4) for the positives and with litharge (PbO) for the negatives. A small percentage of a material such as powdered pumice is added to increase the porosity of the paste. For a given ampere hour capacity (section 23.11), the weight of a pasted plate is only about a third of that of a formed plate.

Students are advised to examine specimen plates or plates taken from disused accumulators.

When weight is of no importance it is common practice to make the positive plates of the 'formed' type and the negative plates of the 'pasted' type. The active material on the positive plates expands when it is subjected to chemical changes; consequently it is found that the greater mechanical stiffness of the 'formed' construction is an important

Fig. 23.8 Arrangement of plates.

advantage in reducing the tendency of the plates to buckle. This tendency to buckle is reduced still further by constructing the cell with an odd number of plates, as shown in fig. 23.8, the outer plates being always negative. This arrangement enables both sides of each positive plate to be actively employed, and the tendency of one side of a plate to expand and cause buckling is neutralized by a similar tendency on the other side.

The plates are assembled in glass, polystyrene, vulcanized-rubber or resin-rubber containers and separated by a special grade of paper or by microporous sheets of a plastic material.

The most suitable relative density of the acid depends upon the type of cell and the state of charge of the cell. An average value, however, is about 1·21.

23.10 Chemical reactions in a lead–acid cell

The chemical reactions taking place during charge and discharge are complicated, and all we can do here is to indicate the most important reactions and to account for the variation in the density of the electrolyte.

When the cell is fully charged, the active material on the positive plate is lead dioxide (PbO_2) and that on the negative plate is spongy or porous lead (Pb). During discharge, the positive hydrogen ions (H^+) travel towards the positive electrode (PbO_2), and after their charges have been neutralized by the absorption of electrons from this electrode, the hydrogen molecules combine with lead dioxide and sulphuric acid to form lead sulphate and water, thus:

$$H_2 + PbO_2 + H_2SO_4 = PbSO_4 + 2H_2O.$$

The negative sulphate ions (SO_4^{--}) travel towards the negative electrode (Pb) and after giving up their surplus electrons, they combine with lead to form lead sulphate, thus:

$$Pb + SO_4 = PbSO_4.$$

During charge, the chemical reactions are reversed and the active material is converted back to lead dioxide on the positive electrode and to spongy lead on the negative electrode.

The above reactions may be summarized thus:

	Positive plate	Electrolyte	Negative plate	
Dis-charge	Lead dioxide (PbO_2)	Sulphuric acid ($2H_2SO_4$)	Lead (Pb)	Charge
	Lead sulphate ($PbSO_4$)	Water ($2H_2O$)	Lead sulphate ($PbSO_4$)	

It will be seen that for every two molecules of sulphuric acid decomposed during discharge, two molecules of water are formed; hence the density of the electrolyte falls as the cell discharges. The reverse process occurs during charging, so that when the cell is fully charged the density of the electrolyte is restored to its initial value.

23.11 Characteristics of a lead–acid cell

When a cell is discharged, its terminal voltage falls at a rate that depends upon the discharge current; and the normal capacity of the cell is taken as the number of ampere hours it can give on a 10-hour discharge at a constant current before its p.d. falls to about 1·85 V. Such a condition is represented by curve A in fig. 23.9. Curve B

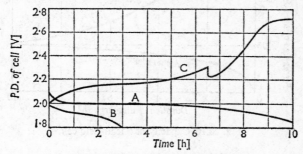

Fig. 23.9 Characteristics of a lead–acid cell.

represents the variation of p.d. when the cell is discharged at about 2·5 times the normal rate. The higher the rate of discharge, the smaller is the number of ampere hours obtainable from the cell before its p.d. falls below the permissible value—a value that depends upon the discharge rate. The lower curve in fig. 23.10 represents the capacity in ampere hours obtainable at different discharge rates, and the upper curve represents the corresponding rates, the values for both curves being expressed as percentages of the 10-hour values.

The reduction in the capacity at high rates of discharge is due to the chemical reactions being at first confined to the outer layers of the active material. Consequently the acid in the pores of these outer layers is used up before fresh acid can take its place. Owing to the relatively high resistance of very weak acid, the voltage falls rapidly, though there may be plenty of unchanged active material still left in the inner layers. If a cell is discharged at a very high rate until the terminal voltage falls to the minimum permissible value and then left on open circuit, the weakened acid becomes strengthened by diffusion of the electrolyte and a further discharge can be obtained from the cell.

The first portion of curve C in fig. 23.9 represents the variation of p.d. during charge at normal rate, but the actual value of the p.d. varies appreciably for different types of cells. It is seen that after about 5 hours the p.d. increases more rapidly, and in order to prevent the temperature exceeding about 40° C, it is necessary to reduce

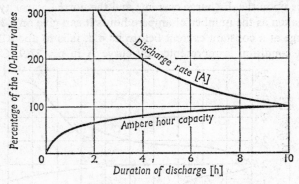

Fig. 23.10 Capacity of a lead–acid cell.

the charging rate; thus the second portion of curve C represents the variation of p.d. with the charging current reduced to half the normal rate.

The ampere hour efficiency of an accumulator is the ratio of the number of ampere hours obtainable during discharge to that required to restore it to its original condition. The value for a lead–acid cell is about 90 per cent. The watt hour efficiency takes the voltage variation into account, and its value is about 75 per cent.

23.12 Alkaline cells

In both the nickel–iron and the nickel–cadmium types, the positive plates are made of nickel hydroxide enclosed in finely perforated steel tubes or pockets, the electrical resistance being reduced by the addition of flakes of pure nickel or graphite. These tubes or pockets are assembled in nickelled-steel plates. In the nickel–iron cell the negative plate is made of iron oxide with a little mercuric oxide to reduce the resistance, the mixture being enclosed in perforated steel pockets, also assembled in nickelled-steel plates. In the nickel–cadmium cell the active material is cadmium mixed with a little iron, the purpose of the latter being to prevent the active material caking and losing its porosity.

In both types of cell, the electrolyte is a solution of potassium hydroxide (KOH) having a relative density of about 1·15–1·2, depending upon the type of cell and the conditions of service. The electrolyte does not undergo any chemical change; consequently the quantity of electrolyte can be reduced to the minimum necessitated by adequate clearance between the plates.

The plates are separated by insulating rods and assembled in sheet-steel containers, the latter being mounted in non-metallic crates to insulate the cells from one another.

23.13 Chemical reactions in an alkaline cell

When the nickel–cadmium cell is in a charged condition, the active material on the positive plates appears to be a hydroxide of nickel having the chemical formula $Ni(OH)_3$ and that on the negative is pure cadmium. During discharge, the $Ni(OH)_3$ is converted into the lower hydroxide $Ni(OH)_2$ and the cadmium is converted into cadmium hydroxide $Cd(OH)_2$. During charge, the chemical reactions are reversed. These reactions may be summarized thus:

Positive plate	*Negative plate*		*Positive plate*	*Negative plate*
$2Ni(OH)_3$	Cd	$\xrightarrow{\text{Discharge}}$ $\xleftarrow{\text{Charge}}$	$2Ni(OH)_2$	$Cd(OH)_2$

In the nickel–iron cell the reactions are exactly similar, except that iron replaces cadmium.

23.14 Characteristics of the alkaline cell

Curve A in fig. 23.11 represents the terminal voltage of a nickel–cadmium cell during discharge at the 10-hour rate, while curve B shows the variation for a 3-hour rate. Owing to the fact that no change occurs in the composition of the electrolyte during discharge, the number of ampere hours obtainable from an alkaline cell is much less affected by the discharge rate than is the case with the lead–acid cell. Fig. 23.12 shows the capacity of a nickel–cadmium cell—expressed as a percentage of the 10-hour capacity—for different rates of discharge.

It is seen that for a 1-hour rate it is possible to obtain about 84 per cent of the 10-hour capacity, compared with about 50 per cent for the lead–acid cell (fig. 23.10).

Curve C of fig. 23.11 represents the variation of the terminal voltage when the cell is charged at 1·5 times the 10-hour discharge rate.

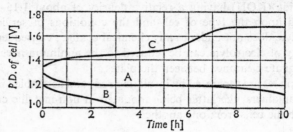

Fig. 23.11 Characteristics of an alkaline cell.

The ampere hour efficiency of the nickel–cadmium cell is about 75–80 per cent, while its watt hour efficiency is about 60–65 per cent.

The advantages of the alkaline accumulator are: (*a*) its mechanical construction enables it to withstand considerable vibration, and (*b*) it is free from 'sulphating' or any similar trouble and can therefore be left in any state of charge without damage. Its disadvantages are: (*a*) its cost is greater than that of the corresponding lead cell; (*b*) its average discharge p.d. is about 1·2 V compared with 2 V for the lead cell, so that for a given voltage the number of alkaline cells is about 67 per cent greater than that of lead cells.

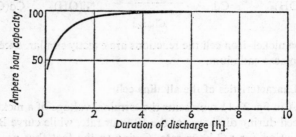

Fig. 23.12 Capacity of an alkaline cell.

Due partly to these disadvantages of the alkaline cell and partly to the improvements made in the construction of the lead–acid cell—especially the portable type—during the past twenty years, the great majority of modern batteries are of the lead–acid type.

23.15 Fuel cells*

The fuel cell has been the subject of much research during the past ten years and received a great deal of publicity when it was used to provide electrical power for a United States spacecraft. The principle of action of the fuel cell was discovered by Sir William Grove as long ago as 1839, when he demonstrated the reversible gas cell. His apparatus consisted of two inverted glass tubes filled with dilute sulphuric acid and partially immersed in this electrolyte as shown in fig. 23.13. Two electrodes, C and D, of platinum black are sealed into the

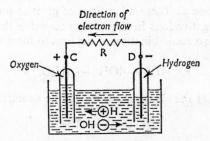

Fig. 23.13 Grove's reversible gas cell.

glass tubes, platinum black being a material that can be prepared by electrolysing a solution of chloroplatinic acid, using a pair of platinum electrodes.

If a battery is connected across terminals CD, with C positive relative to D, as already shown in fig. 23.2, oxygen collects in the anode tube and hydrogen in the cathode tube, as explained in section 23.3. If the battery is then replaced by a resistor R, as in fig. 23.13, the cell generates a current until the hydrogen and the oxygen have all recombined to form water, thereby reproducing the initial conditions.

When the cell is supplying current, the platinum black acts as a catalyst, i.e. a material that assists a chemical reaction but does not itself undergo chemical change. In the vicinity of contact between the negative electrode D and the electrolyte, the effect is to split the hydrogen molecules into positively-charged hydrogen ions and free electrons thus:

$$2H_2 \rightarrow 4H^+ + 4e$$

* The author is indebted to the Thornton Research Centre, Chester, the ASEA Central Laboratory, Sweden, and the Chloride Technical Services, Ltd., Manchester, for information on recent developments in fuel cells.

With the cell terminals connected by a resistor R, as in fig. 23.13, the freed electrons flow through R from D to C, thereby leaving electrode D positively charged and adding a negative charge on to electrode C, with the result that the following reaction takes place at C:

$$O_2 + 2H_2O + 4e \rightarrow 4OH^-.$$

The negative hydroxyl ions, OH^-, are attracted through the electrolyte towards the positively-charged electrode D, while the positive hydrogen ions, H^+, formed at D, are attracted towards the negatively-charged electrode C. These migrations of positive ions in one direction and of negative ions in the reverse direction through the electrolyte result in a recombination of the ions to form water molecules thus:

$$4H^+ + 4OH^- \rightarrow 4H_2O.$$

The net result of the above reactions can be summarized thus:

$$2H_2 + O_2 \rightarrow 2H_2O.$$

The simple cell described above can supply only a very small current, since its action depends upon the area of contact of the platinum with the electrolyte at the surface of the latter. Consequently the reaction rates at the electrodes are very low.

It will be seen from the above description that the principle of action of the fuel cell is very simple. Unfortunately, the construction of an efficient and cheap cell capable of giving appreciable power, without requiring high temperature or high pressure, presents serious difficulties. In one type of cell, developed by Williams and Gregory, the electrode matrix is a microporous plastic on which a thin layer of metal is deposited, followed by a thin layer of catalyst. The electrodes are immersed in an acid electrolyte. With this arrangement it is possible to use air instead of oxygen as the oxidant.

Another form of fuel cell, developed at the Thornton Research Centre, has nickel-plated copper with an alkaline electrolyte, the hydrogen being generated from methyl alcohol, CH_3OH.

A fuel cell developed by the ASEA of Sweden utilizes ammonia for its supply of hydrogen. Its electrode plates are built up of nickel and catalysts, held together in a porous plastic framework with ducts for the supply of the electrolyte and the gases. The catalyst for the oxygen electrode is principally in the form of silver, while the activity of the hydrogen electrode is obtained by means of a special catalyst, essentially of nickel boride.

In a hydrogen fuel cell developed by the Chloride Technical Services, Ltd., each electrode consists of two parallel plates of porous carbon sheets, spaced slightly apart, with the edges of each pair of sheets sealed into an epoxy resin frame. Hydrogen and oxygen are fed into the cavities formed by the carbon sheets of the negative and positive electrodes respectively through hollow terminal posts welded to the top edges of the electrodes. The electrolyte is potassium hydroxide solution.

In the hydrazine fuel cell developed by the same company, the liquid hydrazine fuel, diluted with potassium hydroxide solution, is fed into the cavity in the negative electrode. The positive electrode is the same as that of the hydrogen fuel cell.

At the negative electrode, one molecule of hydrazine (N_2H_4) combines with four hydroxil ions to form one molecule of nitrogen and four molecules of water, and four electrons are released, thus:

$$N_2H_4 + 4OH^- \rightarrow N_2 + 4H_2O + 4e$$

At the positive electrode, the reaction is the same as that already given for the Grove gas cell, namely:

$$O_2 + 2H_2O + 4e \rightarrow 4OH^-$$

It follows that the overall reaction in a hydrazine fuel cell is the oxidation of hydrazine into nitrogen and water, thus:

$$N_2H_4 + O_2 \rightarrow N_2 + 2H_2O$$

This reaction is accompanied by the transfer of four electrons through the electrolyte from the positive to the negative electrode for every molecule of hydrazine decomposed. These electrons flow through the external circuit from the negative to the positive electrode as indicated in fig. 23.13.

A certain hydrazine fuel cell had an output exceeding 50 000 A h when discharged at about 60 A, the terminal voltage being about 0·6 V. The output of a certain hydrogen fuel cell exceeded 100 000 A h when discharged at about 50 A, the terminal voltage being about 0·58 V. In both cells, the operating temperature of the electrolyte is about 60° C.

The great fascination of the fuel cell for scientists is the fact that it converts chemical energy into electrical energy with a relatively high efficiency; for instance, a 1-kW fuel cell can have an efficiency of 30 per cent, compared with about 5 per cent for a 1-kW heat engine such as that used in a lawn mower. The fuel cell has considerable potentialities, especially for such applications as submarines, vehicles and locomotives.

EXAMPLES 23

1. An accumulator is overcharged by 5 A for 20 hours. If the electrochemical equivalents of hydrogen and oxygen are 0·010 45 and 0·082 95 mg/C respectively, find the volume of water in cubic centimetres required to be added to compensate for gassing.

2. If the terminal voltage of a lead–acid cell varies between 2·1 and 1·85 V during discharge, calculate the number of cells required to give 230 V: (*a*) at beginning of discharge; (*b*) at end of discharge.

3. A battery of 50 cells in series is charged through a 4-Ω resistor from a 230-V supply. If the terminal voltage per cell is 2 and 2·7 V respectively at the beginning and end of the charge, calculate the charging current: (*a*) at the beginning, and (*b*) at the end of the charge.

4. A battery of 40 cells in series is to be charged from a 220-V supply. If the average p.d. per cell during charge is 2·2 V, calculate the value of the resistor required in series with the battery to give an average charging current of 5 A.

5. Tabulate the relative advantages and disadvantages of lead–acid and alkaline cells.

Explain the difference between the 'constant current' and the 'constant voltage' methods of charging and give any disadvantages of each system.

A battery of 10 cells is to be charged at a constant current of 8 A from a generator of e.m.f. 35 V and internal resistance 0·2 Ω. The internal resistance of each cell is 0·08 Ω and its e.m.f. ranges from 1·85 V discharged to 2·15 V charged. Determine the maximum and minimum values of series resistor which must be included in the circuit.

(U.E.I.)

6. Explain the terms *ampere hour efficiency* and *watt hour efficiency* used in connection with secondary cells.

A battery of 20 cells is charged through a fixed resistor from a constant 70-V supply. At the beginning of the charge, the e.m.f. per cell is 1·85 V and the charging current is 10 A. When charging is almost complete, the e.m.f. per cell has risen to 2·2 V. Each cell has a constant internal resistance of 0·08 Ω. Calculate the value of the external resistor and the current at the end of the charge.

(U.E.I.)

7. Thirty-five lead–acid secondary cells, each of discharge capacity 100 ampere hours at the 10-hour rate, are to be fully charged at constant current for 8 hours. The d.c. supply is 120 V, the ampere hour efficiency is 80 per cent and the e.m.f. of each cell at beginning and end of

charge is respectively 1·9 V and 2·6 V. Calculate the maximum and minimum values of charging resistor. Ignore internal resistance of cells.

8. A cell is discharged at a constant current of 5 A for 10 hours, the average value of the terminal voltage being 1·92 V. The cell is then recharged at a constant rate of 3 A for 20 hours, with an average terminal voltage of 2·2 V. Calculate: (*a*) the ampere hour efficiency; (*b*) the watt hour efficiency.

9. A battery of 56 cells in series takes a constant charging current of 8 A, the e.m.f. rising from 1·8 to 2·4 V per cell. If each cell has an internal resistance of 0·008 Ω, find the p.d. across the battery at the beginning and at the end of charge.

   If the battery is completely discharged in 40 hours at 4 A at an average voltage of 1·96 V per cell after being charged for 24 hours at 8 A with an average voltage of 2·2 V per cell, find the ampere hour and the watt hour efficiencies of the battery.

10. An alkaline cell is discharged at a constant current of 5 A for 10 hours, the average terminal voltage being 1·2 V. A charging current of 3 A maintained for 21 hours is required to bring the cell back to the same state of charge, the average terminal voltage being 1·48 V. Calculate the ampere hour and the watt hour efficiencies.

11. The 10-hour capacity of a certain accumulator is 60 A h. From figs. 23.10 and 23.12, find the number of ampere hours obtainable on a half-hour discharge rate if the accumulator is of: (*a*) the lead–acid type, and (*b*) the alkaline type.

12. An accumulator has a terminal voltage of 1·9 V when supplying a current of 8 A. The terminal voltage rises to 2·03 V immediately the load is switched off. Calculate the internal resistance of the accumulator.

13. A certain alkaline cell had a terminal voltage of 1·24 V when supplying a load of 3 A. Immediately after the load was disconnected, the terminal voltage was 1·28 V. Calculate the internal resistance of the cell.

14. Describe an experiment in which a slide-wire potentiometer is used to determine the e.m.f. of a primary cell. Give a circuit diagram and derive the equation from which the e.m.f. can be calculated.

   A voltmeter of 100 Ω resistance reads 1·35 V when connected to the terminals of a certain cell. This reading falls to 0·93 V when a 10-Ω coil is also connected across the cell terminals. Deduce the e.m.f. and the internal resistance of the cell. (U.L.C.I., O.1)

15. A separately-excited d.c. generator has an armature resistance of 0·2 Ω. It is used to charge a battery having an open-circuit e.m.f. of 105 V and an internal resistance of 0·1 Ω. The copper conductors between generator and battery have a total length of 100 m and are 2·5 mm in diameter. What will be the value of the generated e.m.f. of the generator when the battery is being charged at 10 A?

   How much power will be wasted in the conductors between generator and battery?

   The resistance of a copper wire, 1 m long and 1 mm^2 cross-section, is $\frac{1}{58}$ Ω. (App. El., L.U.)

APPENDIX

Introduction to the Generalized Theory of Electrical Machines

A.1 Introductory

Electrical machines are, in general, used to convert mechanical energy into electrical energy, as in electric generators, or electrical energy into mechanical energy, as in electric motors. Most electrical machines consist of an outer stationary member and an inner rotating member. The stationary and rotating members consist of iron cores, separated by an airgap, and form a magnetic circuit in which magnetic flux is produced by currents flowing through windings situated on the two members.

A.2 Windings

One member of an electrical machine has a winding, referred to as a *field winding*, the function of which is to produce a magnetic flux. The other member has a winding or group of coils, termed an *armature winding*, in which an e.m.f. is generated by the movement of this winding relative to the magnetic flux produced by the field winding.

In d.c. machines the field winding is stationary and the armature winding is located on a rotating iron core. In a.c. synchronous generators and motors, the field winding is usually on the rotor and the armature winding is stationary, for reasons given in section 15.1.

The types of windings used on electrical machines can be grouped thus:

(*a*) *Concentrated or coil windings* wound around the salient poles of d.c. machines and relatively slow-speed a.c. synchronous generators and motors, as shown in figs. 3.9, 6.1 and 15.2.

(*b*) *Phase or distributed windings* distributed in slots located on the inner surface of a stator, as shown in figs. 15.4 and 15.5, or in slots located on the outer surface of a cylindrical rotor, as in fig. 15.3.

686

The conductors of this type of winding are, in the majority of machines, connected in separate circuits: for instance, figs. 15.7 and 15.9 show how the conductors forming the phase windings of a three-phase stator may be connected. In these diagrams, each group of conductors occupies a phase band of two slots per pole, distributed in regular sequence over the various pole pitches.

(*c*) *Commutator windings* in which the coils are connected to commutator segments as shown in figs. 6.4 and 6.10. This type of winding must be wound on the rotating member of a machine in order that the commutator may act as a switch which automatically reverses the direction of the current in a coil as the latter passes brushes bearing on the commutator surface, as explained in section 6.10.

Commutator windings are used on the armatures of (i) d.c. machines, (ii) single-phase series motors (as in vacuum cleaners) and (iii) three-phase variable-speed commutator motors.

A.3 Electromechanical energy conversion

In the process of converting mechanical energy into electrical energy, or vice-versa, some of the input energy is converted into heat owing to I^2R, iron and friction losses in the machine. There is also an increase or a decrease of energy stored in the magnetic field of the machine if the magnetic flux is increased or decreased. If a change of load on an electric motor is accompanied by a change of speed, some of the input energy is converted into kinetic energy during the period that the machine and its load are being accelerated. On the other hand, if the speed is decreasing, the motor and its load give up some of their kinetic energy, thereby reducing the value of the input energy during the deceleration period.

A machine is said to operate under *steady-state* conditions if its load, speed and excitation remain constant. Under these conditions,

input power = output power + power dissipated as heat.

The losses in a loaded d.c. machine have already been discussed in section 9.1, and a table indicating what happens to the electrical power supplied to an induction motor is given on p. 485. The following table shows what happens to the mechanical power supplied to a loaded a.c. generator having an exciter carried on an extension of its shaft:

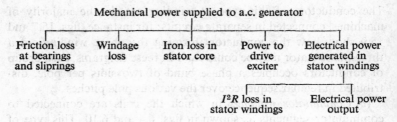

Mechanical power supplied to a.c. generator

| Friction loss at bearings and sliprings | Windage loss | Iron loss in stator core | Power to drive exciter | Electrical power generated in stator windings |

| | I^2R loss in stator windings | Electrical power output |

A.4 Production of torque

It has been explained in sections 6.5, 17.1, 17.6 and 18.1 that in all generators and motors, the torque produced by the magnetic flux on the rotating member is due to the distortion or skewing of that flux in the airgap between the rotating and stationary members. An equal and opposite torque is exerted on the stationary member and is transmitted through the frame of the machine to its foundations.

If a conductor carrying a current I amperes is located on the *surface* of a smooth armature core of length l metres, and if the flux density in the airgap is B teslas, the force on the conductor* is BIl newtons in a tangential direction (see section 2·8). If the distance between the conductor and the axis of rotation is r metres, the torque on the conductor is $BIlr$ newton metres; and the total torque on the armature is the algebraic sum of the torques on the conductors.

With such an arrangement it would be necessary to insert numerous wedges radially into the armature core to prevent the winding moving relative to the core, i.e. the torque would be transmitted to the core via these wedges. Such a necessity is avoided in actual machines by embedding the conductors in slots, as shown in fig. 6.10 for a rotor and in fig. 15.1 for a stator. The flux density in the

---

* When electrons are moving at right angles to a magnetic field, there is a force acting on the electrons in a direction perpendicular to the directions of motion and of the magnetic field, irrespective of whether the electrons are moving in a vacuum, as discussed in section 19.27, or along a conducting wire. In the latter case, the deflection of the electrons is halted by collision with the lattice structure of the conducting material; and the deflecting force is thereby transferred from the electrons to the conductor.

Similarly, when the current-carrying conductors are embedded in slots, the tangential force on the teeth is due to the force exerted on the orientated spinning electrons of the current rings located in the magnetic domains (see section 3.13) which are held rigidly in the structure of the iron adjacent to the airgaps.

slot is very low compared with that in the airgap, so that the torque exerted on the conductors is almost negligible. Practically the whole of the torque is exerted on the teeth. The value of this torque, however, is the same as that which would be produced by the current-carrying conductors if they were located on the surface of the core.

The fact that the conductors are embedded in slots and are therefore situated in regions of low flux density does not affect the magnitude of the e.m.f. generated in the conductors. Thus, when a slot containing a conductor moves towards the right in fig. A.1, the

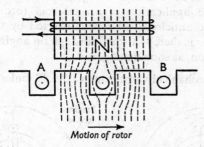

Fig. A.1. Generation of e.m.f.

conductor must pass through the *whole* of the flux entering the armature from the N pole when it moves under the pole face from position A to position B. The conductor therefore generates the same e.m.f. as it would have done had it been located on the surface of the core.

One can imagine the lines of magnetic flux to move comparatively slowly across the teeth where the density is high, and then sweep rapidly across the slots where the density is very low. In this way, the *rate* at which the conductor cuts the flux is the same irrespective of whether the conductor is located on the core surface or embedded in a slot.

A.5 Relationship between torque and $\mathrm{d}M/\mathrm{d}\theta$

Let us consider cylindrical stator and rotor iron cores, the stator having coil AB and the rotor having coil CD wound as shown in fig. A.2. For simplicity, each coil is indicated as a single turn. Suppose the planes of the coils to be displaced by an angle θ radians relative to each other and the coils to carry currents in the directions indicated by the dots and crosses.

The resultant magnetic field is roughly as shown by the dotted lines. (There is practically no flux in the airgaps between A and C and between B and D when the currents in the two coils are equal.) It will be seen that the flux in the airgap is skewed in such a direction as to exert an anticlockwise torque on the rotor and an equal clockwise torque on the stator. Suppose this torque to be T newton metres.

If the friction at the bearings is negligible, we can turn the rotor clockwise by the application of an external torque T equal in magnitude to the anticlockwise torque exerted by the flux on the rotor. If the rotor is thereby turned through an angle $d\theta$ radian in a clockwise direction, as shown in fig. A.3,

$$\text{mechanical work done} = T \cdot d\theta \text{ joules.}$$

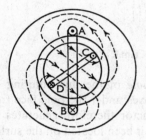

Fig. A.2. Resultant distribution of magnetic flux.

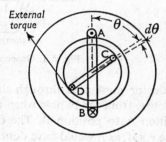

Fig. A.3. Displacement of rotor.

If L_1 and L_2 = self inductances of coils AB and CD respectively,

M = mutual inductance between coils AB and CD

and I_1 and I_2 = currents in coils AB and CD respectively,

then from expression (4.27),

total energy stored in the magnetic field

$$= \tfrac{1}{2}L_1I_1^2 + \tfrac{1}{2}L_2I_2^2 + MI_1I_2 \text{ joules.}$$

The values of the self inductances of coils AB and CD are independent of the position of the rotor and are therefore constant. On the other hand, the value of M decreases from a maximum positive value to zero as the displacement θ is increased from zero to $\pi/2$

radians. If dM henry is the change of mutual inductance when coil CD is turned clockwise through dθ radian, and if the currents in coils AB and CD are maintained constant at I_1 and I_2 respectively, change of energy stored in the magnetic field $= I_1I_2 \,.\, dM$ joules.

Since the value of M has been reduced by the increased displacement dθ, dM is negative. Consequently the expression $I_1I_2.dM$ represents a *decrease* of energy stored in the magnetic field.

The effect of reducing the mutual inductance is to reduce the number of flux-linkages with each coil and thus to induce an e.m.f. in each coil.

If dt = time taken to produce displacement dθ,

$$\text{e.m.f. induced in coil AB} = -I_2.dM/dt \text{ volts}$$
and $$\text{e.m.f. induced in coil CD} = -I_1.dM/dt \text{ volts.}$$

By Lenz's law (section 2.10), the e.m.f. induced in coil AB tries to prevent the reduction in the number of flux-linkages caused by the *increased* displacement dθ. Consequently the direction of this induced e.m.f. is the same as that of the current I_1. This means that the voltage applied to coil AB has to be *reduced* by $I_2.dM/dt$ volts in order to maintain the current in coil AB *constant* at I_1 during the time that the rotor is being turned through dθ radian.

Hence, electrical energy *generated* in coil AB due to this induced

$$\text{e.m.f.} = I_1 \times (I_2.dM/dt) \times dt$$
$$= I_1I_2.dM \text{ joules.}$$

Similarly, electrical energy *generated* in coil CD

$$= I_2 \times (I_1.dM/dt) \times dt$$
$$= I_1I_2.dM \text{ joules.}$$

Hence, total electrical energy *generated* in coils AB and CD

$$= 2I_1I_2.dM \text{ joules.}$$

Since the friction loss at the bearings is being assumed negligible, it follows that the total electrical energy generated in coils AB and CD must come partly from the decrease of energy stored in the magnetic field and partly from the mechanical work done in producing displacement dθ,

i.e.
$$2I_1I_2.\mathrm{d}M = I_1I_2.\mathrm{d}M + T.\mathrm{d}\theta$$
$$\therefore \qquad T = I_1I_2.\mathrm{d}M/\mathrm{d}\theta \text{ newton metres*} \qquad (A.1)$$

Figure A.4 shows coils AB and CD in four different positions relative to each other. The dots and crosses represent the directions of the currents in the coils.

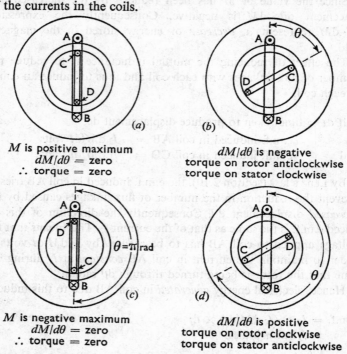

(a)

M is positive maximum
$\mathrm{d}M/\mathrm{d}\theta$ = zero
∴ torque = zero

(b)

$\mathrm{d}M/\mathrm{d}\theta$ is negative
torque on rotor anticlockwise
torque on stator clockwise

(c)

M is negative maximum
$\mathrm{d}M/\mathrm{d}\theta$ = zero
∴ torque = zero

(d)

$\mathrm{d}M/\mathrm{d}\theta$ is positive
torque on rotor clockwise
torque on stator anticlockwise

Fig. A.4. Direction of torque for different relative positions of coils.

Figure A.4 (*b*) shows the coils in the same positions as in fig. A.2, and it has already been explained that for this condition, the torque exerted on the rotor by the magnetic flux is anticlockwise. As the displacement θ of the rotor from the position shown in fig. A.4 (a) is increased from zero up to $\pi/2$ radians, the mutual inductance decreases from a positive maximum to zero, as represented in fig.

* Expression (A.1) can be used to calculate the torque on the moving coil of the electrodynamic (or dynamometer) instrument described in section 22.8. The value of M is determined experimentally for various positions of the moving coil; and from a graph representing the values of M for different values of θ, the values of $\mathrm{d}M/\mathrm{d}\theta$ can be derived for various positions of the pointer over the whole of the scale.

A.5 (a). Further displacement of the rotor through $\pi/2$ radians causes the mutual inductance to vary between zero and a negative maximum. During this displacement of π radians, $dM/d\theta$ is negative and constant in magnitude; hence the torque on the rotor remains constant in an anticlockwise direction, as shown in fig. A.5 (b). It is assumed that the reluctance of the iron cores is negligible compared with that of the airgap, so that the flux density in the airgap due to current in coil AB *alone* would be uniform.

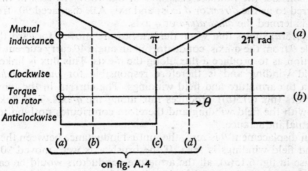

A.5. Variation of mutual inductance and torque with rotor displacement.

While the displacement is varied between π and 2π radians, $dM/d\theta$ is positive and constant in magnitude. Consequently the torque on the rotor remains constant in a clockwise direction.

In a machine with d.c. excitation on the stationary member and a commutator winding on the rotating member, the position of the brushes on the commutator must be such as to reverse the direction of current in each coil at the instant when $dM/d\theta$ for the coil changes from positive to negative and from negative to positive, i.e. at instants (a) and (c) respectively in figs. A.4 and A.5.

If the rotor coil CD is supplied with alternating current through slip-rings, then, in order to maintain the torque unidirectional, the current must reverse its direction at the instant that the sign of $dM/d\theta$ reverses, namely each time the rotor passes through positions corresponding to (a) and (c) in figs. A.4 and A.5. This means that the rotor of a 2-pole machine must rotate through half a revolution in half a cycle of the alternating current; i.e. the rotor must rotate at *synchronous* speed.

If each of the two coils, AB and CD, carries an alternating current, it is necessary for the frequency of the current in the rotor plus the frequency of rotation to be equal to the frequency of the current in the stator coil,

$$\text{i.e. } f_s = f_r + (n_r \times p)$$

as already shown in expression (18.2) for a 3-phase induction motor having p pairs of poles and rotating at n_r revolutions/second.

A.6 Application to commutator machines

Let us consider the *two-pole generator* of fig. 6.14. The axis of the poles is referred to as the *direct* or *d* axis; and axis AB, displaced 90° from the *d*-axis, is termed the *quadrature* or *q* axis.

Fig. 6.15(*a*) shows that when the brushes are moved forward through an angle θ from the q-axis, conductors in an angle 4θ carry current in such a direction as to produce a flux along the *d*-axis. This flux is linked with the field winding and is therefore responsible for *mutual inductance* between the armature and field windings. The current in the remaining conductors (fig. 6.15(*b*)) produces flux along the *q*-axis. This flux is not linked with the field winding and therefore contributes nothing towards the mutual inductance.

When displacement θ is zero, the mutual inductance between the armature and field windings is zero. If the brush axis were moved 90° anticlockwise in fig. 6.15(*a*), all the armature conductors would be carrying current in such a direction that the whole of the armature m.m.f. would be in direct opposition to the m.m.f. of the field winding. Hence the mutual inductance would be at its negative maximum value. For this position of the brushes, half the armature conductors would be carrying current producing a clockwise torque and the other half would produce an anticlockwise torque, so that the net torque would be zero.

If the brushes were moved 90° clockwise from the *q*-axis, the whole of the armature m.m.f. would be acting in the same direction as the field m.m.f. and the mutual inductance would be at its positive maximum value. For this brush position, half the conductors would be exerting a clockwise torque and the other half an anticlockwise torque, so that the net torque would again be zero.

With the brushes on the *q*-axis, i.e. axis AB in fig. 6.11, all the conductors moving in the magnetic field are carrying current in such a direction as to exert a clockwise torque. Hence the torque is a maximum when the brush displacement θ from the *q*-axis is zero.

It is found experimentally (see *The Unified Theory of Electrical Machines* by C. V. Jones, p. 340) that the relationship between the mutual inductance and the brush displacement is practically sinusoidal, i.e. for a brush displacement θ from the *q*-axis.

$$\text{mutual inductance} = M \sin \theta$$
$$\therefore \qquad dM/d\theta = M \cos \theta.$$

In a d.c. machine, the brushes are normally on or very near the *q*-axis, so that $\theta = 0°$ and $dM/d\theta = M$ henrys/radian.

As far as the present discussion is concerned, the d.c. machine could be totally enclosed, with two pairs of wires brought out to external terminals, namely armature terminals AA₁ and field terminals FF₁, as shown in fig. 7.1. If the armature and field currents are represented by I_a and I_t respectively, we have from expression (A.1):

$$\text{torque} = T = I_a I_t dM/d\theta = I_a I_t M \text{ newton metres} \qquad \text{(A.2)}$$

If E_r = rotational e.m.f. generated in the armature

and ω_r = angular speed in radians/second = $2\pi n$

where n = speed in revolutions/second,

then, if the machine is operating as a motor (see section 8.3),

$$\text{mechanical power developed} = E_r I_a.$$

From expression (8.5), $E_r I_a = \omega_r T = \omega_r I_a I_t M$

$$\therefore \qquad\qquad M = \frac{E_r}{\omega_r I_t} = \frac{E_r}{2\pi n I_t} \qquad\qquad \text{(A.3)}$$

The value of M for a given machine can therefore be determined experimentally from the open-circuit characteristic (fig. 7.4) obtained by running the machine as a generator at a known speed; and the torque for given values of I_a and I_t can then be calculated from expression (A.2).

A.7. Application to three-phase synchronous machines

Let us consider the 2-pole, 3-phase synchronous *generator* of fig. 17.1. When the position of the rotor winding relative to the winding of phase R is that shown in fig. 17.1, the mutual inductance between the two windings is zero. If the rotor is displaced 90° *forward* relative to phase R, as in fig. 17.4, the m.m.fs of the two windings are in direct opposition, so that the mutual inductance is at its negative maximum value. This is the rotor position when the current lags the generated e.m.f. by 90°.

If the rotor is displaced 90° *backward* from the position shown in fig. 17.1, the direction of the rotor m.m.f. is the same as that due to phase R, so that the mutual inductance is at its positive maximum value. This is the rotor position when the current leads the generated e.m.f. by 90°.

It is found experimentally that if the rotor is displaced by an angle α* from the position of zero mutual inductance, namely the position shown in fig. 17.1,

$$\left.\begin{array}{l}\text{mutual inductance between rotor}\\ \text{winding and phase R}\end{array}\right\} = \pm M \sin \alpha \qquad \text{(A.4)}$$

$$\text{Hence } dM/d\alpha = M \cos \alpha \text{ (neglecting the signs).} \qquad \text{(A.5)}$$

When the field winding is in the position shown in fig. 17.4 relative to phase R, we can assume that the whole of the flux Φ produced by field current I_t is linked with the winding of phase R. Hence, if the number of conductors per phase is Z, the flux-linkages with phase R is $\Phi Z/2$.

* Symbol α rather than θ is used to avoid confusion with θ when used to represent $2\pi ft$ in a.c. theory.

From expression (4.26), it follows that for a 2-pole machine:
maximum mutual inductance between the field
winding and a stator phase winding $\Big\} = M = \dfrac{\Phi Z}{2I_t}$

If the machine has p pairs of poles,

$$\text{maximum mutual inductance} = M = \frac{\Phi Z p}{2I_t} \tag{A.6}$$

From expression (15.5),

r.m.s. value of e.m.f./phase = $E = 2{\cdot}22k_d k_p Z f \Phi$.

Since $k_d \simeq 0{\cdot}96$ for 3-phase windings and since 3-phase windings are usually full-pitch, so that $k_p = 1{\cdot}0$,

$\therefore$ $\qquad\qquad E \simeq 2{\cdot}22Zf\Phi = 2{\cdot}22Znp\Phi$

where $\qquad\qquad n =$ speed in revolutions/second.

Substituting for $\Phi Z p$ in expression (A.6), we have:

$$M = \frac{E}{2{\cdot}22\,n} \times \frac{1}{2I_t} = \frac{E \times 2\pi}{2{\cdot}22\,\omega_r} \times \frac{1}{2I_t}$$

where $\qquad\qquad \omega_r =$ angular speed of *machine*, in radians/second.

But $\qquad\qquad 2{\cdot}22 = 2 \times 1{\cdot}11 = 2 \times \dfrac{1/\sqrt{2}}{2/\pi} = \dfrac{\pi}{\sqrt{2}}$

Hence $\qquad\qquad M = \dfrac{E \times \pi}{2{\cdot}22\,\omega_r} \times \dfrac{1}{I_t} = \dfrac{\sqrt{2}E}{\omega_r I_t}$ $\qquad$ (A.7)

The value of M can therefore be determined by running the machine on open circuit at synchronous speed and noting the terminal voltage per phase for a given field current. For a synchronous generator,

electrical power *generated* per phase = $I_s E \cos \alpha$

where I_s and E represent the r.m.s. values of the current and generated e.m.f. per phase respectively in the stator winding. Angle α in this expression is the same as angle EOI in fig. 17.8 and also represents the displacement of the rotor relative to phase R from the position shown in fig. 17.1, namely the position of zero mutual inductance between the field winding and phase R.

$\therefore$ total electrical power generated = $P = 3I_s E \cos \alpha$

$$= 3I_s \times \frac{\omega_r I_t M \cos \alpha}{\sqrt{2}}$$

and $\qquad\qquad$ total torque = $T = P/\omega_r = (3/\sqrt{2})I_s I_t M \cos \alpha$

$$= (3/\sqrt{2})I_s I_t \times dM/d\alpha \tag{A.8}$$

Expression (A.8) shows how expression (A.2) has to be modified when applied to a 3-phase synchronous machine.

When the machine is run as a synchronous motor, α is the angle between I and E in fig. 17.16, i.e. $\alpha = 180° - (\theta \pm \phi)$. The plus sign applies when the power factor is leading and the minus sign when it is lagging.

For fuller discussion of this subject, see *Higher Electrical Engineering* by Shepherd, Morton and Spence. (For examples, see page 445.)

Answers to Examples

Examples 1. Page 31

1. 0·4 m/s².   2. 2 kN.
3. 14·715 kN; 220·7 MJ, 61·3 kW h; 368 kW, 26 kW h.
4. 16 370 N m.   5. 212 N m.
6. 94·7 A, 163·7 p.   7. 83·8 per cent, 146 kJ; 67·2 Ω.
8. 12·82 kW, 26·7 A, 102·56 kW h.
9. 59·5 Ω, 0·105 p.   10. 13·93 kW, £2·23.
11. 1·98 g.   12. 241 min.
13. 4·035 V, 780 J.   14. 26·57 V, 13·8 W.   15. 5 Ω.
16. A: 2·4 A, 14·4 V, 9·6 V; B: 1·6 A, 16 V, 8 V; 96 W.
17. 10·96 V; 121·7 A, 78·3 A; 1·33 kW, 0·86 kW.
18. 11·33 Ω, 208 W.   19. 40 Ω, 302·5 W.
20. 0·569 mm.   21. 70°C.
22. 30 Ω.   23. 0·0173 μΩ m, 0·398 Ω.
24. 91·8°C.   25. 4·035 Ω.   26. 99·6°C.
27. 40 Ω; 6 V, 0·9 W; 0·8 W.   28. 1 A, 12 A, 13 A; 104 V.
29. 1·069 A, discharge; 0·1145 A, discharge; 94·66 V.
30. 5; if A is positive with respect to B, load current is from D to C.
31. 72·8 mA, 34·27 mA, 1·927 V.   32. 11·55 A, 3·86 A.
33. I_A, 0·183 A, charge; I_B, 5·726 A, discharge; I_C, 5·543 A, discharge; 108·55 V.   34. 0·32 A.   35. B, 2 volts above A.
36. 2·84 Ω, 1·45 Ω.   37. 20·6 mA from B to E.
38. 0·047 A.   39. 0·192 A.
40. 2·295 A, 0·53 A, 1·765 A, both batteries discharging; 4·875A.
41. AB, 183·3 Ω; BC, 550 Ω; CA, 275 Ω.
42. A, 4·615 Ω; B, 12·31 Ω; C, 18·46 Ω.
43. R_{AB} = 247·8 Ω, R_{BC} = 318·6 Ω, R_{CA} = 223 Ω; 80 Ω.
44. 244·4 V, 240 V, 238·8 V; 2·68 kW; 21·44 kWh.
45. 475 mm², 9·15 p.   46. 261·5 V, 247·6 V, 1·502 kW.
47. AB, 100 A; BC, 20 A; AC, 80 A; V_B, 241 V; V_C, 240·4 V.
48. 106·25 A, 46·25 A, 33·75 A, 63·75 A; 231·5 V, 229·65 V, 232·35 V.
49. 243·43 V, 239·93 V.

Examples 2. Page 55

1. 60 N.   2. 50 A.   3. 0·333 T.   4. 0·452 N.
5. 173 N m, 12·7 kW.   6. 0·667 T.
7. 10 m/s, 0·8 N, 0·48 J.   8. 0·375 T, 0·1406 N.
9. 12 N, 1·2 V, 120 W.   10. 2·33 mV.   11. 0·333 V.
12. 4·42 μV.   13. 37·5 V.   14. 0·32 V.

15. —33·3 V, 0, 100 V.
16. 2700 V, direction same as that of current.   17. 24 V, 224 V.
18. 5·6 V.   19. 1400 rev/min.

Examples 3. Page 97

1. 476 A; $1·19 \times 10^6$ A/Wb, 1120.
2. 720 A/m, 663; 2400 A/m, 398.   3. 553 A, 147 μWb.
4. $1·5 \times 10^6$ A/Wb, 144.
5. 4·5 A; $0·4375 \times 10^6$ A/Wb, 306 A.   6. 121 A, 582.
7. 0·0825 A.   8. 1100 A.   9. 1220 A.   10. 5.65 A.
11. 1520 A.   12. 0·64 T, 358 A/m.
13. 308 A/m, 0·593 T; 616 A/m, 1·04 T; 924 A/m, 1·267 T; 1232 A/m, 1·41 T; 1540 A/m, 1·49 T.
14. 1·22 T, 1620.   15. $1·23 \times 10^{-9}$ C/division.
16. 1800 J, 1·4 T, 11·55 W at 50 Hz.   17. 2·63 W.
18. 0·333 W.   19. 1·11 J.   20. 337 W.
21. 0·37 T; —26 200 A/m.   22. 11·85 cm × 5 cm².
23. 7·95 cm × 0·38 cm².   24. 4·2 cm × 4·4 cm².
25. 9·55 cm × 3·12 cm².   26. 0·218 T.
27. 0·236 T.   28. 578 μV.   29. 0·00764 J.
30. 1·37 A.   31. 13·13 A, 29·85 N.   32. 955 A.
33. 995 A/m, 1·25 mT.
34. 6000 A; 500 000 A/Wb; 2650.   35. 125 N/m.
36. A, 4780 A/m, 6 mT; B, 6370 A/m, 8 mT; C, 2390 A/m, 3 mT; A, O; B, 3185 A/m, 4 mT; C, 2390 A/m, 3 mT.

Examples 4. Page 134

1. 0·375 H.   2. 0·15 H.   3. 160 A/s   4. 3·5 V.
5. 1·25 mH, 0·1 V.
6. 530 A/m, 354 A/m, assuming circular cross-section; 2·66 A; 18·8 μH.
7. 0·24 H, 0·06 H.   8. 1492, 47·1 mH, 1·884 V.
9. 157 μH, 94·2 mH.   10. 15 H; 300 V, assuming zero residual flux.
11. 0·96 H, 182·4 V.   12. 0·15 H, 750 V, 0·1 H.
13. 2·68 A, 13·32 mH, 7·14 V, assuming zero residual flux.
14. 0·09 H, 180 V.   15. 26·67 μH.
16. 100 μH, 6 mV, 0·527.   17. 2·23 mm, 0·125 H.
19. 25 V to 125 V during first second; 100 V during next second; 87·5 V to —12·5 V during last 2 seconds.
20. 0·0833 H, 0·04165 s, 1·667 V.   21. 20 A/s, 1·55, 0·23 s.
22. 20 A/s, 0·1 s, 1·5 A.   23. 0·316 A, 632 Ω, 1·896 H, 0·0946 J.
24. (a) 1 A, (b) 2·5 A, (c) 0, (d) 200 V; (a) 2·5 A, (b) 2·5 A, (c) 700 V, (d) 500 V.   25. 0·316 A, 155·5 V.
26. 40 A/s, 10 J.   27. 12 V.   28. 0·0625 H.
29. —10 V.   30. 15·06 μH.   31. 1·35 mH.
32. 150 μC.   33. 740 μH.
34. $12·8 \times 10^{-10}$ C, 10·65 μH.   35. 1·612 mH, 0·417.
36. 0·0275 J, 0·0175 J; 0·3535.   37. 32 mH, 48 mH, 160 mJ, 96 mJ.
38. 290 μH, 440 μH.   39. 0·6.

Examples 5. Page 177

1. 150 V, 3 mC.   2. 45 μF, 4·615 μF.
3. 3·6 μF, 72 μC, 288 μJ.   4. 1200 μC; 120 V, 80 V; 6 μF.
5. 500 μC; 250 V, 166·7 V, 83·3 V; 0·0208 J.   6. 15 μF in series.
7. 2·77 μF, 18·46 μC.   8. 4·57 μF, 3·56 μF.
9. 3·6 μF, 5·4 μF.
10. 200 V; 1·2 mC, 2 mC, 3·2 mC on A, B and C respectively.
11. 360 V, 240 V; 40 μF, 0·8 J.   12. 267 V; 0·32 J, 0·213 J.
13. 150 V, 100 V, 600 μC; 120 V, 480 μC, 720 μC.
14. 62·5 V.   15. 120 V, 0·06 J, 0·036 J.
16. 590 pF, 0·354 μC, 200 kV/m, 8·85 μC/m².
17. 1593 pF, 0·796 μC, 167 kV/m, 8·85 μC/m².
18. 664 pF, 0·2656 μC, 100 kV/m, 4·43 μC/m².
19. 177 pF, 1062 pF, 33·3 V, 0·0354 μC.
20. 619·5 pF, 31 ms, 0·062 μC.   21. 2212 pF.
22. 1·416 mm.
23. 1290 pF; 1·82 kV/mm (paper), 0·453 kV/mm; 0·0161 J.
24. 0·245 m², 2·3 MV/m in airgap. 25. 8·58 kV/cm in airgap.
26. 20 μC, 15 μC; 1000 μJ, 2250 μJ; 35 μC; 140V; 2450 μJ.
27. 8·8 × 10⁻¹² F/m; 100 kV/m, 0·88 μC/m².
28. 2·81; 30 kV/m, 0·7425 μC/m²; 0·4455 μJ.
29. 397 pF, 8·82 × 10⁻¹² F/m.   30. 0·5 μC, 3·12 × 10¹² electrons.
31. 3·744 × 10²¹ electrons/s.   32. 11·2 mA.
33. 115·2 J, 7·2 × 10²⁰ electronvolts; 8·39 × 10⁶ m/s.
34. 18·72 × 10¹⁵ electrons/s; 6·75 × 10²⁰ electronvolts, 108 J; 6·5 × 10⁶
    m/s.   35. 1·28 × 10⁻¹⁴ N, 6·41 × 10⁻¹⁷ J, 8·43 × 10⁻¹⁰ s.
36. 2·04 mm.   37. 16·4 kV/m, 1·155 × 10⁷ m/s, 16·1°.
38. 5·93 × 10⁶ m/s, 6·74 × 10⁻⁹ s, 10¹⁶ electrons/s.   39. 4 × 10⁻¹⁵ N.
41. 0·8 s, 125 V/s, 12·5 mA, 0·01 C, 0·5 J.   43. 18·4 μA, 4000 μJ.
44. 100 V, 0·1056 ms.   45. 50 mA, 500 mA, 1·25 J.
46. 1342 s.   47. 100 MΩ.
48. 39·7 m (approx.), if wound spirally.
49. 4·75 kV, 1 mJ, 0·4 N.   50. 3·75 μF, 1500 μC.
51. 300 V; 4 μJ, 16 μJ.·   52. 2510 μJ/m³.
53. 20·8 N/m², 20·8 J/m³.
54. 15 930 pF, 3000 kV/m, 159·3 μC/m², 71·7 N.

Examples 6. Page 209

1. 240 A, 498 V; 720 A, 166 V; 119·5 kW.
2. 120 A, 301 V, 36·12 kW; 240 A, 150·5 V, 36·12 kW.
3. 432 V.   4. 0·02 Wb.   5. 946 rev/min.   6. 0·0274 Wb.
7. 1065 rev/min. 8. 140 V.   9. 0·0192 Wb.   10. 0·0333 Wb.
11. 200 At, 3400 At.   14. 4440 At, 555 At; 0·4625 A.
15. 8.   16. 3150 At.   17. 0·443 Ω.

Examples 7. Page 223

1. 0·0219 Wb.   2. 8, 0·0321 Wb.   3. 0·123 Ω, 534·5 V.
4. 282·66 V.   5. 467 V, 373 V, 0, 240 Ω.   6. 400 V, 240 Ω.

7. 354 V; 5 V, 64·5 Ω.   8. 71 Ω, 83 V.
9. 45 A.   10. 236 V, 198 A, 105 V (approx.).
11. 258 V, 100 A (approx.).   12. 6, 0·0296 Ω.   13. 7·5.

Examples 8. Page 245

1. 589 rev/min.   2. 714 rev/min.   3. 190 V.
4. 468 V, 1075 rev/min, 333 N m.   5. 140 A.
6. 208 V, 41 A.   7. 439 rev/min; 400 N m.
8. 730 rev/min, 226 N m.   9. 13·85 mWb, 114·7 N m.
10. 36·2 N m.   11. 35 Ω, 962 rev/min.   12. 5·8 Ω, 60 V.
13. 1 A, 2005 rev/min; 2 A, 1150 rev/min; 3 A, 958 rev/min, etc.
14. 391 rev/min.
15. 5 A, 445 rad/s; 15 A, 148 rad/s; 25 A, 101 rad/s, etc.
16. 5·5 Ω, 23·5 Ω.   17. 356 N m.
18. 210·3 V, 0·0127 Wb, 114·3 N m.   19. 595 rev/min.
20. 40 A, 695 rev/min.   21. 333 N m.
22. 5 A, 2090 rev/min; 10 A, 1035 rev/min; 15 A, 738 rev/min, etc.
23. 1·91 Ω, 80 V.   24. 2·767 Ω, 76·7 V.
25. 18·6 rad/s (or 178 rev/min).   26. 917 rev/min, 2560 N m.
27. 40 Ω.   28. 0·94 Ω, 932 rev/min.
29. 6·53 Ω.
30. 3·05 Ω, 5·67 Ω, 9·2 Ω.

Examples 9. Page 263

1. 800 W.   2. 11·98 kW, 0·8345 p.u.; 8·72 kW, 0·8355 p.u.
3. 7·85 kW, 0·8177 p.u.; 9·44 kW, 0·818 p.u.
4. 1·87 A, 766 rev/min, 10·3 A.   5. 6·49 kW, 0·843 p.u.
6. 4·61 kW, 0·835 p.u.   7. 0·8925 p.u.
8. 6·9 kW, 0·857 p.u., 0·871 p.u., 9·24 kW.
9. 0·914 p.u., 0·896 p.u.   10. 54·1 kW, 0·905 p.u., 146·5 A.
11. 0·914 p.u.   12. Motor, 0·905 p.u.; generator, 0·9205 p.u.
13. Motor, 0·877 p.u.; generator, 0·879 p.u.

Examples 10. Page 283

1. 125·6 V, 33·3 Hz.   2. 20 Hz, 35·5 V, 38·5 V.
3. 28·27 sin 157t volts; 8·74 V.   4. 900 rev/min.
5. 16 poles.   6. 60 Hz.   7. 52·2 V, average; 57·3 V, r.m.s.
8. 14·2 A, 16·4 A, 1·155.   9. 81·5 V, 1·087.
10. 5 V, 5·77 V, 1·154.   11. 30 A, r.m.s.; 20 A, average.
12. 10 V, 11·55 V, 1·155.   13. 0·816, none.
14. 0·637 A, 1·0 A.
15. 0·825 A, moving-coil ammeter; 1·296 A, moving-iron ammeter.
16. 2·61 kV, 3 kV.   17. 7·77 mA, 11 mA.
18. 5·06 A; 1·08, assuming ammeter to read r.m.s. value of sinusoidal current.
19. 500 sin(314t + 0·93) V, 10 sin(314t + 0·412) A, −300 V, −9·15 A, 29·5°, current lagging.
20. 150 Hz, 6·67 ms, 0·68 ms, 7·35 ms, 0·5 kWh.   21. 35 A.
22. 68 V, 48·1 V.   23. 55·9 A, 1·2° lead.

24. 23·1 sin(314*t* + 0·376) A, 16·3 A, 50 Hz.
25. 262 sin(314*t* + 0·41) V, 185·5 V, 50 Hz.
26. 121 V, 9·5° leading; 101·8 V, 101° leading.   27. 72 sin(ω*t* + 0·443).
28. 208 sin(ω*t* − 0·202), 11° 34′ lagging, 18° 26′ leading, 76 sin(ω*t* + 0·528).
29. 10·75 A, 8·5° lagging.

Examples 11. Page 330

1. 125·6 V (max.), 1·256 A (max.), 6·03 N m (max.).
2. 500 W.   3. Zero.   4. 2·65 A, 0·159 A.
5. 6·23 A, 196 V, 156 V, 51·5°.   6. 21·65 Ω, 0·0398 H, 2165 W.
7. 25·12 Ω, 29·3 Ω, 8·19 A, 0·512, 1008 W.
8. 2·64 A, 69·7 W, 0·1147.   9. 0·0435 H, 16 Ω, 58·6°.
10. 170 V, 46°; 3·07 Ω, 0·06 H.   11. 254 Ω, 1·168 mH.
12. 5·34 A, 0·857.   13. 77·5 W, 5·5 Ω.
14. 41·4 A, 4·5°; 41·4 sin(314*t* − 0·078) A; 4·825 Ω, 4·81 Ω, 0·38 Ω.
15. 61·1 Ω, 0·229 H, 34° 18′.   16. 78·6 W, 0·0376 H, 0·716.
17. 174 V, 75·6 V; 109·5 W, 182·5 W; 0·47.
18. 0·129 H, 0·0818 H, 0·706 A or 1·72 A.
19. 49·6 mA, 0·0794; 189 mA, 0·303.   20. 9·95 A.
21. 133 sin(ω*t* − 0·202) V, 6·65 sin(ω*t* + 0·845) A.
22. 25·8 μF, 0·478 leading, 145·5 V.
23. 1·54 A, 1·884 A, 2·435 A, 50° 42′, 308 W, 0·633 leading.
24. 11·67 Ω, 144 μF; 14·55 A, 0·884 leading.
25. 10·3 A, 1060 W, 0·447 lagging, 103 V, 515 V, 309 V.
26. 7·27 A, 5·56 A, 468 V, 0·795 lagging.
27. 7·07 A, 0·707 leading, 5 A, 5 A, 500 VAr.
28. 10 A, 13·42 A, 5·62 A; 20·75 A, 0·95 lagging.
29. 19·75 W, 504 μF.
30. 10·28 A, 10·2 A, 13·75 A, 0·96 leading, 3160 W.
31. 2 A, 4 A, 2 A, 4·56 A, 15° 15′ lagging.
32. 127·4 μF, 56 A, 11·2 kW.
33. 8·5 A, 0·794, 13·53 Ω, 10·74 Ω, 8·25 Ω.
34. 6·05 A, 15·2° lagging.   35. 159 pF, 5 mA; 145 pF, 177 pF.
36. 920 Hz, 173 V.   37. 15·8.
38. 400 V, 401 V, 10 V, 0·5 A, 15 W.
39. 50 Hz, 12·95 Ω, 0·154 H, 207·5 W; 21·85 sin(314*t* + π/4) A.
40. 31·5 V, 60 Ω, 0·167 H, 4220 pF.
41. 84·5 μH, 30 Ω, 17·7.   42. 57·2 Hz, 167 Ω.
43. 125 kΩ.   44. 76·5 pF, 0·87 MHz.   45. 28·7 μF.
46. 5·02 kVA.   47. 10·3 kVA, 0·97 leading; 20·6 kVA, 0·97 lagging.
48. 324 μF.   49. 17·35 A, 4140 W, 390 VAr, 0·996 lagging.
50. 59 kW, 75·5 kVA, 0·78 lagging; 47 kVAr, in parallel.
51. 37·1 A, 0·371, 2750 W; 17·9 A, 0·895, 3200 W; 4000 W.
52. 106·7 W, 0·1433 H.   53. 1·592 A, 11·4 Ω.   54. 21·2 μF, 0·952 A.
55. 0·0126 rad.   56. 1·74 MΩ, 0·061 μF.

Examples 12. Page 355

1. 10 + j0, 10∠0°; 2·5 − j4·33, 5∠ −60°; 34·64 + j20, 40∠ 30°.
2. 20 + j0, 0 + j40, 26 − j15, 5 − j8·66, 51 + j16·34; 53·5, 17° 46′ lead.
3. 11·2∠26° 34′, 8·54∠ −69° 27′.   4. 10 + j17·32, 28·28 − j28·28.

5. 13 − j3, 13·35∠−13°.   6. 7 + j13, 14·77∠61° 42'.
7. 38·28 − j10·96, 39·8∠−16°.   8. −18·28 + j45·6, 49·1∠111° 48'.
9. 50 Ω, 0·0956 H; 30 Ω, 63·7 μF; 76·6 Ω, 0·2045 H; 20 Ω, 92 μF.
10. 0·0308 − j0·0462 S, 0·0555∠−56° 18' S; 0·04 + j0·02 S, 0·0447∠ 26° 34' S; 0·0188 − j0·006 84 S, 0·02∠−20° S; 0·0342 + j0·094 S, 0·1∠70° S.
11. 0·69 − j1·725 Ω, 1·86∠−68° 12' Ω; 10·82 + j6·25 Ω, 12·5∠30° Ω.
12. 4 Ω, 1910 μF; 20 Ω, 31·85 mH; 1·44 Ω, 1274 μF; 3·11 Ω, 8·3 mH.
13. 91·8 + j53 V, 2·53 − j1·32 A; 106∠30° V, 2·85∠−27° 30' A.
14. 0·104 + j2·82 A; 2·82∠87° 53' A.   15. 25 Ω, 50 Ω.
16. 24 Ω, 57·3 mH; 12 Ω, 145 μF; 0·0482∠18° 10' S, 18° 10', current leading.
17. 7·64 A, 8° 36' lagging.
18. 13·7 − j3·2 Ω, capacitive; 2250 W; 13° 9'.   19. 1·192 − j0·538 A.
20. 20 Ω, 47·8 mH; 10 Ω, 53 μF; 0·0347 − j0·0078 S, 12° 40' lagging.
21. 2·83 A, 30° lagging.
22. 2·08 A, lagging 34° 18', from supply; 4·64 A, lagging 59°, through coil; 2·88 A, leading 103° 30', through capacitor.
23. 18·75 A, 0·839 lagging.   24. 21·33 Ω, 93·3 mH.
25. 0·68 μF.   26. 174 Ω, 80·1 mH.
27. 940 W; 550 VAr, lagging.   28. 1060 W; 250 VAr, leading.
29. 4 kVA, 3·46 kW; 8 kVA, 8 kW; 11·64 kVA, 11·46 kW; 44·64 + j37·32 A.

Examples 13. Page 382

1. 4240 V, 70·7 A.   2. 80·6 A.
3. 6·18 A, 5·05 A, 9·87 A, 325 V.   4. 346 V; 346 V, 200 V, 200 V.
5. 11·56 kW, 34·7 kW.   6. 96·2 A, 55·6 A.
7. 8·8 A, 416 V, 15·25 A, 5810 W.
8. 4·26 Ω, 0·0665 H, 18·75 A, 18·75 A, 32·45 A.
9. 1·525 A; 2·64 A, 2·64 A, 4·57 A; 1·32 A, 1·32 A, 0; 210 V.
10. 61·7 A, 0·858, 38 kW.   11. 90·8 A, 240 V.
12. 72·3 μF/phase.
13. 3·81 kVAr, 70·5 μF, 23·5 μF, 0·848 lagging.
14. 9·24 A, 5120 W, 0·8 lagging, 14·3 μF, 7·78 A.
15. 115 Ω, 0·274 H, 1·667 A, 956 W; 1·445 A, 1·445 A, 0.
16. 462 V, 267 V, 16·7 kVAr, 5·87 Ω, 8·92 Ω, 75 A, 33 kW.
17. 43·3 A, 26 A, 17·3 A, 22·9 A.   18. 17·3 A, 31·2 kW.
19. 4 A, 6·67 A, 3·08 A; 6·84 A in R, 10·33 A in Y, 5·79 A in B.
20. 21·6 A in R, 49·6 A in Y, 43·5 A in B.   21. 1·1 kW, 0·715.
22. 8 kW.   23. 0·815.   24. 13·1 kW, 1·71 kW.
25. 21·8 A, 36° 35' lagging; 7 kW; 13 kW.   26. 0·655, 0·359.
27. 15 kW, 0·327.   29. 11·7 kW, 5·33 kW.
30. Line amperes × line volts × sinϕ = 8570 VAr.

Examples 14. Page 422

1. 462, 28, 42 800 mm².   2. 500, 690 V, 4450 mm².
3. 475, 18.   4. 1200 V, 0·675 T.
5. 146·7 V; 22·73 A, 341 A; 0·033 Wb.
6. 60·6 A, 833 A, 0·0135 Wb, 1100 turns.

7. 1·57 kVA, 64 turns.   8. 14, 22, 3500, 805 (primary).
9. 25·9 A.   10. 7 A, 0·735 lagging.   11. 4·43 A, 1·0.
12. 1016 V; 32·9 A, secondary; 8·225 A, 14·25 A, primary.
13. 10, 87.   14. 317·5 V, 952·5 V.
15. Primary and secondary phase currents, 65·8 A; 38·06 A in resistors; 114 A from supply; 65·8 A to load; 87 kW; 29 kW.
16. 5·18 mWb, 345 W, 4·77 A (r.m.s.).   17. 300, 1·44 mm, 88°.
18. 2·3 A, 0·164.   19. 176·5 V.   20. 900 V.
21. 377·6 V.   22. 4 per cent.
23. 2·13 per cent, 225·1 V; 4·22 per cent, 220·3 V; −0·81 per cent, 231·86 V; 0·960 p.u., 0·9585 p.u.; 56·25 kVA.
24. 2·54 per cent.   25. 0·892 p.u.   26. 0·981 p.u.
27. 0·9626 p.u.   28. 0·905 p.u.
29. 0·9598 p.u., 8·65 kVA, 0·9601 p.u. at unity power factor.
30. 0·9868 p.u., 0·9808 p.u.
31. 0·961 p.u., 0·9628 p.u., 23·3 kW.
32. A, 93·77, 96·24, 96·97 per cent; B, 96·42, 97·34, 96·97 per cent; A, 40 kVA; B, 23·1 kVA.
33. A, 0·9553 p.u.; B, 0·9682 p.u.; £3·15.   34. 0·9225 p.u.
35. 44 W, 10 W.   36. 1600 W, 800 W.   37. 362·8 W.
39. 0·80 p.u.
40. 240 turns from one end; 5·45 A in common section, 40 A in end section; 0·88 p.u.

Examples 15. Page 442

1. 3600 rev/min, 0·0227 T.   4. 1·195.   5. 1·24.
6. 4·19 N m.   7. 0·958.   8. 2560 V.   9. 0·960, 503 V.
10. 8 poles, 32·3 mWb.   11. 8.   12. 3·54 kV.
13. 818 V.   14. 0·831, 1600 V.   15. 0·91, 0·0473 Wb.
16. 14.   17. 0·966.   18. 20·5 V.
19. 0·0862 Wb, 525 A.   20. 6·4 kW.

Examples 16. Page 456

1. 500 rev/min.   2. 4 poles.   3. 150 Hz.

Examples 17. Page 476

1. 4·4 per cent, 14·0 per cent, −7·3 per cent.   2. 2412 V.
3. 1585 V.   4. 12·5 per cent increase.
5. 39·9 A, 3290 V, 131·3 kW.   6. 111 A, 1845 V, 615 kW.
7. 71·1 A, 1920 V/phase.   8. 2212 V, 2040 V, 2380 V.
9. 234 V, 198·5 V, 268·5 V.
10. 206 V, 14° 2′; 9·8 A, 0·913 lagging.
11. 500 rev/min, 1165 N m.   12. 1200 kW, 0·3875 leading.
13. 289 kVA, 0·692 leading.   14. 74·8 A, 0·995 lagging.

Examples 18. Page 494

1. 2880 rev/min, 3000 rev/min.   2. 1470 rev/min, 1 Hz.
4. 4·6 per cent, 954 rev/min.   5. 1·585 Hz, 3·17 per cent.

6. 971·7 rev/min.
7. 0·960, 9 mWb.
8. 11·33 A; 0·2, 40·8 A.
9. 10·67 A, 0·376; 21·95 A, 0·303.
10. 0·07 Ω, 37·5 per cent.
11. 960 rev/min, 844 W.
12. 27·5 per cent, 41 per cent.
13. 11 kW, 440 W.
14. 0·857 p.u., 25·57 kW.
15. (i) 1:0·333:0·16; (ii) 1:0·333:0·16.

Examples 19. Page 557

1. 2·96 mA, 61·1 V.
2. 27 kΩ, 11·5 kΩ.
4. 5·45 mA, 118 V.
5. 1·34 mA.
6. 882 Ω, 570 Ω (approx.).
7. 5·4 A, 10·8 A.
8. 45 V.
9. 8·33 kΩ.
10. 3·43 mA.
11. 3 kΩ.
12. 2·6 A, 0·795 A.
13. 38·8 A, 12·03 A, 18·9 A.
14. 32·5 A, 9·13 A.
15. 277·5 V, 0·92 p.u.
16. 5 mA.
17. 1·414 A (moving-iron), 0·9 A (moving-coil).
18. 6·5 mA.
19. For $V_A = 100$ V, $g_m = 2·2$ mA/V, $\mu = 16$ (approx.), corresponding $r_a = 7270 \Omega$.
20. 8200 Ω, 2·5 mA/V (approx.); mutual dynamic characteristic: $I_A = 10·15$ mA for $V_G = 0$; $I_A = 7·2$ mA for $V_G = -4$ V, $I_A = 4·4$ mA for $V_G = -8$ V, etc.
22. 2·6 mA/V, 7800 Ω, 20·3.
23. 4000 Ω, 15, 3·75 mA/V.
24. 12·5 kΩ, 1·667 mA/V, 20·83.
25. 10·5 kΩ, 1·35 mA/V, 14·2.
26. 30 V.
27. 18, 13·85.
28. 1·17 mA, 58 V, 29.
29. 4 mW.
30. 4·9, 1·2 mW.
31. 40·6, 2·9 mA/V, 14 kΩ, 23·9.
32. 6·5, 4, 4·5 mW, 16·7 kΩ.
33. 9·65 mA, 5·36 mA; 105 V, 169·5 V; 16·1; 34·5 mW.
34. 16·9.
35. 14·14:1; 36 mW.
36. 72·6 pF, 43·8 kΩ.
37. 41·7 kΩ, 11·5 mA/V, 480; 27, 580 mW; 27·15.
38. 87 mm, 9·9°.
39. 2·27 $\times$ 10^7 m/s, 2·18 $\times$ 10^{-15} N.
40. 3·75 $\times$ 10^7 m/s, 640 V.
41. 1·74 mT.

Examples 20. Page 599

1. 39, 24.
2. 0·968, 0·986.
3. 0·964, 224, 216; 23·34 dB; 34·4 Ω.
4. 304 kΩ.
5. −26·4, −223, 5890; 37·7 dB; 948 Ω.
6. 80·75 kΩ.
7. 0·297 V, 17·65 μW, 22·1 dB; 2·7 V, 1·46 mW, 38·6 dB.
8. 500 kΩ; 50 kΩ.
9. 21·66 dB, 76·3 Ω; 36·54 dB, 2330 Ω.
10. 906 μW, 4460 μW.
11. 2·83 V, 2·55 V.
12. 53·5, 48·5, 41·5.
13. 2·64 mA, 2·64 V.
14. 6·97 mW.
15. 3·52 V, 1·55 mW.
16. 6·36 mA.

Examples 21. Page 628

1. 11·6 lm/W, 89·2 cd, 0·923 cd/W.
2. 132·6 cd, 4·73 p.
3. 76 500 cd/m^2.
4. 6130 cd/m^2.
5. 99·5 lx.
6. 122·5 cd.
7. 39·06 cd; standard lamp, 1·16 m, test lamp, 1·34 m.
8. 25 lx, 16·67 lx; 0·9 m from 16-cd lamp.
9. 50 lx, 41·9 lx.
10. 10·82 m, 2340 cd.
11. 11·6 lx, 13·1 lx.
12. 9·4 m.

13. 155·5 W, 55·7 lx.
14. 37·3 lx at A; about 38·7 lx at B.
15. 8 lux, 2·87 lux; 76·8 lux.   16. 36.   17. 1550 cd.
18. 60 lamps.   19. 21·5 per cent, −18·5 per cent.
20. 0·606 A, 122·2 V, 97·8 V, 152·3 per cent, 62·5 per cent.
21. 236 cd, 2960 lm.   22. 1·0 V.

Examples 22. Page 660

1. 0·0378 Ω, 6662 Ω.   2. 5005 μΩ, 995 Ω.
3. 8·325 mm.   4. 19·985 kΩ, 0·0378 Ω.
5. 0·015 04 Ω, 9·995 kΩ; 0·231 A.
6. 0·003 753 Ω, 16·662 kΩ; −1·2 per cent, −0·000 36 per cent.
7. $8·1 \times 10^{-5}$ N m.   8. $30·4 \times 10^{-6}$ N m.   9. 25·6 mA.
10. 10.   11. 25.
12. 0·72 per cent, 36 per cent.   13. 0·9, 1·036.
14. 1785 Ω.   15. 33·8 kΩ.   16. 8·13 Ω, 1·6 per cent low.
17. 60·67 Ω, 9·9 per cent high.   18. 1087·5 Ω.   19. 233·5 kΩ.
20. 776 Ω.   21. 5000 Ω, 21·45 mA   22. 900 m.
23. 3·08 mA from D to B.   24. 2·17 per cent high.
25. 1·46 V high.   26. 1·486 V, 1·09 Ω.
27. 4·17 per cent high.   28. 2·04 per cent high.
29. 2·98 per cent low.   30. 1·24 per cent high.
31. 0·41 mA from Z to X.

Examples 23. Page 684

1. 33·6 cm³.   2. 110, 125.   3. 32·5 A, 23·75 A.
4. 26·4 Ω.   5. 1·0625 Ω, 0·6875 Ω.   6. 1·7 Ω, 7·88 A.
7. 3·42 Ω, 1·86 Ω.   8. 83·3 per cent, 72·7 per cent.
9. 104·4 V, 138 V, 83·3 per cent, 74·2 per cent.
10. 79·4 per cent, 64·4 per cent.   11. 21 A h, 42 A h (approx.).
12. 0·016 25 Ω.   13. 0·0133 Ω.
14. 1·414 V, 4·74 Ω.   15. 111·5 V, 35·1 W.

Appendix. Page 445

1. 0·55 μN·m/degree.   2. 33·1°.
3. 1·91 H, 1·01 H.   4. 1·23 H, 295 N·m, 18·5 kW.
5. 1·11 H, 239 V.
6. 0·6 A, 181·5 V, 2605 rev/min, 109 W, 0·883 lagging, 85·8 per cent.
7. 0·225 H, 223 N·m.   8. 16·55 A.

Index